NANOFIBERS – PRODUCTION, PROPERTIES AND FUNCTIONAL APPLICATIONS

Edited by **Tong Lin**

Nanofibers – Production, Properties and Functional Applications
Edited by Tong Lin

Published by InTech
Janeza Trdine 9, 51000 Rijeka, Croatia

Publishing Process Manager Silvia Vlase

Technical Editor Teodora Smiljanic

Cover Designer InTech Design Team

Additional hard copies can be obtained from orders@intechopen.com

Nanofibers – Production, Properties and Functional Applications, Edited by Tong Lin
p. cm.
ISBN 978-953-307-420-7

For EU product safety concerns:
IN TECH d.o.o., Prolaz Marije Krucifikse Kozulić 3, 51000 Rijeka, Croatia,
info@intechopen.com or visit our website at intechopen.com.

Meet the editor

Dr. Tong Lin received his PhD degree in physical chemistry from Chinese Academy of Sciences in 1998. After several years of postdoctoral research in Australia and USA, he joined Deakin in 2003, and is working as the Associate Professor in Fibrous Materials. His main research interest is in functional polymers and fibrous materials, electrospinning and nanofibres. Dr. Lin has published more than 110 peer-reviewed articles in high impact-factor journals.

Contents

Preface

With the rapid development of nanoscience and nanotechnology over the last decades, great progress has been made not only in the preparation and characterization of nanomaterials, but also in their functional applications. As an important one-dimensional nanomaterial, nanofibers have extremely high specific surface area because of their small diameters, and nanofiber membranes are highly porous with excellent pore interconnectivity. These unique characteristics plus the functionalities from the materials themselves impart nanofibers with a number of novel properties for applications in areas as various as biomedical engineering, wound healing, drug delivery and release control, catalyst and enzyme carriers, filtration, environment protection, composite reinforcement, sensors, optics, energy harvest and storage, and many others. More and more emphasis has recently been placed on large-scale nanofiber production, the key technology to the wide usages of nanofibers in practice. Tremendous efforts have been made on producing nanofibers from special materials. Concerns have been raised to the safety issue of nanofibrous materials.

This book is a compilation of contributions made by experts who specialize in their chosen field. It is grouped into three sections composed of twenty-one chapters, providing an up-to-date coverage of nanofiber preparation, properties and functional applications. I am deeply appreciative of all the authors and have no doubt that their contribution will be a useful resource for anyone associated with the discipline of nanofibers.

Dr. Tong Lin
Centre for Material and Fibre Innovation, Deakin University
Australia

Part 1

Production and Assembly

1

Industrial Production Technology for Nanofibers

Stanislav Petrík
ELMARCO s.r.o.,
Czech Republic

1. Introduction

Electrospinning methods for creating nanofibers from polymer solutions have been known for decades (Kirichenko et al., 2007, Ramakrishna et a., 2005). The nozzle-less (free liquid surface) technology opened new economically viable possibilities to produce nanofiber layers in a mass industrial scale, and was developed in the past decade (Jirsak et al., 2005). Hundreds of laboratories are currently active in the research of electrospinning process, nanofiber materials, and their applications. Nanofiber nonwoven-structured layers are ideal for creating novel composite materials by combining them with usual nonwovens. The most developed application of this kind of materials is air filtration (Jaroszczyk et al., 2009). Liquid filters and separators are being developed intensively with very encouraging results. Also well known are several bio-medical applications utilizing nanofiber materials, often from biocompatible/degradable polymers like PLA, gelatine, collagen, chitosan. These developing applications include wound care, skin-, vessel-, bone- scaffolds, drug delivery systems and many others (Proceedings, 2009). Inorganic/ceramic nanofibers attract growing interest as materials for energy generation and storage (solar and fuel cells, batteries), and catalytic materials (Kavan & Grätzel, 2002, Duchoslav & Rubacek, 2008, Rubacek & Duchoslav, 2008, Bognitzki et al., 2001, Guan et al., 2003). To fully explore the extraordinary number of application opportunities of nanofibers, the availability of reliable industrial-level production technology is essential. This chapter intends to demonstrate that the technology has matured to this stage.

2. Theoretical background

The electrospinning process is an interesting and well-characterized physical phenomenon and has been an attractive subject for theoretical investigations of several groups (Bognitzki et al., 2001, Taylor & Van Dyke, 1969, Doshi & Reneker, 1995, Thompson et al., 2007, Shin et al., 2001, Yu et al., 2006, Hohman et al., 2008). Most work concentrates on the essentials of the process – the nanofiber formation from a liquid polymer jet in a (longitudinal) electric field. It has been theoretically described and experimentally proven that the dominant mechanism is whipping elongation occurring due to bending instability (Thompson et al., 2007, Yu et al., 2006, Hohman et al., 2008). Secondary splitting of the liquid polymer streams can occur also (Kirichenko et al., 2007), but the final thinning process is elongation.
In Figure 1, the schematic of bending mechanism derived from physical model (a) is compared with a stroboscopic snapshot (b) (Reneker, 2009).

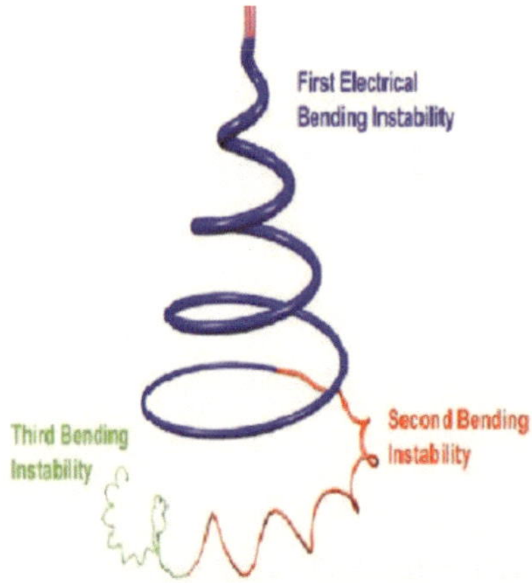

a

b

(Courtesy of Darrell Reneker, University of Akron)

Fig. 1. The path of an electrospinning jet (a - schematic, b - stroboscopic photograph).

A comprehensive analysis (electrohydrodynamic model) of the fiber formation mechanisms published by (Hohman et al., 2008) describes the regions of individual kinds of instability observed during the process. It has predicted and experimentally proven that there is a domain of the process variables where bending instability dominates, as illustrated in Figure 2.

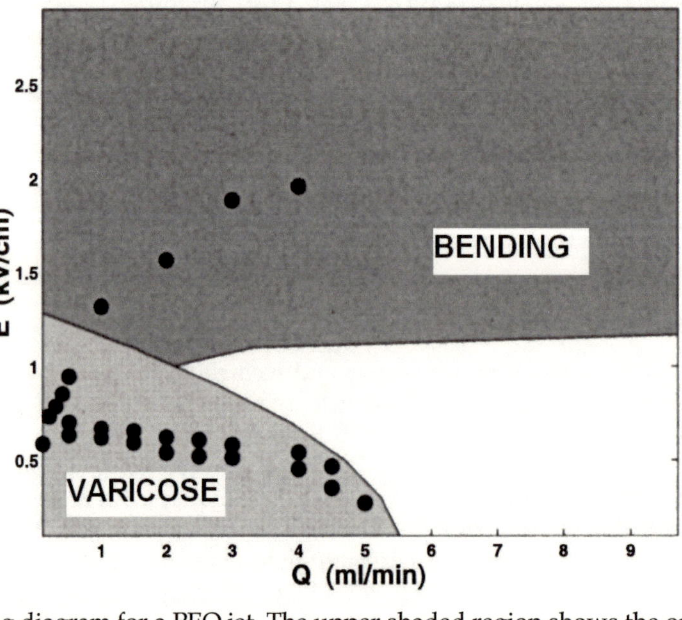

Fig. 2. Operating diagram for a PEO jet. The upper shaded region shows the onset of the whipping instability, the lower one shows the onset of the varicose instability (Hohman et al., 2008).

The efforts to scale up the electrospinning technology to an industrial production level used to be based on multiplication of the jets using multi-nozzle constructions (Kirichenko et al., 2007). In Figure 3, the multi-nozzle spinning head developed by NanoStatics Company is shown. The principle is based on an idea to feed multiple nozzles from a single source of the polymer solution.

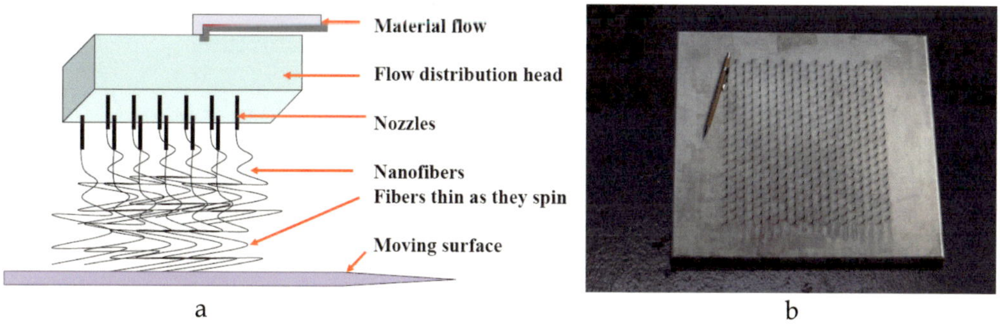

Fig. 3. Schematic (a) and photograph (b) of a multi-nozzle spinning head by NanoStatics (NanoStatics, 2007).

Figure 4 shows the multi-nozzle spinning part of the machine being commercialized by TOPTEC Company. The device uses upwards direction of electrospinning in order to eliminate polymer droplets eventually falling from conventional down-oriented electrospinning elements.

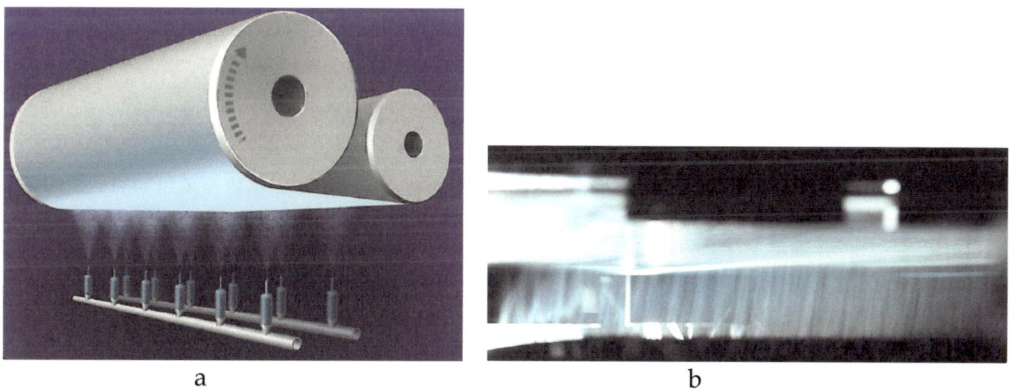

Fig. 4. Schematic (a) and photograph (b) of a multi-nozzle spinning head by TOPTEC (TOPTEC, 2011).

However, the number of jets needed to reach economically acceptable productivity is very high, typically thousands. This brings into play many challenging task, generally related to reliability, quality consistency, and machine maintenance (especially cleaning). The nozzle-less electrospinning solves most of these problems due to its mechanical simplicity, however, the process itself is more complex because of its spontaneous multi-jet nature. The (Lukas et al., 2008) study focused on the process of multi-jet generation from a free liquid surface in an electric field. They showed that the process can be analyzed using Euler's equations for liquid surface waves

$$\nabla\left(\rho\frac{\partial\Phi}{\partial t}+p\right)=0 \tag{1}$$

where Φ is the scalar velocity potential, p is the hydrostatic pressure, and ρ is the liquid density. They derived the dispersion law for the waves in the form

$$\omega^2=(\rho g+\gamma k^2-\varepsilon E_0^2 k)\frac{k}{\rho} \tag{2}$$

where E_0 is electric field strength, γ – surface tension.
The relationship between angular frequency ω and wave number k is in Figure 5, electric field is the parameter. When a critical electric field intensity is reached (E_c, curve 1), ω^2 is turned to be negative, ω is then a purely imaginary value, and hence, the amplitude of the liquid surface wave

$$\xi=Ae^{qt}\exp(ikx) \tag{3}$$

exponentially grows, which leads to an instability.

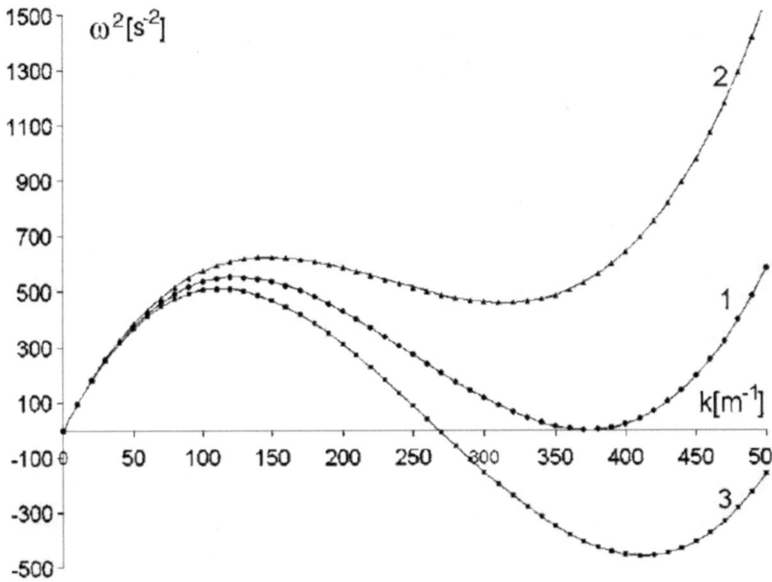

Fig. 5. Relationship between the square of the angular frequency and the wave number for distilled water, electric field is the parameter 1: $E=E_c=2.461\ 945\ 094\times10^6$ V/m, 2: $E=2.4\times10^6$ V/m, and 3: $E=2.5\times10^6$ V/m (Lukas et al., 2008).

Critical field strength can then be expressed

$$E_c=\sqrt[4]{4\gamma\rho g/\varepsilon^2} \tag{4}$$

From this equation, they derived the expression for the critical spatial period („wavelength") - the average distance between individual jets emerging from the liquid surface (Figure 6).

$$\lambda_c = 2\pi / k_c = 2\pi a \tag{5}$$

and

$$\lambda = 12\pi\gamma / [2\varepsilon E_0^2 + \sqrt{(2\varepsilon E_0^2)^2 - 12\gamma\rho g}] \tag{6}$$

a is the capillary length

$$a = \sqrt{\gamma / \rho g} \tag{7}$$

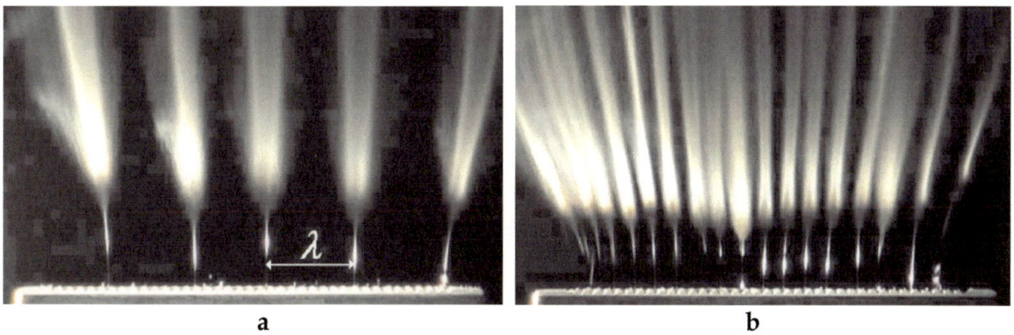

(Courtesy of David Lukas, Technical University of Liberec)

Fig. 6. Free liquid surface electrospinning of Polyvinyalcohol at 32 kV (a) and 43 kV (b).

3. Technical realization

3.1 Description of the nozzle-less technology

The simplest realization of the nozzle-less electrospinning head is in Figure 7a. A rotating drum is dipped into a bath of liquid polymer. The thin layer of polymer is carried on the drum surface and exposed to a high voltage electric field. If the voltage exceeds the critical

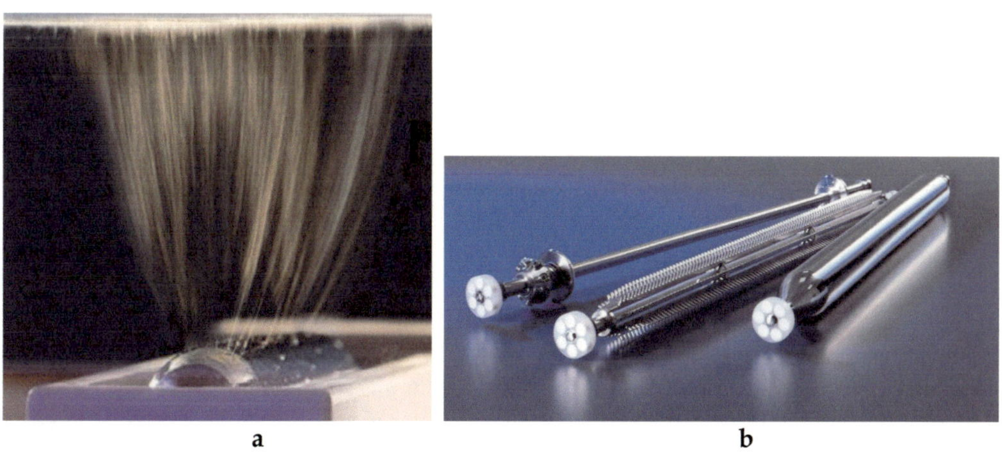

Fig. 7. Free liquid surface electrospinning from a rotating electrode (a) and various types of spinning electrodes (b).

value (4), a number of electrospinning jets are generated. The jets are distributed over the electrode surface with periodicity given by equation (6). This is one of the main advantages of nozzle-less electrospinning: the number and location of the jets is set up naturally in their optimal positions. In the case of multi-needle spinning heads, the jet distribution is made artificially. The mismatch between "natural" jet distribution and the real mechanical structure leads to instabilities in the process, and to the production of nanofiber layers which are not homogenous.

Several types of rotating electrodes for free liquid surface electrospinning for industrial machines have been developed (Figure 7b). However, the drum type is still one of the most productive.

Production variable	**Nozzle**	**Nozzle-Less**
Mechanism	Needle forces polymer downwards. Drips and issues deposited in web.	Polymer is held in bath, even distribution is maintained on electrode via rotation.
Hydrostatic pressure	Production variable - required to be kept level across all needles in process.	None.
Voltage	5 - 20 kV	30 - 120 kV
Taylor cone separation	Defined mechanically by needle distances.	Nature self-optimizes distance between Taylor cones (Eq. (6)).
Polymer concentration	Often 10% of solution.	Often 20% or more of solution.
Fiber diameters	80, 100, 150, 200, 250 and higher. Standard deviation likely to vary over fiber length.	80, 100, 150, 200, 250 and higher. Standard deviation of +/- 30%.

Table 1. Comparison of Nozzle vs. Nozzle-Less Electrospinning.

3.2 Alternative technologies for nanofibers

Research and development centers are very active in their efforts to further improve productivity of the manufacturing process. Novel methods for the production of sub-micron fibers are being developed. The most advanced methods ("Fine Hole" meltblown and "Islets-in-the-sea") are compared with the two current electrospinning approaches in Table 2. The individual methods can be considered to be complementary rather than competing. This is especially true with respect to the fiber diameter distribution and fiber layer uniformity.

Figure 8 shows the extrusion methods being developed by Hills Inc. (HILLS, 2011).

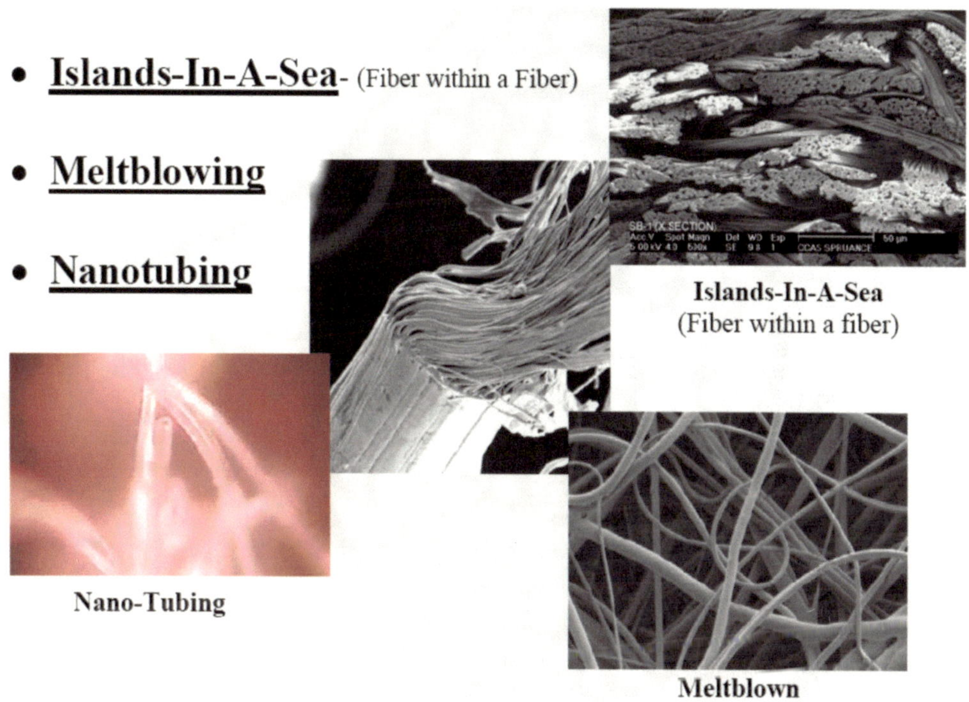

Fig. 8. Fibers made by extrusion methods (HILLS, 2011).

Centrifugal forces for elongation of liquid polymer into thin fibers are used in the approach developed by (Dauner et al., 2008). The productivity of the process they claim is high (up to 1000 $cm^3/m.hr$), however, the fiber diameter distribution and homogeneity of the deposited nanofiber layer is not at the levels achieved by electrospinning.

Individual methods will likely find different areas of application. More productive Nano-meltblown and Islets-in-the-sea technologies compromise fiber diameter and homogeneity and will likely be used in cost sensitive applications like hygiene nonwovens, while high quality electrospun technologies will be used in products where their high added value and need for low amounts of the material can be easily implemented (air and liquid filtration, biomedicine).

a

b

Fig. 9. Centrifugal spinning head (a) and a pilot production line by Dienes/Reiter (b) (Dauner et al., 2008).

	Extrusion		Electrospinning	
Sub-type	Fine hole	Islets-in-the-sea	Nozzle	Nozzle-Less
Description	Fine fiber meltblown process, melted polymer is forced through small holes	Polymer blends are extruded through thicker holes, then separated afterwards, often hydro-entangled	Solvent polymers are forced through a needle and formed through an electrostatic field	Electrostatic field is used to form fibers but is formed without needles using higher voltage on roller-electrodes
Voltage	n/a	n/a	5 - 20 kV	30 - 120 kV
Fiber size (nm)	800 - 2,500 s = +/- 200%	800 - 2,500 s = +/- 200%	?? - 500 s = Varies	80 - 500 s = +/- 30%
Hydrostatic pressure	Yes	Yes	Yes	No
Production ready?	Yes	Yes	No	Yes

Table 2. Nanofiber Production Methods.

The nozzle-less principle using rotating electrodes has been developed into a commercially available industrial scale. A photograph of a modular Nanospider™ machine is in Figure 10.

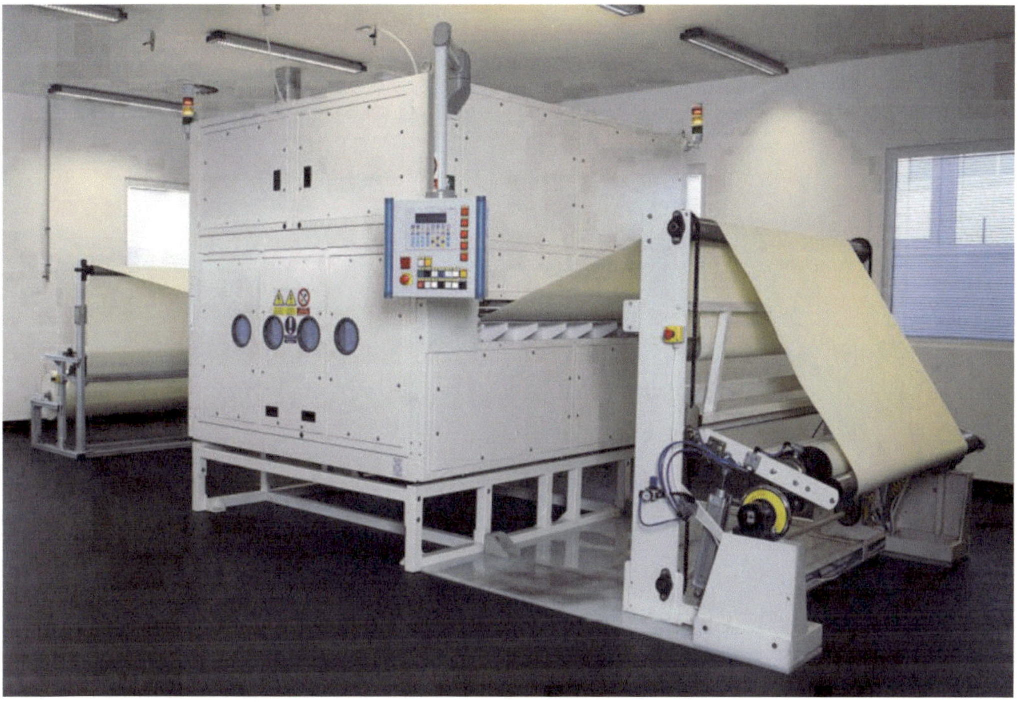

Fig. 10. Nozzle-less production electrospinning line (Nanospider™).

3.3 Performance of the technology

In addition to productivity (or throughput) of the production line, individual industrial applications require certain production consistency. Petrik & Maly illustrated the nozzle-less electrospinning technology performance with the example of air filtration media composed of a regular cellulose substrate and a thin nanofiber layer made from Polyamide 6 (Petrik & Maly, 2009). The product can be characterized by a number of parameters, like fiber diameter distribution (mean value and its standard deviation), basis weight of the nanofiber layer, etc. For the particular application, functional product parameters are more important. Typical values are the initial gravimetric filtration efficiency (IGE), differential filtration efficiency, and pressure drop, measured according to the norms widely accepted within industry.

The correlation between nanofiber diameter and basis weight of the nanofiber layer with differential filtration efficiency is illustrated in Figure 11. To obtain various basis weights, substrate speed was varied from 0.2 m/min to 4 m/min for each series of samples. Polymer solution parameters (concentration, etc.) together with electric field intensity determine the range of nanofiber diameter. Nanofiber diameter distribution has been measured using a scanning electron microscope (SEM). Basis weight values were obtained either by using an analytical scale Mettler (higher values), or by extrapolation from its known dependence on substrate velocity (lower ones).

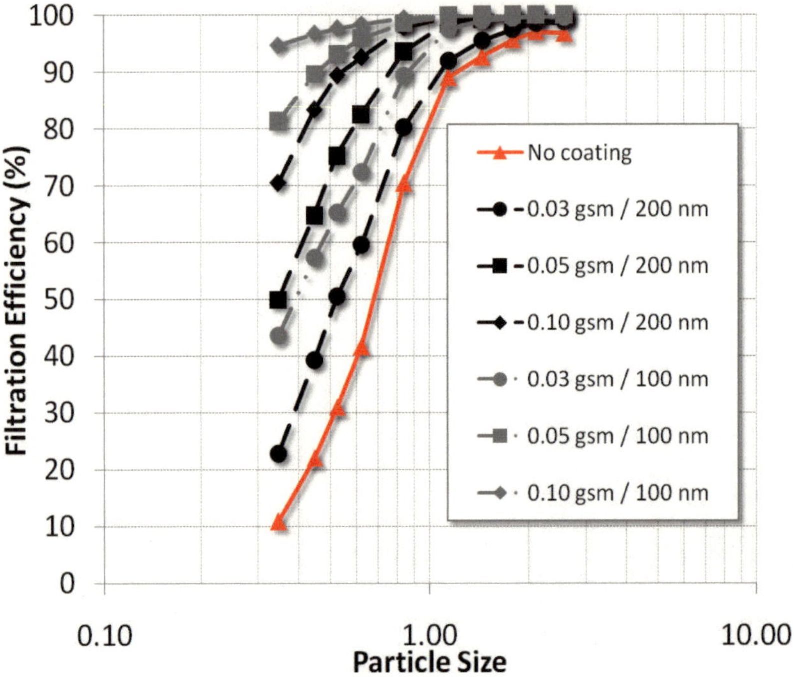

Fig. 11. Filtration efficiency of nanofiber media samples.

Pressure drop for various coatings at 10 m/min face velocity						
Substrate	*Basis Weight*	*NF Diameter*	*Filtration Efficiency*		*Pressure Drop*	
	(g/m^2)	(nm)	at 0.35 micron particle size (%)	D vs. uncoated substrate	(mm of H_20)	D vs. uncoated substrate
Cellulose	No coating	n/a	11	n/a	15.24	n/a
Cellulose	0.03	200	23	108%	17.53	15%
Cellulose	0.05	200	50	357%	19.30	27%
Cellulose	0.10	200	70	545%	24.13	58%
Cellulose	0.03	100	44	298%	18.80	23%
Cellulose	0.05	100	81	644%	22.61	48%
Cellulose	0.10	100	95	766%	29.21	92%

Table 3. Properties of nanofiber filtration media samples shown in Figure 11.

For production technology tests, pressure drop and initial gravimetric filtration efficiency have been chosen as representative product parameters. They were measured using NaCl aerosol at the following settings: air flow speed: 5 m/min, sample area 100 cm², flow rate 50 l/min.

In Figure 12, results of long-term stability and reproducibility of the IGE are presented. It can be seen that the individual runs differ within the standard deviation of the process, and the mean value of the filtration efficiency does not exhibit any significant shift after 16 hours of machine run. Similar consistency exhibits the value of the basis weight of the nanofiber layer, shown in Figure 13.

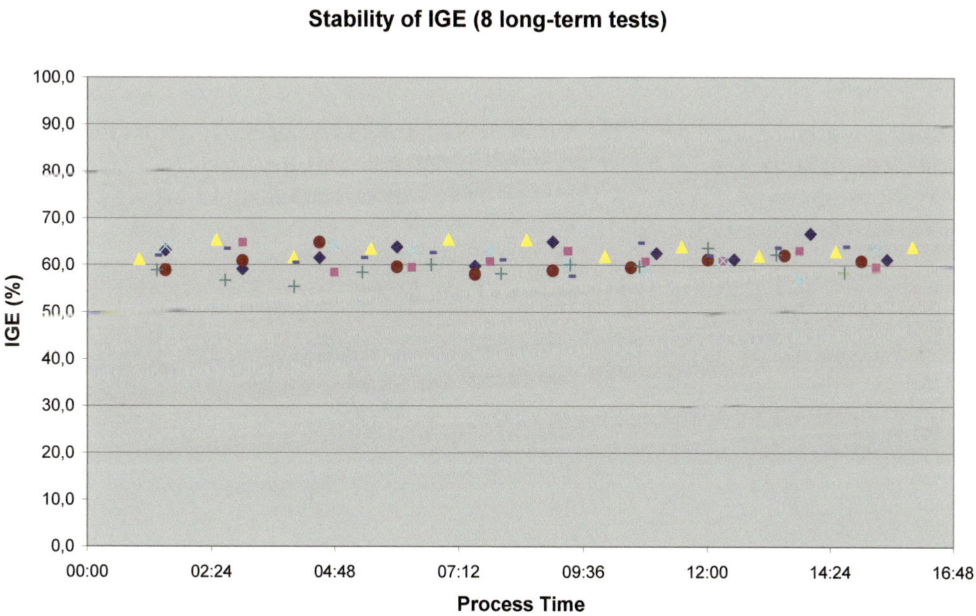

Fig. 12. Stability of initial gravimetric filtration efficiency of the media produced at the industrial nozzle-less electrospinning equipment.

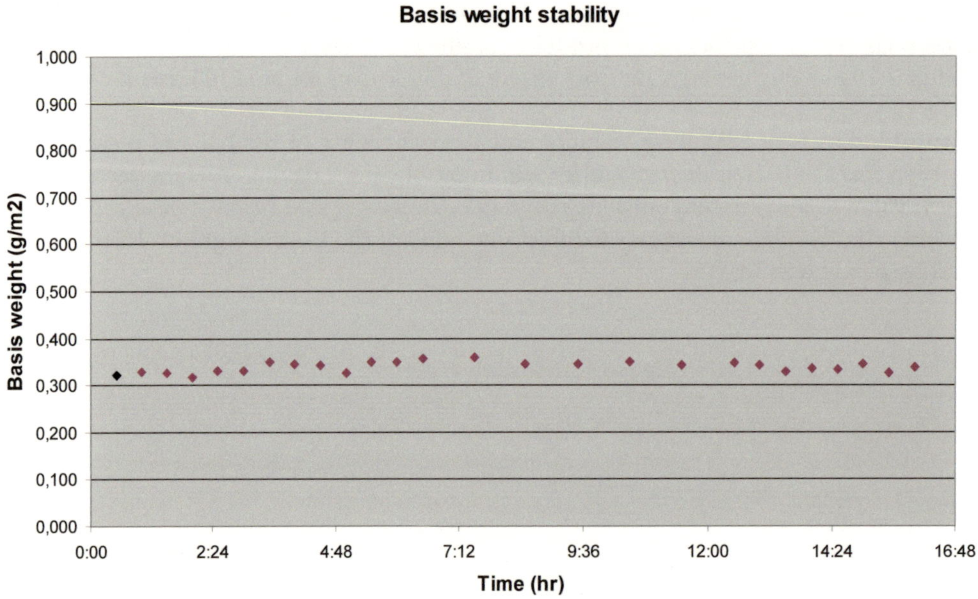

Fig. 13. Consistency of nanofiber layer basis weight produced with industrial nozzle-less electrospinning equipment.

The third important parameter of the filtration media is its homogeneity across the width of the roll. The data for the 1.6 meter wide roll produced with the machine in Figure 6 are shown in the graph in Figure 14. Pressure drop was measured in 10 evenly distributed points at a cross section of the substrate belt.

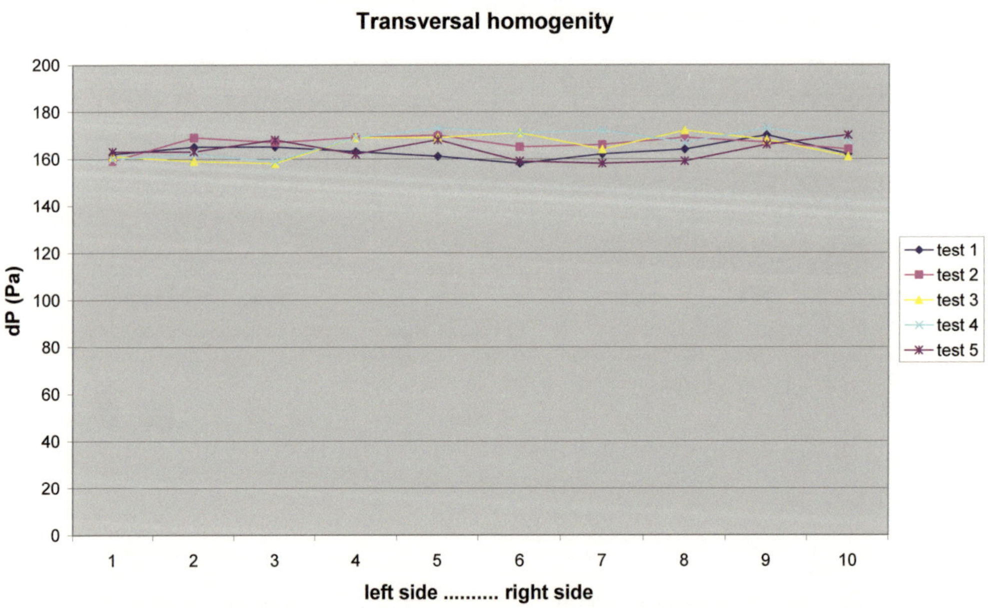

Fig. 14. Transversal homogeneity of the filtration media (1.6 m width, dP = pressure drop).

Production capacity of the industrial electrospinning line for Polyamide 6 is illustrated in Figure 15.

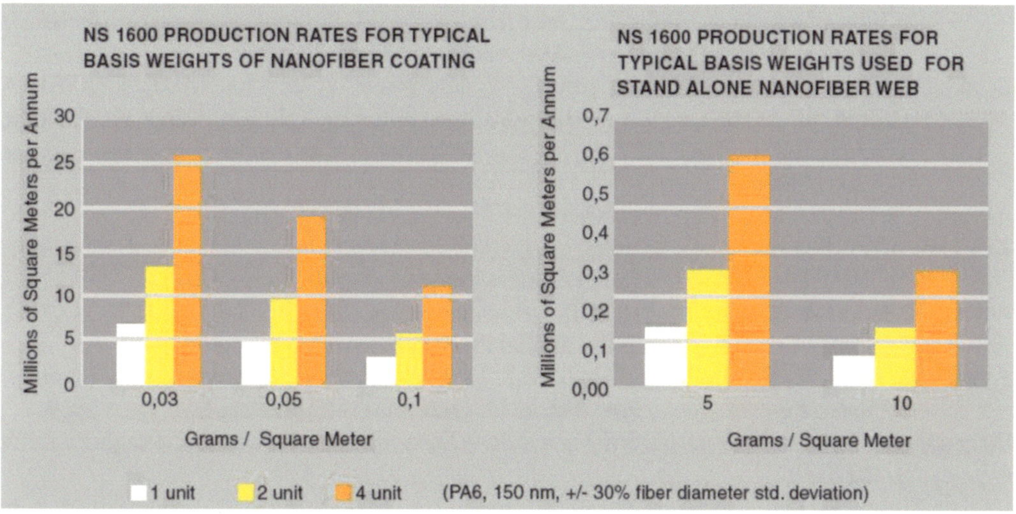

Fig. 15. Production capacity of the nozzle-less electrospinning line with Polyamide 6.

4. Conclusion

High-quality low-cost production of nanofiber layers is essential to support the enormous amount of research results being obtained at many universities and research centers. The described nozzle-less electrospinning technology has matured to a level where large scale production use is common, and can be modified for practically all known polymers soluble in organic solvents and water, as well as for polymer melts. This opens commercial opportunities for hundreds of ideas developed in the academic sphere.

5. References

Bognitzki, M., Czado, W., Frese, T., Schaper, A., Hellwig, M., Steinhard, M., Greiner, A. & Wendorff, J. H. (2001). *Adv. Mater.* 13, 70 (2001)

Dauner, M., Ullrich, A. & Reiter, F. (2008). Nanofibers by Centrifuge Spinning to Improve Filter Media, In: *Proceedings of 10th World Filtration Congress (WFC10)*, Leipzig, Germany, April 14-18, 2008

Doshi, J. & Reneker, D. H. (1995). Electrospinning process and applications of electrospun fibers, *J. Electrost.* 35, 151

Duchoslav, J. & Rubacek, L. (2008). Electrospun TiO2 Fibers as a Material for Dye Sensitizied Solar Cells, In: *Proceedings of NSTI Nanotech Conference*, Boston, MA, June 1-5 (2008)

Guan, H., Shao, C., Wen, S., Chen, B., Gong, J. & Yang, X. (2003). *Inorg. Chem. Commun.* 6, 1302

HILLS Inc. (2011). Hills Nano Technology Brochure, West Melbourne, FL, USA

Hohman, M. M., Shin, M., Rutledge, G. C. & Brenner, M. P. (2001). Electrospinning and electrically forced jets. I. Stability theory, *Phys. Fluids* 13, 2201

Hohman, M. M., Shin, M., Rutledge, G. C. & Brenner, M. P. (2001). Electrospinning and electrically forced jets. II. Applications, *Phys. Fluids* 13, 2221

Jaroszczyk, T., Petrik, S. & Donahue, K. (2009). The Role of Nanofiber Filter Media in Motor Vehicle Air Filtration, In: *Proceedings of 4th Biennial Conference on Emissions Solutions in Transportation*, AFS, Ann Arbor, MI, Oct 5-8 (2009)

Jirsak, O., Sanetrnik, F., Lukas, D., Kotek, V., Martinova, L. & Chaloupek, J. (2005). A method of nanofibers production from a polymer solution using electrostatic spinning and a device for carrying out the method, The Patent Cooperation Treaty WO 2005/024101

Kavan, L. & Grätzel, M. (2002). *Electrochem. Solid-State Lett.*, 5, A39, (2002)

Kirichenko, V., Filatov, Y., & Budyka, A. (2007). *Electrospinning of Micro- and Nanofibers: Fundamentals in Separation and Filtration Processes,* Begell House, USA

Lukas, D., Sarkar A. & Pokorny, P. (2008). *J. Appl. Phys.* 103, 084309 (2008)

NanoStatics (2007). Company Brochure, available at www.nanostatics.com

Petrik, S. & Maly, M. (2009). Production Nozzle-Less Electrospinning Nanofiber Technology, In: *Mater. Res. Soc. Symp. Proc. Vol. 1240, 1240-WW03-07,* Boston, MA, Nov. 2009

Proceedings of *„Nanofibers for the 3rd Millenium – Nano for Life"* Conference, Prague, March 11-12 (2009)

Ramakrishna, S., Fujihara, K., Teo, W., Lim, T., & Ma, Z. (2005). *An Introduction to Electrospinning and Nanofibers,* World Scientific, Singapore

Reneker, D. H., 2009, personal communication.

Rubacek, L. & Duchoslav, J. (2008). Electrospun Nanofiber Layers for Applications in Electrochemical Devices, In: *Proceedings of NSTI Nanotech Conference*, Boston, MA, June 1-5 (2008)

Shin, Y. M., Hohman, M. M., Brenner, M. P. & Rutledge, G. C. (2001). Electrospinning: A whipping fluid jet generates submicron polymer fibers, *Appl. Phys. Lett.* 78, 1149

Taylor, G. F. & Van Dyke, M. D. (1969). *Proc. R. Soc. London, Ser. A* 313, 453

Thompson, C, J., Chase, G. G., Yarin, A. L. & Reneker, D. H. (2007). Effects of parameters on nanofiber diameter determined from electrospinning model, *Polymer,* 48, 6913

Yu, J. H., Fridrikh, S. V. & Rutledge, G. C. (2006). The role of elasticity in the formation of electrospun fibers, *Polymer* 47, 4789

TOPTEC (2011). Company Brochure, available at www.toptec.co.kr

2

Needleless Electrospinning: Developments and Performances

Haitao Niu[1], Xungai Wang[1,2] and Tong Lin[1]
[1]*Centre for Material and Fibre Innovation, Deakin University, Geelong*
[2]*School of Textile Science and Engineering, Wuhan Textile University*
[1]*Australia*
[2]*China*

1. Introduction

Electrospinning technique has attracted a lot of interests recently, although it was invented in as early as 1934 by Anton (Anton, 1934). A basic electrospinning setup normally comprises a high voltage power supply, a syringe needle connected to power supply, and a counter-electrode collector as shown in Fig. 1. During electrospinning, a high electric voltage is applied to the polymer solution, which highly electrifies the solution droplet at the needle tip (Li & Xia, 2004). As a result, the solution droplet at the needle tip receives electric forces, drawing itself toward the opposite electrode, thus deforming into a conical shape (also known as "Taylor cone" (Taylor, 1969)). When the electric force overcomes the surface tension of the polymer solution, the polymer solution ejects off the tip of the "Taylor cone" to form a polymer jet. The charged jet is stretched by the strong electric force into a fine filament. Randomly deposited dry fibers can be obtained on the collector due to the evaporation of solvent in the filament. There are many factors affecting the electrospinning process and fiber properties, including polymer materials (*e.g.* polymer structure, molecular weight, solubility), solvent (*e.g.* boiling point, dielectric properties), solution properties (*e.g.* viscosity, concentration, conductivity, surface tension), operating conditions (*e.g.* applied voltage, collecting distance, flow rate), and ambient environment (*e.g.* temperature, gas environment, humidity).

Electrospun nanofibers exhibit many unique characteristics, such as high surface-to-mass ratio, high porosity with excellent pore interconnectivity, flexibility with reasonable strength, extensive selection of polymer materials, ability to incorporate other materials (*e.g.* chemicals, polymers, biomaterials and nanoparticles) into nanofibers through electrospinning, and ability to control secondary structures of nanofibers in order to prepare nanofibers with core/sheath structure, side-by-side structure, hollow nanofibers and nanofibers with porous structure (Chronakis, 2005). These characteristics enable electrospun nanofibers to find applications in filtrations, affinity membranes, recovery of metal ions, tissue engineering scaffolds, release control, catalyst and enzyme carriers, sensors and energy storage (Fang *et al.*, 2008). In spite of the wide applications, electrospun nanofibers are produced at a low production rate when conventional needle electrospinning setup is used, which hinders their commercialization. Electrospinning with large scale nanofiber production ability has been explored, and some inspiring results have been achieved. This chapter summarizes the recent research progress in large scale electrospinning technologies.

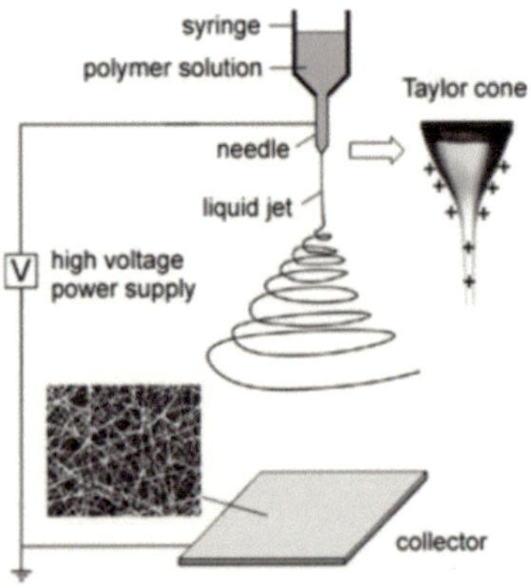

Fig. 1. Schematic illustration of a basic electrospinning setup, the Taylor cone, and a SEM image of electrospun nanofibers (Li & Xia, 2004).

2. Downward multi-jet electrospinning

A straightforward way to increase the electrospinning throughput is to use multi-jet spinnerets as shown in Fig. 2a. The fiber productivity can be simply increased by increasing the jet number (Varesano *et al.*, 2009, 2010; Yang *et al.*, 2010). However, multi-jet electrospinning has shown strong repulsion among the jets and this may lead to reduced fiber production rate and poor fiber quality, which is the main obstacle to practical application. To reduce the jet repulsion, jets have to be set at an appropriate distance, and a large space is required to accommodate the needles for the mass nanofiber production.
To stabilize and optimize the electrospinning process an extra-cylindrical electrode has been used as an auxiliary electrode to cover the multi-jet spinneret (Kim et al., 2006). As shown in Fig. 2b, the presence of the external electrode dramatically reduces the fiber deposition area, thus improving the fiber production rate. Nevertheless, coarser fibers were observed, because the auxiliary electrode shortened the chaotic motion of multi-jets in electrospinning.

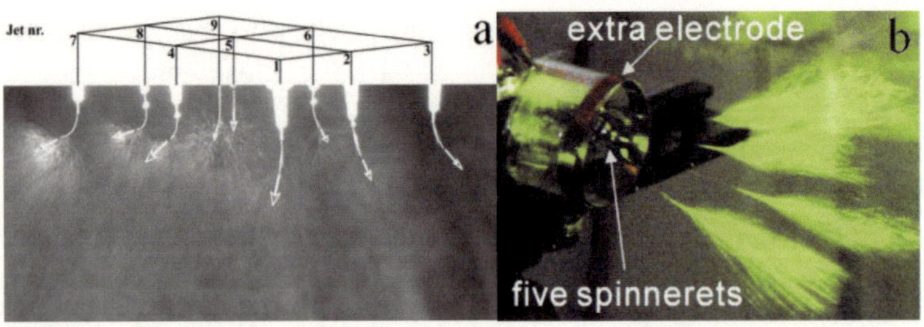

Fig. 2. (a) A multi-jet electrospinning setup (Theron et al., 2005), (b) multi-jet electrospinning with a cylindrical auxiliary electrode (Kim et al., 2006).

In electrospinning, the ejected solution jet carries a large amount of charges, which drive the jet stretching and fiber deposition on the collector. According to the electrospinning mechanism, it can be deduced that interferences in multi-jet electrospinning are unable to be eliminated completely, which will be a barrier to the industrialization of multi-jet electrospinning. Furthermore, the successful operation of multi-jet electrospinning requires a regular cleaning system to avoid the blockage of the needle nozzles. Setting the cleaning device for each needle makes it almost impossible to use multi-jet electrospinning for mass production of nanofibers.

In addition to the multi-jet electrospinning, porous tubes have also been used as electrospinning spinnerets to improve the fiber productivity. This system is herein still classified into multi-jet electrospinning because the electrospinning process is based on conveying solutions inside the tube channels. A porous polyethylene tube with a vertical axis was used to electrospin nanofibers (Fig. 3a) (Dosunmu *et al.*, 2006). The production rate was reported to be 250 times greater than that of single needle electrospinning. However, the SEM results of obtained nanofibers showed large variations in the fiber diameter.

In another example of tube electrospinning (horizontal tube), the polymer solution was pushed through the tube wall with many holes at 1 ~ 2 kPa pressure (Fig. 3b) (Varabhas *et al.*, 2008). This setup can only produce 0.3 ~ 0.5 g/hr of nanofibers due to the small number of holes (fiber generators) that can be drilled per unit area. Although it was mentioned that the production rates can be easily scaled up by increasing the tube length and the number of holes, the space between holes can't be reduced much because of the electric field repulsion between the jets. The strong jet interference in this setup can even result in nanofiber belt instead of fiber web (Varabhas *et al.*, 2008).

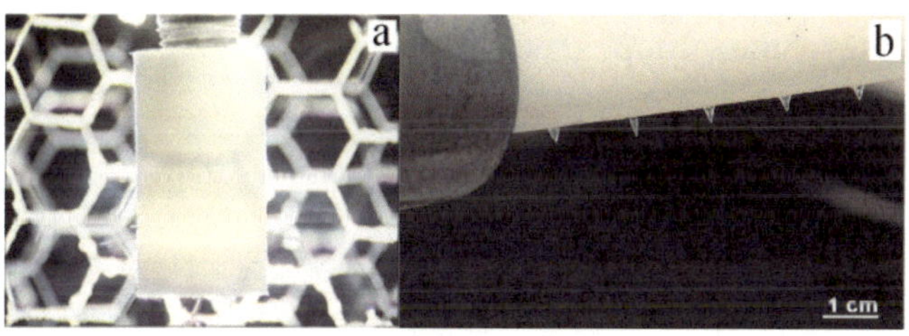

Fig. 3. (a) Electrospinning using a vertical tubular foam spinneret (Dosunmu *et al.*, 2006), (b) horizontal tube electrospinning (Varabhas *et al.*, 2008).

Wang *et al.* reported a conical wire coil electrospinning spinneret, which can work at up to 70 kV without causing corona discharges (Wang et al., 2009). As shown in Fig. 4a~c, the spinning solution was held in the wire cone without using any solution channels. Due to the large surface tension and visco-elasticity, the polymer solution can be retained inside the wire cone. When a high electric voltage was applied to the wire coil, solution was stretched out from the wire surface and the gap between wires to form solution jets. Without using defined solution channels, the fiber ejected independently, eliminating the limitation of jet number that was formed in multi-jet electrospinning and tube electrospinning. In comparison with conventional needle electrospinning, wire coil electrospinning can improve the electrospinning throughput noticeably and produce nanofibers with smaller fiber diameter (Fig. 4d). The results also indicated that fibers prepared by coil electrospinning had a wider diameter distribution than those produced by the needle electrospinning.

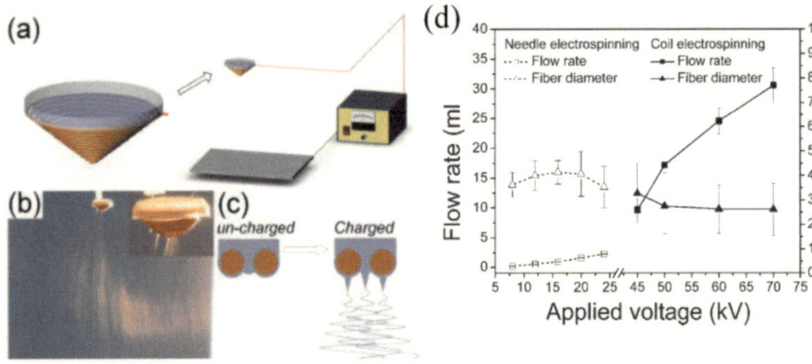

Fig. 4. (a) Schematic illustration of conical wire coil electrospinning setup, (b) photograph of the electrospinning process, (c) illustration of jet formation, (d) comparison between needle electrospinning and coil electrospinning (PVA concentration = 9 wt%; collecting distance = 15 cm) (Wang et al., 2009).

An edge-plate electrospinning setup for improving the fiber productivity was reported by Thoppey *et al* (Thoppey *et al.*, 2010). A plate (Angle with respect to horizontal $\theta = 40°$) for retaining solution was used as a spinneret to electrospin polyethylene oxide (PEO) nanofibers (Fig. 5a). In this method an electrically insulated reservoir connected with one or more plastic pipettes supplied solution to the charged plate, and each pipette supplied a solution stream as jet initiation site. It was found that the production rate was increased by over 5 times even using a single spinning site (one pipette) without getting coarser fibers and wider diameter distribution (Fig. 5 b & c). The surface tension of polymer solution plays a vital role, and the plate angle must be set appropriately according to the solution properties, otherwise solution dripping may occur.

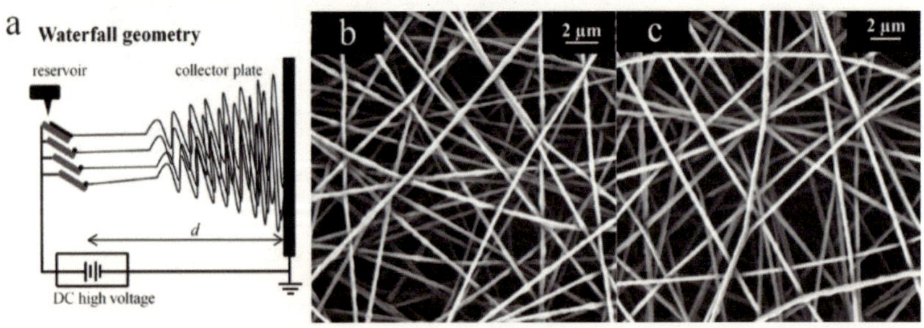

Fig. 5. (a) Schematic illustration of plate edge electrospinning, (b) & (c) SEM images of nanofibers electrospun from (b) conventional needle electrospinning (collecting distance = 15 cm, applied voltage = 11 kV) and (c) edge-plate geometry (collecting distance = 35 cm, applied voltage = 28 kV) (Thoppey *et al.*, 2010).

In another electrospinning design, a solution reservoir was used to provide spinning solution to a metal roller electrospinning spinneret (Fig. 6) (Tang *et al.*, 2010). The polymer solution droplets were splashed onto the surface of a metal roller by a solution distributor, which had a hole at the bottom. When the voltage was applied, solution droplets adhering on the surface of metal roller spinneret were ejected and stretched under the electric force to form nanofibers. This setup was proposed to have the ability to perform electrospinning with improved fiber production rate.

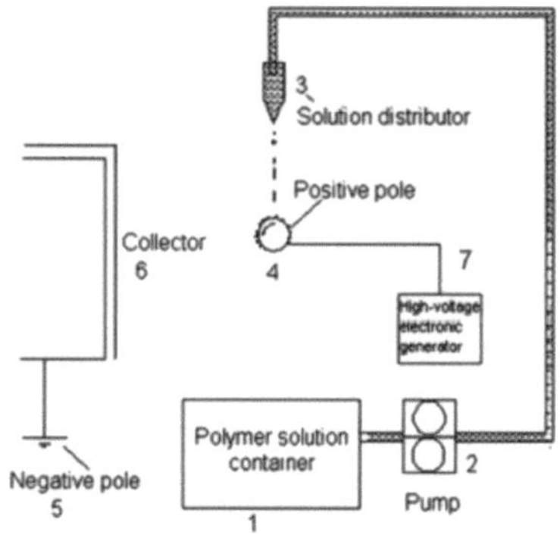

Fig. 6. Schematic illustration of splashing electrospinning setup (Tang *et al.*, 2010).

A rotary cone was used as electrospinning spinneret to perform electrospinning recently, which used a glass pipe to supply the PVP solution to the cone to ensure enough solution for continuous electrospinning (Fig. 7a) (Lu *et al.*, 2010). The electrospinning throughput of this setup was reported to be 1000 times larger than that of conventional needle electrospinning. The morphologies of nanofibers prepared by the cone electrospinning were nearly the same as those produced by conventional needle electrospinning (Fig. 7 b&c).

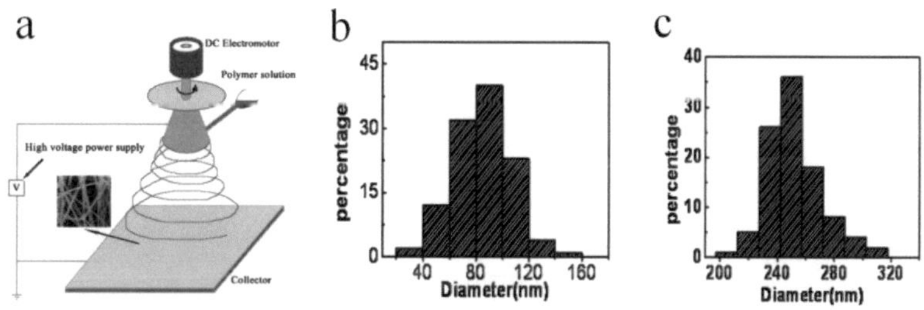

Fig. 7. (a) Schematic illustration of the rotary cone electrospinning setup (inset: SEM image of collected PVP nanofibers, rotational speed of cone = 100 rpm, applied voltage = 30 kV, collecting distance = 20 cm, solution throughput = 10 g min^{-1}), (b) fiber diameter distribution of needle electrospinning, (c) fiber diameter distribution of rotary cone electrospinning (Lu *et al.*, 2010).

3. Upward needleless electrospinning

3.1 Electrospinning techniques

A good electrospinning method suitable for manufacturing nanofibers should have minimal dependences on the fluidic channel numbers to improve the fiber productivity. It should be universal for processing polymer solutions of different properties. Some upward needleless electrospinning setups have good potential.

Yarin and Zussman (Yarin & Zussman, 2004) reported a two-layer-fluid electrospinning setup (Fig. 8a) that could dispose of the problems related to multi-jet electrospinning. In this setup, the lower fluid layer was a ferromagnetic suspension and the upper layer was the polymer solution to be spun. During electrospinning, when a normal magnetic field was applied to the system, steady vertical spikes were formed perturbing the interlayer interface. As a result of applying a high voltage to the fluid at the same time, thousands of jetting ejected upward (Fig.

8b). This upward electrospinning system required a complicated setup and the resultant nanofibers had large fiber diameter and wide diameter distribution (Fig. 8c).

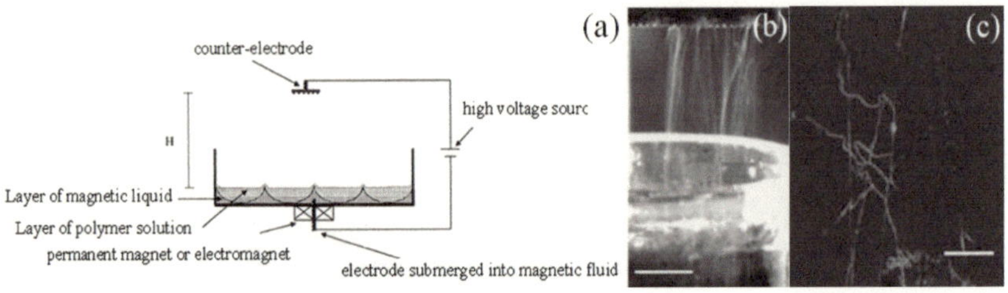

Fig. 8. (a) A two-layer-fluid electrospinning setup, (b) multiple jets ejected toward the counter-electrode, (c) an image of as-spun fibers (scale bar is 50 mm) (Yarin & Zussman, 2004).

In another upward electrospinning work, bubbles or humps were generated on the free surface of a polymer solution to initiate the electrospinning process (Liu et al., 2008). Unlike the previous design, a high pressure gas was used by inserting a gas tube to the bottom of the solution reservoir. A flat aluminum plate was used as collector above the solution. When a high voltage was applied to the solution, Taylor cones were easily formed from the humps. The fiber production rate was reported to depend on the gas pressure, the solution properties and the applied voltage. However, fibers prepared by this method contained large beads.

Jirsak *et al* (Jirsak *et al.*, 2005) invented a needleless electrospinning setup by using a rotating roller as the nanofiber generator. When the roller was partially immersed into a polymer solution and slowly rotates, the polymer solution was loaded onto the upper roller surface. Upon applying a high voltage to the electrospinning system, an enormous number of solution jets can be generated from the roller surface upward (Fig. 9). This setup has been commercialized by Elmarco Co with the brand name "Nanospider™".

Fig. 9. Roller electrospinning process (left), and commercialized Nanospider™ (right) (Jirsak *et al.*, 2005).

Lukas *et al* (Lukas *et al.*, 2008) developed a one-dimensional electrohydrodynamic theory to describe the electrospinning of conductive liquids from an open flat surface based on the phenomenon that nanofibers can be electrospun from linear clefts even without being aided by a magnetic fluid underneath (Fig. 10). This work added a general approach toward studying dynamics of surface waves. During electrospinning from a free liquid surface, due to the electric force the amplitude of a characteristic wavelength boundlessly grew faster than the others. The fastest growing stationary wave marked the onset of electrospinning from a free liquid surface with its jets originating from the wave crests. The proposed theory predicated the critical values of the phenomenon, the critical field strength and corresponding critical inter-jet distance, and the inter-jet distance for field strengths above the critical value. The theory also predicted relaxation time necessary for spontaneous jetting after a high voltage was applied, and explained the fundamental of upward needleless electrospinning.

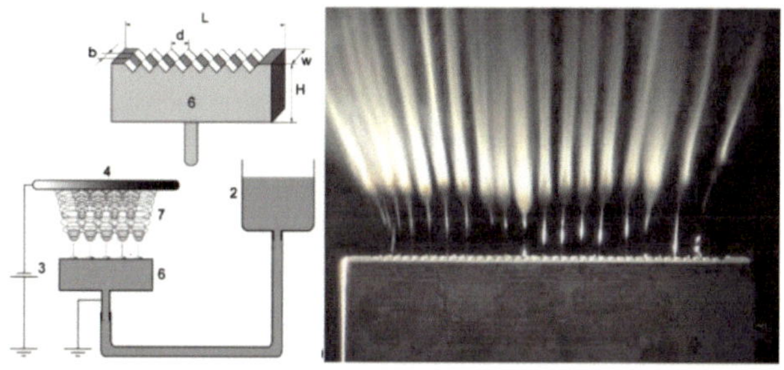

Fig. 10. Schematic illustration of a linear cleft electrospinning setup (left) and electrospinning process (right) (Lukas *et al.*, 2008).

Needleless electrospinning process with rotating spinneret can be summarized as that, the rotation of spinneret loads a thin layer of polymer solution onto the spinneret surface. The rotation and perturbance create conical spikes on the surface of this solution layer. When a high voltage is applied to the spinneret, these spikes tend to concentrate charges and amplify the perturbance and the fluid around the spikes is drawn to these spikes under high electric force. Taylor cones are thus formed. Fine solution jets are then ejected from the tips of these Taylor cones, when the electric force is large enough.

Using a similar design, Niu *et al* systematically compared the needleless electrospinning using different rotary fiber-generators (disc, cylinder or ball) (Niu *et al.*, 2009) (Fig. 11). When PVA solution was charged with a high electric voltage via a copper wire inside the solution vessel, numerous jets/filaments were generated from the spinnerets, which deposited on the collector (*e.g.* rotating drum). With the rotation of the spinneret, PVA solution was loaded onto the spinneret surface constantly, leading to continuous generation of polymer jets/filaments. They also used finite element method to analyze electric field and examine the influence of spinneret shape on the electric field profile. They found that the spinneret with a highly concentrated and evenly distributed electric field is the key to efficient needleless electrospinning of uniform nanofibers.

Based on this understanding, a new needleless electrospinning system using a spiral coil wire as fiber generator was invented (Lin et al., 2010). As shown in Fig. 11, when the applied voltage exceeded a critical value, numerous polymer jets were generated from the wire surface. It was also found that the spiral coil had higher fiber production rate than cylinder spinneret of the same dimension, and the fiber diameter was finer with a narrower diameter distribution (Wang *et al.*, 2009).

High throughput production of nanofibers in the name of "Tip-less Electrospinning" (TLES) has been demonstrated by Wu *et al* (Wu et al., 2010), using a circular cylinder as the fiber generator. This is also an upward needleless electrospinning system using rotating spinneret. The solution ejecting process from the generator surface is shown in Fig. 12. Experimental results showed that the yield of poly(ethylene oxide) nanofibers can be more than 260 times in weight compared to that of a single-jet electrospinning.

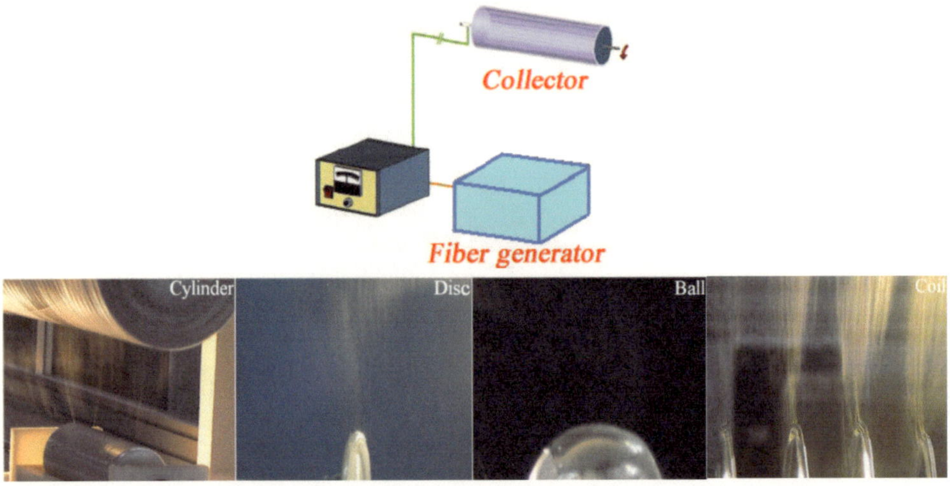

Fig. 11. Schematic illustrations of needleless electrospinning setup, and cylinder, disc, ball, and spiral coil electrospinning processes (Niu *et al.*, 2009; Lin et al., 2010).

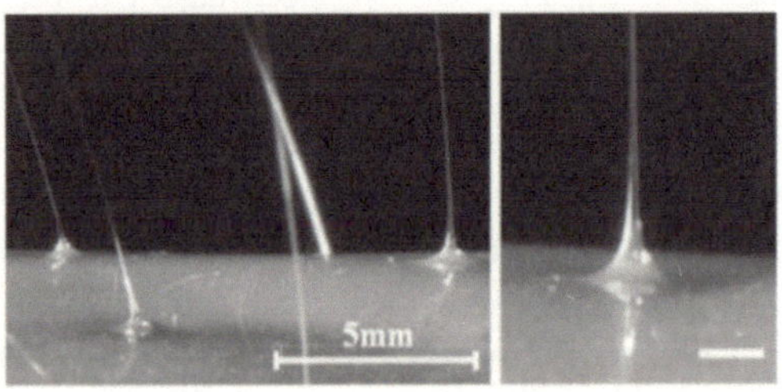

Fig. 12. Fiber ejecting process of a cylinder electrospinning (applied voltage = 70 kV, collecting distance = 15 cm, cylinder diameter = 3 cm) (Wu et al., 2010).

There are significant differences between the needle and upward needleless electrospinning processes. In the upward needleless electrospinning, Taylor cones are created on the surface of polymer solution. If the Taylor cone is stable, it will move together with the surface of rotating roller and produces a solution jet under the strong electric field. Therefore, there must be strong inter-molecular interactions among polymer macromolecules in the solution to stabilize the Taylor cone given that Taylor cone is stretched into a fine jet and deposited on the collector as solid fibers. Different to the conventional needle electrospinning in which Taylor cone is generated and stabilized through constantly feeding polymer solution through the needle, the upward needleless electrospinning forms its Taylor cones by sucking up the solution covering the surrounding fiber generator (Cengiz & Jirsak, 2009). It was observed that the base diameter of Taylor cone in cylinder electrospinning reduced from 1.2 mm to 0.3 mm in the beginning and the end of the electrospinning, respectively (initial polymer film thickness = 1 mm). If the film thickness is further reduced, no Taylor cones or nanofibers can be generated (Wu et al., 2010). Therefore, the solution must have a suitable rheological property. Furthermore, a higher electric voltage is required to initiate the needleless electrospinning, because Taylor cone is formed due to the wave fluctuation.

3.2 Parameters affecting needleless electrospinning

3.2.1 Applied voltage

Applied voltage is a very important parameter affecting both the electrospinning process and fiber properties. A high applied voltage (usually over 40 kV) is usually required to initiate an upward needleless electrospinning. The critical voltage required to initiate electrospinning is closely related to the material properties, ambient environment (*e.g.* humidity, temperature) and collecting distance. High critical voltages are required when either the solution concentration or the collecting distance increases (Fig. 13 a & b). This can be explained that at a high concentration solution, the increased viscosity requires a larger electric force to create Taylor cones. Increasing the collecting distance reduces the electric field strength in the electrospinning zone. It was also found that a solution film in the thickness range of 0.5 mm ~ 2 mm was in favor of forming Taylor cones, and could reduce the critical voltage (Fig. 13c) (Wu et al., 2010).

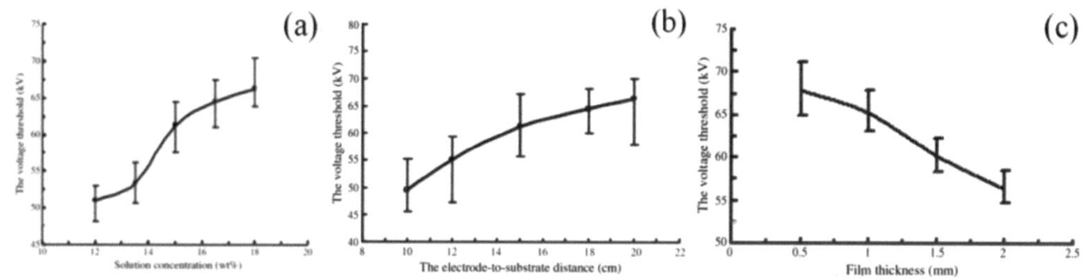

Fig. 13. Critical voltage versus (a) solution concentration (film thickness = 1 mm, electrode-to-substrate distance = 15 cm, (b) electrode-to-substrate distance (film thickness = 1 mm, electrode-to-substrate distance = 15 cm, and polymer solution = 15 wt%), and (c) film thickness (electrode-to-substrate distance = 15 cm and polymer solution = 15 wt%) (Wu et al., 2010).

The spinneret geometry also affects the critical voltage. The spinneret that can generate an intensified electric field (disc spinneret) requires a lower voltage to initiate electrospinning. Niu *et al* found that disc spinneret and cylinder spinneret had different critical voltages for initiating electrospinning, 42 kV and 47 kV, respectively. For the cylinder spinneret at a low applied voltage, the jets were only generated from two end areas, and no jets/filaments were produced from the middle cylinder surface until the applied voltage was above 57 kV. Further increasing the applied voltage led to the generation of jets from the entire cylinder surface. The disc can easily generate high intensity electric field. This was why the disc generated nanofibers regardless of the applied voltage value, as long as the voltage was above the critical value. The fiber morphology was mainly affected by polymer concentration, but little by applied voltage.

Niu *et al* also found that increasing the applied voltage from 47 to 62 kV had little effect on the average fiber diameter in both disc and cylinder electrospinning systems. For the disc electrospinning, the fiber diameter distribution became narrower when the applied voltage was increased (Fig. 14a). For the cylinder electrospinning, the average fiber diameter and diameter distribution showed a very small dependence on the applied voltage. The applied voltage affected the fiber productivity significantly. The electrospinning throughput in both electrospinning systems increased with the increase in applied voltage. When the applied voltage was increased from 57 kV to 62 kV, the fiber productivities of the cylinder electrospinning and disc electrospinning were very similar, indicating that disc spinneret is a high efficiency fiber generator, although the cylinder spinneret is 100 times longer than the disc spinneret (Fig. 14b).

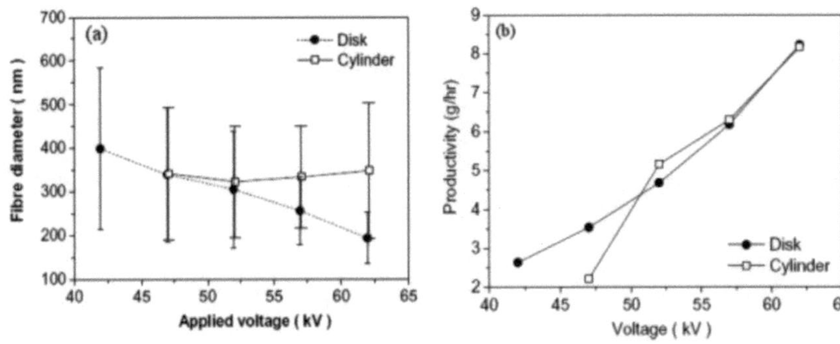

Fig. 14. Influences of applied voltage on (a) fiber diameter, and (b) fiber productivity (collecting distance = 13 cm, PVA concentration = 9 wt%, cylinder diameter = 80 mm, cylinder rim radius = 2 mm, disc diameter = 80 mm, disc thickness = 2 mm) (Niu *et al.*, 2009).

Using a spiral coil, the productivity of electrospun PVA nanofibers increased with the applied voltage (45 kV ~ 60 kV) (Wang *et al.*, 2009). The trends that the fiber diameter decreased but the production rate increased with increasing applied voltage were also reported by other researchers using different polymer systems (Wu *et al.*, 2010). All these results suggest that the applied voltage plays a key role in improving the fiber production rate in the upward needleless electrospinning.

3.2.2 Collecting distance

Reducing the fiber collecting distance has a similar effect to increasing the applied voltage, but the collecting distance can't be reduced infinitely. To collect solid nanofibers, the collecting distance must be large enough to ensure sufficient solvent evaporation from the jet before deposition. The minimal collecting distance is dependent on the solution property and the geometry of fiber generator. For example, the minimal collecting distance for the PEO solution was 10 cm when humidity and temperature are 43% RH and 22 °C, respectively (Wu *et al.*, 2010), and 11 cm for the PVA solution (Niu *et al.*, 2009). However, the collecting distance should not be too large either, since a larger distance would require a higher applied voltage to initiate electrospinning, which may cause corona discharge. A good balance should be maintained between the applied voltage and the collecting distance for successful upward needleless electrospinning.

3.2.3 Rotating speed of fiber generator

The rotating speed of fiber generator in upward needleless electrospinning may be varied over a wide range, which has an influence on electrospinning. A rotating speed of 40 rpm was reported in two studies (Niu et al., 2009, Wang et al., 2009). Cengiz and Jirsak used 3.2~4 rpm for their designs (Cengiz & Jirsak, 2009, Jirsak et al., 2010). Cengiz *et al* (Cengiz et al., 2009) found that the fiber diameter decreased when the rotating speed of a cylinder fiber generator (8 cm in length and 2 cm in diameter) was increased. It is normally believed that increasing rotating speed reduces the life span of the Taylor cone. Very fast spinneret rotating speed is not good for improving the fiber productivity.

3.2.4 Polymer concentration

The polymer solution concentration plays a vital role in upward needleless electrospinning. When the polymer solution concentration is too low, polymer beads rather than nanofibers are usually produced. When the concentration is very high, the polymer solution becomes too thick to be stretched into jets. As long as polymer solutions can be electrospun into nanofibers successfully, the polymer concentration doesn't affect the fiber diameter significantly.

Niu *et al* have studied the effect of PVA concentration on needleless electrospinning. They found that when PVA concentration changed from 8 wt% to 11 wt%, fiber diameter did not change significantly. The nanofibers spun from the disc spinneret had a much narrower diameter distribution than those from the cylinder spinneret. However, the electrospinning throughput was highly dependent on the solution concentration. When the PVA concentration was in the range of 8.0 ~ 11.0 wt%, the productivity of disc electrospinning increased with increasing PVA concentration, while the cylinder electrospinning was highly affected by the PVA concentration. When 9 wt% PVA solution was electrospun with an applied voltage of 52 kV, nanofibers were generated from the whole cylinder surface. 9 wt% PVA also gave the largest electrospinning throughput (Fig. 15). Higher PVA concentration resulted in generation of nanofibers only from cylinder ends.

Fig. 15. Influence of PVA concentration on (a) fiber diameter, and (b) fiber productivity (collecting distance = 13 cm, applied voltage = 57 kV, cylinder diameter = 80 mm, cylinder rim radius = 2 mm, disc diameter = 80 mm, disc thickness = 2 mm) (Niu *et al.*, 2009).

The upward needleless electrospinning has good capability of producing nanofibers from different polymer solution systems. In addition to water solvent system, organic solvent systems, *e.g.* dimethylformamide (DMF), have already been used to prepare nanofibers. The polyimide precursor (polyamic acid) produced from 4, 4′-oxydiphthalic anhydride and 4, 4′-oxydianiline in DMF has been electrospun into nanofibers, on a polypropylene spunbond supporting web, with diameters in the range 143 ~ 470 nm using roller electrospinning (Jirsak *et al.*, 2010). Consequently, these polyamic acid fibers can be heated to convert to polyimide nanofibers. Another example is the polyurethane (PU) nanofibers electrospun using the roller electrospinning from PU/DMF solution (Cengiz & Jirsak, 2009).
Chain entanglements and molecular weight are two important properties that can affect the electrospinning process. Shenoy *et al* (Shenoy *et al.*, 2005) reported that the macromolecule chain entanglement characterized by the entanglement number in solution $(n_e)_{soln}$ can ultimately determine the formation of beads or fibers. Beads are formed when $(n_e)_{soln}$ is below 2 and fibers are produced when $(n_e)_{soln}$ is over 2.5. It was found that the PVA with a molecular weight of 67.000 was not spinnable, while other solutions (molecular weight: 80.000 and 150.000) could be successfully electrospun into nanofibers (Cengiz et al., 2009a). The electrospinning throughputs of both PVA solutions increased with increasing PVA concentration (Fig. 16).

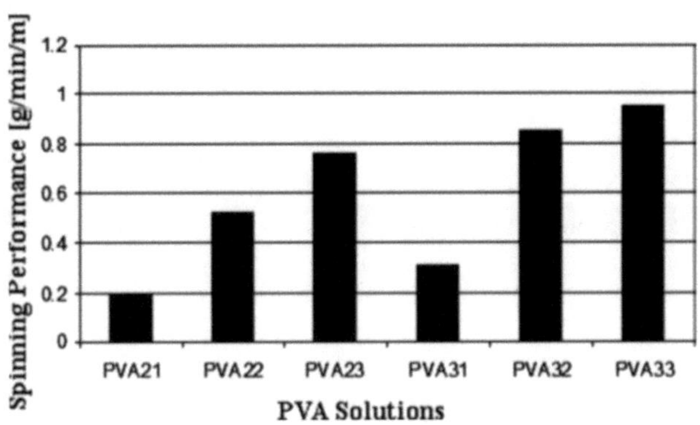

Fig. 16. Influence of polymer solution on the electrospinning performance (cylinder length = 14 cm, cylinder diameter = 2 cm, cylinder rotating speed = 3.2 rpm, collecting distance = 11 cm, applied voltage = 81.2 kV) (Cengiz et al., 2009a).

The addition of salt to the solution can increase the charge density of the solution and improve the electrospinning process. With low or no tetraethylammoniumbromide (TEAB) salt, the PU solution can't be electrospun into nanofibers. When a small amount of TEAB was added into polyurethane solution, the electrospinning was significantly improved. Both fiber diameter and productivity increased with the increase in the TEAB concentration (Cengiz & Jirsak, 2009). The reason for this is that TEAB increases the electric conductivity resulting in the improvement in electrospinning ability.
The upward needleless electrospinning possesses many advantages over the conventional needle electrospinning process. The conglobation of dopants in the electrospinning solution

can easily block the needle spinneret and cease the electrospinning process. In the absence of capillary spinneret, there will be no nozzle blockage in the needleless electrospinning. PVA nanofibers containing carbon nanotubes (CNTs) have been successfully prepared using a roller electrospinning process (Kostakova *et al.*, 2009).

It was also found that the spinneret geometry significantly affected the electrospinning process and fiber property (Fig. 17). Under the same electrospinning conditions, the disc produced the finest nanofibers with the narrowest fiber diameter distribution. Cylinders produced coarse nanofibers with the largest fiber productivity. Compared to disc and cylinder spinnerets, ball spinneret produced coarser nanofibers with lower productivity. The spiral coil electrospinning combined the advantages of cylinder and disc spinnerets, by producing fine and uniform nanofibers at a high productivity (Lin *et al.*, 2010).

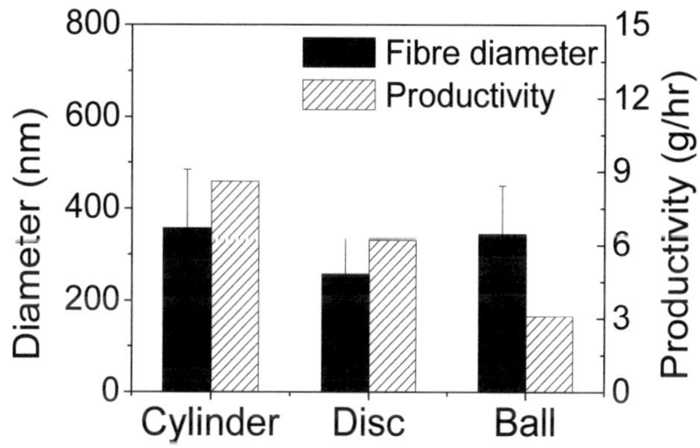

Fig. 17. Comparisons among cylinder, disc and ball spinnerets in fiber diameter and productivity (applied voltage = 57 kV, collecting distance = 13 cm, cylinder diameter = 80 mm, cylinder rim radius = 5 mm, disc diameter = 80 mm, disc thickness = 2 mm, ball diameter = 80 mm).

4. Fiber collection in upward needleless electrospinning

The fiber collection in needleless electrospinning can generally be classified into three types: grounded plate, rotating drum and moving substrate. The plate collector is normally used for electrospinning at a moderately high production rate, *e.g.* multi-jet electrospinning and wire coil electrospinning. However, these fiber collectors are not suitable for upward needleless electrospinning systems, because the deposition of numerous nanofibers accumulates a large amount of charges on the collector, which can finally disturb or even stop the electrospinning process if they are not dissipated quickly. Moving collectors are therefore necessary in these systems. There are two types of moving collectors: rotating drum collector and moving substrate. The drum collector was used by Niu *et al* (Fig. 11).

Drum collector can also collect aligned nanofiber mat. When the mat reaches a certain thickness, it can be peeled off from the collector. Jirsak *et al* (Jirsak et al., 2010) used a moving substrate to collect nanofibers in their systems. The advantage of this type of fiber collection is that nanofibers can be collected continuously and the nanofiber web does not have to be very strong; such a collection system also allows the incorporation of nanofibers into the fiber collecting substrate itself, if required.

5. Electric field analysis

One of the vital conditions for initiating an electrospinning process is the high electric field intensity and the strong interactions between electric field and polymer fluid. The electric force has been identified as the driving force for electrospinning. Although the driving force in all electrospinning is the same, the electrospinning process could be influenced by the spinneret geometry. In needle electrospinning, only the polymer solution at the needle tip is under electric force. Induced by a high electric voltage, charges accumulate on a liquid surface, giving rise to electric forces. At a low applied voltage, under the influences of electric force, the droplet reduces its size so that the force balance is maintained. With an increase in the applied voltage, the shape of solution droplet evolves from the hemi-sphere to a cone shape (Taylor cone) with a high electric force concentrated at the tip of the Taylor cone. When the electric field reaches a critical value, the droplet at the cone tip overcomes its surface tension, to eject into the electric field formed between the tip and collector, and a solution jet is thus generated. Through the above analysis, one can conclude that jet initiation is determined by the applied voltage. In needle electrospinning, the critical applied voltage for electrospinning was proposed in equation (1) (Taylor, 1969):

$$V_c^2 = 4ln\left(\frac{2h}{R}\right)(1.3\pi R\gamma)(0.09) \tag{1}$$

where h is the distance from the needle tip to the collector, R denotes the needle outer radius, and γ is the surface tension. The factor 0.09 was inserted to predict the voltage. Ludas *et al* (Lukas *et al.*, 2008) have explained the self-organization of jets happening on a free liquid surface in needleless electrospinning process. The critical electric field intensity for electrospinning nanofibers was proposed as:

$$\mathrm{E_c} = \sqrt[4]{4\gamma\rho g/\varepsilon^2} \tag{2}$$

where ρ is liquid mass density, g is gravity acceleration, γ is surface tension, ε is the permittivity. In both the models, electric force plays a crucial role in the jet initiation.
In a static electric field, it is well known that the relationship between voltage and electric field intensity can be expressed as:

$$\mathrm{E} = -\nabla \mathrm{V} \tag{3}$$

Therefore, a surface with a higher curvature will have higher electric field intensity. In needleless electrospinning, because the surface electric field is highly determined by the spinneret shape, the electrospinning process is also influenced by the shape of the spinneret. The region with higher surface curvature can generate high intensity electric field, and electrospinning can therefore be initiated easily from this region.

The electric field of needleless electrospinning spinneret is quite different from needle spinneret. The electric fields of coil-wire and needle electrospinning spinnerets were calculated and compared to examine the relationship between electric field intensity and electrospinning performance (Wang et al., 2009). The intensified electric field is generated on the wire surface (Fig. 18 a & b), the needle spinneret generates intensified electric field at its tip, but with a lower intensity due to the lower applied voltage used (Fig. 18 c & d). This should be the reason as to why a coil-wire electrospinning can produce finer PVA nanofibers than a needle electrospinning (Wang et al., 2009).

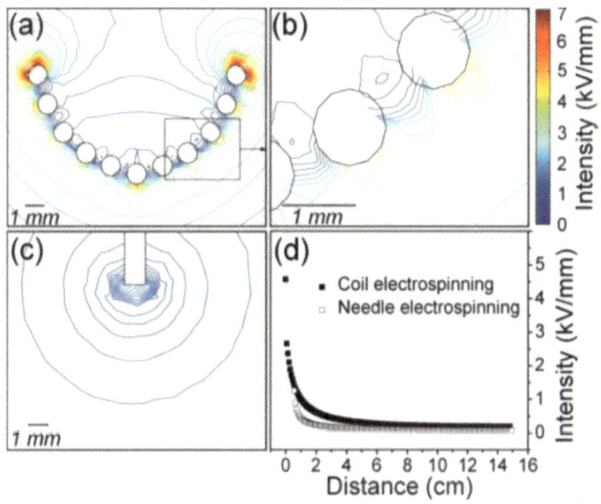

Fig. 18. Cross-sectional view of electric field intensity profiles on (a) & (b) conical coil spinneret (applied voltage = 60 kV), (c) needle spinneret (applied voltage = 22 kV), and (d) electric field intensity profiles along the electrospinning direction. Spinneret (0, 0), collector (0, 15) (Wang et al., 2009).

In the plate edge electrospinning, the sharp edge can generate strong electric field close to the edge, which decays rapidly toward to the ground, while the electric field profile is similar to that of a needle spinneret (Fig. 19 a & b). Many spinning sites can be formed along the entire edge, generating much more solution jets than employing an array of needles (Thoppey *et al.*, 2010). In the plate stack spinneret, more spinning sites can still be formed along these plate edges for generating more nanofibers, although the electric field intensity gradient close to the plate edge is reduced (Fig. 19 c).
Our recent study indicated that the difference in geometry between the cylinder, disc, ball, and coil spinnerets led to totally different electric field intensity profiles. As shown in Fig. 20, much higher electric field intensity is formed at the cylinder top ends than the middle top surface. As a result, the cylinder ends produced nanofibers more easily than the middle area and the nanofibers produced from the whole cylinder had a wide diameter distribution. If the rim radius of the cylinder ends was reduced, stronger electric field tended to be generated. However the electric field in the cylinder middle area was little affected by the rim radius. When the rim radius was 5 mm, the cylinder spinneret resulted in the highest production rate, while the cylinder length and diameter remained unchanged.

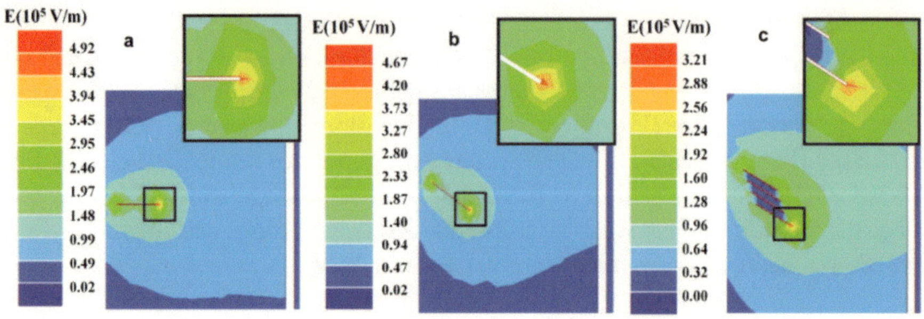

Fig. 19. Electric field intensity distribution of different electrospinning setups (collecting distance = 15 cm, applied voltage = 15 kV, insets are the magnifications of the indicated square areas in each figure), (a) conventional needle electrospinning, (b) edge-plate electrospinning, and (c) waterfall electrospinning (Thoppey *et al.*, 2010).

When the cylinder length and the rim radius were kept unchanged, the area with high electric field intensity shrank with reducing cylinder diameter, and the electric field intensity in the middle surface increased with a decrease in the cylinder diameter (Fig. 20a). Reducing cylinder diameter can contribute to the improvement in electrospinning throughput. It was also reported that when cylinder diameter increased from 1 cm to 6 cm, the critical voltage for electrospinning a PEO solution increased from about 25 kV to 61 kV (Wu et al., 2010).

Cylinders can produce nanofibers from the whole surface, but only when the applied voltage is high enough. This makes the cylinders less efficient compared with other spinnerets such as disc and spring coil, in terms of power consumption. The electric field is narrowly distributed on the disc top edge in Fig. 20b, and the intensity around the disc surface shows a high dependence on the disc thickness. Thinner disc produced higher electric field intensity around the disc circumference. Compared with the cylinders, discs required a lower critical voltage to initiate the electrospinning and produced fibers with a narrower fiber diameter distribution. The thinnest disc showed the largest fiber productivity.

High electric field intensity is mainly generated on the top half of the ball spinneret (Fig. 20c). The generated electric field was more evenly distributed along the ball surface compared to the cylinder spinnerets but had lower intensity. The experimental results also verified the theoretical calculation results that ball spinneret had low production rate and required a higher voltage to initiate electrospinning (Fig. 20c).

The coil spinneret can generate intensified electric field on each spiral. The coil length, coil distance, coil diameter, and wire diameter all showed significant influence on the generated electric field and electrospinning throughput. The electric field intensity decreased with the increase in coil length, thus less polymer jets were produced on each spiral. When the coil distance decreased from 8 cm to 1 cm, the productivity on each spiral decreased gradually since the electric field intensity decreased evidently. The spirals interfered each other when the pitch was too small (*e.g.* 2 cm), very strong electric field was formed on the side area of the coil but much weaker field in the center. The electric field intensity increased with increasing coil diameter, which increased the fiber productivity. The electric intensity increased with the decrease in wire diameter, and the productivity increased slightly. When wire diameter was higher than 6.35 mm, very low electric field intensity was formed, resulting in an evident drop in fiber productivity.

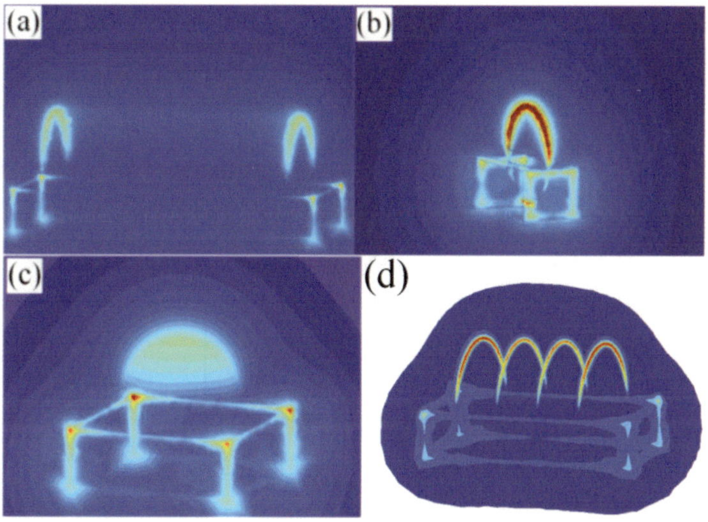

Fig. 20. Electric field intensity profile of (a) cylinder, (b) disc, (c) ball, and (d) coil spinnerets (Lin *et al.*, 2010).

6. Issues associated with needleless electrospinning

The upward needleless electrospinning has been proven to be the most successful needleless electrospinning system. The solution bath is normally open to air. The evaporation of solvent from solution can increase the solution viscosity and decrease the solution uniformity. To ensure good electrospinning ability, the solution in the bath must be calculated precisely.

Due to the formation of a large number of solution jets in a small space (from the spinneret to the collector), the concentration of organic solvent in the electrospinning zone could reach a high value during electrospinning. How to recycle the organic solvent efficiently has became a major issue in designing an upward needleless electrospinning system. An air ventilating system has been used to address this issue in the Nanospider™ (Fig. 21).

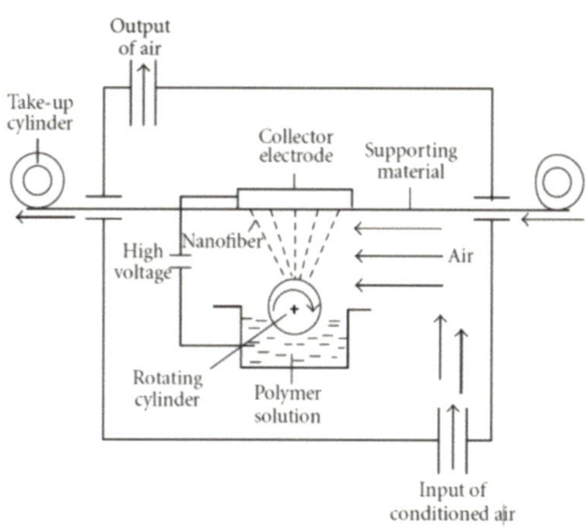

Fig. 21. Air ventilating system used in Nanospider™ (Jirsak *et al.* 2010).

7. Concluding remarks

Electrospun nanofibers have numerous applications in various fields. Developing an electrospinning technique for large-scale nanofiber production has become more and more important, as the conventional needle electrospinning has limited productivity and is only suitable for research purpose. Upward needleless electrospinning has been shown the ability to mass produce nanofibers and it is also the most successful design for practical applications. It is expected that the solvent can be recycled effectively so that the fibers are produced with minimal impact on the environment. An efficient needleless electrospinning system to produce nanofibers from thermoplastic polymers is yet to be developed.

8. References

Anton F. (1934). Process and apparatus for preparing artificial threads. *United States patent application 1975504.*

Cengiz F., Dao T. A. & Jirsak O. (2009). Influence of solution properties on the roller electrospinning of poly(vinyl alcohol). *Polymer Engineering & Science,* Vol. 50, No. 5, (May 2010), pp. 936-43, ISSN 1548-2634

Cengiz F. & Jirsak O. (2009). The effect of salt on the roller electrospinning of polyurethane nanofibers. *Fibers and Polymers,* Vol. 10, No. 2, pp. 177-84, ISSN 1229-9197

Cengiz F., Krucinska I., Gliscinska E., Chrzanowski M. & Goktepe F. (2009). Comparative Analysis of Various Electrospinning Methods of Nanofiber Formation. *FIBERS & TEXTILES in Eastern Europe,* Vol. 1, No. 72, (January/March 2009), pp. 13-19.

Chronakis I. S. (2005). Novel nanocomposites and nanoceramics based on polymer nanofibers using electrospinning process - A review. *Journal of Materials Processing Technology,* Vol. 167, No. 2-3, (August 2005), pp. 283-93, ISSN 0924-0136

Dosunmu O. O., Chase G. G., Kataphinan W. & Reneker D. H. (2006). Electrospinning of polymer nanofibers from multiple jets on a porous tubular surface. *Nanotechnology,* Vol. 17, No. 4, pp. 1123-27, ISSN 1361-6528

Fang J., Niu H., Lin T. & Wang X. G. (2008). Applications of electrospun nanofibers. *Chinese Science Bulletin,* Vol. 53, No. 15, pp. 2265-86, ISSN 1001-6538

Jirsak O., Sanetrnik F., Lukas D., Kotek V., Martinova L. & Chaloupek J. (2005). A method of nanofibers production from a polymer solution using electrostatic spinning and a device for carrying out the method. *WO 2005/024101 A1.*

Jirsak O., Sysel P., Sanetrnik F., Hruza J. & Chaloupek J. (2010). Polyamic acid nanofibers produced by needleless electrospinning. *Journal of Nanomaterials,* Vol. 2010, pp. 1-7, ISSN 1687-4110

Kim G., Choa Y. S. & Kim W. D. (2006). Stability analysis for multi-jets electrospinning process modified with a cylindrical electrode. *European Polymer Journal,* Vol. 42, No. 9, (September 2006), pp. 2031-38, ISSN 0014-3057

Kostakova E., Meszaros L. & Gregr J. (2009). Composite nanofibers produced by modified needleless electrospinning. *Materials Letters,* Vol. 63, No. 28, (November 2009), pp. 2419-22, ISSN 0167-577X

Li D. & Xia Y. 2004. Electrospinning of Nanofibers: Reinventing the Wheel? *Advanced Materials,* Vol. 16, No. 14, (July 2004), pp. 1151-70, ISSN 1521-4095

Lin T., Wang X., Wang X. & Niu H. (2010). Electrostatic spinning assembly. *WO/2010/043002*

Liu Y., He J. H. & Yu J. Y. (2008). Bubble-electrospinning: a novel method for making nanofibers. *Journal of Physics: Conference Series,* Vol. 96, pp. 012001, ISSN 1742-6596

Lu B., Wang Y., Liu Y., Duan H., Zhou J., Zhang Z., Li X., Wang W., Lan W. & Xie E. (2010). Superhigh-Throughput Needleless Electrospinning Using a Rotary Cone as Spinneret. *Small,* Vol. 6, No. 15, (August 2010), pp. 1612-16, ISSN 1613-6829

Lukas D., Sarkar A. & Pokorny P. (2008). Self-organization of jets in electrospinning from free liquid surface: A generalized approach. *Journal of Applied Physics,* Vol. 103, No.8, pp. 084309, ISSN 0021-8979

Niu H., Lin T. & Wang X. (2009). Needleless electrospinning. I. A comparison of cylinder and disk nozzles. *Journal of Applied Polymer Science,* Vol. 114, No. 6, (December 2009), pp. 3524-30, ISSN 1097-4628

Shenoy S. L., Bates W. D., Frisch H. L. & Wnek G E (2005). Role of chain entanglements on fiber formation during electrospinning of polymer solutions: good solvent, non-specific polymer-polymer interaction limit. *Polymer,* Vol. 46, No. 10, (April 2005), pp. 3372-84, ISSN 0032-3861

Tang S., Zeng Y. & Wang X. (2010). Splashing needleless electrospinning of nanofibers. *Polymer Engineering & Science,* Vol. 50, No. 11, (November 2010), pp. 2252-57, ISSN 1548-2634

Taylor G. (1964). Disintegration of Water Drops in an Electric Field. *Proceedings of the Royal Society of London. Series A. Mathematical and Physical Sciences,* Vol. 280, No. 1382, (July 1964), pp. 383-97, ISSN 1471-2946

Taylor G. (1969). Electrically driven jets. *Proceedings of the Royal Society of London. Series A, Mathematical and Physical Sciences,* Vol. 313, No. 1515, (December 1969), pp. 453-75, ISSN 1471-2946

Theron S. A., Yarin A. L., Zussman E. & Kroll E. (2005). Multiple jets in electrospinning: experiment and modeling. *Polymer,* Vol. 46, No. 9, (April 2005), pp. 2889-99, ISSN 0032-3861

Thoppey N. M., Bochinski J. R., Clarke L. I. & Gorga R. E. (2010). Unconfined fluid electrospun into high quality nanofibers from a plate edge. *Polymer,* Vol. 51, No. 21, (October 2010), pp. 4928-36, ISSN 0032-3861

Varabhas J. S., Chase G. G. & Reneker D. H. (2008). Electrospun nanofibers from a porous hollow tube. *Polymer,* Vol. 49, No. 19, (September 2008), pp. 4226-29, ISSN 0032-3861

Varesano A., Carletto R. A. & Mazzuchetti G. (2009). Experimental investigations on the multi-jet electrospinning process. *Journal of Materials Processing Technology,* Vol. 209, No. 11, (June 2009), pp. 5178-85, ISSN 0924-0136

Varesano A., Rombaldoni F., Mazzuchetti G., Tonin C. & Comotto R. (2010). Multi-jet nozzle electrospinning on textile substrates: observations on process and nanofiber mat deposition. *Polymer International,* Vol. 59, No. 12, (December 2010), pp. 1606-15, ISSN 1097-0126

Wang X., Niu H. T., Lin T. & Wang X. (2009). Needleless electrospinning of nanofibers with a conical wire coil. *Polymer Engineering & Science,* Vol. 49, No. 8, (August 2009), pp. 1582-86, ISSN 1548-2634

Wang X., Niu H. T., Wang X. & Lin T. (2009). Large-scale electrospinning of polymer nanofibers using needleless nozzle. *Proceedings of the 38th Textile Research Symposium,* (3-5 September 2009), pp. 117-122

Wu D., Huang X., Lai X., Sun D. & Lin L. (2010). High Throughput Tip-Less Electrospinning via a Circular Cylindrical Electrode. *Journal of Nanoscience and Nanotechnology,* Vol. 10, No. 7, (July 2010), pp. 4221-26, ISSN 1533-4899

Yang E., Shi J. & Xue Y. (2010). Influence of electric field interference on double nozzles electrospinning. *Journal of Applied Polymer Science,* Vol. 116, No. 6, (June 2010), pp. 3688-92, ISSN 1097-4628

Yarin A. L. & Zussman E. (2004). Upward needleless electrospinning of multiple nanofibers. *Polymer,* Vol. 45, No. 9, (April 2004), pp. 2977-80, ISSN 1548-2634

3

Electrospinning of Metal Doped Alumina Nanofibers for Catalyst Applications

Sneha Swaminathan and George Chase
The University of Akron,
USA

1. Introduction

The world is turning into a vast technological jungle and the amount of pollution has increased proportionally to the number of industries and world population. The majority of the pollution is created by anthropogenic activities that include burning of fossil fuels, exhaust of automobiles and other combustion processes. Beginning with the 1940s, some urban cities started experiencing air pollution because of the increasing number of automobiles (Haagen-Smit, 1970). By the 1960s, cars had been manufactured in large quantities and had become popular in providing personal mobility to an increasing range of people. But in automobiles, the process of combustion of gasoline in the engine to carbon dioxide (CO_2) and water was not a very efficient process (Twigg, 2007). The byproducts of combustion include residual uncombusted hydrocarbons (HCs), partial combustion product carbon monoxide (CO), nitric oxides (NO_x) formed from atmospheric nitrogen during combustion, and particulate matter (PM), especially carbonaceous particulate formed in diesel engines (Shelef & McCabe, 2000). In the engine exhausts the CO level ranged from 1 to 2 vol. % while unburnt HCs ranged between 500- 1000 vppm. During the process of combustion, very high temperatures are produced by the burning of gasoline droplets that result in the thermal fixation of N_2 in the air to form NO_x with concentrations in the range of 100-300 vppm (Patterson& Henein,1972; Armor, 1994; Heck & Farrauto, 1995; Silver et al., 1992). NO_x (where $1<x<2$), consisting mainly of nitrogen monoxide (NO) and nitrogen dioxide (NO_2) are one among many toxic pollutants in air. NO is the most abundant nitrogen derivative present in the atmosphere, and represents 95% of the total NO_x emissions (Degobert & Marshall, 1995). The total nitrogen oxides emitted in the world from automotive exhausts was 6.4 million in 2005 (EPA,2005). NO_x undergoes photochemical reactions leading to the generation of smog. It has been proven to cause serious health impacts such as respiratory troubles, circulation problems, and premature death (Patterson& Henein,1972; Meulenbelt, 2007). Another major pollutant is carbon monoxide. CO poisoning is the most common type of fatal poisoning in many countries. CO adversely effects human health leading to cardiovascular and neurobehavioral problems, developmental disorders and in some cases even death (Raub et al., 2000). CO primarily occurs in urban traffic where gasoline engines often idle resulting in high concentration levels of up to 100 mg/m^3 (Degobert & Marshall, 1995).

To prevent environmental pollution, it is essential to eliminate these toxic air pollutants by converting them into less harmful substances. Hence their abatement using catalysts is a very important topic of research and application. Noble metals are most commonly used as catalysts for reducing these air pollutants and to meet the stringent federal regulations (Patterson & Henein, 1972) The global market for energy and environmental catalysts was worth $12.2 billion in 2006, $13.0 billion in 2009 and an estimated $18.5 billion in 2012 at a compound annual growth rate (CAGR) of 7.4 % from 2007 to 2012 (BCC Research, 2007).

Noble metals can be used as nanocatalysts to eliminate these toxic substances from the various exhausts. Nanosize materials possess high surface energy which result in specific catalytic activity and selectivity (Zhou et al., 2004). The catalytic activity of supported metal particle catalysts is strongly dependent on the size and shape of the particles. Hence, on a mass basis nanocatalysts are very active due to their high surface areas and high specific catalytic activity (Gai et al., 2002). Nanocatalysts also optimize the use of precious metals thus minimizing the specific cost per function.

Due to high surface area to volume ratio, noble metal nanoscale metals differ in their properties from bulk materials (Narayanan & El-Sayed, 2005). These materials have potential applications in catalysis and as sensors (Takashi et al.,2005). Supported nanoparticles have been used for a variety of heterogeneous catalysis reactions such as acylation, alkane isomerisation, benzene hydrogenation, Suzuki-Miayura reaction and Heck's reaction. Nanoparticles have properties in between bulk materials and molecules and are potentially useful for new generations of chemical, optical, magnetic, and electronic devices. Well distributed nano sized noble metal particles show better potential as catalysts than bulk materials [Gniewek et al., 2008; Luo et al., 2005; . Tribolet & Kiwi-Minsker, 2005).

Metal oxide nanofibers are potential candidates as catalyst supports for heterogeneous catalytic reactions (Patel et al., 2007; Park et al., 1998; Park et al.,2008). Nanofibers can be produced by electrospinning. Electrospinning is a process in which fibers of nanoscale or submicron scale are generated, as the electrified jet (composed of a highly viscous polymer solution) is continuously stretched due to electrostatic repulsions between the surface charges. The jet continues to stretch until it either encounters a grounded surface or enough of the solvent has evaporated that the jet is no longer fluid enough to stretch (Li & Xia, 2004). The fiber diameter can be controlled by varying the process parameters such as polymer concentration, viscosity, electric field, solvent composition, flow rate, diameter and angle of spinneret (Doshi & Reneker, 1995). As catalyst supports, nanofibers can provide a means of constructing catalytic systems that have large surface area, are lightweight, are thermally and chemically stable, and result in optimization of precious metals (Takashi et al., 2005). The formation of nanoscale diameter fibers from electrospinning have been explored for high-performance filters, wound dressing materials, and biomaterial scaffolds for tissue engineering (Wei & Ma, 2004; Filatov, 2007). The structures thus produced have a high surface area to volume ratio, high length to diameter ratio, high porosity and interconnected pores, good surface functionalities, and superior mechanical performance (e.g. stiffness and tensile strength) (Doshi & Reneker, 1995; Sukigara et al., 2003; Huang et al., 2003). Figure 1 shows the schematic diagram of an electrospinning setup.

Traditionally, the catalysts are coated onto commercial monolithic supports by the wetness impregnation technique. In this method, the catalyst precursor salt is dispersed on the support surface by the impregnation of the solution of the precursor. The precursor particles are formed in the drying process, calcined to form metal-oxides and subsequently reduced to form fine supported metal particles (Zhou et al., 2004).

In the present work, an alternative approach has been developed wherein the noble metal nanocatalysts are incorporated into a ceramic nanofiber. The catalysts on ceramic nanofibers increase the overall exposed catalyst area and simultaneously immobilize the catalyst to

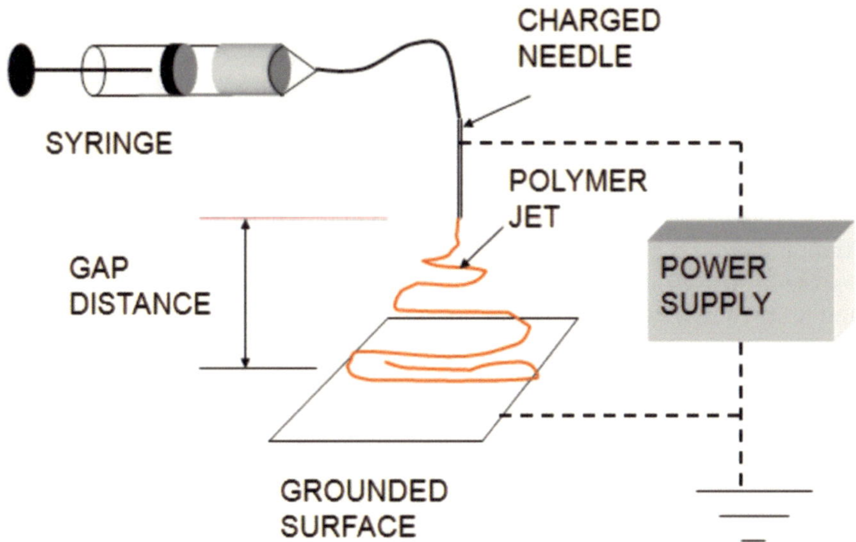

Fig. 1. Electrospinning setup.

minimize catalyst loss. A small amount of the catalyst incorporated ceramic nanofibers are mixed with microfibers to fabricate a filter disk by a vacuum molding technique. This filter disk, termed as 'catalytic filter', is a combination of catalytic elements and filter. The catalyzed ceramic nanofiber augmented microfiber filter media can be used for two applications: reduction of NO_x and oxidation of CO and for enhanced particulate removal. This chapter describes the process of fabrication of the catalytic filters and its applications in reducing NO_x and CO emissions.

The combination of two operations, catalysis and filtration, in one operation improves performance and reduces processing, investment and maintenance costs (Heidenreich et al., 2008). The individual nanoparticles are dispersed throughout the entire fiber and firmly anchored to the walls of the pores of the alumina support; hence their tendency to sinter and coalesce is minimized. In addition, there is free diffusional access of reactants and exit of products from the nanocatalyst, thereby facilitating the catalytic turnover of bulky organic molecules such as unburnt hydrocarbons (Hermans et al., 2001; Raja et al., 2001; Thomas et al., 2001). Nanomaterials are more effective than the monolithic support structures for two reasons. First, their extremely small size yields a tremendous surface area-to-volume ratio. Second, when materials are fabricated on the nanoscale, they achieve properties not found within their macroscopic counterparts. Both of these reasons account for the versatility and effectiveness of nanocatalysts. They also help in reducing the cost of processes which are indispensable without them (Schlogl & Hamid, 2004).

Fibrous materials in filter media have advantages which include high filtration efficiency and low air resistance (Tsai et al., 2002). One method to develop highly effective and efficient filters is by the addition of nanofibers. When the channels and structural elements of a filter are matched to the size of the particles or droplets to be captured, the capture efficiency improves considerably (Graham et al., 2002). Nanofibrous media have low basis weight, high permeability and small pore size that make them appropriate for a wide range of filtration applications. Nanofibers offer unique properties of high specific surface area, high porosity and good interconnectivity of pores, potential to incorporate active chemistry or functionality on nanoscale and high permeability for gases (Barhate & Ramakrishna, 2007; Ahn et al., 2006). The nanofibrous media have advantages over mesoporous media by avoiding the mass transfer limitation of substrates due to reduced thickness and intrafiber porosity. In particular, they are highly successful for the development of high-performance air filters. The electrospinning process in particular can be used to tailor the porosity, fiber diameter, and pore size to achieve the desired structural configuration (Barhate & Ramakrishna, 2007).

From filtration theory, submicron fibers have better filter efficiency than larger fibers at the same pressure drop in the interception and inertial impaction regimes. The interception and inertial impaction efficiencies will increase faster, more than compensating for the pressure drop increase (Graham et al., 2002). A second factor for submicron fibers is the effect of slip flow on the fiber surface. One assumption in filtration theory is that of continuous flow around the fiber with no-slip condition at the fiber surface. But, when the scale of the fiber becomes small enough that the molecular movements of the air molecules are significant in relation to the size of the fibers and the flow field, this assumption becomes invalid and slip-flow model is added to extend the filtration theory.

2. Experimental: fabrication of catalytic filters

The manufacture of the catalytic filter media involves the steps shown in Fig.2. The active ingredient in the catalytic filter is the ceramic nanofibers i.e. metal incorporated alumina nanofibers which is responsible for the conversion of NO_x and CO into safer products.

Over the years, ceramic nanofibers have evolved with a multitude of applications which can be categorized primarily in four ways (Ramaseshan et al., 2007):

1. Chemistry based applications: catalysis, storage batteries, membranes, fuel cells, sensors,
2. Physical change based applications: optical communications, data and magnetic storage,
3. Materials based applications: filtration, paints, packaging, aerospace, and
4. Biomedical applications: biosensors, bone and tissue engineering, biomechanical devices

Hence, in this research the immense potential of ceramic nanofibers is exploited for catalysis and filtration applications.

The main steps in the fabrication of a catalytic filter are:

1. Preparation of the electrospinning solution,
2. Electrospinning to form metal oxide incorporated alumina nanofibers,
3. Vacuum molding to form fibrous filter, and
4. Reduction to convert the catalyst metal oxides in the alumina nanofibers to their catalytic metal state.

The synthesis of ceramic nanofibers involves three main steps:

1. Preparation of sol with suitable inorganic precursor and carrier polymer with suitable concentration, mixing and temperature conditions,
2. Electrospinning the sol by adjusting and optimizing conditions such as viscosity, electric field, concentration, flow rate etc. to obtain composite fibers, and
3. Calcination of the composite mat to decompose the polymer template and to convert the metal precursors to their respective oxides.

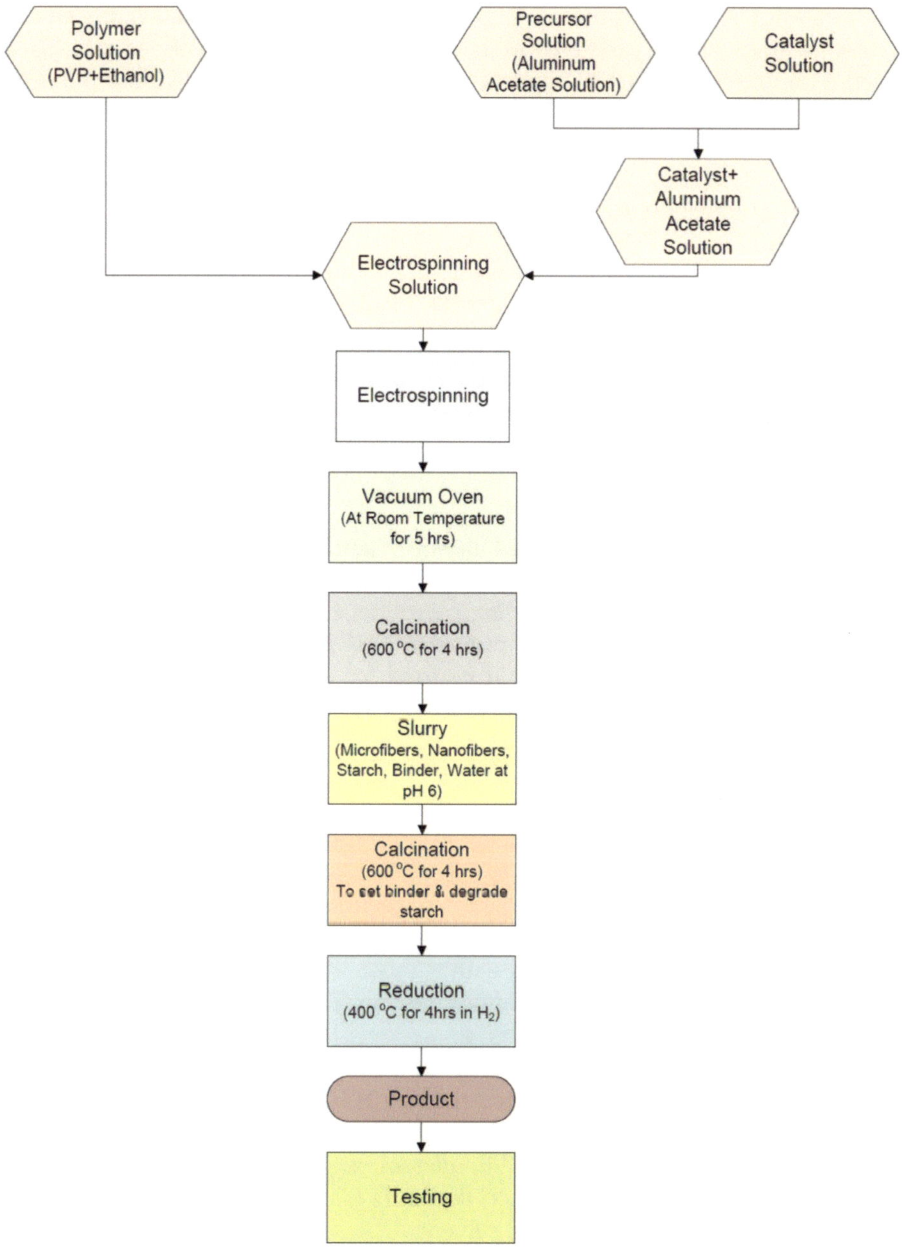

Fig. 2. Process flow diagram for the manufacture of catalytic filter.

The purposes of a carrier polymer are to assist with the electrospinning process and to form uniform colloidal dispersions of catalyst salt particles (Esumi et al., 1994). In this case, polyvinylpyrrolidone (PVP) was chosen as the carrier polymer because PVP can form a complex with the metal ions or metal particles, which can prevent the agglomeration of the cluster, and synthesis monodisperse nanoparticles with uniform size (Qian et al., 2001). PVP also plays an important role in protecting and stabilizing the colloidal dispersions of noble metals. Since soluble polymers can cover the whole surface of the catalyst, the effect of the supporting polymer may be promoted in the case of the polymer- protected colloidal metals compared with that by inorganic supports (Hirai et al., 1985). The precursor for alumina is basic aluminum acetate. The oxidation of basic aluminum acetate produces alumina. Aluminum acetate is calcined in air at 600 °C to form alumina. At this temperature, the carrier polymer degrades primarily to CO_2 and H_2O. The stepwise conversion is as follows: From 170 to 220 °C, condensation of the OH groups on the basic acetate occurs by the reaction (Ayral & Droguet, 1989).

$$2\ Al(OH)(OAc)_2 \longrightarrow Al_2O(OAc)_4 + H_2O$$

From, 220 to 475 °C the two carboxylates groups decompose to form amorphous alumina. Other phases of the alumina can be obtained by calcining at higher temperatures (Ayral & Droguet, 1989). This is described by the reaction

$$\text{Amorphous } Al_2O_3 \xrightarrow{800\,^{\circ}C} \gamma Al_2O_3 \longrightarrow \delta Al_2O_3 \longrightarrow \theta Al_2O_3 \xrightarrow{1200\,^{\circ}C} \alpha\ Al_2O_3.$$

2.1 Electrospinning of palladium doped alumina nanofibers

To optimize the electrospinning process, combinations of several parameters have to be considered. 9 wt. % solution of polyvinylpyrrolidone (PVP, Aldrich, M_W 1,300,000) in ethanol (Pharmco-AAPER alcohol) was prepared. This molecular weight of PVP was used because low concentrations of PVP could be used for electrospinning when diluted with ceramic precursor solution to obtain nanofibers with small diameters. Since the carrier polymer is generally sacrificed in the end to obtain ceramic nanofibers, lower concentrations of the polymer are preferred to increase the yield of ceramic nanofibers. Aluminum acetate precursor solution (AA solution) was prepared by mixing aluminum acetate (Alfa Aesar, Basic hydrate), water and formic acid in the weight ratio of 1:2.5:1, respectively. The PVP solution and AA solution were mixed in the weight ratio of 1:2, respectively. Palladium chloride (Sigma-Aldrich, 60 % Pd) was mixed in the resulting mixture in the ratio of 5 wt% $PdCl_2$ with respect to aluminum acetate and stirred overnight at 40 °C using a magnetic stirrer. The solution was further diluted with ethanol. The solution was electrospun with a flow rate of 2 µl/min and an electric field of 1 kV/cm. The sample was placed in a vacuum oven at room temperature for 5 hours to remove trapped solvents. The electrospun nanofibers were calcined to 600 °C to degrade the polymer and convert the metal precursors to metal oxides. From Thermogravimetric Analysis (TGA) there was significant mass loss as the temperature was ramped up to 500 °C. Above 500°C, the sample mass became constant, indicating that the organic materials (PVP, ethanol, acetate group) had degraded and the volatiles had evaporated (Figure 3). Hence, the fibers calcined to 600 °C were relatively pure metal oxide and free of PVP and residual solvent.

To obtain the fiber diameter distribution, the diameters of 100 fibers were measured from 10 Scanning Electron Microscope (SEM) images similar to those in Figure 4. The distribution was calculated using Equations (1) and (2) (Dosunmu et al., 2006; Varabhas et al., 2008):

$$\ln(\overline{x_g}) = \frac{1}{L_t}\sum L_i \ln(x_i) \quad (1)$$

$$\ln(\sigma_g) = \sqrt{\frac{\sum(L_i(\ln(x_i) - \ln(\overline{x_g}))^2)}{L_t}} \quad (2)$$

where, L_i is the length of the fiber segments with diameter x_i and $L_t = \sum L_i$. The distributions are asymmetric and may be fitted with a log-normal distribution.

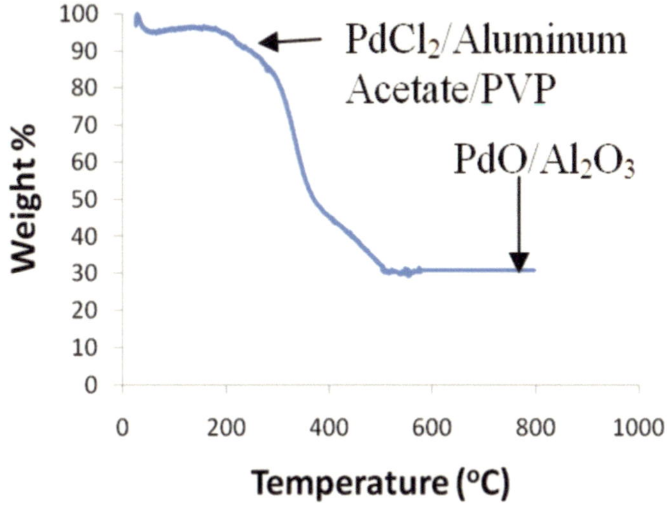

Fig. 3. TGA of $PdCl_2$/Aluminum Acetate/PVP nanofibers.

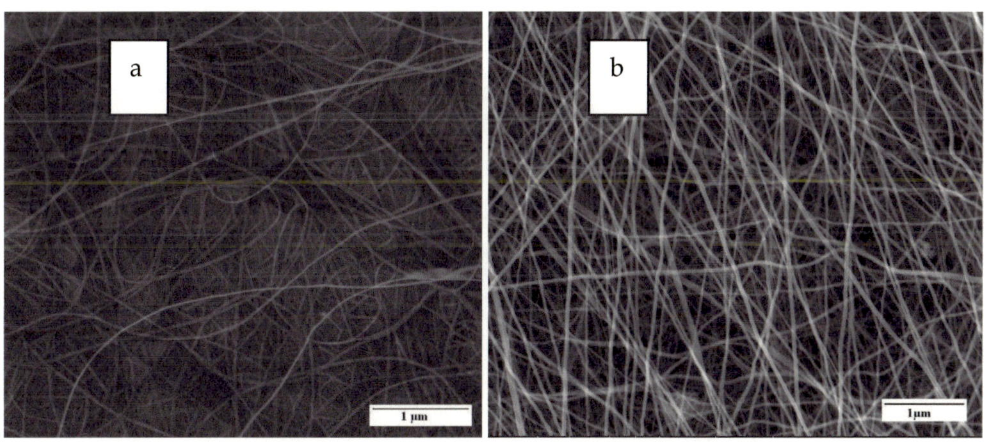

Fig. 4. SEM images of (a) $PdCl_2$/Aluminum Acetate/PVP nanofibers and (b) PdO/Al_2O_3 nanofibers.

The frequency distributions are determined from Equation (3)

$$f_i = \frac{1}{d_i} \frac{1}{\ln\sigma_g \sqrt{2\pi}} exp\left(-\frac{(\ln(x_i)-\ln(\overline{x_g}))^2}{2\left(\ln \sigma_g\right)^2}\right) \quad (3)$$

and plotted in Figure 5. The average fiber diameter was $\overline{x_g}$ = 54.59 nm and standard deviation was σ_g = 1.28 for the log-normal distribution.

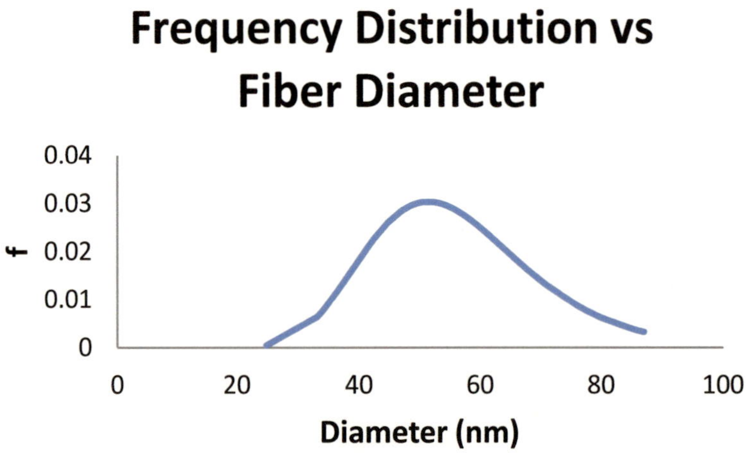

Fig. 5. Length weighted log-normal frequency distribution of PdO/Al_2O_3 fiber diameters.

From Transmission Electron Microscope (TEM) images (Figure 6) the average particle size of 50 measurements of the PdO nanoparticles on the Al_2O_3 nanofiber was 2.43±0.70 nm.

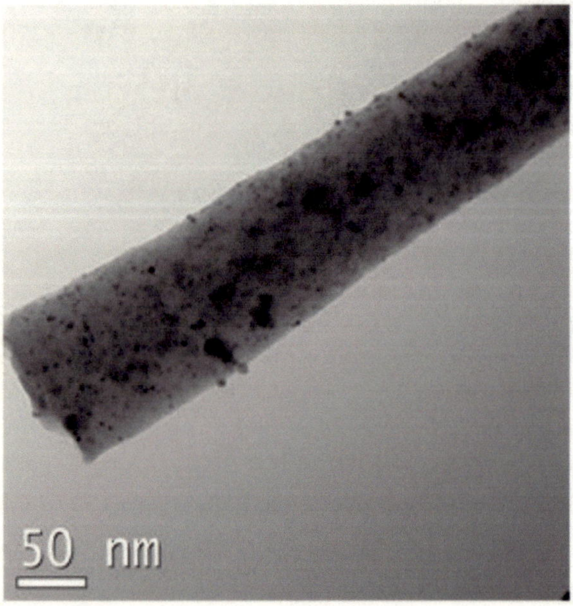

Fig. 6. Transmission electron micrographs of PdO/Al_2O_3 nanofibers.

2.2 Electrospinning of platinum incorporated alumina nanofibers

The 9 wt. % PVP solution and AA solution was mixed in the ratio of 1:3 by weight, respectively. Platinum chloride ($PtCl_2$, 99.9 %, Sigma Aldrich) was dissolved in a mixture of water and ethanol. This solution contained 3.8 wt. % of platinum chloride with respect to aluminum acetate. The solution was mixed overnight at 50 °C. Then the solution was electrospun at a flow rate of 2 µl/min and an electric field of 1 kV/cm. The sample was placed in a vacuum oven at room temperature for 5 hours to remove trapped solvents. The electrospun nanofibers were calcined to 600 °C for 4 hours to obtain PtO/Al_2O_3. Similar to Figure 3 for PdO/Al_2O_3, the TGA analysis of PtO/Al_2O_3 showed significant mass loss occurred as the temperature was ramped up to 450 °C. Above 450 °C the sample mass became constant, indicating that the organic materials (PVP, ethanol, acetate group) had degraded and the volatile components had evaporated. Hence, the fibers calcined to 600 °C were relatively pure PtO/Al_2O_3 and free of PVP and residual solvents.

Figure 7a shows scanning electron micrographs of $PtCl_2$/Aluminum Acetate/ PVP composite nanofibers. From Figure 7b, the average fiber diameters of PtO/Al_2O_3 were in the range of 35-65 nm. After these composite nanofibers were calcined to 600 °C, they were converted to PtO/Al_2O_3. Figure 8 shows a sample TEM image of PtO/Al_2O_3 of PtO nanoparticles on the alumina nanofiber. From the TEM images the average particle size of 100 measurements of the PtO nanoparticles on the Al_2O_3 nanofiber was calculated to be about 2.87±0.82 nm.

The length weighted log-normal frequency fiber diameter distributions for PtO/Al_2O_3 are shown in Figure 11 with average fiber diameter of $\overline{x_g}$ = 53.05 nm and standard deviation of σ_g = 1.53 for the log-normal distribution.

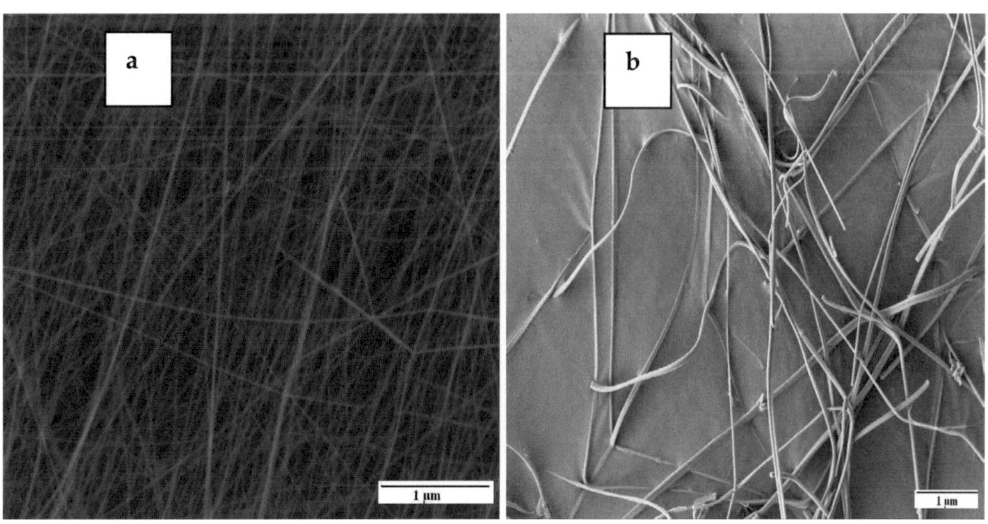

Fig. 7. Scanning electron micrographs of (a) $PtCl_2$/Aluminum Acetate/ PVP composite nanofibers (b) PtO/Al_2O_3 nanofibers.

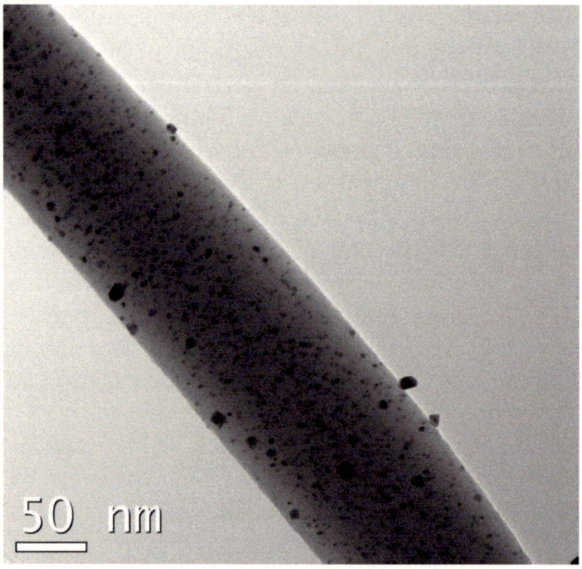

Fig. 8. Transmission electron micrographs of PtO/Al_2O_3 nanofibers.

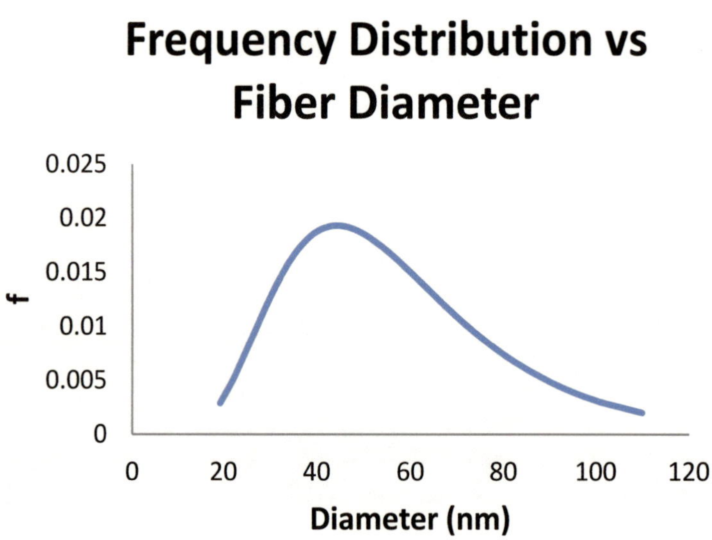

Fig. 9. Length weighted log-normal frequency distribution of PtO/Al_2O_3 fiber diameters.

2.3 Electrospinning of rhodium incorporated alumina nanofibers

9 wt % PVP solution and AA precursor were added in the ratio of 1:4 by weight, respectively. Rhodium chloride ($RhCl_3$, 99.9 %, Sigma Aldrich) was dissolved in ethanol in the ratio of 3.7 wt % of rhodium chloride respect to the aluminum acetate. This solution was mixed overnight at the 50 °C. Using the rotating drum collector electrospinning setup, nanofiber mat was produced with a flow rate flow rate of 2 μl/min and an electric field of 1 kV/cm. The nanofibers were calcined at 600°C for 4 hrs to form Rh_2O_3/Al_2O_3 nanofibers.

The TGA analysis of Rh_2O_3/Al_2O_3 precursor fibers, similar to the TGA analysis of PdO/Al_2O_3 in Figure 3, showed significant mass loss as the temperature ramped up to 450 °C. Above 450 °C, the sample mass became constant, indicating degradation of the PVP and evaporation of volatiles.

Figure 10(a) shows SEM images of $RhCl_3$/Aluminum Acetate/ PVP composite nanofibers. After these composite nanofibers were calcined to 600 °C, they were converted to Rh_2O_3/Al_2O_3 (Figure 10b). Figure 11 shows an example TEM image of Al_2O_3 fibers with the Rh_2O_3 nanoparticles on the nanofiber. The average particle size of the Rh_2O_3 nanoparticles was 1.16±0.35 nm.

The length weighted log-normal frequency fiber diameter distribution for Rh_2O_3/Al_2O_3 is shown in Figure 12. The average fiber diameter is $\overline{x_g}$ = 61.57 nm and the standard deviation is $\sigma_g = 1.33$ for the log-normal distribution.

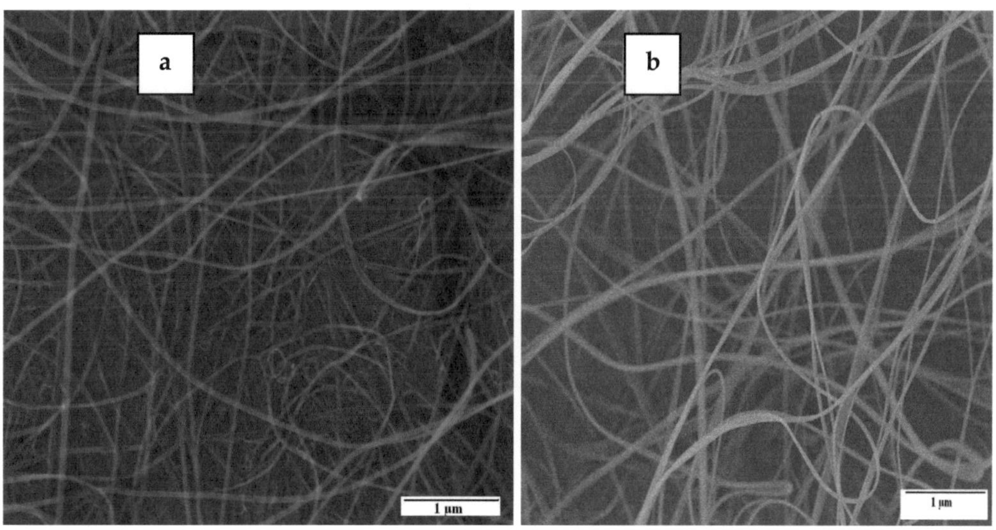

Fig. 10. (a) Scanning electron micrographs of of $RhCl_3$/Aluminum Acetate/ PVP composite nanofibers (b) Rh_2O_3/Al_2O_3 nanofibers.

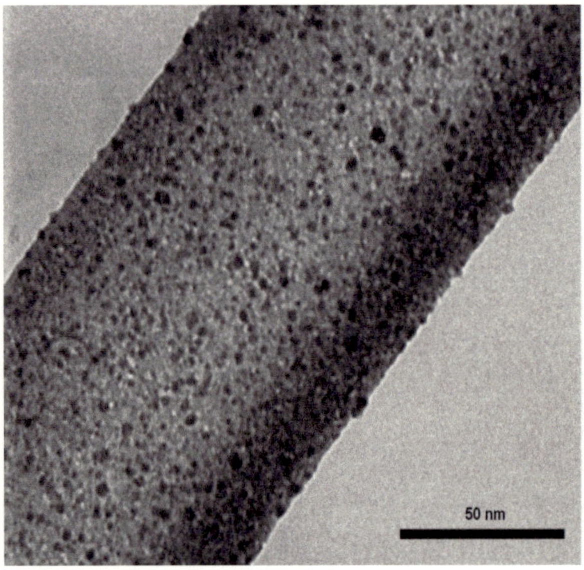

Fig. 11. Transmission electron micrograph of Rh_2O_3/Al_2O_3 nanofibers.

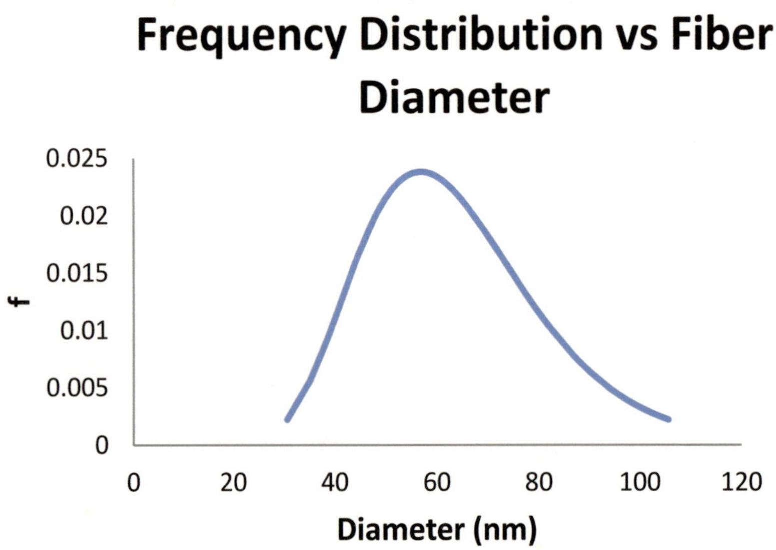

Fig. 12. Length weighted log-normal fiber diameter frequency distribution of of Rh_2O_3/Al_2O_3 nanofibers.

2.4 Formation of nanofiber based filter media

Filter media were prepared by vacuum molding an aqueous slurries of fibers and binder. The vacuum molding apparatus consists of a mixing tank, filter mold, collection tank and vacuum pump. The slurry was prepared using a mixture of microfibers, ceramic nanofibers, corn starch and alumina binder in water. Diluted acid was added to maintain the slurry at a pH ~ 6.0 and starch was added to enhance the binding capability of the binder and fibers. The slurry was mixed overnight using a stirrer to obtain a uniform dispersion of materials. It was poured into the mixing tank and agitated by bubbling into the bottom of the tank. The Plexiglas mold was located at the bottom of the mixing tank. The mold had an internal diameter of 2.3 cm, equal to the diameter of the filter. The bottom of the mold had a fine steel wire mesh to capture the fibers. A vacuum was applied to the collection tank. A hose from the collection tank to the underside of the mold transferred the vacuum pressure to the mold to pull the slurry from the mixing tank through the steel mesh. The fibers were retained on the steel mesh forming a wet fibrous filter cake. The wet cake was heated to dry the cake and to thermally set the binder.

The slurry in the mixing tank consisted of 4 L of water, 0.5 g of alumina microfibers (SAFFIL HA, Thermal Ceramics), 0.05 g catalytic fibers, 0.02 g corn starch, 1 mL alumina binder (alumina rigidizer, Zircar Ceramics) and about 20 drops of dilute H_2SO_4 acid. The slurry was stirred for 8 hours before it was vacuum molded at 5 psi to form a filter wet filter cake. The wet cake was dried at 120 °C for 2 hours to remove moisture and calcined to 600 °C for 4 hours to set the binder. The filter disk was heated at 400 °C with hydrogen gas (20 % H_2, 80 % N_2) for 4 hours to convert the precious metal oxides to their native metal form.

Figure 13 is a sample SEM image of the alumina microfibers. The average fiber diameter range determined by log-normal frequency distribution (Figure 14) was calculated to be $\overline{x_g}$ = 3.20 μm and the standard deviation was σ_g = 1.29 for the log-normal distribution. Alumina microfibers were used to structurally and functionally support the catalytic fibers because they can withstand high temperatures and they have a similar coefficient of expansion in comparison to the catalytic nanofibers.

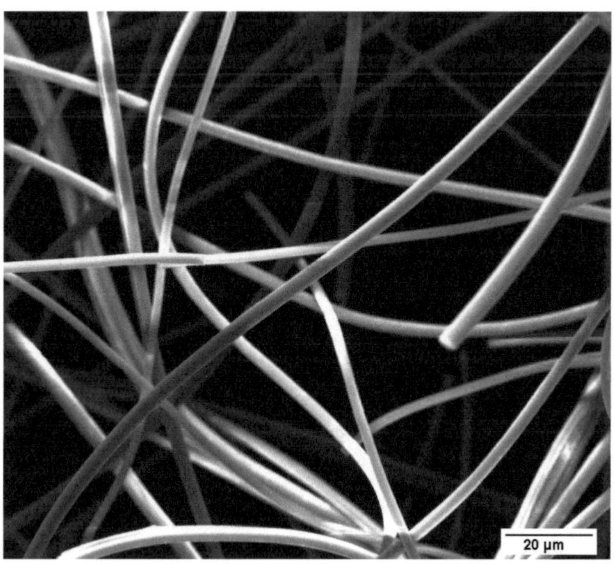

Fig. 13. Scanning electron micrographs of Al_2O_3 microfibers.

Figure 15 (a) shows picture of a catalytic filter having diameter of 2.3 cm and height of 1.5 cm. The filter can be molded to the desired shape and dimensions using molds and meshes with the desired diameter. Figure 15 (b) are scanning electron micrographs of sections of catalytic filters. The nanofibers are entrapped by the large microfibers. This structure results in higher capture efficiencies and quality factor for a very small increase in pressure drop in comparison to a filter with only microfibers (Srinivasan, 2005).

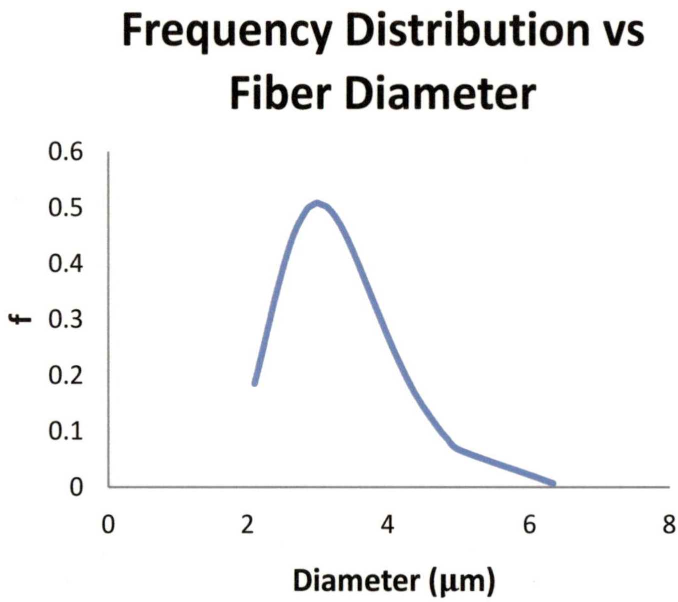

Fig. 14. Length weighted log-normal frequency of fiber diameter of Al_2O_3 microfibers.

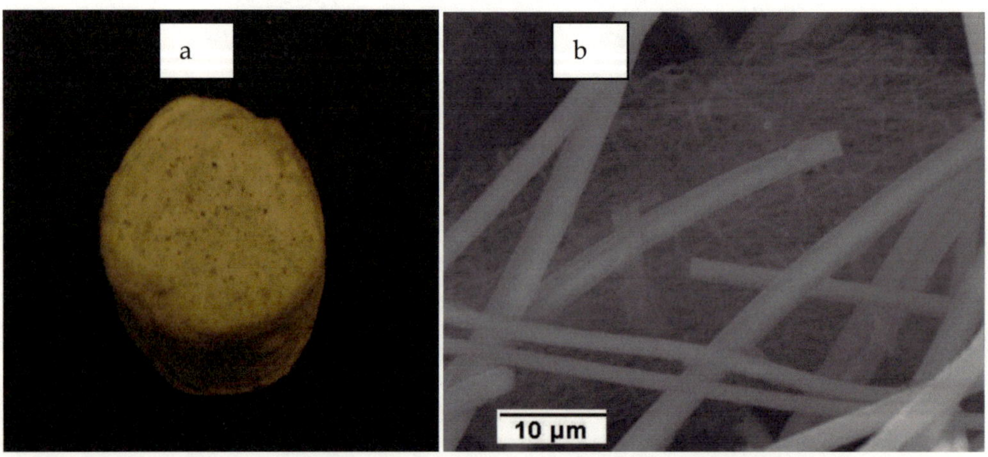

Fig. 15. (a) Disk shaped catalytic filter approximately 2.3 cm in diameter and 1.2 cm thick. (b) SEM image of nanofibers mixed among microfibers in the catalytic filter.

2.5 Characterization of filter media

The catalytic filters were characterized for porosity, permeability and hardness.

2.5.1 Porosity

Porosity is a 3-dimensional measurement of the void volume of a filter material. Porosity is defined as the ratio of the volume of all of the pores of a material to the volume of the filter and is one of the parameters that describes the structure of the filter material. A special made pycnometer was used to measure the bulk porosity of the filter media by a gas expansion method. The pycnometer consists of two chambers, one (with a removable gas-tight lid) and the second chamber of a fixed, known internal volume - referred to as the reference volume or added volume. The device also comprises a valve to admit a gas under pressure to the second chamber, a pressure measuring device and a valved vent from the second chamber.

The porosities of three of the disk shaped catalytic filters are listed in Table 1. Each listed porosity is the average of three filters of the same type. The accuracy of the special made pycnometer is at best ±0.01.

Type of catalytic filter	Porosity±0.01
Pd/Al_2O_3	0.97
Pt/Al_2O_3	0.97
Rh/Al_2O_3	0.96

Table 1. Porosity of the catalytic filters.

2.5.2 Permeability

Permeability is defined as the volume of a fluid that will move across a known area of filter media at a constant pressure drop. It is used to describe how easily a fluid flows through a material and is measured by a Frazier Air Permeability tester (Frazier Precision Instrument Co.) modified to hold the disk shaped filters. The Frazier test is based on Darcy's law which relates the pressure drop to flow through a packed bed with the permeability coefficient as given by Equation (4):

$$\frac{k}{\mu}\left(\frac{P_O-P_L}{L}\right)=\frac{Q}{A} \tag{4}$$

where k= Permeability in m^2
P_o= Initial pressure, in Pa
P_L= Final pressure, in Pa
L=Thickness of media, in m
Q=Volumetric flow rate, in m^3/s
μ=Kinematic viscosity, in Ns/m^2
A=Area, in m^2

Table 2 lists the permeabilities of the catalytic filters for measurements done in triplicate for each type of catalytic filter.

Type of catalytic filter	Permeability (m^2)
Pd/Al_2O_3	$1.25\pm0.05*10^{-11}$
Pt/Al_2O_3	$1.15\pm0.08*10^{-11}$
Rh/Al_2O_3	$1.4\pm0.1*10^{-11}$

Table 2. Permeability of the catalytic filters.

2.5.3 Hardness

A durometer (Rex Gauge Co. Model 1600 Type O) was used to measure the hardness of the filter. Type O durometers are used for very soft elastomers, textile windings, soft granular materials (Rex Durometers, 2006) and is suitable for measuring the hardness of fibrous filters. In this durometer, the load deflection curve is a straight line where 100 durometer O points equals 822 grams and zero durometer O points equals 56 g. The durometer was pressed down and held in a vertical position till the entire foot was in contact with the filter. The dial hand gave the durometer reading. The reading after 15 seconds was also noted while maintaining firm contact of the durometer foot against the filter to make sure that the value was consistent and there was no creep or cold-flow of the filter. The hardness values of the catalytic filters are given in Table 3, as the average of triplicate measurements.

Type of catalytic filter	Durometer O Points	Load(g)
Pd/Al_2O_3	10 ±2	110
Pt/Al_2O_3	14±1	140
Rh/Al_2O_3	13±1	120

Table 3. Hardness of the catalytic filters

2.6 Reaction test setup

The reaction test setup was constructed to test the activity of the catalytic filter when challenged with NO and CO gases. He, CO and NO were forced through the catalytic filter cell with controlled flow rates using mass flow controllers. To increase the velocity and to maintain the stoichiometry of NO and CO, He was used as the carrier gas. In the construction of the filter cell, two stainless steel cylindrical blocks were machined with a central hole of diameter 2.3 cm and three holes in the corners for screwing the two parts together. These two parts were bolted together. A gasket made of ceramic fibers was used for sealing the reactor. Bottom and top parts of the reactor were connected with 1/8″ stainless steel tubing to the rest of the setup (Figure 16). To control the reactor temperature, the reactor was wrapped with heating tape connected to a temperature controller and insulated.

The pressure drop across the reactor was measured using a pressure transducer. The concentration of the inlet and the outlet gases were measured using a gas chromatograph (GC). The GC was connected to a computer where the peaks of the gases were displayed and their respective concentrations were calculated. Figure 17 shows the flow diagram of the reactor setup.

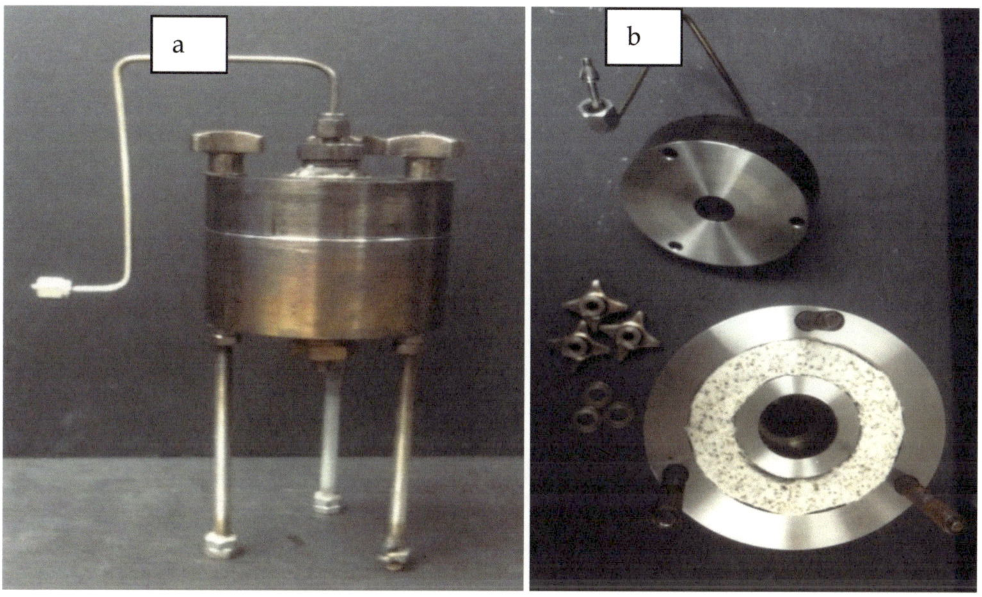

Fig. 16. (a) Reactor (b) Internal view of the reactor and screws.

The ceramic filter medium was positioned in the filter holder for measuring its activity. The gases were continuously flowed from the bottom of the reactor and exhausted from the top. The inlet concentration and the exhaust concentration were monitored by the GC. The temperature of the reactor was gradually increased to observe the light-off temperature of the NO and CO gases. The concentration profile of the inlet and outlet gases was obtained with respect to different temperatures. The maximum temperature the setup was taken to was 450 °C.

For the Pd/Al_2O_3 (Figure 18) catalytic filter which contained 0.05g of the PdO/Al_2O_3 nanofibers the decomposition temperature of NO was 350 °C. In the case of Pt/Al_2O_3 (Figure 19) the filter contained 0.05g of PtO/Al_2O_3 nanofibers and the decomposition temperature of NO was 325 °C. The Rh/Al_2O_3 (Figure 20) filter decomposed NO at a much lower temperature of 250 °C. This is because NO readily dissociates on Rh {111} and N_2 is adsorbed from the Rh surfaces from 200 to 300 °C (Lambert & Bridge, 1984; Hendrickx & Nieuwenhuys, 1986). On the other hand, Pt and Pd are good for oxidation reactions and but are less able to dissociate NO than Rh (Thomas & Thomas, 1997).

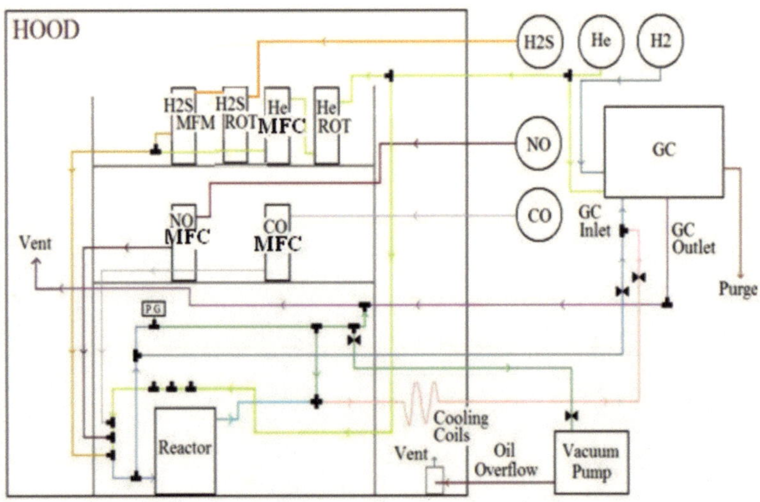

MFC - Mass Flow Controller; MFM - Mass Flow Meter; ROT - Rotameter;
P.G. - Pressure Gage; GC - Gas Chromatograph

Fig. 17. Flow diagram of the reactor setup.

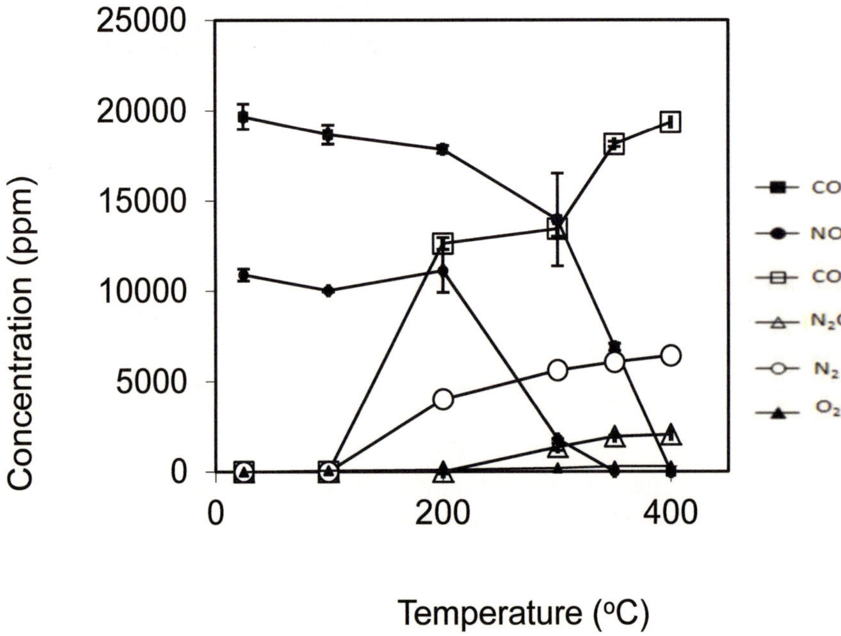

Fig. 18. Concentration vs. temperature for the reaction over 0.05g of Pd/Al_2O_3.

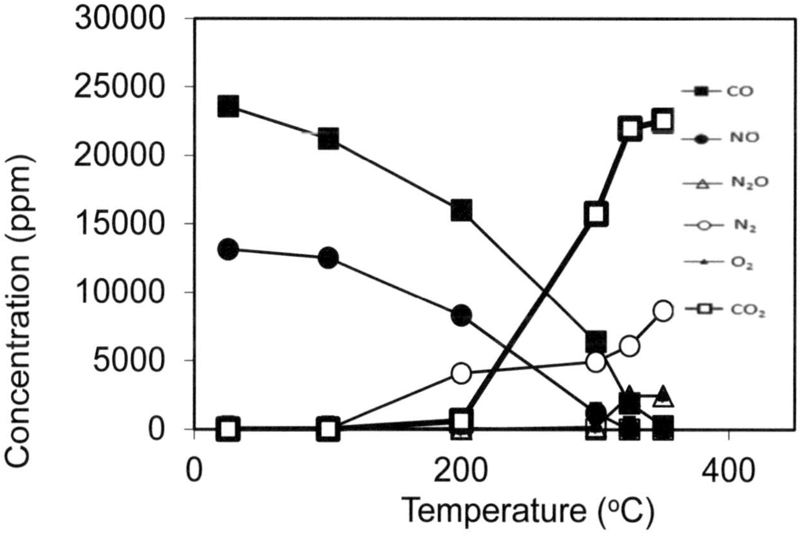

Fig. 19. Concentration vs. temperature for the reaction over 0.05g of Pt/Al_2O_3.

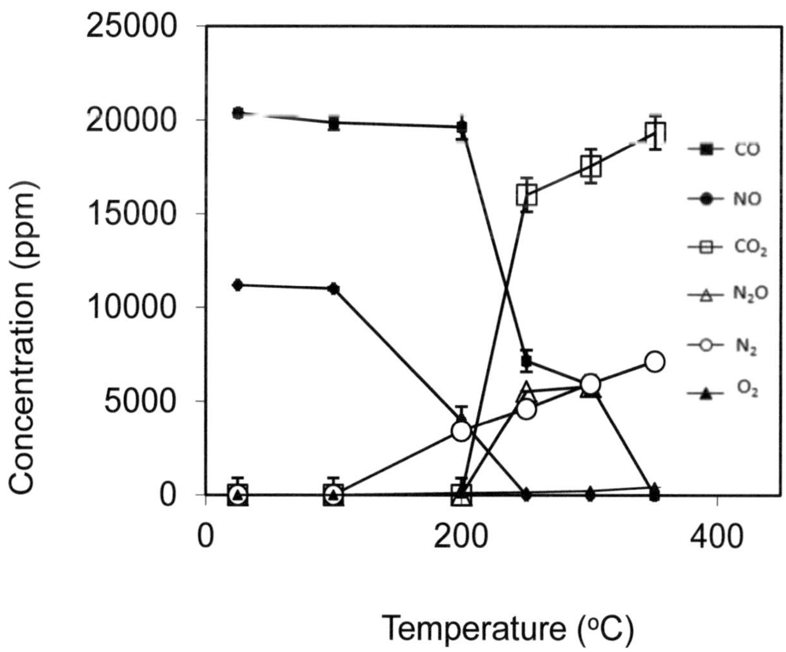

Fig. 20. Concentration vs. temperature for the reaction over 0.05g of Rh/Al_2O_3.

3. Conclusion

In this research, the noble metals (palladium, platinum and rhodium) have been incorporated into alumina nanofiber support by sol gel processing and electrospinning techniques. The diameter of the nanofibers and the particle sizes of the noble metals are summarized in Table 4.

Material	Fiber Diameter (nm)	Catalyst Particle Diameter (nm)
PdO/Al_2O_3	40-70	2.43±0.70
PtO/Al_2O_3	35-65	2.87±0.82
Rh_2O_3/Al_2O_3	50-80	1.16±0.35

Table 4. Diameter of the nanofibers and catalyst particles.

These nanofibers were mixed with alumina microfibers to form a filter using vacuum molding. The catalytic filter was reduced under flowing hydrogen to reduce the noble metal oxide into their native metal form. This filter was tested in a stream of NO and CO. The NO gas was completely decomposed in the temperature range 225 ~ 400 °C depending on the type and amount of catalyst and reaction conditions. Nitrogen, carbon dioxide, and nitrous oxide gases were detected as reaction products. The electrospinning solutions for fabricating the catalyst supporting nanofibers were optimized by varying the ratio of aluminum acetate precursor solution and carrier polymer solutions to obtain maximum yield of ceramic nanofibers.

4. References

Ahn, YC; Park, SK; Kim, GT; Hwang, YJ; Lee, CG; Shin, HS; Lee, JK. (2006) Development of high efficiency nanofilters made of nanofibers. *Current Applied Physics.* Vol. 6, pp. 1030–1035.

Armor JN. (1994) *Environmental Catalysis.* American Chemical Society, Washington DC.

Ayral, A; Droguet, JC. (1989). Alumina powders via a controlled precipitation of aluminum acetate. *Journal of Materials Research.* Vol. 4, pp. 967-971.

Barhate, RS; Ramakrishna, S. (2007) Nanofibrous filtering media: Filtration problems and solutions from tiny materials. *Journal of Membrane Science.* Vol. 296, pp. 1–8.

Degobert, P; Marshall, N. (1995) Society of Automotive Engineers. Automobiles and Pollution. s.l. : Editions TECHNIP.

Doshi J; Reneker DH.(1995) Electrospinning process and applications of electrospun fibers. *Journal of Electrostatics.*Vol. 35, pp. 151-160 .

Dosunmu, OO; Chase. GG; Kataphinan, W; Reneker, DH. (2006) Electrospinning of polymer nanofibres from multiple jets on a porous tubular surface. *Nanotechnology.*Vol. 17, pp. 1123–1127.

EPA. National Summary of Nitrogen Oxides Emissions, May 10, 2010, Available from <://www.epa.gov/air/emissions/nox.htm>

Esumi, K; Itakura, T; Torigoe, K.(1994) Preparation of organo palladium sols from palladium complexes in various alcohols. *Colloids and Surfaces A: Physicochemical and Engineering Aspects,* Vol. 82, pp. 111-113.

Filatov, Y; Budyka, A; Kirichenko, V; (2007). *Electrospinning of Micro- and Nanofibers Fundamentals and applications in separation and filtration processes.* Begell House,Inc, Connecticut.

Gai, PL; Roper, R; White, MG. (2002) Recent advances in nanocatalysis research. *Current Opinion in Solid State and Materials Science.* Vol. 6, pp. 401-406.

Gniewek, A; Ziolkowski, JJ; Trzeciak, AM; Zawadzki, M; Grabowska, H; Wrzyszcz, J. (2008) Palladium nanoparticles supported on alumina-based oxides as heterogeneous catalysts of the Suzuki–Miyaura reaction. *Journal of Catalysis.* Vol. 254, pp. 121–130.

Haagen-Smit; AJ. (1970) A Lesson from the Smog Capital of the World. *Proceedings of the National Academy of Sciences.* Vol. 67, pp. 887-897.

Heck RM, Farrauto RJ. *Catalytic Air Pollution Control Commercial Technology.* New Jersey : John Wiley & Sons, Inc., 1995.

Heidenreich; S, Nacken; M, Hackel; M, Schaub; G. (2008) Catalytic filter elements for combined particle separation and nitrogen oxides removal from gas streams. *Powder Technology.* Vol. 180, pp. 86– 90.

Hendrickx; HACM, Nieuwenhuys; BE. (1986) Surface structure effects in the adsorption and desorption of nitric oxide on rhodium. *Surface Science.* Vol. 175, pp. 185-196 .

Hermans, S; Raja, R; Thomas, JM; Johnson, BFG; Sankar G, Gleeson D. (2001) Solvent-Free, Low-Temperature, Selective Hydrogenation of Polyenes using a Bimetallic Nanoparticle Ru - Sn Catalyst. *Angewandte Chemie International Edition.* Vol. 40, pp. 1211-1215.

Hirai, H; Chawanya, H; Toshima, N. (1985) Colloidal palladium protected with poly(N-vinyl-2-pyrrolidone) for selective hydrogenation of cyclopentadiene. *Reactive Polymers.* 1985, Vol. 3, pp. 127-141.

Huang, ZM; Zhang, YZ; Kotaki, M; Ramakrishna, S. A review on polymer nanofibers by electrospinning and their applications in nanocomposites. *Composites Science and Technology.* 2003, Vol. 63, pp. 2223-2253.

Lambert, RM; Bridge, ME. (1984). Chemistry and Physics of Solid Surfaces and Heterogeneous Catalysis. [ed.] Woodruff DP King DA. Elsevier, Amsterdam. Vol. 3, pp. 83-101.

Li, D; Xia, Y (2004). Electrospinning of Nanofibers: Reinventing the Wheel? *Advanced Materials.* Vol. 16, pp. 1151-1170.

Luo, C; Zhang, Y; Wang, Y. (2005) Palladium nanoparticles in poly(ethyleneglycol): the efficient and recyclable catalyst for Heck reaction. *Journal of Molecular Catalysis A: Chemical.* Vol. 229, pp. 7-12.

Meulenbelt; J. Nitrogen and nitrogen oxides. (2007). Medicine, Vol. 35, pp. 638

Narayanan; R, El-Sayed; MA. Effect of colloidal nanocatalysis on the metallic nanoparticle shape: The Suzuki Reaction. *Langmuir.* 2005, Vol. *21,* pp 2027–2033.

Park; C, Baker; RTK. (1998) Catalytic behavior of graphite nanofiber supported nickel particles. The influence of the nanofiber structure. *Journal of Physical Chemistry B.* Vol. 102, pp. 5168-5177.

Park; SJ, Bhargava; S, Bender; ET, Chase; GG, Ramsier; RD. (2008) Palladium nanoparticles supported by alumina nanofibers synthesized by electrospinning. *Journal of Materials Research.* Vol. 23, pp. 1193-1196.

Patel; C, Li; S, Wang; C, Zhang; W,Wei; Y. (2007) Electrospinning of porous silica nanofibers containing silver nanoparticles for catalytic applications. *Chemistry of materials.* Vol. 19, pp. 1231-1238.

Patterson; DJ, Henein; NA. (1972) *Emissions from Combustion Engines and their Control.* Ann Arbor : Ann Arbor Science Publishers, Inc.

Graham; K, Ouyang; M, Raether; T, Grafe; T, McDonald; B, Knauf; P. (2002) *Polymeric Nanofibers in Air Filtration Applications.* Fifteenth Annual Technical Conference & Expo of the American Filtration & Separations Society.Galveston.

Qian; XF, Yin; J, Feng; S, Liu; SH, Zhu; ZK. (2001). Preparation and characterization of polyvinylpyrrolidone films containing silver sulfide nanoparticles. *Journal of Materials Chemistry*. Vol. 11, pp. 2504-2506.

Raja; R, Khimyak; T, Thomas; JM, Hermans; S, Johnson; BFG. (2001) Single-Step, Highly Active, and Highly Selective Nanoparticle Catalysts for the Hydrogenation of Key Organic Compounds. *Angewandte Chemie International Edition*. Vol. 40, pp. 4638 - 4642.

Ramaseshan; R, Sundarrajan; S, Jose; R, Ramakrishna; S. (2007). Nanostructured ceramics by electrospinning. *Journal of Applied Physics*. Vol. 102, pp. 111101-111101-17.

Raub; JA, Nolf; MM, Hampson; NB. (2000). Carbon monoxide poisoning — a public health perspective. *Toxicology*. Vol. 145, pp. 1-14.

Research, BCC. April 10, 2011. Catalysts for Environmental and Energy Applications. Available from < http://bccresearch.com/report/catalysts-environment-energy-chm020c.html>

Rex Gauge Company, Inc. April 10, 2011. Available from <http://www.rexgauge.com/content/tech_reference_guide>.

Schlogl; E, Hamid; SBA. (2004). Nanocatalysis: Mature science revisited or something really new? *Angewandte Chemie International Edition*. Vol. 43, pp. 1628 –1637.

Shelef; M, McCabe; RW. (2000). Twenty-five years after introduction of automotive catalysts: what next? *Catalysis Today*. Vol. 62, pp. 35–50.

Silver; RG, Sawyer; JE, Summers; JC. (1992). *Catalytic Control of Air Pollution*. American Chemical Society, New York.

Srinivasan; P. (2005). M.S. Thesis, Nanofiber incorporated glass fiber filter media, The University of Akron.

Sukigara; S, Gandhi; M, Ayutsede; J, Micklus; M, Ko; F. (2003). Regeneration of Bombyx mori silk by electrospinning - part 1: processing parameters and geometric properties. *Polymer*. Vol. 44, pp. 5721-5727.

Takashi; M, Masaru; H, Yuichi; I, Kyoko; KB, Nobuyuki; M, Makoto; T, Yuji; Y. (2005) Effect of noble metal particle size on the sulfur tolerance of monometallic Pd and Pt catalysts supported on high-silica USY zeolite. *Applied Catalysis A*. Vol. 286, pp.249-257.

Thomas; JM, Johnson; BFG, Raja; R, Sankar; G, Midgley; PA. (2004). High-Performance Nanocatalysts for Single-Step Hydrogenations. *Accounts of Chemical Research*. Vol. 36, pp. 20–30.

Thomas; JM, Thomas; WJ. (1997). *Heterogeneous Catalysis*. VCH Publishers Inc.,New York.

Tribolet P, Kiwi-Minsker L. Palladium on carbon nanofibers grown on metallic filters as novel structured catalyst. *Catalysis Today*. 2005, Vol. 105, pp. 337–343.

Tsai; PP, Gibson; HS, Gibson; P. (2002) Different electrostatic methods for making electret filters. *Journal of Electrostatics*. Vol. 54, pp. 333-341 .

Twigg; MV.(2007). Progress and future challenges in controlling automotive exhaust gas emissions. *Applied Catalysis B: Environmental*. Vol. 70, pp. 2–15.

Varabhas; JS, Chase; GG, Reneker; DH. (2008). Electrospun nanofibers from a porous hollow tube. *Polymer*.Vol. 49, pp. 4226–4229.

Wei; G, Ma; PX. (2004). Structure and properties of nano-hydroxyapatite/polymer composite scaffolds for bone tissue engineering. *Journal of Biomaterials*.Vol. 25, pp. 4749-4757.

Zhou; B, Hermans; S, Somorjai; GA. (2004). *Nanotechnology in Catalysis: Roadmap to new catalyst systems: Palladium nanoparticles*. American Chemical Society, Springer, New York.

4

Chitin Nanofibers with a Uniform Width of 10 to 20 nm and Their Transparent Nanocomposite Films

Shinsuke Ifuku, Antonio Norio Nakagaito and Hiroyuki Saimoto
Graduate School of Engineering, Tottori University, Koyama-cho Minami, Tottori, Japan

1. Introduction

Nanofibers have been paid much attention in research community due to the recent developments in nanotechnology. Generally, nanotechnology deals with structures sized 1-100 nm in at least one dimension. Nanotechnology may create many new materials and devices with a vast range of applications such as in medicine, electronics, biomaterials and energy production. As described in some reports, the fabrication of nanofibers is the subject of much attention because of their unique characteristics such as a very large surface area to mass ratio and other properties, which allow many promising properties for advanced applications. Nanofibers have been produced by a variety of methods. Most nanofibers are artificially prepared by electrospinning process, which produces fibers from polymer solution using interactions between fluid dynamics, electrically charged surfaces and electrically charged liquids. However, the processability, biocompatibility, and biodegradability of the obtained composites are limited compared to those of natural polymers.

There is a growing interest in producing nanofibers from natural polymers, since they introduce additional functionalities such as biodegrability, biocompability, renewability and availability. A variety of nanofibers are produced in nature, such as collagen fibrils, fibroin in silk, deoxyribonucleic acid (DNA) with double helix, and so on.

Among the variety of nanofibers, cellulose microfibrils are the most abundant natural nanofiber in nature. Recently, Abe et al. succeeded in isolating cellulose nanofibers with a uniform width of approximately 15 nm from wood (Abe et al., 2007). Wood cell walls consist of stiff cellulose microfibrils, embedded in soft matrix substances such as hemicelluloses and lignin which behave just as adhesive between microfibrils. Due to the extended crystalline structure, the microfibrils have efficient physical properties; their Young's modulus is 138 GPa (Nishino et al., 1995), their tensile strength is at least 2 GPa, based on experimental results for kraft pulp, and their thermal-expansion coefficient in the axial direction is as small as 0.1 ppm/K (Nishino et al., 2004) . These structural features of wood cell walls make the wood stiff and tough, and are helpful in supporting the huge body of a tree. Since cellulose nanofibers are embedded in the matrix phase, they can be isolated by first removing the matrix and subsequently extracted as nanofibers by a very simple grinding treatment.

After cellulose, chitin is the second most abundant biomacromolecule in nature, existing mainly in exoskeletons of crabs, shrimps, and insects, and is synthesized in the amount 10^{10} to 10^{11} tons every year (Gopalan & Dufresne, 2003). Owing to its linear (1→4)-*β*-*N*-acetyl glycosaminoglycan structure with two reactive hydroxyl groups and an acetamide group per anhydro-*N*-acetylglucosamine unit, chitin has broad potential in the design of advanced polymeric biomaterials (Figure 1). However, most of chitin is discarded as industrial food residue without effective utilization. Thus, it is important to develop efficient use of chitin as a natural and eco-friendly material.

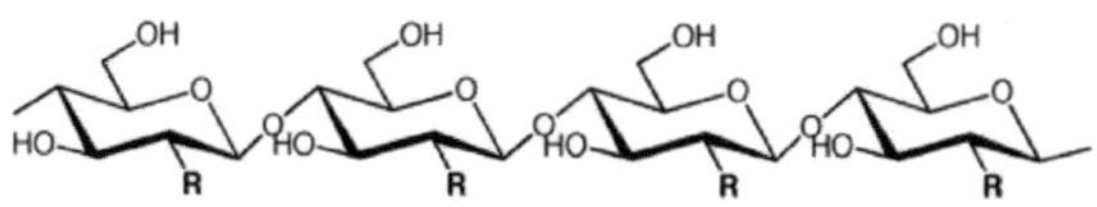

Cellulose: **R** = OH
Chitin: **R** = NHAc

Fig. 1. Chemical structure of cellulose and chitin.

Native chitin in crab shells has highly crystalline structure, arranged as *α*-chitin in an antiparallel fashion forming microfibrils strongly connected by hydrogen bonds. The chitin microfibrils consist of nanofibers about 2–5 nm in diameter and about 300 nm in length embedded in several protein matrices (Raabe et al., 2006; Chen et al., 2008). Since chitin nanofibers are considered to have great potential, various methods have been employed for their preparation, such as acid hydrolysis (Gopalan & Dufresne, 2003; Revol & Marchessault, 1993), ultrasonication (Zhao et al., 2007), and electrospinning (Min et al., 2004). However, the nanofibers obtained by these processes were different from the native chitin nanofibers in terms of width, aspect ratio, crystallinity, chemical structure, and/or width distribution. Recently, Isogai et al, have developed an efficient method to prepare completely individualized cellulose nanofibers by 2,2,6,6-tetramethylpiperidine-1-oxy radical (TEMPO) mediated oxidation of native cellulose, followed by ultrasonic treatment. Although TEMPO mediated oxidation was applicable to crab shell *α*-chitin, the average nanocrystal length was considerably low (Fan et al., 2008a). In addition, the nanocrystals are actually chitin derivatives. They reported a procedure for preparing chitin nanofibers 3–4 nm in width from squid pen *β*-chitin by ultrasonication treatment under acidic conditions, however, the crystallinity of the nanofibers from the squid pen is relatively low (Fan et al., 2008b). Moreover, the biomass quantity of the pen is considerably lower than those of crab shells.

Similar to wood cell walls, the exoskeletons of crab and prawn have a strictly hierarchical structure consisting of crystalline *α*-chitin nanofibers and various types of proteins and minerals (mainly calcium carbonate) (Chen et al, 2008; Raabe et al, 2006; Giraud-guille, 1984). Thus, just like cellulose nanofibers, chitin nanofibers are embedded in matrix components. As Abe et al. described a preparation method can be used to isolate cellulose nanofibers from any natural plant containing lignin and hemicellulose, we considered that this isolation procedure would be applicable to any natural nanofiber source consisting of bionanofibers and other embedding matrixes. Accordingly, we here studied the extraction of natural *α*-chitin nanofibers with a uniform width of 10–20 nm from crab shells.

2. Preparation of chitin nanofibers

2.1 Chitin nanofibers from crab shell

In general, the crab shell has a strictly hierarchical organization which reveals various structural levels, as shown in Figure 2 (Chen et al., 2008; Giraud-guille, 1984; Raabe et al., 2005, 2006). Chitin molecules are aligned in an antiparallel manner that gives rise to α-chitin crystals in the form of thinner nanofibers. These nanofibers are wrapped in protein layers. The next level in the scale consists in the clustering of some of these nanofibers into chitin/protein fibers of about 50-300 nm in diameter. The next step is the formation of a planar woven and branched network of such chitin-protein fibers with a variety of thicknesses. These strands are embedded in a variety of proteins and minerals. The minerals mainly consist of crystalline calcium carbonate. The thickness of strands vary widely among crustaceans. Furthermore, these woven and networked planes form a twisted plywood pattern. This structure is formed by the helicoidal stacking sequences of the fibrous chitin-protein layers. The thickness of the twisted plywood layer corresponds to a certain stacking density of planes, which are gradually rotated about their normal axis.

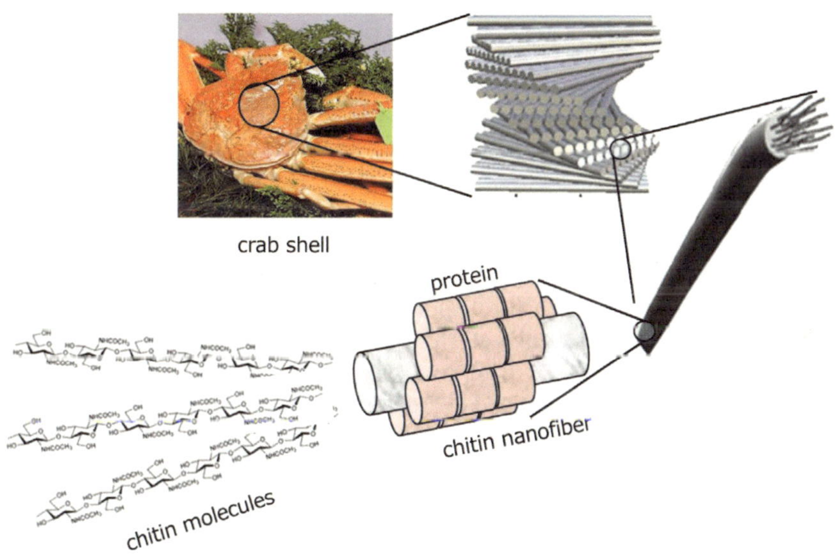

Fig. 2. Schematic presentation of the exoskeleton structure of crab shell.

To extract chitin nanofibers from crab shell, the samples were purified by a series of chemical treatments according to the flowchart shown in Figure 3. Dried crab shell flakes of *Paralithodes camtschaticus* (Red king crab), which is commercially available as a fertilizer at low cost were used for this study as the starting material. Crab shell powder was purified according to the general method. First, crab shell flakes were treated with 2 N HCl for 2 days at room temperature to remove the mineral salts. The suspension was filtered and washed thoroughly with distilled water. The sample was refluxed in 2 N NaOH for 2 days to remove the various proteins (Gopalan & Dufresne, 2003; BeMiller & Whistler, 1962, Shimahara & Takiguchi, 1998).

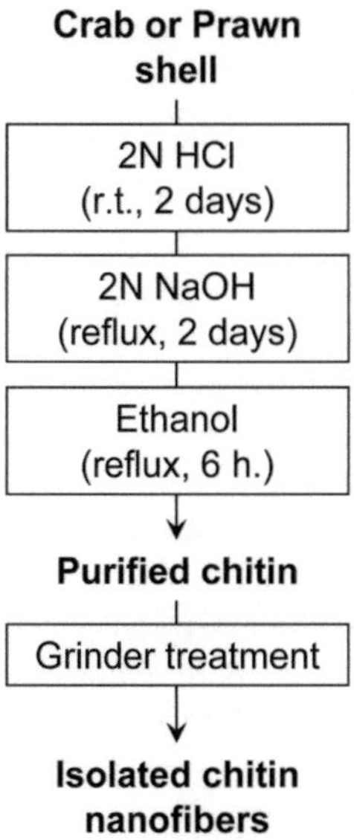

Fig. 3. Preparation procedure of chitin nanofibers from crab and prawn shell.

After filtration and rinsing with distilled water, the pigments and lipids in the sample were extracted by ethanol followed by filtration and rinsing with water (Shimahara & Takiguchi, 1998). It is known that almost all proteins and minerals (mainly calcium carbonate) can be removed by treatment with NaOH and HCl solutions each <0.1%. On the other hand, since complete removal of hemicelluloses from wood fiber is difficult, considerable amount of hemicelluloses remain in cellulose by alkali treatment. The yield of chitin from the crab shells was estimated to be 12.1 wt %. Since the process of drying chitin nanofibers generates strong hydrogen bonding between the nanofiber bundles, making difficult to obtain thin and uniform nanofibers, the material was kept wet after the removal of the matrix. Figure 4 shows field emission scanning electron microscope (FE-SEM) images of the crab shell surface after the removal of the matrix. The image represents the endocuticle, which makes up approximately 90 vol.% of the crab exoskeleton. We could observe that the chitin was made up of regularly structured chitin fiber networks, and the networks were found to be comprised of bundles of chitin nanofibers.

The purified wet chitin from dry crab shells was dispersed in distilled water at 1 wt %, and the slurry was passed through a grinder (MKCA6-3; Masuko Sangyo Co., Ltd.) under a neutral pH condition, with a careful adjustment of the clearance between the grinding stones. In principle, there is no direct contact between the grinding stones due to the presence of chitin suspension. The suspension of chitin fibrils was subjected to oven-drying after replacement of water by ethanol, and the obtained sheets were coated with an approximately 2 nm layer of platinum by an ion sputter coater and observed with FE-SEM. The widths of the fibers derived from crab shell after grinder treatment were widely distributed over a range 10 to 100 nm (Figure 5a). The chitin fiber network structure seems to have been broken completely by the grinder treatment after removal of the matrix substances. However the thicker fibrils of approximately 100 nm, which were bundles of nanofibers of 10-20 nm in width, still remained after grinder treatment even though the protein layers were removed under a never-dried condition.

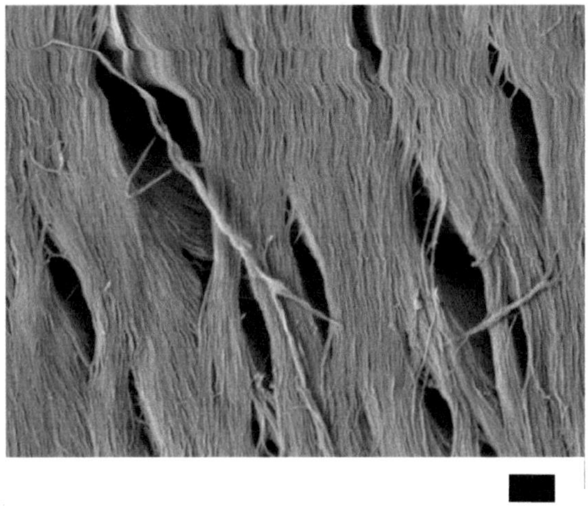

Fig. 4. FE-SEM image of crab shell surface after the removal of matrix. The length of scale bar is 300 nm.

Fan et al. reported a preparation method of chitin nanofibers from squid pen β-chitin in water at pH 3–4 (Fan et al., 2008a). Cationization of C2 amino groups in β-chitin under an acidic condition is important to maintain the stable dispersion state by electrostatic repulsions. Therefore, we considered that the purified α-chitin from crab shell would also be well dispersed under an acidic condition by the cationization of the amino groups on the fiber surface, which would facilitate fibrillation into 10-20 nm chitin nanofibers. Thus, the pH value of the purified chitin suspension was adjusted to approximately 3 by the addition of AcOH, and then the chitin was subjected to a grinder treatment. It should be emphasized that the chitin slurry thus obtained formed a gel after a single pass through the grinder, as the extremely large surface area leads to high dispersion property in water, suggesting that fibrillation was successfully accomplished, (Abe et al., 2007). Figure 5b and 5c show FE-SEM micrographs of the dried chitin gel. The isolated chitins were highly uniform nanofibers with a width of 10–20 nm, suggesting that the fibrillation process was facilitated in acidic solution as expected. Since broken fibers are not observed even over a wide observation area, the aspect ratios of the nanofibers are very high. These results indicate that chitin nanofibers were successfully isolated from crab shells without altering their natural shape. Thus, the never-dried process that has previously been used to extract the cellulose nanofibers from wood cell walls was here proven applicable for the preparation of 10-20 nm chitin nanofibers from crab shells by mechanical grinding at pH 3–4 condition.

2.2 Characterization of chitin nanofibers

The degree of *N*-acetylation of the obtained chitin nanofibers were calculated from the C and N contents from the elemental analysis data by using an elemental analyzer, and was found to be 95%. Thus, the degree of substitution of amino group was just 5%, indicating that deacetylation hardly occurred after the removal of proteins with 2N NaOH, and chitin nanofibers were obtained as natural fibers. Thus, even though the ratio of amino group was only 5%, cationization of amino groups facilitated the fibrillation of chitin fibers into nanofibers and maintained the stable dispersion state in the water by electrostatic repulsions between the chitin nanofibers with cationic surface charges.

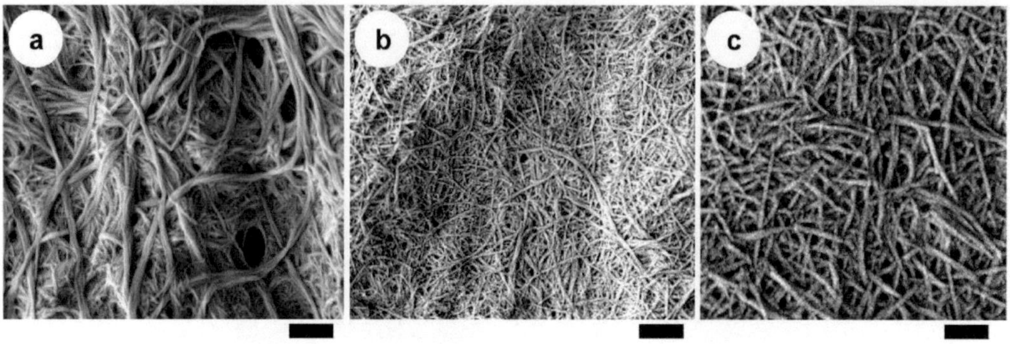

Fig. 5. SEM images of chitin nanofibers from crab shell after one pass through the grinder (a): under neutral pH condition, (b and c): under acidic condition at pH 3. The length of the scale bar is (a) 400 nm, (b) 400 nm, and (c) 200 nm, respectively.

Infrared spectra (FT-IR) recorded using potassium bromide pellets of (a) commercially available pure chitin, (b) newly prepared chitin nanofibers from crab shell, and (c) dried crab shell flakes without purification are shown in Figure 6. The spectrum of the dried crab shell flakes is very different from the other two spectra because of the matrix components included in the shell. On the other hand, the spectrum of newly prepared chitin nanofibers is in excellent agreement with the spectrum of commercial pure α-chitin, indicating that the matrix, protein and minerals were well removed by a series of processing steps. In particular, the absorption band at 1420 cm^{-1} derived from protein has completely disappeared, suggesting that these treatments were sufficient to eliminate all the proteins. Moreover, the OH stretching band at 3482 cm^{-1}, NH stretching band at 3270 cm^{-1}, amide I bands at 1661 and 1622 cm^{-1}, and amide II band at 1559 cm^{-1} of the chitin nanofibers are observed. These characteristic absorption peaks in the carbonyl region corresponds to anhydrous α-chitin (Gopalan & Dufresne, 2003).

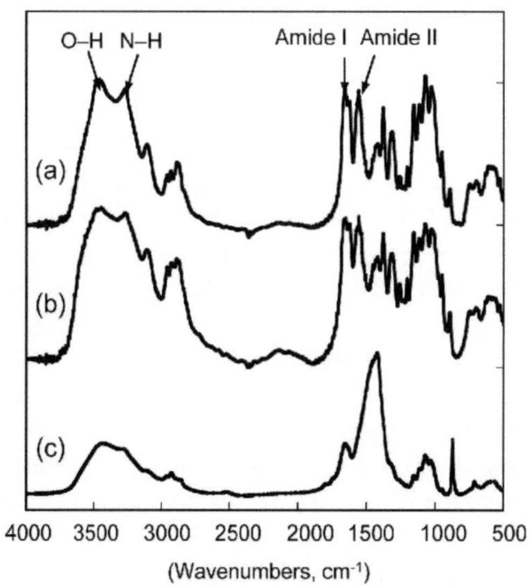

Fig. 6. FT-IR spectra of (a) commercially available chitin powder, (b) chitin nanofibers, and (c) crab shell flakes.

X-ray diffraction profiles obtained with Ni-filtered Cu Kα of (a) commercially available pure *α*-chitin, (b) newly prepared chitin nanofibers from crab shell, and (c) dried crab shell flakes derived from red king crab are shown in Figure 7. The diffraction peak at 29.6°, which is typical of calcium carbonate, completely disappeared from the profile of chitin nanofibers, indicating that the mineral component was entirely removed from the newly prepared chitin nanofibers. The four diffraction peaks of chitin nanofibers observed at 9.5°, 19.5°, 20.9°, and 23.4°, which corresponds to 020, 110, 120, and 130 planes, respectively, are typical crystal patterns of *α*-chitin, and are closely coincident with commercial *α*-chitin (Minke & Blackwell, 1978). Thus, the chitin nanofibers were extracted from the crab shell, and the original molecular structure and *α*-chitin crystalline structure were maintained even after the removal of the matrix and the grinder treatments.

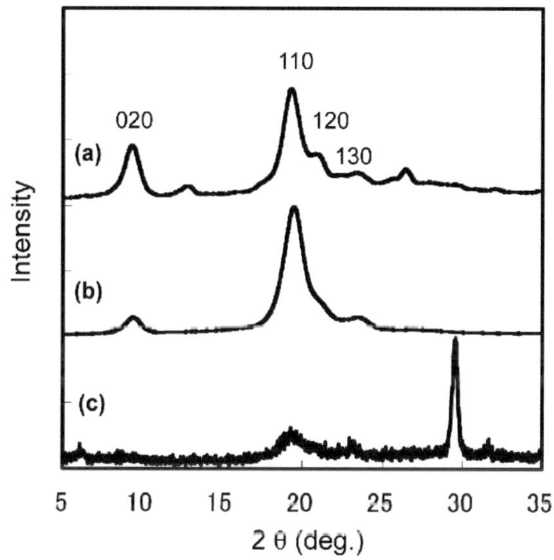

Fig. 7. X-ray diffraction profiles of a) commercially available chitin powder, (b) chitin nanofibers, and (c) crab shell flakes.

2.3 Chitin nanofibers from prawn shell

We succeeded in preparing *α*-chitin nanofibers from crab shells with a uniform width of approximately 10–20 nm (Ifuku et al., 2009). The grinder treatment under acidic conditions is the key to preparing chitin nanofibers. Cationization of amino groups in the chitin by the addition of an acid is important to maintain a stable dispersion state by electrostatic repulsions, which facilitate nano-fibrillation into chitin nanofibers. However, in some cases, the acidic condition may cause significant problems for application of chitin nanofibers such as in biomedical materials, nanocomposites, electronics devices, and so on, because in general, these materials are sensitive to acid. Indeed, it is difficult to remove acidic chemicals from chitin nanofiber suspension, since chitin nanofibers are homogeneously dispersed in water to give high viscosity. Therfore preparation of chitin nanofibers under a neutral pH condition is required to expand the application of nanofibers. Here, we show the extraction of chitin nanofibers from prawn shells under neutral conditions without using any acid.

Fresh shells of *Penaeus monodon* (black tiger prawn) was purified to prepare the chitin nanofibers. It is a widely cultured prawn species in the world and some of its shell is thrown away as industrial waste without effective utilization. Proteins and minerals were removed according to the conventional method using aqueous NaOH and HCl, respectively to extract the chitin component from the prawn shell (Shimahara & Takiguchi, 1998). The pigment component in the sample was then removed using ethanol. The yield of dry chitin from the wet prawn shells was approximately 16.7%. The degree of deacetylation (DDA) of the samples determined by the C and N content in the elemental analysis data was 7%. The SEM micrograph of the black tiger prawn shell surface after removal of the matrix components (without grinding treatment) is shown in Figure 8. This image is from the exocuticle, which is the main part of the prawn shell. Although the appearance of the prawn shell is still intact after removal of the matrix, we could observe uniform chitin nanofibers with an elaborate design.

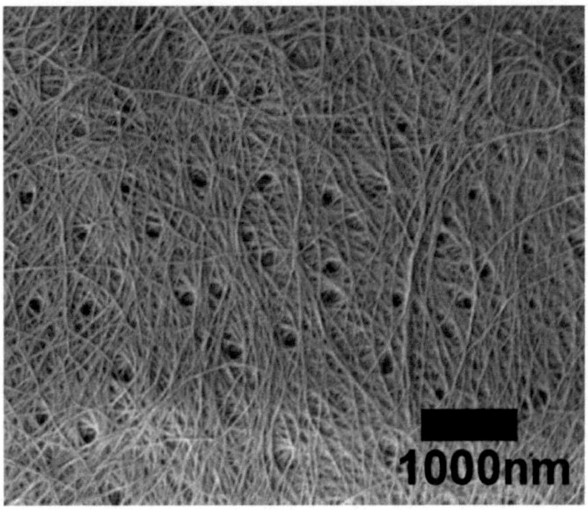

Fig. 8. FE-SEM image of the surface of the black tiger prawn after removing matrix components.

The purified chitin suspension with a concentration of 1 wt% was crushed by a domestic blender, and passed through a grinder for nano-fibrillation without using any acid. The chitin slurry thus obtained was viscous after a single grinding treatment similar to the chitin nanofibers from crab shell. The obtained chitin slurry was observed using FE-SEM (Fig. 9). In the case of crab shell, the widths of the fibers were widely distributed over a range from 10 to 100 nm by grinder treatment under a neutral condition as described above (Fig. 5a, Ifuku et al., 2009). On the other hand, in the case of prawn shell, we could see a uniform shape of chitin nanofibers using the same extraction treatment. The chitin nanofibers were highly uniform over an extensive area (Fig. 9a), and the width of the nanofibers was approximately 10 - 20 nm, including a 2 nm thick platinum coating layer (Fig. 9b). The fiber thickness of 10 - 20nm was similar to nanofibers from crab shell fibrillated under an acidic condition (Ifuku et al., 2009). Thus, chitin nanofibers were successfully prepared from prawn shell with a uniform width under a neutral pH condition.

The plausible explanation of successful fibrillation is following. The exoskeleton of crab and prawn is made up of mainly two parts, the exocuticle and the endocuticle. Although the exocuticle has a very fine twisted plywood-type structure, the endocuticle has a much coarser structure with a thicker fiber diameter (Fig. 4). In general, approximately 90% of a crab shell is made of endocuticle (Chen et al., 2008). In contrast, the exoskeleton of *Natantia,* which has a semitransparent soft shell, including the black tiger prawn, is made up primarily of a fine exocuticle, as shown in Figure 8 (Yano, 1972, 1975 & 1977). As a result, due to the differences in the cuticle structure, fibrillation of prawn shell is easier than that of crab shell.

The preparation method for chitin nanofibers from prawn shells under a neutral pH condition was applicable to other prawn shells. Figure 10 shows SEM images of the chitin nanofibers derived from (a) *Marsupenaeus japonicus* (Japanese tiger prawn) and (b) *Pandalus eous* (Alaskan pink shrimp), both of which are important food sources. These chitin nanofibers were prepared by the same preparation procedure, that is, removal of matrix components and subsequent grinding treatment under a neutral pH condition. Both chitins are also observed as uniform nanofibers with a width of 10–20 nm, which is similar to the nanofibers from black tiger prawn. This result suggests that chitin nanofibers can be obtained from other prawn species having a very fine exocuticle structure by nano-fibrillation under neutral pH conditions. Since many materials are sensitive to acid chemicals, this study will expand the application of chitin nanofibers.

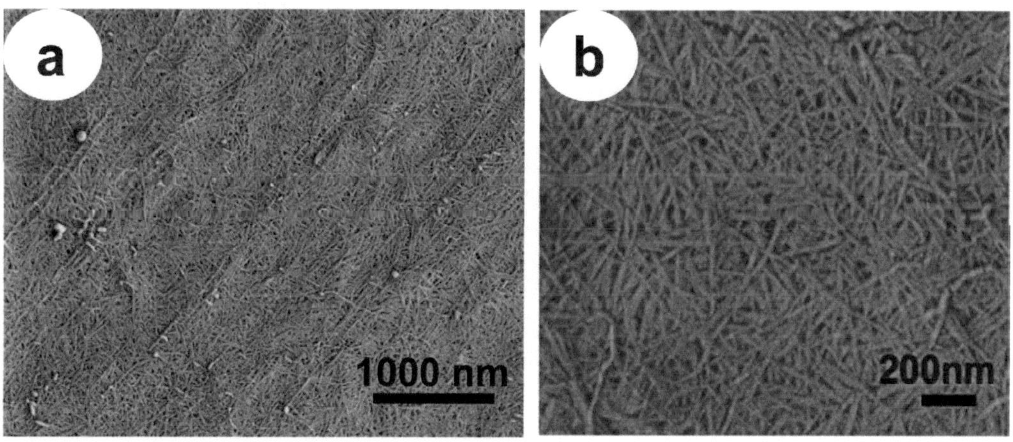

Fig. 9. FE-SEM images of chitin nanofibers from black tiger prawn shell after one pass through the grinder.

2.4 Facile preparation of chitin nanofiber from dry chitin

We have succeeded in isolating α-chitin nanofibers from crab and shrimp shells with a uniform width of 10–20 nm and a high aspect ratio (Ifuku et al., 2009). In general, the drying process of chitin and cellulose fibers generates strong hydrogen bonding between these fibers after removal of the matrix, which makes it difficult to fibrillate them into nanofibers. Therefore, chitin and cellulose must be kept wet after removal of the matrix for nanofiber preparation (Abe et al, 2007; Fan et al, 2009; Ifuku et al, 2009; Iwamoto et al., 2008; Saito et al., 2006). However, this requirement causes a strong disadvantage in the commercial

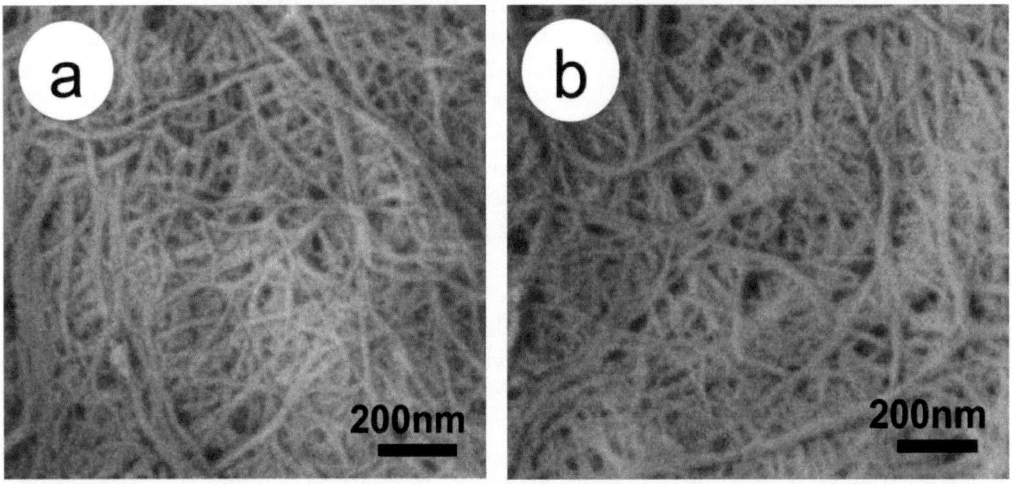

Fig. 10. FE-SEM images of chitin nanofibers (a) from Japanese tiger prawn shell, and (b) from Alaskan pink shrimp shell.

production of nanofibers. If chitin nanofibers could be obtained from dry chitin, we could prepare nanofibers easily and immediately whenever required. Therefore, preparation of nanofibers from a dried pure chitin is an important goal for expanding the use of chitin nanofiber. As described above, for the preparation of nanofibers, grinder treatment under an acidic condition is the key to fibrillating the chitin effectively (Fan et al., 2008b). A small number of amino groups in the chitin are cationized by the addition of an acid, which promotes the fibrillation of chitin into nanofibers by electrostatic repulsion. Indeed, electrostatic repulsion force is applied for the fibrillation of cellulose to obtain nanofibers through TEMPO-catalyzed oxidation too (Saito et al., 2006). If the electrostatic repulsion force of amino cations can break the strong hydrogen bonds between the chitin bundles, chitin nanofibers could be obtained from pure dry chitin aggregate. Accordingly, we here show the fibrillation of dried chitin into nanofibers by a grinding method under acidic conditions (Ifuku 2010b).

First, we tried to prepare cellulose nanofibers from dry pulp fibers, as a preliminary experiment. Figure 11 shows FE-SEM micrographs of cellulose fiber (a, b) before and (c) after single pass through the grinder with a concentration of 1 wt % for fibrillation of pulp fibers into cellulose nanofibers. In Figs. 11a and 11b, it can be seen that pulp fibers consist of a bundle of cellulose nanofibers. Figure 11c shows that the pulp fibers were not fibrillated into nanofibers at all after the grinder treatment. This is obviously because the drying process causes strong hydrogen bonding between the cellulose nanofibers after the removal of matrix substances hemicelluloses and lignin, thus making it difficult to obtain cellulose nanofibers. Thus, in general, the sample should never dry out for bionanofiber preparation after the removal of matrix (Abe et al., 2007; Iwamoto et al., 2008).

Next, we tried to fibrillate dried chitin into nanofibers. Dried chitin was prepared from crab shell flakes with an α-chitin crystal structure. The chitin was purified by the removal of matrix components of proteins and calcium carbonate according to the conventional method (Shimahara & Takiguchi, 1998), followed by drying in an oven to promote hydrogen bonding interaction between chitin nanofibers. Figure 4 is FE-SEM image of a crab shell after

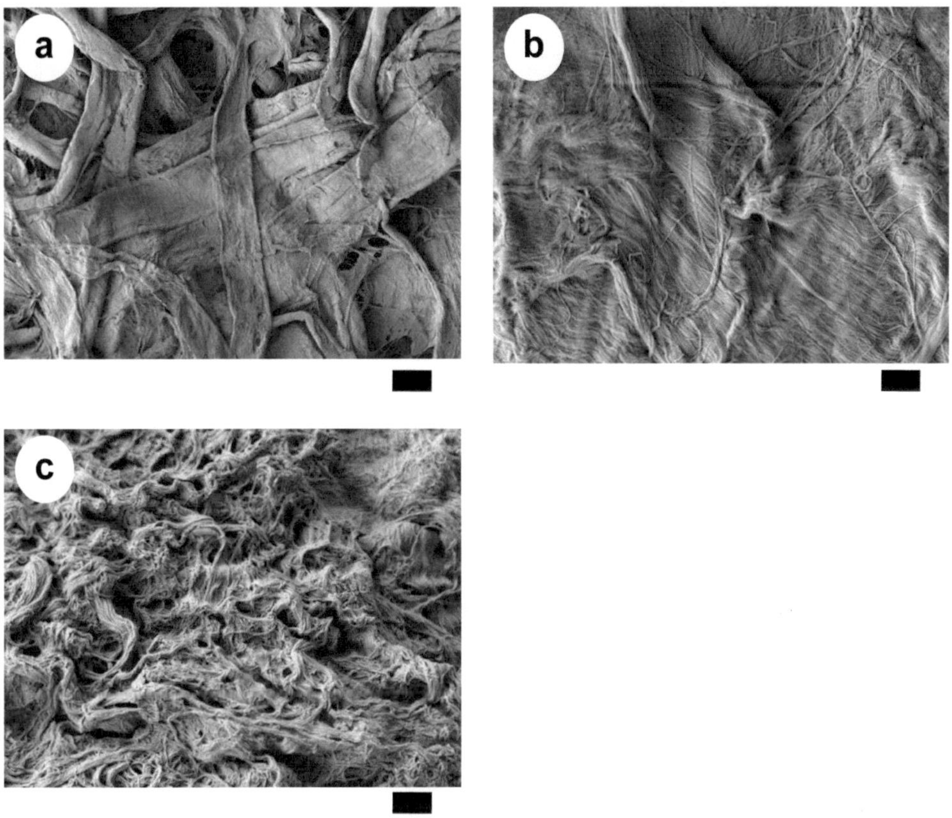

Fig. 11. FE-SEM images of cellulose fibers from wood pulp (a and b) before and (c) after one pass through grinder. The length of the scale bar is (a) 30,000 nm, and (b, c) 1000 nm, respectively.

removing matrices, followed by drying. It can be observed that the chitin was made up of structurally well organized bundles of nanofibers, as mentioned above. The slurry of purified dry chitin was passed through a grinder with a concentration of 1 wt % with and without acetic acid to fibrillate the bundles of chitin nanofibers. In the case of treatment under an acidic condition, the obtained chitin slurry had high viscosity, similar to the chitin nanofibers prepared by the previous method, suggesting that dried chitin was successfully fibrillated and homogeneously dispersed in acidic water with a very high surface ratio (Ifuku et al., 2009). Figure 12 shows FE-SEM images of chitin fibers after one pass through the grinder treated (a) without and (b, c) with acetic acid. In the case of grinder treatment under neutral pH condition, dried chitin could not be fibrillated at all, and aggregates of chitin nanofibers were observed (Fig. 12a). This lack of fibrillation also occurred because the drying process of purified chitin causes strong hydrogen bonding between hydroxyl, acetamide, and amino groups with strong dipole moments on the chitin fiber surface, which make it hard to fibrillate bundles of chitin nanofibers, as in the case of dry pulp fibers. On the other hand, chitin slurry treated at pH 3 was completely fibrillated into nanofibers (Fig. 12b and c).

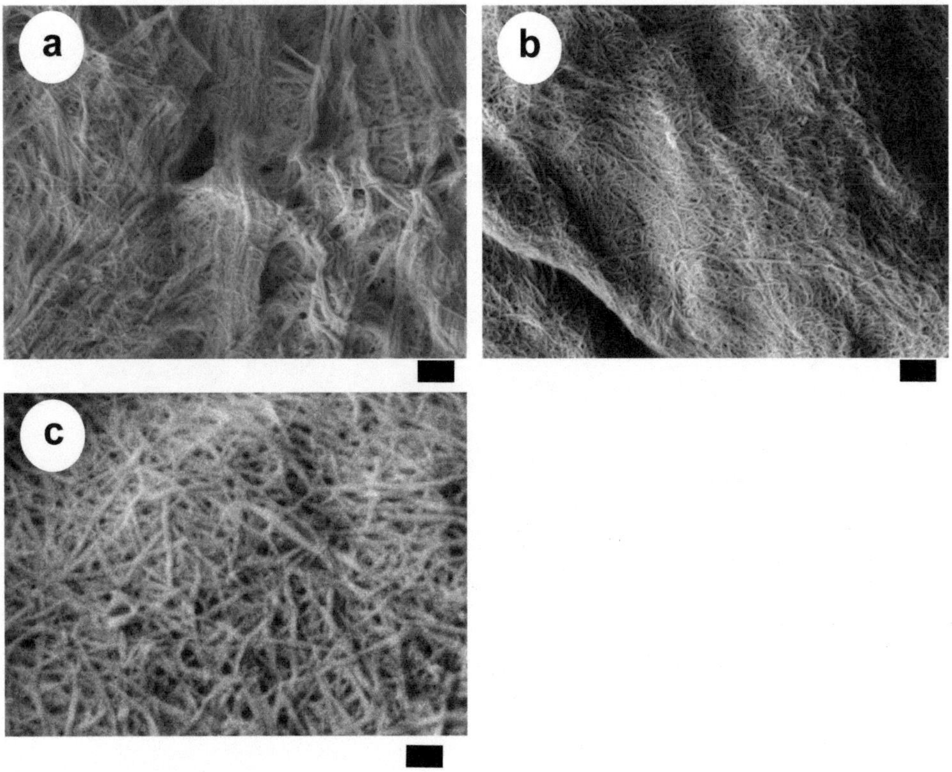

Fig. 12. FE-SEM images of chitin fibers after one pass through the grinder treated (a) without and (b, c) with acetic acid. The length of the scale bar is (a, b) 300 nm, and (c) 100 nm respectively.

The fibrillated chitin has a very fine nanofiber network; the structure is highly uniform with a width of 10 to 20 nm and a high aspect ratio, and we cannot see thicker fibers within the extensive area. The appearance of the fibers was very similar to nanofibers prepared with the never-dry process (Figs. 5b and c). The success of this new method is obviously owing to the electrostatic repulsions between the nanofibers. A few C2 amino groups on the chitin fiber surface were cationized under acidic conditions, which facilitated the fibrillation into nanofibers by electrostatic repulsions. Even though the degree of substitution of the amino groups calculated from the elemental analysis data was only 3.9 %, the electrostatic repulsion force caused from the cationic surface charges was enough to break the strong hydrogen bonds between the nanofibers. Thus, grinder treatment under acidic condition allowed us to obtain chitin nanofibers with a uniform width of 10–20 nm from purified dry chitin easily and immediately. This method could provide a significant advantage for industrial production of nanofibers from the viewpoints of stable supply, storage stability, transportation costs, storage space, and so on, since chitin nanofibers can be prepared by a simple method from light, low-volume, and nonperishable dried chitin. Use of the chitin nanofibers could be quite different from that of cellulose, which does not have ionic functional group to cause electrostatic repulsions. Since other acidic chemicals are also available to facilitate fibrillation, including ascorbic acid, lactic acid, citric acid, and so forth, several acidic additives are available for number of application.

The preparation method for chitin nanofibers was also found to be applicable to dry chitin powder purchased from a chemical reagent company. In Figure 13a, we can see well that commercially available dry chitin powder from crab shell is also made up of nanofibers (Chen et al., 2008; Raabe et al., 2005). Figure 13 shows FE-SEM micrographs of chitin fibers after one pass through the grinder (b) without and (c) with acetic acid, respectively. In Figure 13b, it can be seen that the chitin powder was not fibrillated at all because of the strong interfibrillar hydrogen bonding. On the other hand, in Figure 13c, the aggregates have clearly been fibrillated into homogeneous nanofibers with a width of 10 to 20 nm by grinder treatment at pH 3, even though the degree of substitution (DS) of amino groups was only 3.9 % and the slurry formed a gel. This is also owing to the electrostatic repulsion caused by cationic charges on the chitin fiber surface. In this way, since commercially available dry chitin powder is also made up of bundles of nanofibers, chitin nanofibers could easily be prepared from the dry chitin powder by applying electrostatic repulsion to break hydrogen bonds between chitin fibers' bundles. Chitin nanofibers from commercial prepurified dry chitin is advantageous for laboratory-scale investigations and commercial production because a large amount of chitin could be easily obtained within few hours by a simple fibrillation process under acidic conditions without any purification processes (removal of proteins, minerals, lipids, and pigments), which generally require at least 5 days (Shimahara & Takiguchi, 1998).

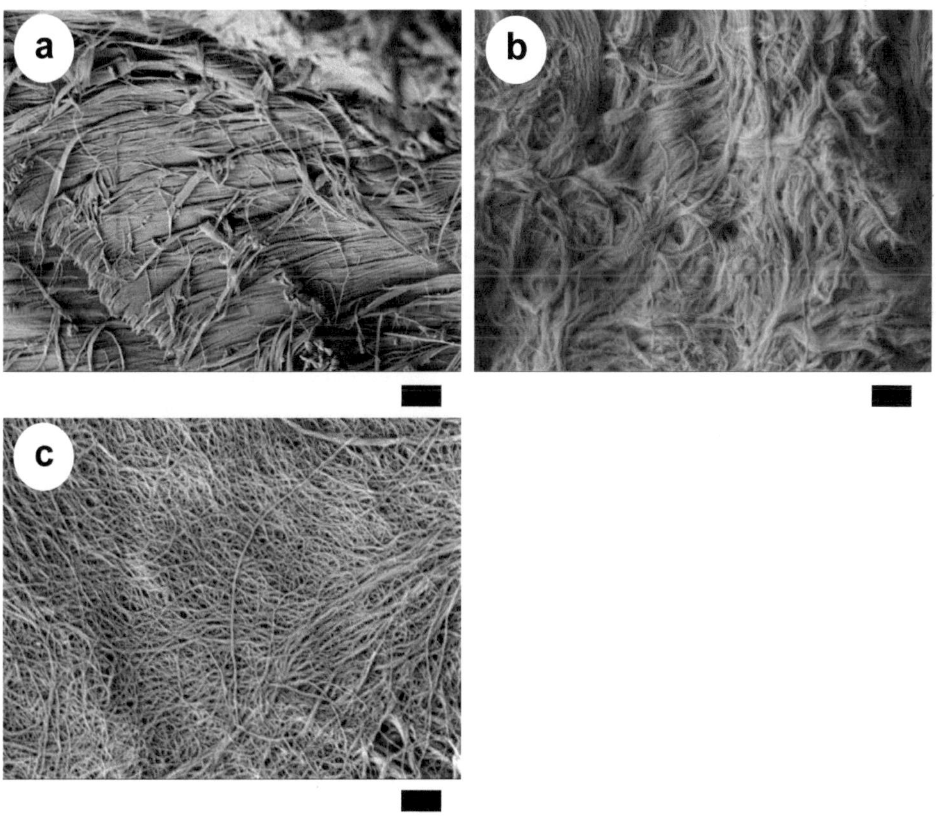

Fig. 13. FE-SEM images of (a) commercially available dry chitin powder, and chitin fibers after one pass through the grinder (b) without and (c) with acetic acid. The length of the scale bar is (a) 1000 nm and (b and c) 300 nm, respectively.

3. Preparation of optically transparent nanocomposites

3.1 Chitin nanofiber/acrylic resin composites

Since the chitin nanofiber consists of an antiparallel extended crystalline structure, the nanofiber has excellent mechanical properties: high Young's modulus and fracture strength, and low thermal expansion (Vincent & Wegst, 2004; Wada & Saito, 2001). Given the characteristic nano-form and efficient physical properties, the chitin nanofibers could be strong candidates for a reinforcement agent to create high-performance nanocomposites.

Ahead of the chitin nanofibers, cellulose nanofibers have attracted much attention in recent years as a reinforcement agent to make high performance nanocomposite (Nakagaito & Yano, 2008; Nakagaito et al., 2009). In particular, Yano et al. have fabricated a unique cellulose nanofiber composite film (Yano et al., 2005). Due to the nano-size effect, the nanocomposite film was optically transparent. This finding is applicable for chitin nanofibers to obtain optically transparent nanocomposite with acrylic resin as well (Ifuku, et al., 2010; Shams et al., 2011). We expect that the transparent chitin nanofiber/acrylic nanocomposite with improved mechanical and optical properties encourages the development of an optical device such as a flexible display. Additionally, chitin has different characteristics from cellulose such as biocompatibility (Tokura et al., 1983), wound healing activity (Okamoto et al., 1995), high purity (Shimahara & Takiguchi, 1998), and hydrophobicity (Shams et al., 2011), which strongly distinguish it from cellulose nanofiber reinforced composite film. In this study, we prepared chitin nanofiber composite with two different types of acrylic resins, and characterized their transparency, Young's modulus, mechanical strength, and thermal expansion. Specific characterization of the effects of chitin nanofibers on a variety of resins would be especially valuable for designing advanced nanocomposite materials.

Chitin nanofibers were prepared from commercially available dry chitin powder by fibrillation using a grinding apparatus. The fibrillated chitin samples were found to have a fine nanofiber network. The structure was highly uniform with a width of 10-20 nm and a high aspect ratio. Fibrillated chitin nanofibers were dispersed in water at a fiber content of 0.1 wt%. The suspension was vacuum-filtrated using a hydrophilic polytetrafluoroethylene membrane filter. The obtained chitin nanofiber sheets were hot-pressed to obtain a dried sheet. The chitin nanofiber sheets were cut into 3 cm × 4 cm sheets with approximately 45 μm thickness and a weight of 40 mg. The sheets were impregnated with bi-functional acrylic resins with 2 wt% of 2-hydroxy-2-methylpropiophenone photoinitiator under reduced pressure for 24 hours. Poly(propylene glycol) diacrylate (A-600) and tricyclodecanedimethanol dimethacrylate (DCP) were used in this study as acrylic resins (Fig. 14). The resin-impregnated sheets were polymerized using UV curing equipment. The chitin nanofiber composite films thus obtained were approximately 60 μm thick, and the fiber contents were approximately 40%.

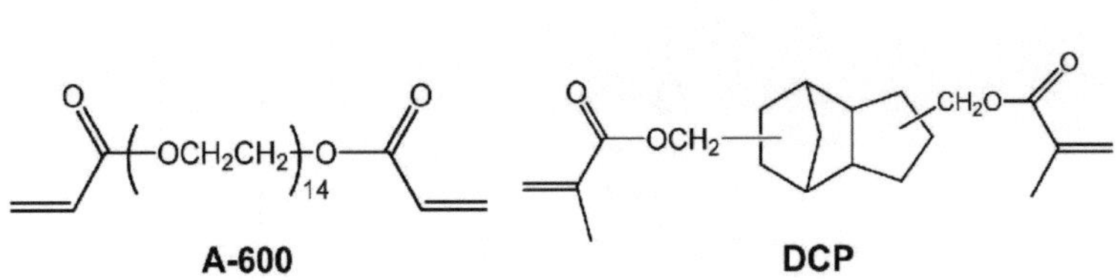

Fig. 14. Chemical structures of A-600 and DCP.

Both nanocomposite films were optically transparent despite the high fiber content of 40 wt%, due to the nanofiber size effect (Fig. 15, Yano et al., 2005). Figure 16 shows the regular light transmittance spectra of nanocomposites reinforced with chitin nanofibers. Transmittance of DCP nanocomposite was higher than that of A-600. This transparency is attributed to the difference in the refractive index (RI) of the resins. The RI of chitin nanofiber was reported to be 1.56 (Vigneron et al., 2005), which was closer to the RI of DCP than that of A-600, as shown in Table 1. The gap in the RI between chitin nanofiber and resins resulted in transparency difference.

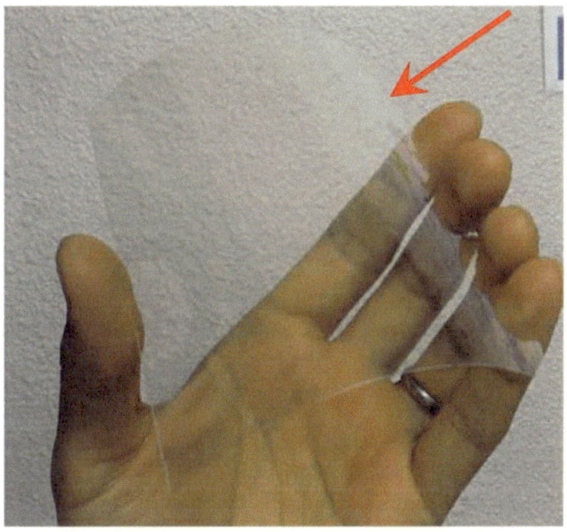

Fig. 15. Appearance of DCP films reinforced with chitin nanofibers.

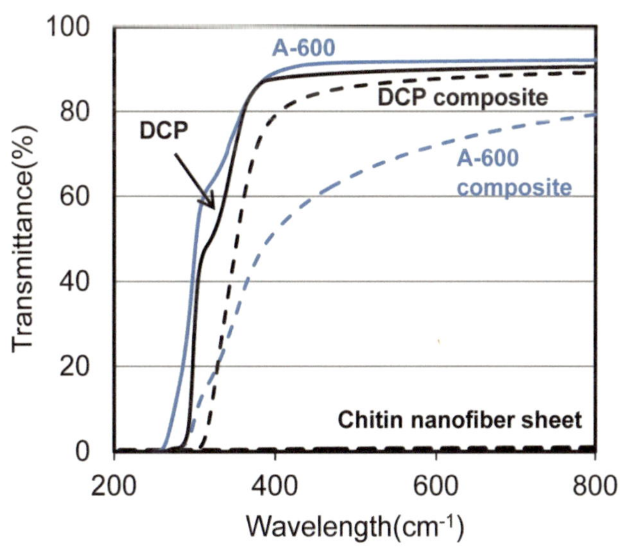

Fig. 16. Regular light transmittance spectra of neat acrylic resins (solid lines) and their nanocomposites (dashed lines).

	A-600		DCP		chitin nanofiber sheet
	neat resin	composite	neat resin	composite	
refractive index	1.468	-	1.500	-	-
transmittance (%)	91.8	72.1	89.8	87.6	0.4

Table 1. Regular light transmittance of neat acrylic resins and their nanocomposites at 600 nm.

Generally, thermal expansion has an inverse relationship with Young's modulus. Since the Young's modulus of chitin crystal is estimated to be at least 150 GPa (Vincent et al., 2005), the coefficient of thermal expansion (CTE) of chitin nanofiber sheet is only 10 ppm K^{-1}. Although the CTEs of A-600 and DCP were very high (184 and 100 ppm K^{-1}), the CTEs of their nanocompsites decreased drastically to 19 and 24 ppm K^{-1}, which corresponded to 90 and 76% decrease (Fig. 17 and Table 2). Thus, chitin nanofibers with low CTE and high Young's modulus could effectively decrease the CTE of acrylic resins by a reinforcement effect.

Since chitin nanofiber has high fracture stress and Young's modulus due to the antiparallel extended crystalline structure, it could be available as a reinforcing element to improve mechanical properties of a resin. The stress-strain curves of A-600 and DCP resin, and their nanocomposites are shown in Figure 18, and their mechanical properties are listed in Table 2. The Young's moduli of both resins effectively increased. Especially, Young's modulus of soft A-600 resin increased hundredfold, obviously due to the reinforcement with 40 wt% of chitin nanofiber. The tensile strengths of the two nanocomposites also successfully increased. These outstanding enhancements of mechanical properties strongly support the fact that chitin nanofiber with high Young's modulus and high tensile strength works effectively as a reinforcing element. Furthermore, chitin nanofiber can make a fragile resin ductile. That is, the DCP resin was extremely brittle due to its higher crosslinking density. However, the fracture strain of the DCP nanocomposite increased from 0.6 to 1.2%.

In conclusion, two types of acrylic resins reinforced with chitin nanofibers were highly transparent and flexible, and had low CTE, high Young's modulus, and high tensile strength, owing to the unique nano-size structure and excellent mechanical properties of chitin nanofiber.

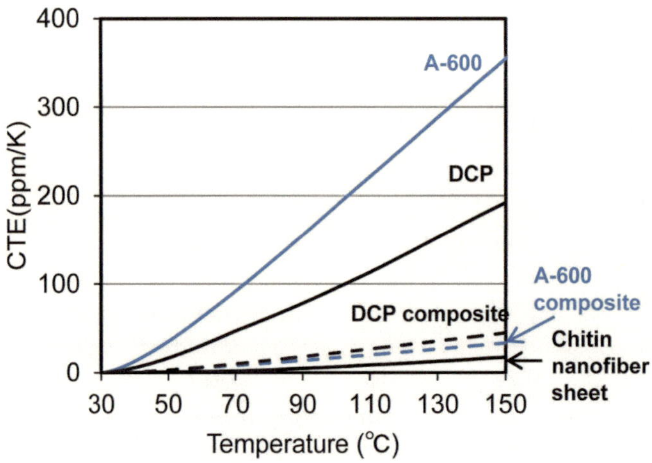

Fig. 17. Coefficients of thermal expansion of acrylic resins (solid lines) and their nanocomposites (dashed lines).

	A-600		DCP		chitin nanofiber sheet
	neat resin	composite	neat resin	composite	
Young's modulus (GPa)	0.02	2.03	2.30	5.34	2.50
fracture stress (MPa)	4	41	11	56	45
fracture strain (%)	16.5	9.0	0.6	1.2	7.8
CTE (ppm/K)	184	19	100	24	10

Table 2. Mechanical properties and CTEs of neat acrylic resins and their nanocomposites.

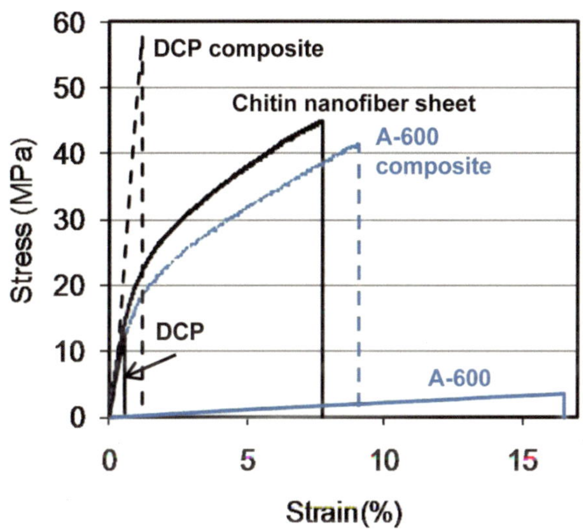

Fig. 18. Stress-strain curves of neat acrylic resins and their nanocomposites.

3.2 Acetylation of chitin nanofibers

We expect that chitin nanofibers with a uniform width and very high surface area will be developed into novel green nanomaterials. To expand the applications of chitin nanofibers, chemical modification of the chitin nanofiber surface is very important. By the introduction of hydrophobic functional groups into hydrophilic hydroxyl groups on chitin fibers, it is expected that dispersibility in nonpolar solvents, hygroscopicity, adhesion properties with hydrophobic matrices for fiber-reinforced composite materials are improved (Glasser et al., 1994). In chemical modification, acetylation is considered to be a simple, popular, and inexpensive approach to change the surface property (Kim et al., 2002; Ifuku et al., 2007). However, there have been no reports regarding acetylation of chitin nanofibers. Hence, the reaction behavior of highly crystalline chitin nanofibers, and the relationships between acetyl DS values and the various properties of the nanofibers remain unclear. We have showed that chitin nanofibers have very promising characteristics as a reinforcing material. The nanocomposites are optically transparent even with a high fiber content. We consider that acetylated chitin nanofiber would also be available as a filler to reinforce plastics. It is

therefore desirable to investigate acetyl DS dependency on the properties of chitin nanofiber composites, including transparency and hygroscopicity. We here prepared chitin nanofibers with different acetyl DS values and their nano-composites, and characterized them.
For acetylation of chitin nanofibers, the suspension of chitin nanofibers was vacuum-filtered using a membrane filter. The obtained chitin nanofiber sheets were dried and were cut into 3 x 4 cm sheets 60 μm thick, with a weight of 55 mg. The sheets were placed in a Petri dish containing a mixture of acetic anhydride and perchloric acid. The mixture was stirred for the desired time at room temperature. After acetylation, the chitin sheets were washed by Soxhlet extraction with methanol overnight Figure 19 shows the effects of reaction time on the DS of acetyl groups.

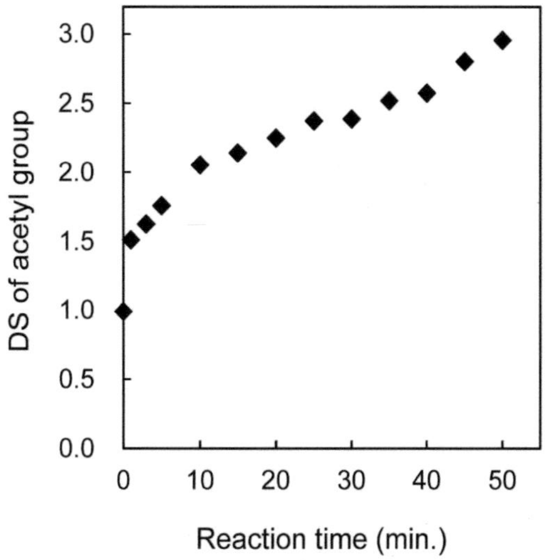

Fig. 19. Effect of reaction time on the acetyl DS.

The DS values were estimated by C and N content in the elemental analysis data. In general, since chitin has poor solubility in typical solvents, the reaction rate of acetylation of chitin is very small. However, in the case of chitin nanofibers, the acetyl DS ranged from 0.99 to 2.96 after 50 minutes reaction time, indicating an almost complete substitution of acetyl groups in the chitin nanofibers. The graph shows that the DS reached 1.51 from 0.99 after only 1 minute of acetylation. The high reaction rate is obviously due to the very high surface area, which works effectively for the liquid-solid phase reaction. The acetyl DS values then increased slowly and almost proportionally up to DS 2.96. The observed nonlinearity may have been caused by a heterogeneous reaction. That is, first, the chitin nanofiber surfaces were acetylated, and then the insides of the nanofibers were acetylated more gradually, as we describe later. Thus, the DS appears to be strictly adjustable by changing the acetylation time. This result could be applicable to other esterifications of chitin nanofibers using the corresponding acid anhydride.
The FT-IR spectra of acetylated chitin nanofibers with DS values of 0.99, 1.81, and 2.96 recorded using potassium bromide pellets are shown in Figure 20. As the DS of acetyl groups increased, two major bands at 1231 cm^{-1} and 1748 cm^{-1} increased, corresponding to the C-O and C=O stretching vibration modes of the acetyl group, respectively. In contrast,

the O-H stretching band at 3972 cm^{-1} decreased with increasing acetyl DS and almost disappeared with a DS of 2.96, indicating a complete substitution of acetyl groups in the chitin nanofibers.

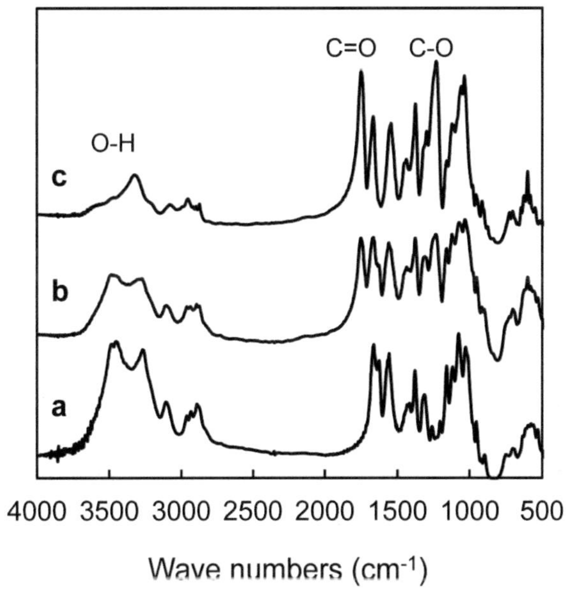

Fig. 20. FT-IR spectra of acetylated chitin nanofibers of (a) DS 0.99, (b) DS 1.81, and (c) DS 2.96.

Figure 21 shows the X-ray diffraction profiles of a series of acetylated chitin nanofibers. In original chitin nanofiber (DS 0.99), the four diffraction peaks of chitin nanofibers observed at 9.5, 19.4, 20.9, and 23.4°, which corresponds to 020, 110, 120, and 130 planes, respectively, show the typical antiparallel crystal pattern of α-chitin (Minke, 1978). At DS 2.96, the diffraction pattern from α-chitin had completely disappeared, and the sample showed a well-defined uniform pattern of di-*O*-acetylated chitin (chitin diacetate) at $2\theta = 7.4$ and 17.7°. On the other hand, the diffraction from α-chitin still remained even at DS 1.81, which indicates that approximately 50% of OH groups were substituted. Moreover, diffraction from chitin diacetate was also observed at 7.4°, and a small shoulder peak was also observed at 17.7°. This profile clearly shows that chitin nanofibers are acetylated heterogeneously from the surface to the core (Kim et al., 2002; Ifuku et al., 2007).

Figure 22 shows the FE-SEM images of acetylated chitin nanofibers with a series of DS values. Nanofiber shapes remain even in the DS 2.96 sample, which indicates that chitin diacetate is insoluble in the reaction mixture (mainly acetic anhydride), although cellulose triacetate is soluble in acetic anhydride. In all the images of the chitin nanofiber sheets, individual nanofibers seem to exist separately without aggregation, and as the DS values increase, the thickness of the nanofibers is increased. The average thicknesses of nanofibers with DS: 0.99, 1.81, and 2.96 were 21.6, 28.9, and 32.1 nm, respectively. This change in thickness is obviously due to the introduction of bulky acetyl groups into chitin nanofibers.

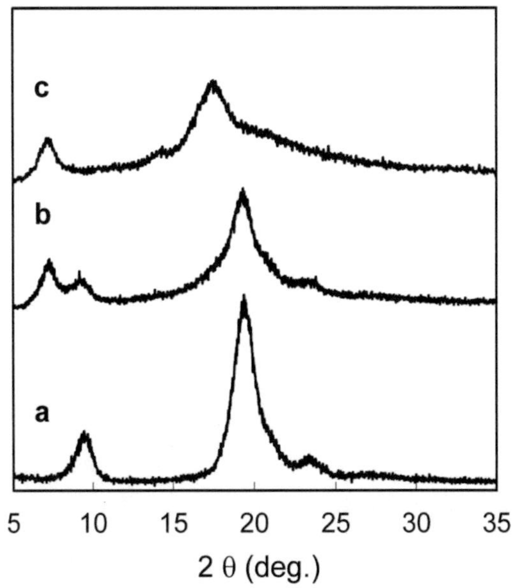

Fig. 21. X-ray diffraction profiles of acetylated chitin nanofibers of (a) DS 0.99, (b) DS 1.81, and (c) DS 2.96.

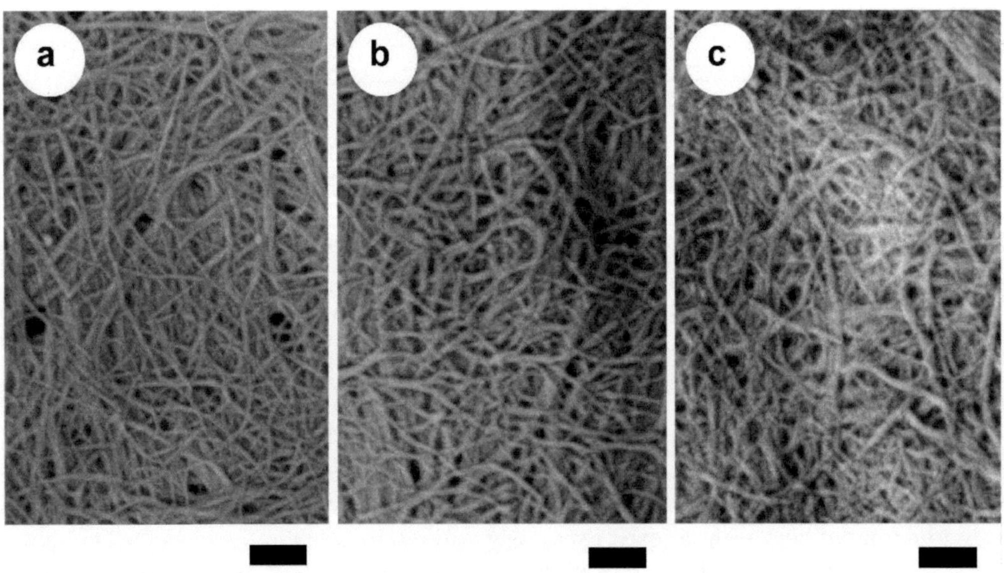

Fig. 22. FE-SEM micrographs of acetylated chitin nanofiber samples of (a) DS 0.99, (b) DS 1.81, and (c) DS 2.96 .The length of scale bar is 200 nm.

Thermogravimetric analysis (TGA) of a series of acetylated chitin nanofibers was carried out under nitrogen atmosphere to evaluate their degradation profiles and thermal stability. Figure 23 shows the derivative TGA curves of acetylated chitin nanofibers with DS 0.99, 1.81, and 2.96. The thermal degradation temperature of the original fibers (DS 0.99) was 388 °C according to the derivative curves. The derivative TGA curves drastically changed in the DS 2.96 sample, with no evidence of the peak derived from the original chitin nanofibers, and there were two peaks at 242 and 326 °C, which are attributed to the thermal decomposition of chitin diacetate. Interestingly, at DS 1.81, the TGA curve showed two kinds of thermal decomposition behaviors derived from chitin and chitin diacetate. This result obviously also indicates that acetyl groups were introduced from the surface to the core of the nanofibers.

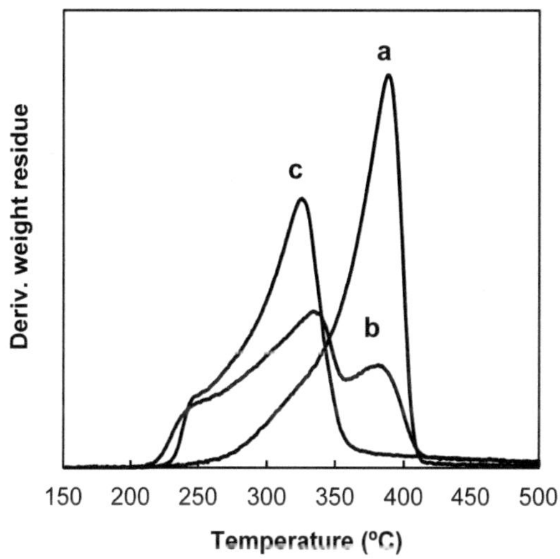

Fig. 23. Derivative TGA curves of acetylated chitin nanofibers of (a) DS 0.99, (b) DS 1.81, and (c) DS 2.96.

3.3 Characterization of acetylated chitin nanofiber composites

Acetylated chitin nanofiber composites were fabricated with acrylic resin DCP. The chitin nanofiber composite films thus obtained were approximately 150 μm thick, and the fiber content was approximately 25%. Since the diameters of the chitin nanofibers are sufficiently smaller than the visible light wavelength, all nanocomposites have high transparency despite a variety of acetyl DS values ranging from 0.99 to 2.96. Figure 24 shows the regular transmittance of a series of acetylated chitin nanofiber composites at 700 nm wavelength. At 0.99, the transparency of the nanocomposite was 77%. However, as the acetyl DS values increased, the transmittance decreased slightly and proportionally up to 73% at DS 2.96. This decrease can likely be attributed to the change in the refractive index of the nanofibers. The optimum refractive index of resin to obtain the highest transparency nanocomposite with chitin nanofibers is known to be approximately 1.56, which is close to that of the DCP resin (1.532, Vigneron, 2006). However, since acetylation decreases the refractive index, the gap in the refractive index between chitin nanofiber and resin became wider with acetylation, resulting in a reduction in transparency. The transparency of chitin nanofiber composite was less sensitive to acetylation than that of bacterial cellulose (BC) nanofiber composite (Ifuku et al., 2006). The optical loss of the acetylated BC nanocomposite was approximately 20% with changes in the acetyl DS from 0.74 to 1.76. The difference of the optical loss is obviously due to the size effect. Since thickness of the chitin nanofiber is approximately 20 nm, which is considerably smaller than that of BC nanofibers (50 nm width), the chitin nanofibers are freer from light scattering than BC nanofibers (Nogi et al., 2005).

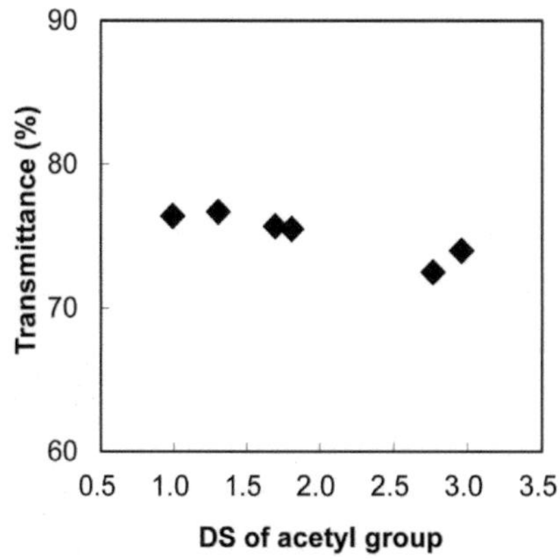

Fig. 24. Regular light transmittance of a series of acetylated chitin nanocomposites.

Although the chitin nanofiber reinforced composite seems to have promising characteristics as an optically functional composite, it is quite hygroscopic. Absorption of moisture causes deformation of the composite. Acetylation seems to reduce the moisture content of the nanocomposite. We therefore investigated the effects of acetylation on the hygroscopicity of the nanocomposites. Figure 25 shows the moisture content of a series of acetylated chitin nanofiber sheets and their composites. Moisture content was evaluated by exposing samples under a constant relative humidity. After measuring the sample weight equilibrated at 75.1% RH and 30 °C with a saturated aqueous solution of NaCl, the sample was oven-dried at 105 °C for 24 h, and the moisture content was then determined on the basis of the ovendried weight. Three samples were used to determine the moisture content. Interestingly, acetylation of chitin nanofiber sheet decreased its moisture content just slightly. The slight difference from 7.8% to 7.3% may be due to the nano-size effect. Since chitin nanofiber sheet with very high surface area was exposed to the air, water molecules were easily adsorbed onto the large nanofibers' surface. In the case of their composites, the moisture content of the original sample with DS 0.99 was 4.0%, which is considerably higher than the 0.33% of acrylic resin due to the hygroscopic chitin nanofiber filler. However, the moisture absorption of the nanocomposite with DS 1.30 drastically decreased to 2.2%. Clearly, this can be attributed to the introduction of hydrophobic acetyl groups into the hydrophilic hydroxyl moiety of chitin molecules. Acetylation of the nanofibers improved the compatibility with the acrylic resin, and improvement of miscibility at the filler/matrix interface decreased water adsorption onto the interface. However, with further acetylation, the moisture content showed little change, with a range of 2.0-2.5%. This result suggests that most hydroxyl groups on the chitin nanofiber surface were acetylated after 1 minute reaction time, as mentioned above.

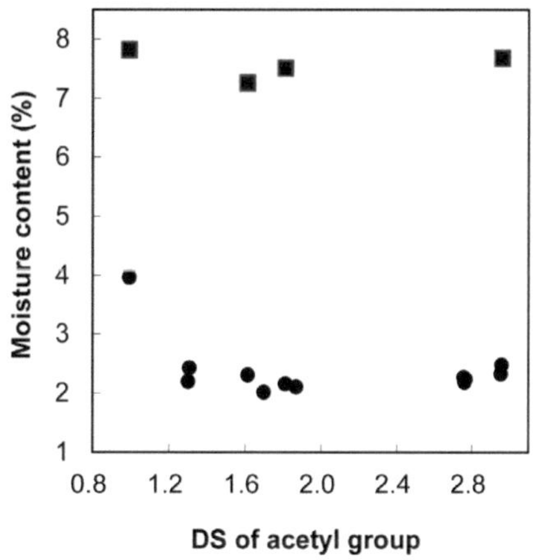

Fig. 25. Moisture contents of a series of acetylated chitin nanofiber sheets (squares) and their nanocomposites (circles).

Since chitin nanofibers have an extended antiparallel crystalline structure, the CTE of the nanofibers is small. We therefore thought that the chitin nanofibers act as a reinforcement agent to reduce the thermal expansion of a resin. However, the CTE of nanofibers and its composites will be changed by the acetylation. Figure 26 shows the CTE of chitin nanofiber sheets and DCP resin nanocomposites. The measurements were carried out from 30 to 165 °C by elevating the temperature at a rate of 5 °C min^{-1} in a nitrogen atmosphere in tensile mode. Although the CTE of the DCP resin was 64 ppm K^{-1}, the CTE of the chitin nanofiber/DCP composite was measured to be 23 ppm K^{-1}. It is easy to see that the chitin nanofibers with a low thermal expansion of 92 ppm K^{-1} drastically reduced the CTE of the DCP by reinforcement, with a fiber content of 25%. On the other hand, the CTE of chitin nanofiber samples and their composites increased proportionally and gradually with increases in the acetyl DS. This increase is due to the acetylation reducing the degree of crystallinity of the chitin nanofibers, as shown in Figure 21, which increases the CTE of chitin nanofibers and the nanocomposites.

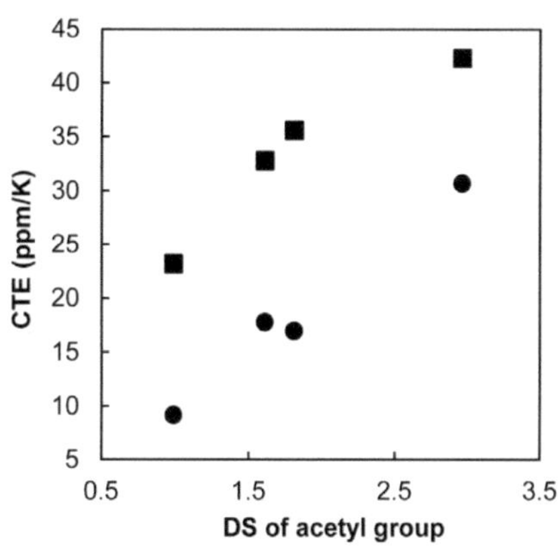

Fig. 26. Coefficient of thermal expansion of a series of acetylated chitin nanofibers (circles) and their nanocomposites (squares).

4. Conclusion

Chitin nanofibers were prepared from dried-crab shell which has a complex hierarchical structure with a uniform width of approximately 10-20 nm by conventional chemical treatment followed by mechanical treatment. This study demonstrates that the grinding treatment in a never-dried state after the removal of the matrix is applicable to not only wood but also crab shell to isolate nanofibers. Chitin nanofibers could be prepared from dried chitin as well. Mechanical treatment under acidic conditions is the key to fibrillating dried chitin. The cationization of amino groups on the chitin fiber surface breaks the strong hydrogen bonds between the nanofibers by electrostatic repulsion. These results indicate that the never-dried process is not necessary for chitin nanofiber preparation; this finding is quite different than that for cellulose nanofibers, since cellulose does not have ionic charges. Nanofibers from dry chitin are advantageous for commercial production in terms of storage, supply, transportation. This method allows us to effectively obtain homogeneous chitin nanofibers with a high surface-to-volume ratio. Moreover, chitin nanofibers could be prepared from a variety of prawn shells too. Since prawn shell has a finer structure than crab shell, chitin nanofibers were isolated without any acid.

Two different types of acrylic resins reinforced with chitin nanofibers were prepared. Due to nano-sized structure and excellent mechanical properties of chitin nanofiber, the composites were highly transparent and flexible, and had low thermal expansion, high Young's modulus, and high tensile strength. This study will expand the application of chitin nanofiber as a novel nanofiber reinforcement and encourage the use of chitin, most of which is thrown away as industrial waste, as a natural and environmentally friendly material.

Chitin nanofibers were chemically modified by acetyl groups. Acetylated chitin nanofibers with a variety of DS values were obtained by adjusting the reaction time. Acetyl groups were heterogeneously introduced inside the nanofibers after the surface reaction. X-ray diffraction profiles, SEM images, and TGA curves show that the acetylation of chitin nanofibers changes their crystal structure, fiber thickness, and thermal degradation temperature, respectively. Chitin nanofiber-reinforced plastics have high transparency with a variety of acetyl DS values. Acetylation reduced the moisture absorption of chitin nanocomposite by the introduction of hydrophobic acetyl groups.

As the thermal expansion is a property intrinsic to the material itself, its restriction depends on the reinforcing elements incorporated to the matrix. These reinforcements must not interfere with light transmittance in the case of transparent polymers. The use of natural nanofibers like chitin has proved to be a viable way to reduce the high thermal expansion of polymers without significant losses in transparency. These nanofibers are obtained from sustainable resources through environmentally friendly processes, and development of new extraction methods is attaining progressive cost reduction. As a result of this research, flexible transparent polymers with very low CTEs, low moisture adsorption, high Young's moduli, and high tensile strengths are emerging as potential candidates for optical functional devices.

5. References

Abe, K.; Iwamoto, S.; Yano, H. (2007). Obtaining cellulose nanofibers with a uniform width of 15 nm from wood. *Biomacromolecules*, 8: 3276 - 3278.

BeMiller, J. N.; Whistler, R. L. (1962). Alkaline Degradation of Amino Sugars. *J. Org. Chem.* 27: 1161 - 1164.

Chen, P.-Y.; Lin, Y.-M.; McKittrick, J.; Meyers, M. A. (2008). Structure and mechanical properties of crab exoskeletons. *Acta Biomateralia*, 4: 587 - 596.

Fan, Y.; Saito, T.; Isogai, A. (2008a). Chitin nanocrystals prepared by TEMPO mediated oxidation of α-chitin. *Biomacromolecules*, 9: 192 - 198.

Fan, Y.; Saito, T.; Isogai, A. (2008b). Preparation of chitin nanofibers from squid pen β-chitin by simple mechanical treatment under acid conditions. *Biomacromolecules*, 9: 1919 - 1923.

Fan, Y.; Saito, T.; Isogai, A. (2009). TEMPO-mediated oxidation of β-chitin to prepare individual nanofibrils. *Carbohydrate Polymers*, 77: 832 - 838.

Giraud-guille, M.-M. (1984). Fine structure of the chitin-protein system in the crab cuticle. *Tissue Cell*, 16: 75 - 92.

Glasser, W. G.; Taib, R.; Rajesh, R. K.; Kander, R. (1999), Fiber-reinforced cellulosic thermoplastic composites. *J. Appl. Polym. Sci.* 73: 1329 - 1340.

Gopalan Nair, K.; Dufresne, A. (2003) Crab shell chitin whisker reinforced natural rubber nanocomposites. 1. Processing and swelling behavior. *Biomacromolecules*, 4: 657 - 665.

Ifuku, S.; Nogi, M.; Abe, K.; Handa, K.; Nakatsubo, F.; Yano, H. (2007). Surface modification of bacterial cellulose nanofibers for property enhancement of optically transparent composites: Dependence on acetyl-group DS. *Biomacromolecules*, 8: 1973 - 1978.

Ifuku, S.; Nogi, M.; Abe, K.; Yoshioka, M.; Morimoto, M.; Saimoto, H., et al. (2009). Preparation of chitin nanofibers with a uniform width as α-chitin from crab shells. *Biomacromolecules*, 10: 1584 - 1588.

Ifuku, S.; Morooka, S.; Morimoto, M.; Saimoto, H. (2010a). Acetylation of chitin nanofibers and their transparent nanocomposite films. *Biomacromolecules*,11: 1326-1330.

Ifuku, S.; Nogi, M.; Yoshioka, M.; Morimoto, M.; Yano, H.; Saimoto, H. (2010b). Fibrillation of dried chitin into 10–20 nm nanofibers by a simple grinding method under acidic conditions. *Carbohydr. Polym.*, 81: 134-139.

Iwamoto, S.; Abe, K.; Yano, H. (2008). The effect of hemicelluloses on wood pulp nanofibrillation and nanofiber network characteristics. *Biomacromoles*, 9: 1022 - 1026.

Kim, D.-Y.; Nishiyama, Y.; Kuga, S. (2002). Surface acetylation of bacterial cellulose *Cellulose*, 9: 361 367.

Min, B. M.; Lee, S. W.; Lim, J. N.; You, Y.; Lee, T. S.; Kang, P. H.; Park, W. H. (2004). Chitin and chitosan nanofibers: Electrospinning of chitin and deacetylation of chitin nanofibers. *Polymer*, 45: 7137 - 7142.

Minke, R.; Blackwell, J. (1978). The structure of □-chitin. *J. Mol. Biol.*, 120: 167 - 181.

Nakagaito, A. N.; Yano, H. (2008). Toughness enhancement of cellulose nanocomposites by alkali treatment of the reinforcing cellulose nanofibers *Cellulose*, 15, 323-331.

Nakagaito, A. N.; Fujimura, A.; Sakai, T.; Hama, Y.; Yano, H. (2009). Production of microfibrillated cellulose (MFC)-reinforced polylactic acid (PLA) nanocomposites from sheets obtained by a papermaking-like process. *Compos. Sci. Technol.*, 69, 1293-1297.

Nishino, T.; Takano, K.; Nakamae, K. (1995). Elastic-modulus of the crystalline regions of cellulose polymorphs. *J. Polym. Sci., Part B: Polym. Phys.*, 33, 1647-1651.

Nishino, T.; Matsuda, I.; Hirano, K. (2004). All cellulose composite. *Macromolecules*, 37, 7683-7687.

Nogi, M.; Handa, K.; Nakagaito, A. N.; Yano, H. (2005). Optically transparent bionanofiber composites with low sensitivity to refractive index of the polymer matrix. *Applied Physics Letters*, 87: 243110.

Okamoto, Y.; Shibazaki, K.; Minami, S.; Matsuhashi, A.; Tanioka S.; Shigemasa, Y. (1995). Evaluation of Chitin and Chitosan on Open Wound Healing in Dogs. *J. Vet. Med. Sci.*, 57, 851-854.

Raabe, D., Sachs, C., & Romano, P. (2005). The crustacean exoskeleton as an example of a structurally and mechanically graded biological nanocomposite material. *Acta Biomaterialia*, 53, 4281 - 4292.

Raabe, D.; Romano, P.; Sachs, C.; Fabritius, H.; Al-Sawalmih, A.; Yi, S. B.; Servos, G.; Hartwig, H. G., (2006). Microstructure and crystallographic texture of the chitin-protein network in the biological composite material of the exoskeleton of the lobster Homarus americanus. *Mater. Sci. Engin. A.*, 421, 143-153.

Revol, J.-F., & Marchessault, R. H. (1993). In oifro chiral nematic ordering of chitin crystallites. International Journal of Biological Macromolecules, 15, 329 - 335.

Saito, T., Nishiyama, Y., Putaux, J.-L., Vignon, M., & Isogai, A. (2006). Homogeneous suspensions of individualized microfibrils from TEMPO-catalyzed oxidation of native cellulose. *Biomacromolecules*, 7, 1687 - 1691.

Shams, M. I.; Ifuku, S.; Nogi, M.; Oku, T.; Yano, H., (2011). Fabrication of optically transparent chitin nanocomposites. *Appl. Phys. A*, 102, 325-331.

Shimahara, K.; Takiguchi, Y. (1998). Preparation of crustacean chitin, In: *Methods in enzymology Vol. 161*, Wood, W. A.; Kellogg, S. T. (Eds.), pp. 417-423, Academia Press, 0121820629, CA.

Tokura S.; Nishi, N. Nishimura, S.; Somorin, O. (1983). Lysozyme-accessible fibers from chitin and its derivatives. *Sin-i Gakkaishi*. 39: 45-49.

Vincent, J.; Wegst, U. (2004). Design and mechanical properties of insect cuticle. *Arthropod Struct. Dev.*, 33: 187-199.

Vigneron, J. P.; Rassart, M.; Vandenbem, C.; Lousse, V.; Deparis, O.; Biro, L. P.; Dedouaire, D.; Cornet, A.; Defrance, P. (2006). Spectral filtering of visible light by the cuticle of metallic woodboring beetles and microfabrication of a matching bioinspired material. *Phys. ReV. E.* 73: 041905.

Wada, M.; Saito, Y., (2001). Lateral Thermal Expansion of Chitin Crystals. *J. Polym. Sci. B*, 39: 168-174.

Yano, H., Sugiyama, J., Nakagaito, A. N., Nogi, M., Matsuura, T., Hikita, M., et al. (2005). Optically transparent composites reinforced with networks of bacterial nanofibers. *Advances Materials*, 17, 153 - 155.

Yano, I. (1972). A historical study on the exocuticle with respect to its calcification and associated epidermal cells in a shore crab. *Bull. of Jap. Soc. Sci. Fisheries*, 38: 733 - 739.

Yano, I. (1975). An electron microscope study on the calcification of the exoskeleton in a shore crab. *Bull. of Jap. Soc. Sci. Fisheries, Gakkaishi*, 41: 1079-1082.

Yano, I. (1977). Structures of exoskeletons of prawn and crab shells. *Kagaku to Seibutsu*, 15, 328 - 336.

Zhao, H.; Feng, X.; Gao, H. (2007). Ultrasonic technique for extracting nanofibers from nature materials. *Appl. Phys. Lett.* 90: 073112.

5

Advances in Electroactive Electrospun Nanofibers

Paulo H. S. Picciani[1], Eliton S. Medeiros[2], William J. Orts[3] and Luiz H. C. Mattoso[4]
[1]*Instituto de Macromoléculas, Universidade Federal do Rio de Janeiro, Centro de Tecnologia, Ilha do Fundão, Rio de Janeiro, RJ*
[2]*Departamento de Engenharia de Materiais, Universidade Federal da Paraíba, Centro de Tecnologia, Cidade Universitária, João Pessoa – PB*
[3]*Bioproducts Chemistry and Engineering Unit, Western Regional Research Center, Agricultural Research Center, USDA, Albany, CA*
[4]*Laboratório Nacional de Nanotecnologia Aplicada ao Agronegócio, Embrapa Instrumentação Agropecuária, São Carlos-SP,*
[1,2,4]*Brazil*
[3]*USA*

1. Introduction

The first detailed description of the technique known nowadays as electrospinning dates from 1930's when Anton Formhals filed a patent entitled "*Process and Apparatus for Preparing Artificial Threads*". His patent described the formation of artificial filaments when a polymer solution or melt is submitted to a high strength electric field. Recently, with the emerging of nanotechnology, electrospinning was rediscovered and became an important method for producing ultrathin polymer fibers with diameters down to the nanometer scale. Many classes of polymers have been electrospun, opening up new promising applications such as in filtration, protective clothing, scaffolds for tissue engineering and a myriad of other possibilities.

Contrasting with the most common fiber processing techniques like melt spinning, dry spinning, wet spinning and extrusion; electrospinning is able to produce ultrathin fibers with very high surface area. This is of particular interest in sensors, actuators and other electroactive devices, especially when conducting polymers are used. Many kinds of conducting polymers have been used to produce electroactive polymer devices based on nanosized electrospun fibers as polyaniline, polypyrrol, polythyiophene, among others. Although the available literature on electrospinning of polymer mats is extensive, there are few reports on electrospinning of electroactive polymers and their related applications.

In this chapter, special topics involving the production of electroactive nanofibers by electrospinning and their use in electroactive devices such as sensors, actuators, electroluminescent devices, conductive membranes and others are reviewed. Especial attention to processing parameters and effects of electroactive polymer structure and properties in solution will be given. Finally, some perspectives on electrospinning of electroactive nanofibers obtained by electrospinning and their uses will be considered.

1.1 Historical background and current technologies

Electrospinning has emerged as one of the techniques that can be successfully used to produce polymeric fibers, mats and scaffolds down to the nanometer scale. Besides, electrospinning has been currently cited in an enormous amount of scientific papers and reports that show how it can be used to produce a vast number of morphologies with many potential applications. The development of this technique started with the studies of William Gilbert in the late 1600's, with the attempt to describe the deformation of water droplets under magnetic and electric fields. Later, in the 1880's, Lord Rayleigh was the first to estimate theoretically the pressure resulting from a charge q on a droplet of spherical radius r and surface tension σ. When the amount of charge q overcomes the Rayleigh limit q_r the natural quadrupolar oscillation becomes unstable and the liquid spread out in fine jets **[1]**. The Rayleigh limit can also be reached by the liquid evaporation. The ejecting event is frequently called Rayleigh discharge or Coulomb fission and more commonly electrospray **[2]**.

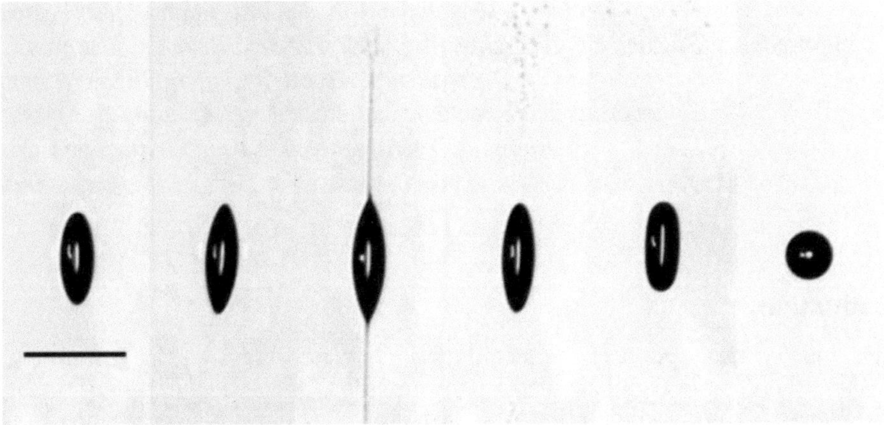

Fig. 1. The droplet (radius, 24 micron) is imaged on a vertical charge-coupled-device array to determine and control its vertical position in the levitator, and on a photomultiplier that detects instability-onset, quadrupole-shaped oscillations of the droplet. These oscillations trigger a signal to a flashlamp that fires after a predefined delay, Δt **[1]**.

Later in 1914, John Zeleny described that electrical discharge of liquid points consisted of acidulated water placed in a metal bar **[3]**. By the use of a refined experimental technique, Zeleny was able to measure the electric intensity at the surface of liquid droplets before and during the electrical discharge, leading to a better understanding of liquids surface instabilities due to the presence of electrical charges. In a paper published in 1917, Zeleny also observed that when electrification is increased to a certain limit value, depending on liquid surface tension and the curvature of its surface, the liquid surface becomes unstable, due to the inside pressure becomes equals to the outside pressure, and any slight accidental displacement may result in a rapid and intense deformation of liquid surface. Above this critical condition, in case of further increase in electric density, the liquid is ejected under forms of fine threads which can eventually be broken into small drops **[3]**.

At that time, some speculations on applying electrospray for viscous fluids arose. Later, this process became known as electrospinning and the process of spinning fine threads by the use of electrical charges was firstly patented by J. F. Cooley in February 1902 (U.S. Patent 692,631) and by W. J. Morton in July of the same year (U.S. Patent 0,705,691). In fact, the patent filed in 1934 by Anton Formhals (U.S. Patent 1,975,504) entitled "Process and Apparatus for Preparing Artificial Threads" is widely known as the first document which effectively consider the production of fiber threads by means of electric fields. This work was succeed by a sequence of patents from 1934 to 1944 (U.S. Patent 2,349,950) and Formhals established a very precise experimental setup for producing electrospun fibers.

It is worth noting the fabulous work of Sir Geoffrey Taylor, from 1964 to 1969, who published various papers detailing the theoretical underling on disintegration of water drops under electric fields, including electrical force calculation on fine jets under strong electric fields and the deformation of liquid surfaces by an electrical field into a characteristic shape that is currently known as Taylor cone **[4]**.

The term electrospinning was mainly popularized by Dr. Darrell H. Reneker in early 1990 in demonstrating that many types of polymers can be spun by means of electrostatic forces into fine threads known as micro or nanofibers depending of their diameters. Today, thousands of scientific papers and patents have been published on electrospinning. Besides fascinating many scientists around the world, this topic involves both scientific and technologic developments and includes many fields of the human knowledge such as physics, chemistry, material science and engineering, biology, biochemistry.

2. Nanofibers of conducting polymers

2.1 Introduction to conducting polymers

Conducting polymers have been used in a variety of applications. These materials exhibit the ability to allow the passage of electric current, store charge or display redox activities. Common polymers are mostly dielectric and can be modified in order to conduct electric current by the addition of conducting fillers, as metal powders, carbon black and other conductive materials. These filled polymers are commonly called Extrinsic Conducting Polymers - ECP's and can be used as, pressure sensors, actuators, electromagnetic interference (EMI) shielding etc. **[5]**

On the other hand, polymers formed by chains that contain conjugated chemical double bond, with a minimal molecular size able to produce electrical conductivity, are usually called Intrinsically Conducting Polymers - ICP's or simply conducting polymers. These materials formed by organic molecules containing conjugated chemical bonds such as aromatic rings and (or) conjugated oligomeric (or) polymeric conjugated oligomeric/polymeric chains have attracted enormous attention from scientists because of the possibility of their utilization in the development of electronic devices. Depending on the size of the molecule and its conjugation extension, interesting physical and chemical properties may arise - electrical conductivity, electroluminescence, electrochromism - which may be used in several practical applications, such as static charge dissipation, diodes and light emitters, memory storage, rechargeable batteries, chemical and biological sensors **[6-9]**.

The first attempt in obtaining polymers with conjugate chemical bonds (polyenes) dates from 1958 through the polymerization of acetylene by Natta and coworkers**[10]**. The obtained polyene was widely accepted as consisted of a long conjugated polymer chain. Theoretical speculations on electrical properties of Polyacetylene - PAc, indicated a metallic conducting behavior in an alternate double bonded linear chain. However, at that time there was no clear idea if a long polyene chain would be energetically more stable than that with bonds of equal lengths (delocalization).

Using an adapted synthetic route proposed by Natta, Shirakawa and coworkers **[10]** synthesized polyacetylene in the 1970's. This polymer was obtained as a smooth film with metallic brightness and relatively low conductivity, ranging from 10^{-8} to 10^{-7} S.cm^{-1}, for *cis*-polyacetylene, to 10^{-3}-10^{-2} S.cm^{-1}, for trans-polyacetylene. Simultaneously, MacDiarmid and Heeger **[11]** were investigating an inorganic polymer, poly(sulfur nitride) - SNx, which has exhibited electrical conductivity between 10 and 1,700 S cm^{-1} at room temperature and superconductivity at low temperatures (~0.26K). Later, in 1977, Shirakawa, MacDiarmid and Heeger **[12]** performed a modification in the previous preparation procedure of *trans*-polyacetylene, introducing small amounts of iodine, which is known for its high oxidizing strength. The polymer obtained through this new procedure exhibited high electrical conductivity, 10^3 S.cm^{-1}, that was attributed to the presence of charge carriers formed upon oxidation of the conjugated chemical bonds of the polymer backbone. The oxidation was called "doping" in analogy to semiconductors.

Conjugate polymers are considered one of the most innovative materials and different types are available nowadays, since a great number of compounds is used as monomers and different synthetic routes can be adopted for this purpose **[13,14]**. All the synthetic routes developed so far can be grouped into two main types: chemical and electrochemical synthesis. Enzymatic **[15]** and plasma synthesis **[16]** of conjugated polymers have also been reported. Processing in conducting polymers is another important issue that needs to be considered. Such materials can mostly be obtained in an easy and controllable way. They are chemically and thermally stable even in the conducting state, and exhibit high electrical conductivity which can be controlled according to different doping methods **[13,17]**. Since the discovery of conducting polymers, this research area has witnessed the increasing number of conjugated polymers with improved solubility that can be processed by different techniques. Table 1 shows some examples of these polymers.

Polymer name	Chemical structure	Polymer name	Chemical structure
Trans-polyacetylene (*trans*-PAc)		Polypyrrole (PPy)	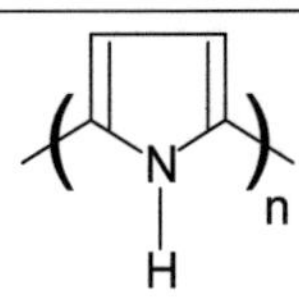
Poly(*p*-phenylene) (PPP)	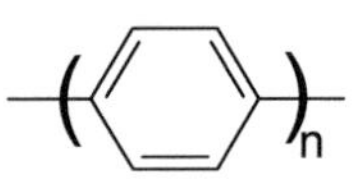	Polythiophene (PTh)	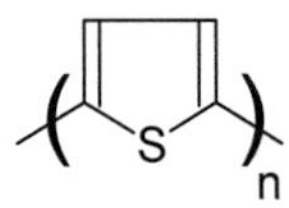
Poly(*p*-phenylene-vinylene) (PPV)	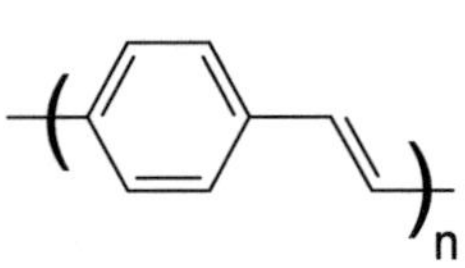	Poly(3,4-ethylene dioxythiophene) (PEDOT)	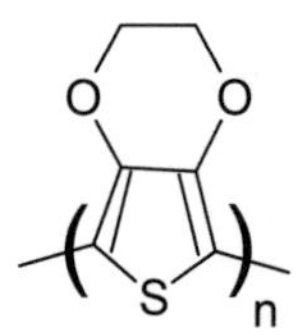
Poly(*p*-phenylene-sulphide) (PPS)	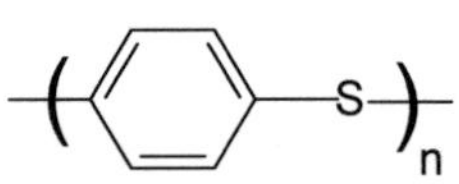	Polyaniline (PAni)	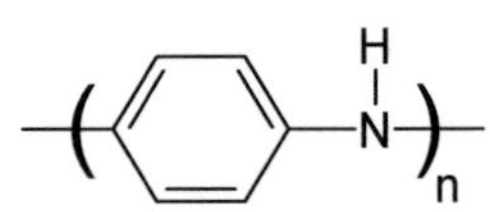
Polyfurane (PFu)	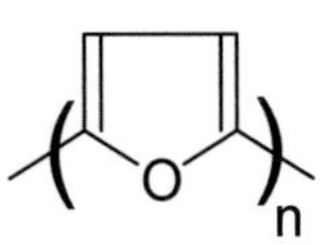	Polycarbazole (PCbz)	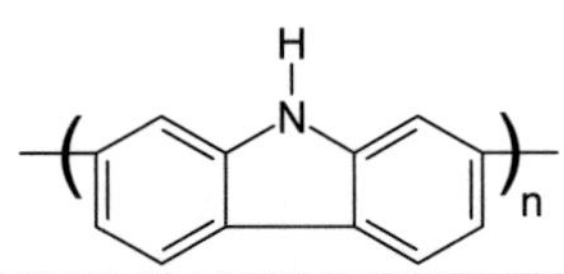

Table 1. Examples of the most common intrinsically conducting polymers (ICP's).

Techniques used for processing conjugated polymers should take advantage of their physical and chemical properties in order to produce structures with elevated organization and transparency depending on the method of fabrication and end use.

2.2 Electrical conduction and doping in conjugated polymers

Electrical conductivity is a physical property that can vary several orders of magnitude, ranging from 10^{-22} S cm^{-1} for the best insulators up to 10^{10} S cm^{-1} at 1K for the best conductors. Semiconductors exhibit intermediary conductivity, between 10^{-7} and 10^{-3} S cm^{-1}, once the energy gap between conduction and valence band is of the order of few electrons volt (eV). Depending on the type of polymer, doping level and dopant, conductivity of conjugated polymers varies over a broad range of conductivity as can be seen in Figure 2.

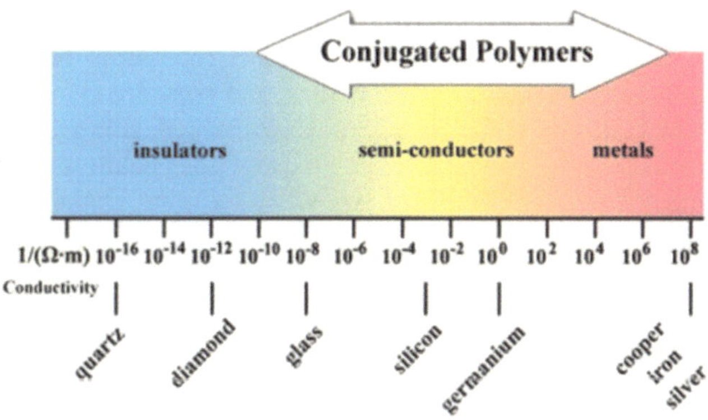

Fig. 2. Schematic illustration of conductivity range of conjugated polymers compared with most common materials. (source: The Nobel Prize in Chemistry, 2000, www.nobelprize.org)

As already mentioned, conjugated polymers are formed by a main backbone containing alternating single and double chemical bonds known as conjugation. Whereas a single bond (or sigma bond, σ) is strong and contains strongly localized electrons under a sp^3 orbital symmetry, a double bond (or pi bond, π) is weaker and electrons are less localized following a sp^2 orbital symmetry. This means that π electrons may exhibit higher mobility when compared to σ electrons and can move along the backbone since the conjugation leads to the formation of extended delocalized orbitals.

One can imagine that because of this conjugated configuration polymers are electrical conductors. However, for conjugated polymers to become conducting, an extra electron needs to be removed from the backbone in order to create a vacancy, through a redox process called doping. In fact, the first results obtained by Shirakawa's group **[5]** showed that pristine polyacetilene had conductivity values around 10^{-3} S cm^{-1}, a semiconductor. The vacancy or "hole" is a position where an electron is missing in the valence band. If a neighboring electron jumps into the hole a new hole is generated from where the electron jumped and this process perpetuates allowing charge to migrate a long the polymer backbone, which is in fact the electrical conduction **[18]**.

Many different theoretical models **[19]** have been already proposed to address the electrical conduction exhibited by conjugated polymers after doping. However, a band model is usually referred to because it is able to explain many of the phenomena related in such systems. According to the band model, it was initially believed that during doping valence electrons were removed from the valence band (the highest occupied molecular orbitals - HOMO) and added to the conduction band (the lowest unoccupied molecular orbitals - LUMO) like in a HOMO-LUMO transition. Nevertheless, this hypothesis did not predict the existence of charge carriers with a null spin and it had to be reformulated.

In the case of polyacetylene, structural defects (or free radicals, in a chemistry terminology) are generated during the synthesis of the polymer due to oxidation/reduction reactions. Such defects were called solitons, but they are still unable to perform the electrical conduction due to the energy associated to its transition to conduction band to be around 1.4 eV and, therefore, thermal energy is not enough to promote electrons to the conduction band. After the doping process, which is basically a process where electrons are removed from or added to the polymer, charged solitons are generated; these are the carriers responsible for the electrical conduction. Charged solitons are species with a null spin and, hence, one concludes that the electrical conduction in polyacetylene is based on fully occupied bands in the ground state **[18,19]**. Figure 3 represents band energy levels of possible solitons configuration, i.e., positive, neutral and negative soliton according to their spin properties. In fact, solitons are not created separately. A soliton / anti-soliton pair is always created after redox process in a conjugated polymer chain. Figure 4 illustrates the soliton soliton / anti-soliton separation process when polymer chains are submitted to an external electric field.

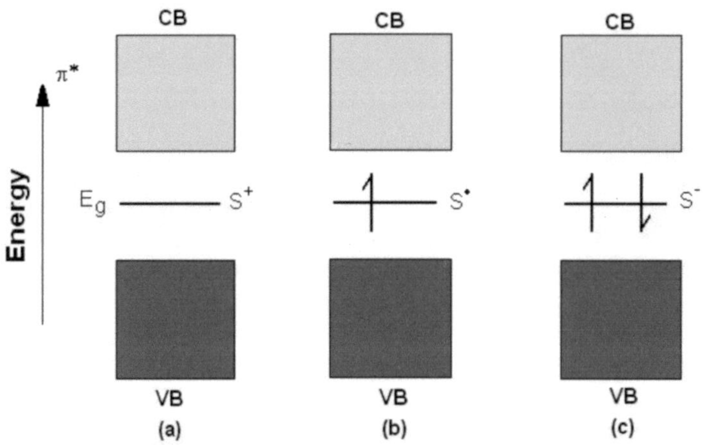

Fig. 3. Positive (a), neutral (b) and negative (c) solitons. Solitons are located in an energy level equivalent to the half of the gap size. Positive and negative solitons are formed through p-doping and n-doping process, respectively. Both are null spin entities **[5]**.

The proposed theoretical model for polyacetylene is, however, unable to explain the electrical conduction in aromatic and heteroaromatic conjugated polymers, mainly due to the existence of other types of carriers in such polymers. The removal of an electron from these polymers is accompanied by a local distortion in the crystal lattice, which may be represented by a conversion of the aromatic rings from the benzenoid to the quinoid form. As the quantized states of crystal structures can be expressed in terms of phonons, this electron-phonon coupling is called polaron and represents a different type of charge carrier in conjugated polymers. The quinoid form has a lower formation energy and higher electron affinity than the benzenoid form. Therefore, during the doping, the charge accommodation

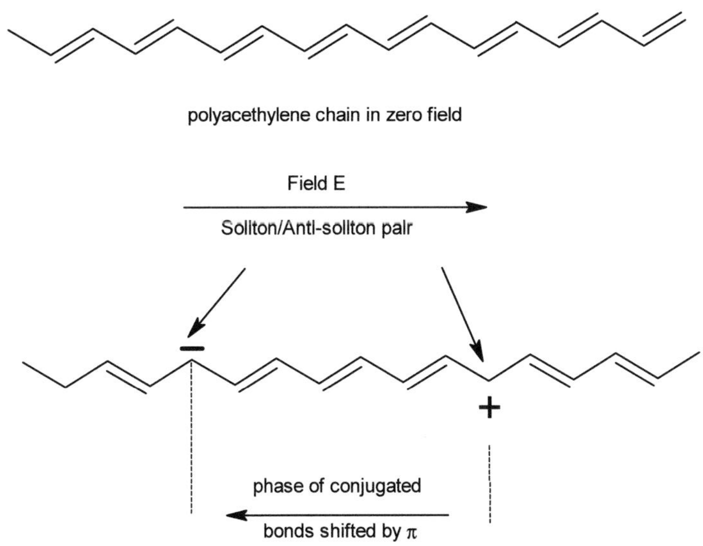

Fig. 4. Schematic representation of soliton - anti-soliton separation and electrical conduction in Polyacetilene chain. (source: http://electrons.wikidot.com).

in the polymeric chain is energetically favored by a combination of oxidation reaction and chain distortion **[20]**.

As mentioned before, a polaron is considered as a positive charge carrier formed by the electron-phonon coupling and, in a chemistry definition, is a radical-cation with ½ spin moment and two electronic states. Both states are located around the Fermi level of the polymer with the lower energy state being occupied by one electron. When the doping level increases such states are broadened forming semi-filled conduction bands. The removal of a second electron can be performed and leads to the formation of a bipolaron which is defined as a pair of positive charges associated to a stronger lattice distortion. Bipolarons also form two bands but both are empty resulting in a Fermi level close to the maximum of the valence band **[18-21]**. Figure 5 represents the energy levels of doped polypirrol with respective levels of polaron and bipolaron charge carriers.

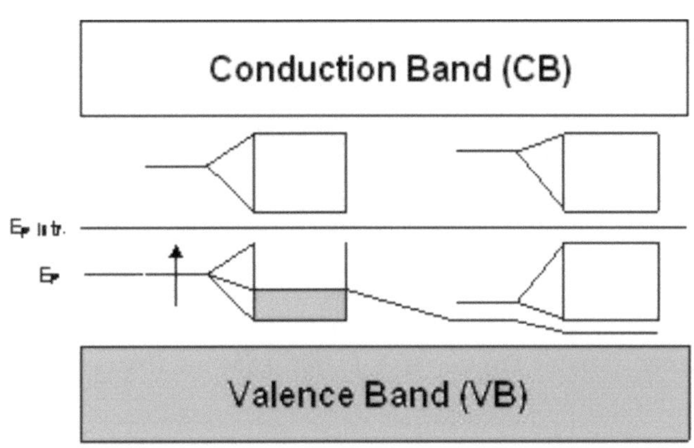

Fig. 5. Scheme of energy levels of doped polypirrol with polarons and bipolarons **[21]**.

Assuming those models to explain the electrical conduction in conjugated polymers, doping methods have been developed in order to produce conjugated polymers with high conductivity. Although there is still a current search for doping methods able to produce polymers with elevated conductivity, many of the technological applications proposed for conjugated polymers do not demand highly conducting materials and hence, the semiconducting behavior of these polymers has been more often exploited **[13,22]**.

Doping is the term used to define the process where a material, insulator or semiconducting, is converted into an electrical conductor. In a general sense, the doping process is based upon the removal or addition of electrons from/to a material which can be carried out in different ways. In the case of inorganic semiconductors, doping is represented by the addition of small amounts (10^{-5}-10^{-8} atoms of dopant per atoms of the material) of a doping agent or dopant, which replaces atoms from the original material. Differently from that, doping of conjugated polymers demands larger amounts of dopants (10^{2}-10^{-1} moles of dopant per moles of monomer) but there is no replacement of atoms. The doping process in conjugated polymers is based on chemical reactions between the dopant and the polymer, usually oxidation/reduction reactions in which the initially neutral polymer is converted into a charged macromolecule whose charge is compensated by a counter-ion original from the dopant. According to MacDiarmid **[13]** there are three main types of doping process that can be applied to conjugated polymers: (a) redox doping; (b) doping without dopant ions and (c) non-redox doping.

In the first type, redox doping, the π bonds in a conjugated polymer undergo chemical or electrochemical oxidation/reduction reactions in which the number of electrons is changed. All conjugated polymers can be doped by this type of doping. If we consider the oxidation of the π bonds, electrons are removed and therefore the polymer backbone exhibits a positive net charge which is balanced by the presence of a counter-ion of opposite charge (negative). Usually, this type of doping is referred as p-doping in analogy to inorganic semiconductors. This process can be carried out by a chemical dopant, such as iodine and other halogen compounds (Figure 6), or electrochemically by anodic oxidation of the polymer in a medium containing a salt, for instance $LiClO_4$. The counter-ions are, respectively, the reduced iodine (I^-) and the perchlorate ion ($ClO4^-$). In the n-doping, the polymer backbone is reduced and electrons are incorporated either chemically or electrochemically. The n-doping was first achieved in trans-polyacetylene by treating it with sodium amalgam or sodium naphthalide. The electrochemical reduction can be performed in the presence of $LiClO_4$ where Li^+ is the counter-ion **[13]**.

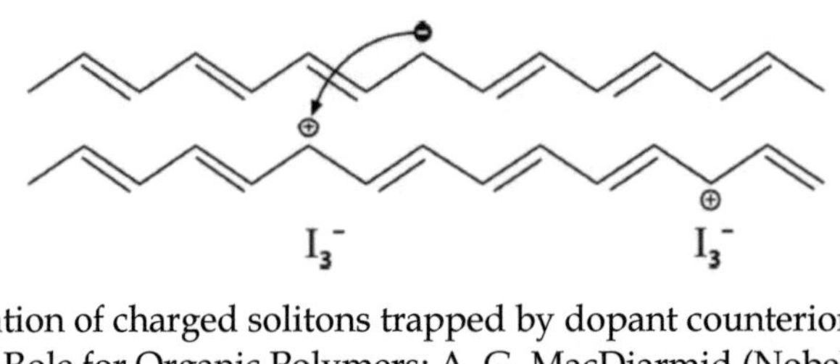

Fig. 6. Representation of charged solitons trapped by dopant counterions. (source: "Synthetic Metals": A Novel Role for Organic Polymers; A. G. MacDiarmid (Nobel Lecture, 2000); www.nobelprize.org).

In a second type of doping, counter-ions are not incorporated into the polymer matrix where photo-doping and charge-injection doping may be considered. In the photo-doping conjugated polymer is exposed to a radiation source with energy greater than its band-gap and as consequence, electrons are promoted across the gap. If the polymer is kept under radiation and an electrical potential is applied, electrons and holes generated during photo-doping separate from each other and photoconductivity is observed. In order to succeed the charge-injection doping, a film of conjugated polymer is assembled onto the surface of a metal with a separating layer made of a dielectric material. When an appropriate electrical potential is applied to this configuration, a surface charge layer is formed on top of the polymeric film which induces the formation of charges on it, however, without the presence of a dopant ion **[13]**.

The third type of doping, called protonation, is based on non-redox process what means that the number of electrons associated with the polymer backbone remains unchanged. This mechanism is based on acid-basic reactions and is observed exclusively in polyaniline and its derivatives. Briefly, the emeraldine oxidation state of polyaniline can undergo an acid-base reaction where the imine nitrogen atoms are protonated by a protonic acid, or Lewis acids, generating positively charge carriers, or polarons, without changing its oxidation state **[13, 22-27]**.

2.3 Processing of electroactive polymers

When dealing with conjugated polymers, processing becomes a very important issue. For many years, these materials were considered intractable, since they often degrade before melting and show poor solubility in many common solvents, mostly due to their elevated aromaticity, high hydrogen bond density and low flexibility of polymer chains **[28]**. Many efforts have been concentrated to improve conjugated polymer processability **[29]**. Monomer functionalization is one of the approaches used to obtain processable conducting polymers **[30]**. In order to be used in practical applications, a conjugated polymer must be cost-effective to be synthesized and purified, have good chemical and electrical stability, and easily processed from either solution or melt. The latter is the most problematic aspect of conjugated polymers.

Solubility in conjugated polymers has been improved with the synthesis of derivatives and also by using different combinations of dopants and solvents. For example, poly(o-ethoxyaniline (POEA) is a polyaniline derivative with improved solunility and processability due to addition of an ethoxy group at the ortho position of polyaniline aromatic ring. These two approaches have enabled, for example, the fabrication of thin films of conjugated polymers by different techniques. In the form of thin films, conjugated polymers are very adequate to be employed as sensitive layers, transducers and support for immobilization of different molecules in different device systems configurations **[31]**. Moreover, their performance might be enhanced to unimaginable levels when films with thickness around hundreds of nanometers are fabricated.

Many efforts have started to be made in order to produce materials at very low physical dimensions since many properties can be enhanced or even new properties can arise at the nanoscale level **[32]**. Some of the methods used to obtain these structures are Lanmguir-Blodgett, layer-by-layer (LbL) deposition and spinning techniques.

2.3.1 Langmuir-blodgett

One of the most used techniques to produce nanostructures based on conducting polymers is the Langmuir-Blodgett approach. Langmuir-Blodgett technique (LB) was developed in

1935 by I. Langmuir and K. Blodgett pointing the production of monomolecular layers of fatty acids onto solid supports from an aqueous subphase **[9,33]**. Lately, this technique has been employed to fabricate films of many different materials, including conjugated polymers **[34-36]**. Among interesting features presented by the LB technique one can cite its ability in constructing ultra-thin films (of the order of nanometers) with structures whose molecular architecture could be manipulated and appropriately designed. Mono and multilayer films can be produced with a high degree of control of the film thickness and may even provide a high degree of orientational order **[33-35]**. In LB technique, generally, a solution of an amphiphilic molecule is spread over an aquous subphase allowing the solvent evaporation. Thus, the molecules are self-organized with their hydrophilic parts pointing to the water surface. After, a moveable barrier is able to compress the molecules in a solid phase oriented film. Figure 7 represents Langmuir film orientation and Langmuir-Blodgett film deposition **[37]**.

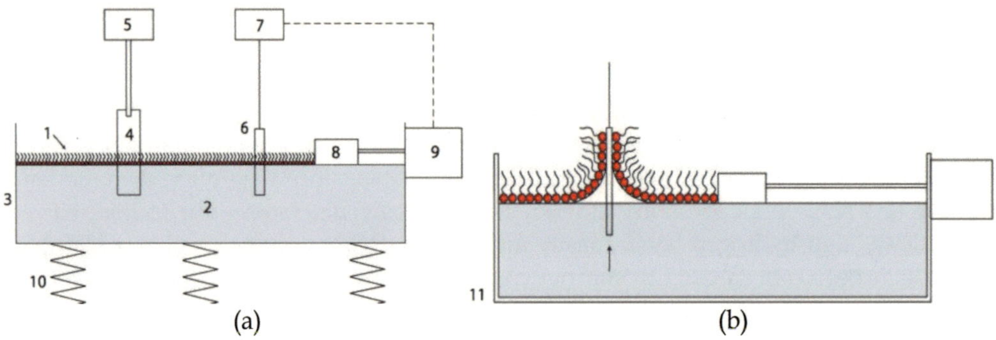

Fig. 7. (a) a schematic of a Langmuir Blodgett trough: 1. Amphiphile monolayer 2. Liquid subphase 3. LB Trough 4. Solid substrate 5. Dipping mechanism 6. Wilhelmy Plate 7. Electrobalance 8. Barrier 9. Barrier Mechanism 10. Vibration reduction system 11. Clean room enclosure; (b) monolayer transfer onto a substrate after film compression (source: http://en.wikipedia.org/).

For the reasons exemplified above, several approaches have been used to process conjugated polymers in the form of LB films since they can be produced only in specific conditions. Among these approaches, mixing the conjugated polymer with a surface-active compound, use of polymer derivatives and polymerization in the Langmuir trough are the most investigated for conjugated polymer processing **[34,35]**.

In the first approach, mixing, the conjugated polymer is mixed with a surface-active compound such as a fatty acid or a plasticizer. The resulting film is characterized by polymer molecules embedded into a well-organized matrix of fatty acid molecules (or plasticizer). Poly(alkyl thiophenes) and PAni derivatives, whose monolayers were too rigid to be transferred, have been processed by mixing **[34,35]**.

Conjugated polymer derivatives are usually more soluble in common organic solvents and hence can be spread over an aqueous subphase. Good quality LB films from poly(o-alkoxyanilines) and pure poly(3,4-dibutoxythiophene) have been fabricated with no need for addition of surface-active compounds or processing aids **[34-36]**.

In a third approach, Langmuir monolayers from the monomer are deposited followed by a subsequent in situ polymerization. Often, amphiphilic derivatives monomers are spread over the subphase which contains the polymerizing agent. Polymerization then occurs by interaction between the agent in the subphase and the film forming molecules. The polymeric monolayer would be subsequently transferred onto a solid support. LB films from PAni and PPy derivatives have been fabricated in this way. This approach can be further modified where the polymerization is carried out after the LB film deposition. The LB film containing monolayers of a specific monomer is polymerized by thermal treatment, as in the case of poly(p-phenylenevinylenes) and polyacetylenes, electrochemical oxidation for films of pyrrole derivatives or exposure to UV light **[34,35]**.

2.3.2 Layer-by-layer (LBL)

There are other interesting and promising approaches in producing nanometrically organized structures with conjugated polymers, one of them is called layer-by-layer technique or self-assembly. The layer-by-layer (LBL) deposition technique was originally proposed by Iler in 1966 **[38]** when dealing with the adsorption of colloidal particles. However, this technique was reformulated in the 80's by Sagiv **[39]** and extended to polyelectrolytes by Decher **[40]** in 1990. Although the LB technique is a promising and elegant way of producing thin films, its elevated cost and experimental procedure represent disadvantageous concerning the fabrication of films in large scale. In this sense, the LBL represents an alternative to the LB method mainly due to its low cost and easier experimental procedure.

LBL films are formed by the spontaneous adsorption of molecules from their solutions onto a solid support through different kinds of adsorption mechanisms. Basically, a chosen support is immersed into the deposition solution containing the material of interest for a pre-determined period of time and the molecules adsorb forming a layer. This layer has a thickness ranging from 10 to 100 Å. Later, the support is removed, washed and dried and, in a next step can be immersed again in the same solution or in a different one, if multilayer films are desired (see Figure 8). The solution is kept in a beaker and there is no need of a clean room. Any type of material like metals, glass or plastics and in different shapes such as spheres, slides or rods can be used as support to film deposition **[41,42]**.

LBL films from different types of conjugated polymers have been fabricated and examples include PAni and its alkoxy derivatives **[43-47]**, sulfonated-PAni **[44]**, Polypyrrole **[43]**, poly(thiophene acetic acid) **[44]** and poly(p-phenylene vinylene) **[44]**. In general, the adsorption of each layer contributes with the same amount of material to the film, which means that the film thickness can be controlled by the number of deposited layers. Moreover, the layer thickness can be varied by manipulating the polymeric solution parameters, such as pH, ionic strength and polymeric concentration. LBL films of conjugated polymers exhibit low roughness and their morphology can be controlled by the addition of different polyanions, such as sulfonated polystyrene, sulfonated lignin, poly(vinyl pyrrolidinone) and poly(allylamine hydrochloride). The polyanion presence can also play an important role in the stability of the conjugated polymer in the conducting state, since polyelectrolyte complexes are formed between them, which stabilize their charge and prevent proton loss in the case of PAni and its derivatives **[48]**.

2.3.3 Spin coating

Spin coating and casting techniques have also been employed, mainly for soluble conjugated polymers and for their blends with conventional polymers. Spin coating

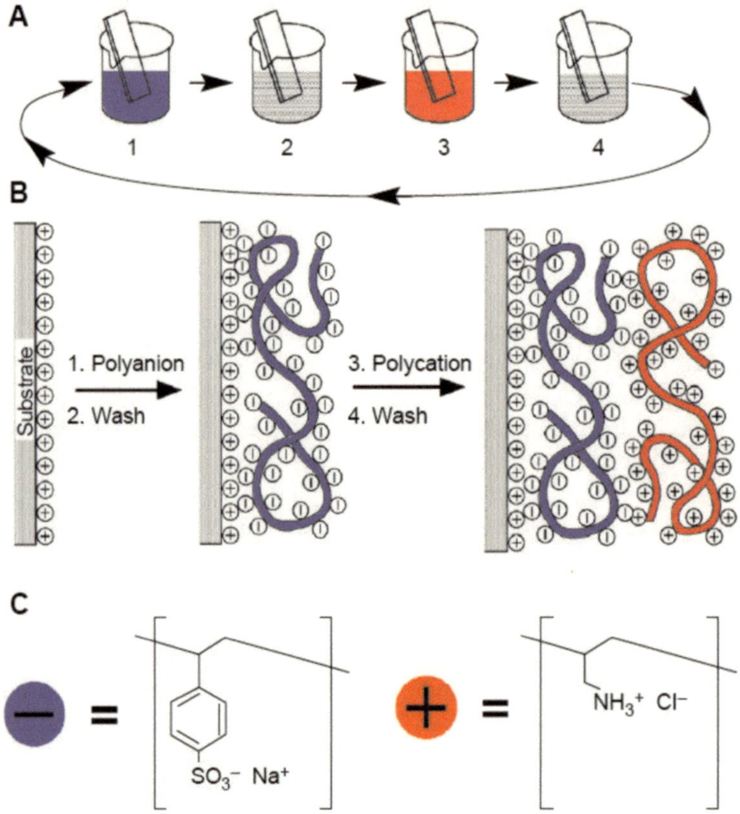

Fig. 8. Schematic of the film deposition by the LBL process according to Decher **[41]**.

can be considered as an improvement of casting film technique where a polymer solution is spread over an appropriate substrate and solvent is allowed to evaporate. A typical process involves deposition of small droplets of a polymer solution onto the center of a substrate and then spinning the substance at high speed. Thus, centripetal force will cause the spread of the substance over the substrate in the form of thin film. Evidently, the final thickness and other film properties will depend of the physical properties of the solution and process parameters. Spin coated, films can usually be obtained with thickness range around 0.1-200 µm **[49]**. Figure 9 exemplifies the basic steps in spin coating process.

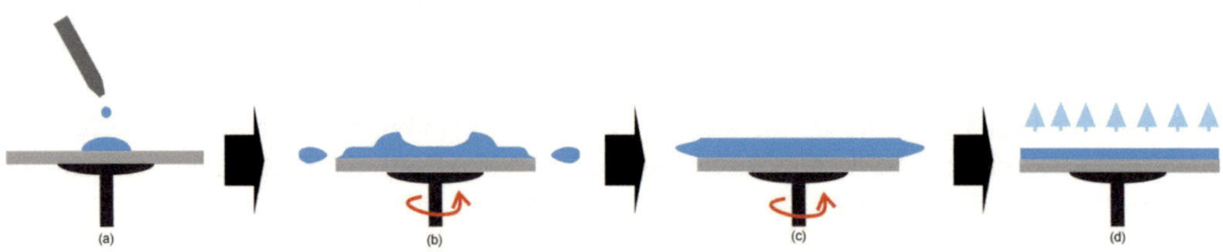

Fig. 9. Schematic illustration of spin coating process; (a) solution deposition, (b) substrate acceleration, (c) constant spinning rate (d) drying.
(http://www.freephotos.biz/images/science/chemistry/solgel_spincoating.jpg).

Many studies have been done by using spin coating to produce thin films of conducting polymers, mainly for electroluminescent materials and devices **[50-52]**. Meier and coworkers reported the influence of film thickness on the phase separation mechanism in poly[(1-methoxy)-4-(2-ethylhexyloxy)-p-phenylenevinylene] (MEH-PPV) and poly(Nvinylcarbazole) (PVK) spin coated films **[53]**. According to the authors, films with thickness from 16 to 61 nm where obtained by spin coating solutions with concentrations from 1 to 9 g L^{-1} at 2000 rpm.

Spin coating is relatively simple and cost-effective and, furthermore, thin films with highly molecular orientations can be readily produced. Furthermore, the thickness of spin coated films are still greater than that obtained by LB and LBL, and the constant presence of solvents used for processing must be considered. Films with very low thickness, around hundreds of nanometers, are usually preferred in recent technological applications in order to assure new and unrelated properties due to size reduction **[54]**.

2.3.4 Fiber production

Another very interesting and useful way to produce nanostructures of conducting polymers is the production nanofibers. Conductive materials in fibrillar shape may be advantageous comparing to films due to their inherent properties as anisotropy, high surface area and mechanical strength. Fibrous conductive materials are, for example, of particular interest in electroactive composites.

Fine metal wires, carbon fibers and carbon nanotubes have been efficiently distributed in an insulating polymer matrix in order to improve both electrical and mechanical properties **[5]**. Combination of electrical properties with good mechanical performance is of particular interest in ECP's technology. Fibers have intrinsically high structure factor which results in lower percolation threshold values avoiding material facture with low filler content. Also, the use of mechanically stronger fibers will result in stronger composites.

Some authors compare the electrical performance of conducting fiber composites to gelation process in a polymer after crosslinking, where the conductive network corresponds to the gel fraction. If a fiber segment is able to conduct it must be connected to the gel in both ends. In the case of low filler concentration, there is no conductive network. When the fiber content is increased until a three-dimensional network is firstly formed, the gel point is achieved and conductivity increases by a large factor **[5]**. If some processing technique is able to spatially orient the fibers, the conductive network will be formed at even lower concentrations, and the ECP will have good electrical performance associated to the plastic mechanical properties of the matrix.

Currently, there are many techniques able to produce polymeric nanofibers as, drawing, template synthesis, phase separation, self-assembly, solution blow spinning and electrospinning **[55,56]**. Ondarçuchu and Joachim demonstrated the possibility of drawing polymer fibers down to the nanometer scale with the tip of a micropipette and a microdroplet of a polymer solution. The authors compared this process to dry-spinning at molecular level and nanofibers with dimensions comparable to carbon nanotubes can be drawn by this easy and inexpensive fashion **[57]**. Obviously, the drawing process requires a polymer with appropriate viscoelastic properties which are able to being deformed and kept connected by cohesive forces. Besides being simple and inexpensive, this technique is very limited for conjugate polymers, once most of them have lower solubility and form solutions with small viscous modulus.

In phase separation approach, a polymer is solubilized and then undergoes to gelation process. Due to the physical incompatibility of the gel and solvent, this is able to be removed and the left structure, after freezing, is obtained in nanofibrilar structures **[55]**. Template synthesis basically implies in the use of a template or mold to obtain a desired structure. Commonly metal oxide membranes with nanoporous are used, where a polymer solution is forced to pass through to a non-solvent bath originating nanofibers depending of the pores diameter **[55]**.

A similar method was developed to obtain polyaniline nanofibrils in such way that growing polymer chains separate from solution according to their molecular size. According to Huang and Kaner **[58]**, polyaniline nanofibers are observed to be formed spontaneously during the chemical oxidative polymerization of aniline. The authors observed that the key of the nanofibril formation is the suppression of the secondary chain growth that leads to agglomerated particles. Depending of the doping acid nanofibers with diameters values between 30 and 120 nm can be obtained by this aproach. These nanofibrils can be used in as a template to grow inorganic/polyaniline nanocomposites that lead to exciting properties such as electrical bistability that can be used for nonvolatile memory devices **[59]**.

Medeiros and coworkers successfully produced nanostructured films of poly(o-ethoxyaniline) (POEA) alternated with cellulose nanofibrils (CnF) by layer-by-layer assembly (LBL) at different pH values. According to the authors it was possible to build up films by alternating POEA and CnF layers with relatively precise architectural control by controlling the number of layers and pH. Film thickness had a dependence on pH which is a combination of the effects of the deposited amount for each POEA layer and the pH at which the absorption of the cellulose nanofibrils was carried out **[60]**.

The same authors also produced conductive nanofibrils by *in situ* coating cellulose nanowhiskers with different thickness of polyaniline. One of the advantages of using these coated whiskers instead of pristine conjugated polymers is the inherent strong nature of cellulose allied with the conductive nature of polyaniline **[61]**.

2.3.5 Solution blow spinning

A new technique called solution blow spinning has been developed as an alternative method for making non-woven webs of micro- and nanofibers with down to the nanometer scale with the advantage of having a fiber production rate several times higher **[56]**. This solution blow spinning method is based on the use of a syringe pump to deliver a polymer solution connected to an apparatus consisting of concentric nozzles whereby the polymer solution is pumped through the inner nozzle while a constant, high velocity gas flow is sustained through the outer nozzle (See Figure 10). The aerodynamic forces are able to stretch the solution exiting the inner nozzle to produce long filaments with diameter down to the nanoscale.

With this technique the authors have been able to produce a variety of morphologies such as smooth and porous fibers as well as beaded fibers. Moreover, polymers such as PLA, PMMA, PEO, PS and PVC have been successfully used to produce micro and nanosized fibers with diameters as low as 40 nm **[56, 62]**.

2.3.6 Electrospinning

A very innovative and promising technique to obtain electroactive fibers is through the application of a static electric field on a polymer solution or melt. This technique known as

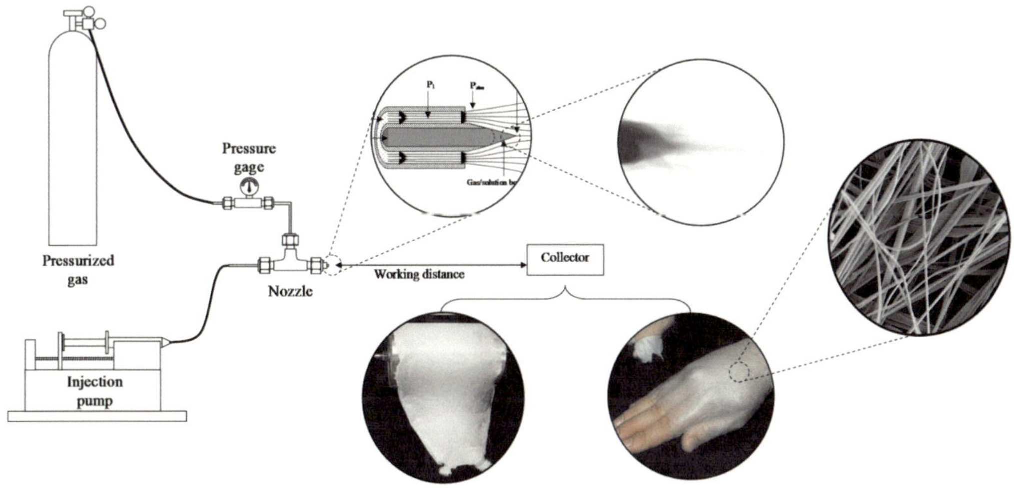

Fig. 10. Schematic illustration of the Solution Blow Spinning technique **[56]**.

electrospinning can be used to produce polymer fibers down to the nanometer. Contrasting with the most common fiber processing techniques as melt spinning, dry spinning, wet spinning and extrusion, electrospinning is able to produce ultrathin fibers with low cost and elevated surface area. This point is of particular interest for producing sensors, actuators and other electroactive devices, especially possible when conducting polymers are used in electrospinnnig. MacDiarmid and his group firstly verified the possibility to electrospun fibers of conducting polymers once no chain degradation was observed after electrical field application **[63, 64]**. Since then, many conducting polymers such as polyaniline, polypyrrol and polythyiophene have been used to produce nanosized electrospun fibers **[65-68]**. Although the available literature on electrospinning of polymer mats is extensive, there are still few reports on electrospinning of electroactive polymers and its related applications[1]. Details of the electrospinning process will be given in the next section.

3. Electrospinning process

3.1 Basics of electrospinning

Electrospinning has emerged as an experimental technique based on the application of a static electric field on a polymer solution or melt through a spinneret. The electrostatic spinning of polymer fibers matches both basic science and technological development pointing the development of new materials with unrelated properties that will cause revolution in the way of producing polymer materials **[2]**.

On a first approach, electrospinning appears to be a simple and easily controllable technique able to produce polymeric nanofibers. A typical experimental setup is based on a capillary injection tip, a high voltage source able to apply electric fields of 100-500 KVm^{-1} and a metallic collector, or counter-electrode. Electric current in electrospinning experiments are usually in the order of few miliamperes **[2]**. A typical apparatus for electrospinning of polymers consists of an injection pump with hypodermic syringe to pump solution though

[1] A total of 98 scientific papers and patents were found at SciFinder and 56 at Web of Science until May, 2011 under the key words: "conjugated polymer electrospinning".

the needle/nozzle, a grounded collector that can be either stationary or rotating, and a high voltage supply. Experiments can also be carried out in a box in order to precisely control environmental conditions such as temperature and relative humidity (see Figure 11).

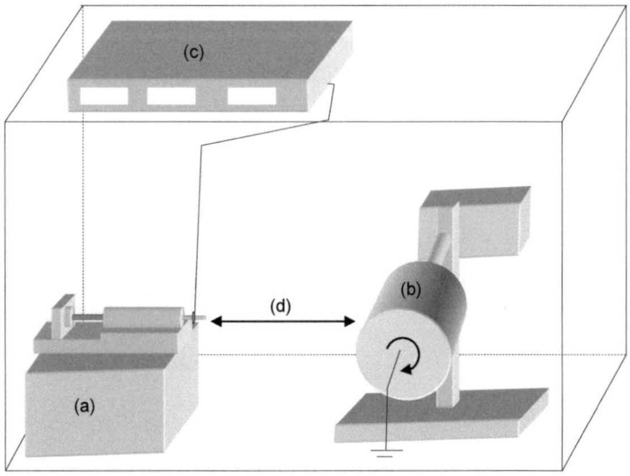

Fig. 11. A typical setup used to produce nanofibers by electrospinning. (a) Injection pump with hypodermic syringe; (b) grounded collector; (c) high voltage supply; and (d) working distance.

The electric potential causes the deformation of the fluid drop and, when the applied voltage develops enough force and balances with the fluid surface tension of the polymer solution, the drop is deformed under a cone shape with a semi-vertical angle of approximately 30°. Recently, Reneker and cowoerkers **[69]** demonstrated that beyond this critical value (Raylegh limit), the electrostatic forces generated by the charge carriers present in fluid (which move toward the surface of the polymer solution) overcome the surface tension and the deformed droplet undergoes to a transition zone just before the fiber jet is initiated to the collector screen **[69-72]**. At this moment, fluid is submitted to an expressive stretching, but inside tension is still small and the authors proposed a Newtonian flux behavior in this transition zone. Experimental measurements have shown that typical stretching rates in transition zone are around 100 - 1000 s^{-1} **[69]**. Thus, any change in jet format will imply a dynamic redistribution of charges on its surface, leading to instability due to bending caused by this redistribution of electric charges. After, a linear segment takes place and the pre-stretched jet is submitted to rates of 20 s^{-1}. In linear segment, the flow is basically controlled by effects of electrical field and the longitudinal tension of the viscoelastic fluid. Due to the high electric fields commonly used in electrospinning, fluid jet is kept stable under small distances (around 2-4 cm) before reach the scattering region, where longitudinal instabilities take place **[73-77]**. Polymer jets can be ejected at velocities up to 40 $m.s^{-1}$ **[2]**, Figure 12 represents the different regions of a polymer jet in electrospinning process.

A closer look to the electrospinning process reveals its intricate complexity. Recent studies compare the electrical instability on electrospinning process to aerodynamic instabilities **[77]** where partial differential equations of the aerodynamics of the fluid jet with electric field equations are used to detailed describe the phenomenon **[78,77,79]**. Many theoretical models have been used to describe the combination of a viscoelastic fluid driven by an electrical field.

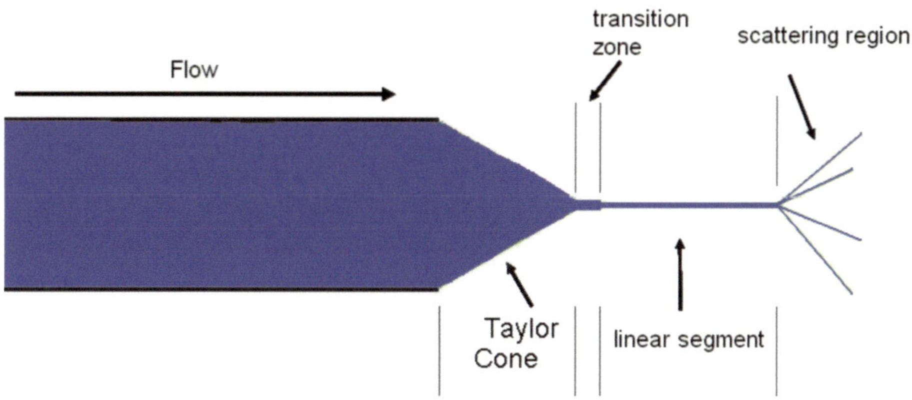

Fig. 12. Representation of fluid deformation in electrospinning process.

Based on Taylor studies and with the use of high velocity cameras, Baumgarten proposed that the electrospinning process occurs in two moments: (1) geometry deformation of the fluid droplet by the electric field and (2) the formation of a continuous jet from the top of deformed droplet **[80]**. A one dimensional model was proposed by Hohman e Gañán based on numerical calculations of fluid atomization in a steady state. Based on this approach, the authors were able to reasonably describe the intermediate region between the cone and scattering region of electrified jets **[81,82]**. Later, Holman and coworkers studied the fluid jet stability under an electric field **[70, 83]**. The studies searched for evidences of the experimental parameters on the beginning of electrospinning process. The authors proposed that the nanofiber formation is governed by the scattering region, where the surface of charged jets interacts to the electric field leading to the scattering region (Figure 9). Many others theoretical models have been proposed to electrospinning. Recently, Doshi and Reneker considered the electrospinning jet as a mass-spring system divided into four distinct regions containing beads of electrical charge (e) and mass (m) connected by viscoelastic elements **[71,85-87]**.

For many applications a precise diameter control is required. Fiber dimensions and morphology depend strongly on process parameters as, for example, polymer properties: molar mass, molar mass distribution, glass transition temperature, solubility; solution properties as viscosity, viscoelasticity, concentration, surface tension, electrical conductivity, dielectric constant, vapor pressure and ambient conditions as humidity and temperature. Basically, electrospinning process parameters can be classified into three different topics; solution parameters, process parameters and ambient parameters.

Solution parameters as viscosity, electrical conductivity, and surface tension affect directly the fiber dimensions and morphology. Viscosity is one of the most important solution parameters. Beadless fibers are commonly obtained when the polymer solution develops a minimum polymeric chain network, the entanglement concentration. In fact, both solution viscosity and concentration are related by Berry Number (*Be*). Experimental findings show that the diameter of electrospun fibers is dependent of solution concentration and polymer chain conformation in solution. Thus, the Berry Number is defined by equation 1;

$$Be = [\eta]^*C \qquad (1)$$

where, [η] is intrinsic polymer viscosity and C is the polymer solution concentration. As the degree of polymer chain entanglements can be represented by the *Be* number, one can say that *Be* values determines the electrospun fiber diameter. When *Be* < 1, polymer molecules in solution are sparsely distributed and there is a low probability of an individual molecule to bind with another. As a result, only beads and beaded fibers are formed. Figure 13, compares typically beaded and beadless fibers.

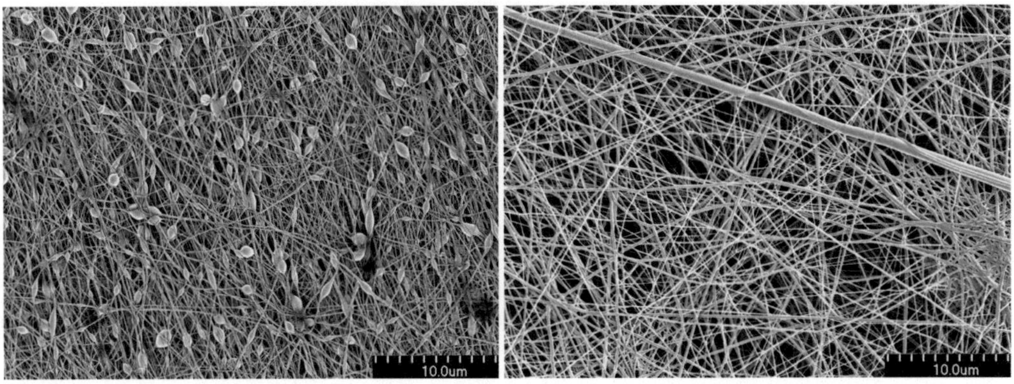

Fig. 13. (a) Poly(lactic acid) electrospun fibers with beaded morphology, (b) beadless Poly(lactic acid) electrospun fibers.

At 1 ≤ *Be* ≤ 2.7, entanglement probability increases and favorable conditions for fiber production takes place. This *Be* number range is convenient for nanofiber production. At *Be* ≥ 2.7, polymer chain entanglements probability increases and the average fiber diameter goes above the micrometer range **[88, 89]**.

Surface tension is directly related to the Taylor cone formation and this is related to the electric field strength applied over the fluid droplet able to deform its shape. This tension value is called critical tension and polymer solutions with different solvents will have different the critical values. Also, it was observed that the decrease in surface tension values will favor a beadless morphology **[90, 91]**.

The electrical conductivity of solution also plays an important role on fiber morphology. Higher solution electrical conductivity is associated to a greater number of charges in solution, which favors the electrospinning process. Generally, both electrical conductivity of solvents and polymers are small and in some cases, inorganic salts are added to solution in order to favor the spinning process. This methodology is successfully used in production of nanofibers with uniform diameter values with beadless morphology **[92,93]**. Moghe and coworkers **[94]** produced bead-free ultrafine fibers with narrow fiber diameter distribution from poly(ε-caprolactone) (PCL) via electrospinning. The high quality product was achieved with the use of a new solvent system that involves an acid–base reaction to produce weak salt complexes, which serve to increase the conductivity of the polymer solution.

In electrospinning, process parameters are typically considered as the applied electric field, working distance, flow rate and, in some cases, rotor (collector) velocity.

According to Doshi e Reneker **[71]**, there is a range of applied voltage values where a stable jet is obtained for poly(ethylene oxide) solutions. For example, solutions at 6 wt.%, a stable jet is formed between 5 and 15 KV, with a working distance of 12.5 cm. The authors also concluded that the diameter of the jet decreases as it moves away from the needle tip until it reaches a minimum value dependant of the Taylor cone initiated. These values were also corroborated by Medeiros and collaborators for poly(vinyl alcohol) **[95]**. Also, according to Deitzel and coworkers **[96]**, applied voltage values are directly related to beads formation and the monitoring of electric current during the process is able to indicate the electric field values where beads density significantly increases.

Carroll and Joo **[97]** reported a study considering axisymmetric instabilities of highly conducting viscoeleastic solutions of poly(ethylene oxide). In this theoretical study, a linear stability analysis combined with a model for stable electrospun jet was used to calculate the expected bead grow rate and the wave number for a given electrospinning conditions. According to the authors, the analysis reveals that the unstable axisymmetric mode for electrically driven, highly conducting jets is not a capillary mode, but is mainly driven by electrical forces due to the interaction of charges on the jet. The authors observed that both experiments and stability analysis elucidated that the axisymmetric instability with a high growth rate can be seen in practice when the electrical force is effectively coupled with viscoelastic forces.

Unlike the applied electric field, the working distance, i.e., distance between needle tip and the collector, seems less important in the formation and morphology of fibers. However, a value of minimum working distance is needed to ensure complete solvent evaporation, and a maximum value for the electric field is effective in forming the Taylor cone and consequently the formation of nanofibers **[92]**. As observed by Gomes and coworkers **[93]**, fiber diameter decreases when the working distance increase from 2 to 14 cm. Further increase in distance, from 14 to 20 cm, has no effect on fiber diameter. Also, in small working distance condition, solvent is not completely evaporated when fibers reach the collector and porous morphologies are obtained.

The environmental parameters, temperature, humidity and air composition, can affect the formation and morphology of nanofibers **[98]**. Medeiros and coworkers **[99]** also found that electrospun fibers of PVA, PLA, PVC and PS had their morphology strongly dependent on the relative humidity surrounding the spinning process. Depending on the relative humidity used, fibers with different sizes and porosities were obtained. With the same aim, Vrieze and coworkers **[99]** studied the effect of humidity and temperature in the nanofibers of cellulose acetate and PVP (polyvinyl pyrrolidone). They found that for PVP increased humidity resulted in a decrease in the average fiber diameter, while for cellulose acetate fiber diameter increased. The authors attributed this behavior to the chemical nature of the polymer. However, the dependence of the diameter with temperature was not linear for both polymers, since for lower temperatures, 283 and 293K, initially there was an increase in diameter and with increasing temperature, 303K, there was a decrease in diameter **[99]**.

Chuangchote and coworkers **[100]** promoted an extensive study on the effect of different solvents on the electrospinnability of PVP fibers. The authors confirmed that dielectric constant, viscosity, and surface tension of the solvents affect the electrospinnability, morphological appearance, and fiber size. The authors also observed that small and uniform PVP fibers can be obtained using solvents with high dielectric constants, low surface tension, and low viscosity. Furthermore, diameters of PVP fibers decreased with the dielectric constant, dipole moment, and density of the solvents.

3.2 Processing parameters of electroactive polymers

Electroactive polymers have a very specific chemistry which may limit the obtention of purely conjugated polymers. Because of limitations on molecular weight and solvents suitable for electrospinning, only a few conjugated polymers such as polyaniline, poly(dodecylthiophene), and poly[2-methoxy-5-(2′-ethylhexyloxy)-p-phenylene vinylene] have been electrospun **[101]**. Thus, electrospinning of conjugated polymers is very limited due to the absence of chain entanglement (conjugated backbones are stiffer and offer low or almost no entanglements), which is considered a prerequisite in electrospinning technique.

Some authors have proposed several modified process approaches to obtain electroactive electrospun nanofibers **[102]**. One first practical and easy way is to spin a nonconductive polymeric web and after, polymerize conductive polymers onto fiber surface. For example, conductive Polyamide-6 (PA-6) nanofibers were prepared by polymerizing pyrrole molecules directly on the fiber surface of PA-6. Firstly, a solution of PA-6 added with ferric chloride in formic acid was electrospun with average diameter values around 260 nm. Secondly, fibers were then exposed to pyrrole vapor and a compact coating of polypyrrole was formed on the fiber surface. According to the authors, polypyrrole coating on the fibers turned out to be conductive with a pure resistive characteristic **[102]**.

A similar approach was used by Ketpang and Park **[103]** to electrospin PVDF/PPy composites which were prepared by spinning a nowoven web from a solution of PVDF and $CuCl.2H_2O$ in DMAc and then exposing the spun fibers to pyrrole vapors in order to produce the conductive composites. According to authors, the electrical conductivity of the PPy composites was affected by the fabrication method and oxidant content in the non-woven web.

Lee and coworkers **[104]**, produced uniform poly(3-hexylthiophene), P3HT nanofibers by electrospinning. As the solubility of P3HT is limited in chloroform the authors adapted the nozzle for being clogged by using a coaxial electrospinning setup where polymer solution is fed through the inner nozzle and pure chloroform is provided through the outer nozzle in order to retard solvent evaporation. According to the authors, this continuous method can be employed to produce organic-based devices on a massive scale.

A very interesting adaptation on electrospinning methodology was developed by Sundarrajan and coworkers for producing a P3HT/PCBM electroactive solar cloth **[105]**. As noted by the authors, electrospinning of pure conjugated polymers is not possible due to the absence of polymer chain entanglements. Since, they proposed the co-electrospinning of poly(3-hexyl thiophene) (P3HT) (a conducting polymer) or P3HT/PCBM as the core and poly(vinyl pyrrolidone) (PVP) as the shell. This approach reveled successful once the PVP shell could be washed giving rise to conductive P3HT or P3HT/PCBM cloth.

On the other hand, MacDiarmid's group was the first to report the production of pure polyaniline (PAni) fibers by electrospinning **[63]**. According to the authors, 100% PAni fibers with average diameter of 139 nm and conductivity value of a single fiber ~0.1 S/cm were produced placing a 20 -wt% solution of polyaniline in a 98% sulfuric acid in a glass pipette above a copper cathode immersed in pure water at 5000 V potential difference. Later, Chronakis and coworkers produced pure PPy electrospun fibers with diameters of ~ 70 nm by dissolving $[(PPy3)^+(DEHS)^-]_x$ in DMF, where the doping agent $(DEHS)^-$ derives from di(2-ethylhexyl) sulfosuccinate sodium salt (NaDEHS) **[106]**.

In order to improve polyaniline processability, a first approach to obtain PAni.HCSA doped nanofibers blended with usual polymers by electrospinning was done by MacDiarmid's group **[64]**. In this study a non-woven mat was obtained by using a polyaniline/ polyethylene oxide PEO solution dissolved in chloroform. By controlling the ratio of polyaniline to PEO in the blend, fibers with conductivity values comparable to that of PAn.HCSA/PEO cast films were produced.

Another study on production of PAni(CSA) nanofibers dispersed in poly(methyl methacrylate) (PMMA) solution in chloroform were produced by Veluru and coworkers

[107]. According to the authors very good aligned fibers with diameters in the range of 500nm to 5μm were obtained and dc conductivity was estimated to be around 0.28 S/m.
Gizdavic-Nikolaidis and coworkers **[108]** used a mixture of dimethyl sulfoxide/ tetrahydrofuran in order to obtain homogeneous blended nanofibers of HCl-doped poly(aniline-co-3-aminobenzoic acid) (3ABAPANI) copolymer and poly(lactic acid) (PLA) for tissue engineering. Once solvent system DMSO/THF (50:50) is quite difficult to be removed from the nonwoven mats, the authors used a heated collector to facilitate the solvent removal. This procedure can be considered essential in cases where nonwoven fiber mats will be used in cell growth. Besides composite electroactive fibers achieve lower conductivity values comparing to pure conjugated polymer fibers, this lower conductivity is quite appropriate to tissue engineering.
As observed earlier, the addition of ions in electrospinning solution can improve fiber spinnability due to the increase of charge carriers in solution. It is also known that conjugated polymers may have high density of charge carriers and this can also affect the electrospinning process. It was recently observed that the addition of PAni.TSA to PLA solution in HFP caused similar effect of inorganic salts addition **[65]**. According to the authors, the average fiber diameter reduced around 400 nm after addition of 0.2 -wt% of PAni.TSA. Also, diameter distribution narrowed and a beadless morphology was observed. Furthermore, electrorheological effects can be observed in polymer solution with high density of charge carriers **[66]**.
Composite nanofibers of Poly(vinylidene fluoride-trifluoroethylene)/polyaniline-polystyrene sulfonic acid with diameters of ~ 6 nm were reported by Abreu and coworkers **[109]**. As observed before, the addition of the conjugated polymer PANi-PSSA also increased the charge density of the solution and assisted the fabrication of homogeneous nanofibers at lower than normal PVDF concentrations in DMF.
Attout and coworkers used a very interesting experimental setup to produce aligned polyaniline based nanowires and nanotubes based on electrostatic steering **[110]**. As demonstrated by the authors, electrospun nanofibers can be aligned on a substrate using an alternative electrostatic field generated between two collectors. This technique suggests promising strategies to achieve fiber alignment and counting with an "immobile" experimental setup (Figure 14).

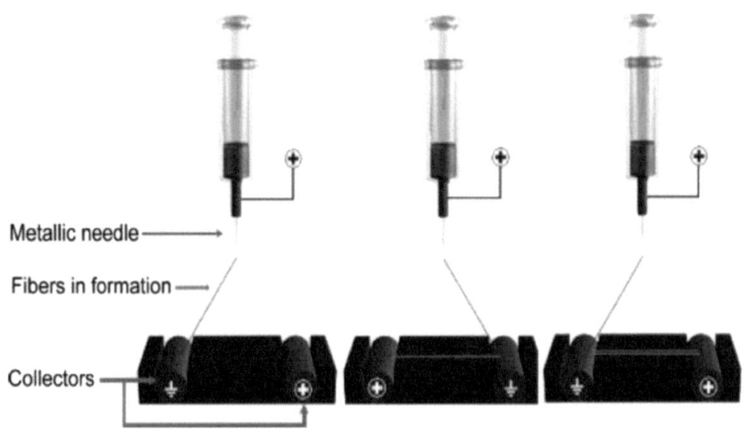

(a)

Fig. 14. (Continued)

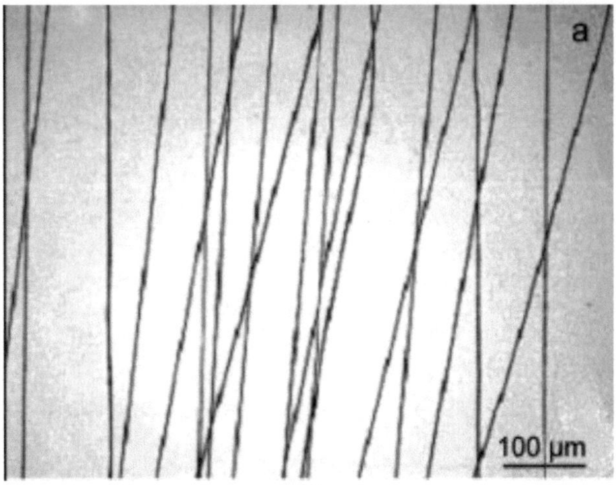

(b)

Fig. 14. (a) Electrospun fibers alignment set up developed by Attout and coworkers and optical pictures of PMMA aligned fibers **[110]**.

3.3 Electrospun electroactive polymer fibers: novel properties due to the size reduction, uses and devices

Nanofibers of electroactive polymers have received great attention recently because of their unique and useful properties **[63]**. New and unrelated properties may arise from size confinement which may be important for several applications in electronic devices, optics and biomedical materials **[111-113]**, protective clothing **[114]**, filtration media **[115]**, charge storage devices **[116,117]**, sensors and actuators **[118-120]**.

Recently Shin and coworkers fabricated conducting nanofibers by blending multiwalled carbon nanotubes (MWNTs) and polyaniline (PAni)/poly(ethylene oxide) (PEO) using electrospinning. The authors observed an unexpected transition in the electrical conductivity of the conducting composite while measuring the *I–V* characteristics of the nanofibers aligned on an electrode when they were exposed to an applied high voltage. This unexpected transition in the electrical conductivity was attributed to the interactions between the MWNTs and the conducting polymer inside the fiber due to an annealing effect of the PANi/PEO matrix from the thermal dissipation of the CNTs. The authors also related this unusual transition to the self-heating effect of the MWNTs incorporated into the conducting polymer which will be very helpful in enhancing the electrical properties of nanoscale conducting composite fibers **[121]**. Figure 15 represents the interactions between carbon nanotubes and polymer chain.

Jeong and coworkers **[122]** prepared a conductive composite based on multiwall carbon nanotubes and nylon 6,6 by electrospinning. In this work a methodology was developed in order to produce stable dispersions of MWNT's functionalized with $-NH_2$ terminations in formic acid. After, nylon 6,6 solution in formic acid was electrospun with different filler concentration. According to the authors, the *I–V* characteristics were found to be non-ohmic and improved with increasing filler concentration in the nylon nanofiber. This increase was attributed by the authors to the enhancement of the electron conduction process by the increase of MWNT content. Figure 16 shows the nylon 6,6/MWNT´s composite electrospun fiber followed by its I-V characteristics **[122]**.

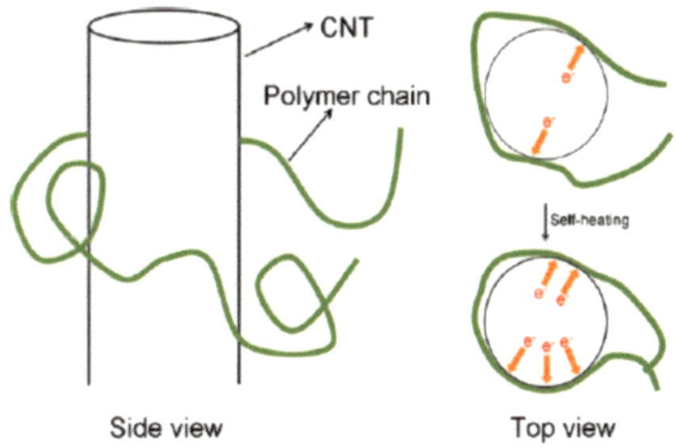

Fig. 15. Schematic representation illustrating the self-heating phenomena between a CNT and a polymer chain inside a conducting composite electrospun nanofiber **[121]**.

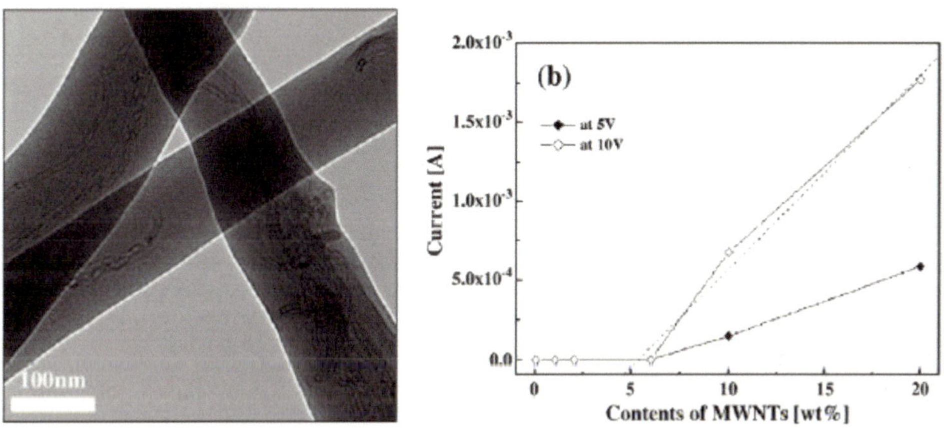

Fig. 16. TEM image of the MWNTs/nylon composite nanofibers with 10 wt.% and its I–V characteristics plot **[122]**.

Babel and coworkers electrospun nanofibers of two series of binary blends of poly[2-methoxy-5-(2-ethylhexoxy)-1,4-phenylenevinylene] (MEH-PPV) with regioregular poly(3-hexylthiophene) (PHT) and (MEH-PPV) with poly(9,9-dioctylfluorene) (PFO) aiming the production of fibers with tunable, composition dependent, optical, and charge transport properties that could be exploited in nanoscale devices.The authors were able to produce fibers with diameters ranging from 100 to 500 nm. It was observed phase-separated morphology by SEM images. However, because of confinement of the liquid jets during electrospinning, the length scales of the phase separation in these blend fibers are much smaller than those of the MEH-PPV/PHT blend thin films prepared by spin coating where the length scales of the phase-separated domains were on the order of 100-150 nm **[100]**. Furthermore, the red shift in electronic absorption peaks suggests that the polymer chains in the fibers are more extended, which may lead to the increase of π-conjugation length. Moreover, the extended polymer chains should be oriented along the fiber axis due to the strong stretching of the liquid jet during electrospinning better π-electron delocalization. An interesting feature in the absorption spectra of MEH-PPV/PFO blend nanofibers is a 20- 30 nm red shift of the PFO absorption band to 400- 410 nm, suggesting that the PFO chains are also extended and oriented along the fiber axis.

PAni.TSA/PLA blended electrospun nanofibers have been recently produced **[67]**. The authors observed no phase segregation of PAni in PLA matrix in electrospun fibers while phase segregation was observed in cast films with the same composition. According to the obtained results, the authors concluded that due to rapid solvent evaporation in electrospinning process, no crystalline structures in fiber mats were formed compared to cast films. Highly homogeneous electroactive fibers can be useful in construction of electronic devices and sensors. Similar behavior was observed to the (PVDF-TrFE/PANi-PSSA) electrospun nanofibers **[123]**.

Laforgue reported the production of flexible supercapacitors using electroactive fibers obtained by electrospinning **[124]**. In this work the author demonstrated that Polyvinylpyrrolidone (PVP) fibers covered with poly(3,4-ethylenedioxythiophene) (EDOT) by vapor phase polymerization. The conductive mats presented elevated electrical conductivity (60±10 S cm^{-1}) and were separated by a layer of PAN in order to assembly an all flexible capacitor device. Figure 17 represents the assembly process of capacitor device using electrospun electroactive fibers.

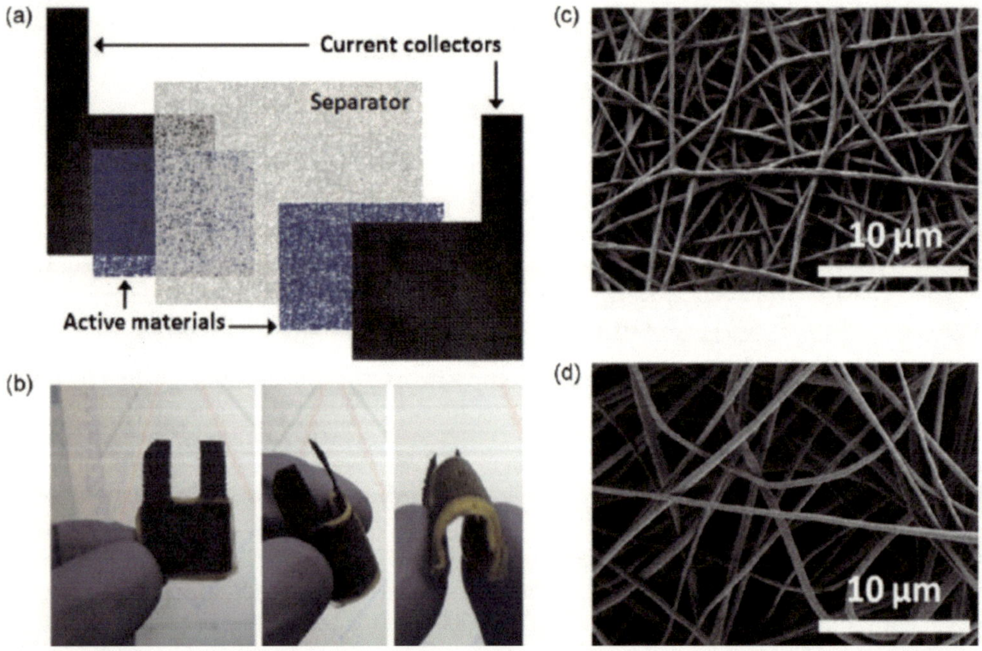

Fig. 17. Assembly of a textile supercapacitor. (a) Schematic representation of the stacked textile layers; (b) photographs of a 3cm^2 all-textile flexible supercapacitor; (c) SEM image of the active materials (PEDOT nanofibers); (d) SEM image of the separator (PAN nanofibers) **[124]**.

According to the authors, the electrochemical performances of the solid-state supercapacitors were very similar to the ones obtained in liquid electrolyte. Owing to the nanostructure nature of the active materials, an effective wetability by the electrolyte and a limited diffusion length of the doping ions within the polymer structure were observed **[124]**.

A similar approach was used by Sundarrajan and coworkers to produce a solar cloth by electrospinning technique as presented previously in this chapter **[105]**, as represented in Figure 18. The authors have found an efficiency of fiber cloth around 8.7×10^{-8}, however environmental parameters and fiber diameter reduction can be improved in order to produce more efficient materials.

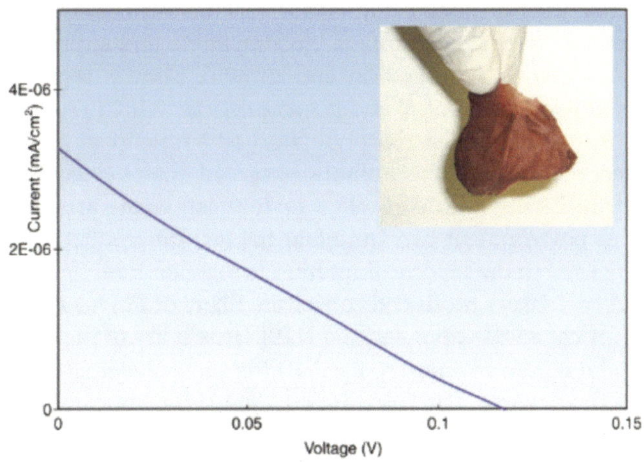

Fig. 18. Jsc–V graph of the P3HT/PCBM solar cloth measured under 1 Sun conditions. Inset shows a picture of the solar cloth fabricated using electrospinning **[105]**.

Electrospun conducting fibers are also used in production of biocompative systems for tissue engineering and biosensors. Lee and coworkers produced PLGA electrospun nanofibers coated with polypyrrole for neural tissue applications **[125]**. The authors used PPy–PLGA electrospun meshes to support the growth and differentiation of rat pheochromocytoma 12 (PC12) cells and hippocampal neurons comparing to non-coated PLGA meshes. It was suggested that PPy–PLGA may be suitable as conductive nanofibers for neuronal tissue scaffolds.

Also, conductive polymers can be blended with other polymers to provide an electrical current to increase cell attachment, proliferation, and migration. McKeon and coworkers **[126]** electrospun several polyaniline and poly(D,L-lactide) (PANi/PDLA) mixtures at different weight percents. Interestingly only the 75/25 electrospun scaffold was able to conduct a current of 5 mA with a calculated electrical conductivity of 0.0437 S cm^{-1}. Later, primary rat muscle cells were cultured on scaffolds and on tissue culture polystyrene as a positive control. The authors observed that, although the scaffolds degraded during this process, cells were still able to attach and proliferate on each of the different scaffolds. The cellular proliferation measurements showed no significant difference between the four groups measured and the conductivity and cellular behavior demonstrate the feasibility of fabricating a biocompatible, biodegradable, and electrically conductive PDLA/PANi scaffold.

As cited before, HCl-doped poly(aniline-co-3-aminobenzoic acid) (3ABAPANI) copolymer and poly(lactic acid) (PLA) blend were electrospun in the form of three-dimensional networks with a high degree of connectivity, onto glass substrates **[108]**. The authors evaluated the ability to promote proliferation of COS-1 fibroblast cells over this conductive scaffold. According to the authors, this new class of nanofibrous blends can potentially be used as tissue engineering scaffolds and showed promise as the basis of a new generation of functional wound dressings that may eliminate deficiencies of currently available antimicrobial dressings.

Li and coworkers investigated the potential applications of polyaniline containing gelatin nanofibers for producing conductive scaffolds for tissue engineering purposes **[127]**. The authors have found that the polyaniline addition affects the physicochemistry of PAni-gelatin blend fibers and that this kind of substrate is biocompatible and support the cells (H9c2 rat cardiac myoblast) attachment, proliferation and growth. Similar work was developed by Borriello and coworkers on testing PAni and polycaprolactone (PCL) electrospun membranes as platforms for to mimic either the morphological and functional features of the cardiac muscle tissue regeneration **[128]**. The authors observed that development of PANi/PCL membranes by electrospinning with controlled texture can create an electrically conductive environment and this environment can stimulate the cell differentiation to cardiomyocites, and can successful be used in the myocardium muscle regeneration.

Aussawasathien and coworkers produced composite fibers of Poly(o-anisidine)-polystyrene via electrospinning for chemical vapor sensing **[129]**. Sensibility of the composite fibers were tested under water and ethanol vapor (being that) the sensors elements responded better to the high polarity of the solvent. The CSA doped POA/PS composition seems to be stable under the submitted ambient one to ethanol. The sensor could be reused several times without any change in sensing behavior and/or damage to the sensing materials.

A flexible nanotube membrane of Poly(3,4-ethylenedioxythiophene) (PEDOT) was produced by Kwon and coworkers electrospinning mediated for ammonia gas detection **[130]**. Initially, Poly(vinyl alcohol) (PVA) solution was electrospun and further treated with $FeCl_3$ solution to adsorb Fe ions on the nanofibers surface. Later, EDOT monomer was evaporated and polymerized on the PVA surface leading to coaxial PVA/PEDOT fibers which were after washed with distillated water giving 140 nm PEDOT tubes. According to the authors, PEDOT nanotubes achieved electrical conductivity values of 61 S cm^{-1}, higher than the usual PEDOT nanomaterials produced with $FeCl_3$. PEDOT nanotubes revealed owing faster recovery times than PVA/PEDOT coaxial fibers due to the elevated surface area, demonstrating the possibility of this methodology to produce 1D nanomaterials for sensors applications.

Pinto and coworkers evaluated the electric response of isolated polyaniline fibers to vapors of aliphatic alcohols **[131]**. According to the authors, the large surface to volume ratio, the uniform diameter and small quantity of active material used in the sensor construction are comparable to or faster than those prepared from nanofiber mats of the same polymer. Also, the sensors made from individual fibers exhibit larger responses, especially for bigger alcohol molecules, and also show true saturation upon exposure and removal of the alcohol vapor. As observed by the authors, the response of sensors made from electrospun nanofibers to small alcohol molecules is opposite to that observed for nanofiber mats. This effect was related by the authors to the doping process used in the preparation of the polymer in either case **[135]**.

4. Conclusions and perspectives

Nanotechnology has recently emerged as a unique and fruitful area in modern science that basically encompasses all aspects of the human knowledge. Nevertheless, the use of nanostructured materials dates back to ancient Rome, where artifacts such as wine glasses that changed their color according to the incidence of light were made with gold nanoparticles. Obviously, the knowledge about atomic and molecular manipulation did not exist in those ancient times and only recently with the perception of Richard Feynman and the work of many scientists, nanotechnology has become such a broad field of science with enormous implications and advancements. In fact, emergence of nanotechnology began in the 1980's caused by the convergence of experimental advances such as the development of the scanning tunneling microscope in 1981 by Gerd Binnig and Heinrich Rohrer at IBM Zurich Research Laboratory, and the discovery of fullerenes in 1985 by Harry Kroto, Richard Smalley, and Robert Curl, and others discoveries of materials and techniques. Furthermore, the elucidation and popularization of the conceptual framework of nanotechnoloy began in 1986 with the publication of the book "Engines of Creation: The Coming Era of Nanotechnology" by K. Eric Drexler.

One of the greatest objectives in current research in nanotechnology is based on observations that a number of physical phenomena may become more pronounced as the size of the system decreases. These effects are more significant in case of statistical mechanics and quantum effects usually, especially in case of electronic and optical properties of some solids. Quantum effects become more pronounced, for example, when the nanometer size range is reached, typically at distances of 100 nanometers or less. Another important effect when dealing with nanosized materials is the increase in surface area to volume ratio which is able to produce significant changes in mechanical, thermal and catalytic properties of many materials, including polymers and polymer composites.

Various effects can be considered regarding to the research in nanoscience and nanotechnology of electroactive polymers, especially with regard to electronic conduction in disordered materials of low-dimensional conductors and also with respect to the significant increase in surface area. The latter is considered of especial interest in research and development of sensors based on electroactive polymers.

Electrospinning technique allows us to obtain electroactive polymeric structures of high surface area and organization at the nanoscale. As we have seen before, electrospinning is easy to operate, relatively low cost, and is capable to produce a wide range of electroactive polymer nanofibers and nanomats that can be uses in many applications, ranging from sensors and actuators to solar clothes, supercapacitors and bioactive materials. However, there is only a limited number of papers published in this field and there is still need for further investigations, particularly with the regard to understanding the electrospinning process within a single theoretical framework, capable of accurately predict fiber dimensions and morphologies experimentally obtained. This point can be considered of fundamental importance for the development of electrospinning and its applications at larger scales.

In fact, this is a topic that deserves most attention from researchers in electrospinning. Despite the ease of obtaining and operating, fibers productivity by electrospinning is considered very small when compared with other techniques for producing fibers. In this regard, considerable efforts should be made to increase the productivity of spun fibers while precisely controlling dimensions and morphology of the fibers. Some efforts have already been done towards this direction indeed since the time of Formhals, but to date there are no records of large-scale production of electroactive polymer fibers by electrospinning despite the fact that this technique has given its first steps towards mass production of micro and nanofibrous mats. In addition, the emergence of new effects and phenomena related to electronic and ionic transport in conductive nanofibers can be studied in depth which certainly will lead to materials and devices with unknown properties.

5. References

[1] Duft, D. Achtzehn, T.; Müller R.; Huber, B. A.; Leisner T. (2003). Nature, 421, p. 128.
[2] Greiner A., Wendorff , J.H. (2007) Angewandte Chemie International Edition, 46, p. 5670.
[3] Zeleny, J. (1914). Physical Review, 3, p. 69.
[4] Taylor, G. (1965). Proceedings of the Royal Society of London A: Mathematical, Physical & Engineering Sciences, 291, p 145.
[5] Blythe T.; Bloor D. (2005). Electrical Properties of Polymers, Cambridge University Press, Cambridge, UK.
[6] Heeger, A. J. (2001). Angew. Chem. Int. Ed. 40, p. 2591.
[7] Mullen K., Wegner, G. (1998). Electronic Materials: The Oligomer Approach, Wiley-VCH, Weinhem.
[8] Guiseppi-Elie, A.; Wallace, G. G.; Matsue, T. (1998) in Handbook of Conducting Polymers edited T. A. Skotheim, R. L. Elsenbaumer, and J. R. Reynolds, Marcel Dekker, New York, p.963.
[9] Ulman, A. (1991). An Introduction to Ultra-thin Films: from Langmuir-Blodgett to Self-assembly, Academic Press, Boston.
[10] Hideki Shirakawa, Nobel Lecture, 2000. (www.nobelprize.org.)
[11] Heeger, A. J.; Smith, P. (1991). Solution Processing of Conducting Polymers: Opportunities for Science and Technology, in: Brédas, J. L.; Silbey, R.; Conjugated Polymers: The Novel Science and Technology of Highly Conducting and Non-linear Optically Active Materials. Dordrecht (The Nederlands): Kluwer Academic Publishers.
[12] Chiang, C. K.; Fincher, C. R.; Park, Y. W.; Heeger, A. J.; Shirakawa, H.; Louis, E. J.; Grau, S. C.; MacDiarmid, A. G. (1977). Phys. Rev. Lett. 39, p. 1098.
[13] MacDiarmid, A. G. (2002). Synth. Met. 125, p. 11.
[14] Skotheim, T. A.; Elsenbaumer, R. L.; Reynolds, J. R. (1998). Handbook of Conducting Polymers edited, Marcel Dekker, New York, p. 467.
[15] Liu, W.; Kumar, J.; Tripathy,S.; . Senecal, K. J; Samuelson, L. (1999). J. Am. Chem. Soc. 121, p. 71
[16] Shi, F. F. (1996) Surf. Coat. Tech. 82, p. 1.
[17] MacDiarmid, A. G. (1997). Synth. Met. 84, p. 27.
[18] Nordén, B.; Krutmeijer, E. (2000). The Nobel Prize in Chemistry, 2000: Conductive polymers, Kungl. Vetenskapsakademien (The Royal Swedish Academy of Sciences), p.1.
[19] Skotheim, T. A. (1986). Handbook of Conducting Polymers, Marcel Dekker, New York.
[20] Brédas J. L.; Street, G. B. (1985). Acc. Chem. Res. 18, p. 309.
[21] Appel, G.; Böhme, O.; Mikalo, R. (1999). Chem. Phys. Lett., 313, p. 411.
[22] Hadziioanou G.; Van Hutten, P. F. (2000). Semiconducting Polymers - Chemistry, Physics and Engineering, Wiley-VCH, Weinheim.
[23] Huang, W. S.; Humphrey, B. D.; MacDiarmid, A. G. (1986). J. Chem. Soc. Faraday Trans. 1, 82, p. 2385.
[24] Letheby, H. J. (1862). Am. Chem. Soc. 15, p. 161.
[25] Ray, A.; Asturias, G. E.; Kershner, D. L.; Richter, A. F.; MacDiarmid, A. G.; Epstein, A. J. (1989). Synth. Met. 29, p. 141.
[26] Chiang, J. C.; MacDiarmid, A. G. (1986). Synth. Met. 13, p.193.

[27] Picciani, P.H.S.; Souza Jr., F.G.; Comerlato, N.M.;, Soares, B.G. (2007). Synthetic Metals, 157, p. 1074.
[28] Heeger, A. J. (2001) Nobel Lecture: Semiconducting and metallic polymers: The fourth generation of polymeric materials, Reviews of Modern Physics, v. 73, n.3, p. 681.
[29] Paul, R.K., Pillai, C.K.S. (2001). Polymer International, 50, 4, p.381.
[30] Wessling, B. (1998). Handbook of Conducting Polymers edited T. A. Skotheim, R. L. Elsenbaumer, and J. R. Reynolds, Marcel Dekker, New York, p. 467.
[31] Medeiros et al. (2005). In. Encyclopedia of Sensors, edited by Grimes, C. A.; Dickey, E.C.; Pishko, M.V. Encyclopedia of Sensors. Volume X, p. 36.
[32] Taniguchi, N. (1996). Ed., Nanotechnology: Integrated Processing Systems for Ultra-Precision and Ultra-Fine Products, Oxford University Press, Oxford.
[33] Petty, M. C. (1996). Langmuir-Blodgett Films - An Introduction, Cambridge University Press, Cambridge.
[34] Mattoso, L. H.C.; Oliveira Jr., O. N.; Ferreira, M. (1986). in: Polymeric Materials Encyclopedia edited J. C. Salamone, CRC Press, Boca Raton, p.1432.
[35] Dhanabalan, A.; Mattoso L. H. C.; Oliveira Jr., O. N. Curr. (2000). Trends in Polym. Sci. 5, p. 19.
[36] Faria, R. M.; Mattoso, L. H. C.; Ferreira, M.; Oliveira Jr., O. N.; Gonçalves, D.; Bulhões, L. O. S. (1992). Thin Solid Films 221, p. 5.
[37] Oliveira Jr, O.N.; Raposo, M.; Dhanabalan, A. (2001). Langmuir-Blodgett (LB) and Self-assembled (SA) polymeric films, Handbook of Surfaces and Interfaces of Materials, ed. H.S. Nalwa, Vol. 4, pp. 1-63, Academic Press.
[38] Iler, R. (1966). J. Colloid Interf. Sci. 21, p. 569.
[39] Sagiv, J. (1980). J. Am. Chem. Soc. 102, p.92.
[40] Decher, G. Hong, J. D. (1991). Ber. Bunsenges. Phys. Chem. 95, p.1430.
[41] Decher, G. (1997). Science 277, p.1232.
[42] Oliveira Jr., O. N.; He, J.-A.; Zucolotto, V.; Balasubramanian, S.; Li, L.; Nalwa, H. S.; Kumar, J.; Tripathy, S. K. (2002).In Handbook of Polyelectrolytes and their Applications edited S. K. Tripathy, J. Kumar, H. S. Nalwa, American Scientific Publishers, Los Angeles, p.1.
[43] Cheung, J. H.; Fou, A. C.; Rubner, M. F. (1994). Thin Solid Films 244, p. 985.
[44] Ferreira, M.; Cheung, J. H.; Rubner,M. F. (1994) Thin Solid Films 244, p. 806.
[45] Cheung, J. H.; Stockton, W. B.; Rubner, M. F. (1997). Macromolecules 30, p. 2712.
[46] Pontes, R. S.; Raposo, M.; Camilo, C. S.; Dhanabalan,A.; Ferreira, M.; Oliveira Jr., O. N. (1999). Phys. Stat. Sol. A 173, p. 41.
[47] Raposo, M.; Oliveira Jr., O. N. (2000). Langmuir 16, p. 2839.
[48] Paterno,L. G.; Constantino, C. J. L.; Oliveira Jr., O. N.; Mattoso, L. H. C. (2002). Colloids and Surf. B: Biointerfaces 23, p. 257.
[49] http://www.cpmt.org/mm/pkglab/theory/spin_theory.html
[50] He , X.; Gao, F.; Tu, G.; Hasko, D. G.; Hüttner, S. Greenham, N. C.; Steiner,U.; Friend, R. H.; Huck, W. T. S. (2011). Adv. Funct. Mater., 21, p. 139.
[51] Zuoquan, J.; Tengling, Y.; Yang, C.; Yang, D.; Zhu, M.; Zhong, C.; Qin, J.; Ma,D. (2011). Chem. Mater., 23, 771-777.
[52] Palewicz, M.; Iwan, A.; Doskocz, J.; Strek, W.; Sek, D.; Kaczmarczyk,B.; Mazurek, B. (2011). POLYM BULL 66 (1), p. 65.

[53] Meier, R. Ruderer, M. A.; Diethert, A.; Kaune,G.; Krstgens, V.; Roth, S. V.; Muller-Buschbaum, P. (2011). J. Phys. Chem. B, 115 (12), p. 2899.
[54] Salleo, A.; Kline, R.; DeLongchamp, M.; Chabinyc, M. L. (2010). Advanced Materials, 22, p. 3812.
[55] Ramakrishina, S.; Fujihara, K.; Teo, W-E. (2005). An introduction to electrospinning and nanofibers, Singapore: World Scientific, 396 p.
[56] Medeiros, E. S.; Glenn, G. M.; Klamczynski, A. P.; Orts, W. J.; Mattoso, L. H. C. (2009). Journal of AppliedPolymer Science, 113, p. 2322.
[57] Ondarcuhu, T.; Joachim, C. (1998). Europhys. Lett., 42, p.215.
[58] Huang, J. Kaner, R. B. (2006). Chem. Commun., 2006, 367–376.
[59] Huang, J. (2006). Pure Appl. Chem., 78, p. 15.
[60] Medeiros, E. S.; Mattoso, L. H. C.; Bernardes-Filho, R.; Wood, D. F.; Orts, W.J. (2008). Colloid Polym Sci 286, p. 1265.
[61] Mattoso, L.H.; Medeiros, E.S.; Baker, D.A.; Avloni, J.; Wood, D.F.; Orts, W.J. (2009). J. Nanosci Nanotechnol., 9(5), p.2917.
[62] Oliveira, J. E.; Moraes, E. A.; Costa, R. G. F.; Afonso, A. S.; Mattoso,L. H. C.; Orts, W. J.; Medeiros,. E.S. (2009). J. Appl. Polym., 113, p. 2322.
[63] MacDiarmid, A.G.; Jones Jr., W.E. Norris, I.D.; Gao,J.; Johnson Jr., A.T.; Pinto, N.J.; Hone, J. B.; Han, Ko, F.K.; Okuzaki, H.; Llaguno, M. (2001). Synthetic Metals, 119, p. 27.
[64] Norris, I. D.; Shaker, M. M.; Ko, F. K.; MacDiarmid; A. G. (2000). Synthetic Metals, 114, p. 109.
[65] Picciani, P. H. S.; Medeiros, E. S., Pan, Z.; Orts, W. J.; Mattoso, L. H. C.; Soares, B. G. (2009). Journal of Applied Polymer Science, 112, p. 744.
[66] Picciani, P. H. S.; Soares, B. G.; Medeiros, E. S.; Souza Jr., F. G.; Wood, D. F.; Orts, W. J.; Mattoso, L. H. C. (2009). Macromolecular Theory and Simulations, 18, p. 528.
[67] Picciani, P. H. S.; Medeiros, E. S.; Pan, Z.; Wood, D. F.; Orts, W. J.; Mattoso, L. H. C.; Soares, B. G. (2010). Macromolecular Materials and Engineering, 295, p. 618.
[68] SUBBIAH, T., BHAT, G.S., TOCK, R.W. PARAMESWARAN, S, RAMKUMAR, S.S. (2005). Journal of Applied Polymer Science, 96, p. 557.
[69] Reneker, D.H. & Yarin, A.L. (2008). Polymer, 49, p.2387.
[70] Fomhals, A. (1938). "Artificial Fiber Construction", Patent US 2109333.
[71] Doshi, J. & Reneker, D.H. (1995). J. Electrostat., 35, p.151.
[72] McCann, J.T.; Li, D. & Xia, Y.N. (2005). J. Mater. Chem., 15, p.735.
[73] Feng, J.J. (2002). Phys. Fluids, 14, p. 3912.
[74] Hohman, M.M. et al. (2001). Phys. Fluids, 13, p.2201.
[75] Yarin, A.L. et al. (2006) Aiche J., 52, p.217.
[76] Yarin, A.L.; Koombhongse, S. & Reneker, D.H. (2001). J. Appl. Phys., 90, p.4836.
[77] Feng, J.J. (2003). J. Non-Newton. Fluid, 116, p.55.
[78] Brannonpeppas, L. (1995). Int. J. Pharm., 116, p.1.
[79] Shin, Y.M. et al. (2001). Appl. Phys. Lett., 78, p.1149.
[80] Reneker, D.H. et al. (2000). J. Appl. Phys., 87, p.4531.
[81] Taylor, G. (1969). Proc. R. Soc. Lond. A, 313, p.453.
[82] Baumgarten, P.K. (1971). J. Colloid Interf. Sci., 36, p.71.
[83] Hohman, M.M. et al. (2001). Phys. Fluids, 13, p.2201.
[84] Han, T.; Reneker, D.H. & Yarin, A.L. (2007). Polymer, 48, p.6064.

[85] Han, T.; Reneker, D.H. & Yarin, A.L. (2008). Polymer, 49, p.2160.
[86] Han, T.; Yarin, A.L. & Reneker, D.H. (2008). Polymer, 49, p.1651.
[87] Ko, F., Aufy, A., Hoalam, G., MacDiarmid, A.G. (2005). Eletrostatically generated nanofibers for wearable electronics. In Tao, X. Wearable Electronics and Photonics. Cambridge: CRC Press, 250 p. p. 13-41.
[88] Ko, F.K. (2006). Nanofiber Technology. In. Gogotsi, Y. Nanomaterials Handbook. London: CRC Press, p.533.
[89] Lin, Z. et al. (2010). J. Appl. Polym. Sci., 116, p.895.
[90] Guerrini, L.M. et al. (2009). J. Appl. Polym. Sci., 112, p.1680.
[91] Moghe, A.K. et al. (2010). Polymer, 50, p.3311.
[92] Wang, C.; Chien, H.S.; Yan, K.W.; Hung, C.L.; Hung, K.L.; Tsai, S.J.;Jhang, H.J. (2009). Polymer, 50, p.6100.
[93] Gomes, D.S.; da Silva, A. N. R.; Morimoto, N. I.; Mendes, L. T. F.; Furlan, R.; Ramos, I. (2007). Polímeros: Ciência e Tecnologia, 17, p.206.
[94] Moghe, A.K.; Hufenus,R.; Hudson, S.M.; Gupta, B.S. (2009). Polymer, 50, 14, p. 3311.
[95] Medeiros, E. S. Mattoso, L. H. C.; Ito, E. N.; Gregorski, K. S.; Robertson, G. H.; Offeman, R. D.;Wood, D. F.; Orts, W.J. Imam, S. H. (2008). Journal of Biobased Materials and Bioenergy, 2, p. 1.
[96] Deitzel, J.M.; Kleinmeyer, J.; Harris,D.; Beck Tan, N.C. (2001). Polymer, 42, p. 261.
[97] Carroll, C. P.; Joo, Y. L. (2009). Phys. Fluids 21, p. 103101.
[98] Medeiros, Eliton S. ; Mattoso, Luiz H.C. ; Offeman, Richard D. ; Wood, Delilah F. ; Orts, William J. (2008) Canadian Journal of Chemistry, v. 86, p. 590.
[99] Vrieze et al. (2009). Mat Sci 44(5), p. 1357.
[100] Chuangchote, S.; Sagawa, T:; Yoshikawa, S. (2009). Journal of Applied Polymer Science. 114, 5, p. 2777.
[101] Babel, A.; Li, D. Xia, Y.; Jenekhe, S.A. Macromolecules 2005, 38, 4705-4711
[102] Granato, F.; Bianco, A.; Bertarelli, C.; Zerbi, G. (2009) Macromol. Rapid Commun., 30, p. 453.
[103] Ketpang, K.; Park, J. S. (2010). Synthetic Metals, 160, p. 1603.
[104] Lee, S.; Moon, G. D.; Jeong, U. J. (2009). Mater. Chem. 19, p. 743.
[105] Sundarrajan, S.; Murugan, R.; Nair, A. S.; Ramakrishna, S. (2006). Materials Letters, 64, p. 2369.
[106] Chronakis, I. S.; Grapenson, S.; Jakob, A. (2006). Polymer 47, p. 1597-1603
[107] Veluru, J.B.; Satheesh, K.K.; Trivedi, D.C. (2007). JOURNAL OF ENGINEERED FIBERS AND FABRICS, 2, p. 25.
[108] Gizdavic-Nikolaidis, M.; Ray, S.; Bennett, J. R.; Easteal, A.J.; Cooney, R.P. (2010). Macromol. Biosci., 10, p. 1424
[109] Abreu, M.; Montanez, S.; Pinto, N. J. (2011). Journal of Applied Polymer Science, 119, p. 3640.
[110] Attout, A.; Yunus, S. P.; Polymer Engineering & Science, 48, p. 00.
[111] Yun, M.H., Myung. N.V., Vasquez. R.P.; Lee, C.S.; Menke, E.; Penner, R.M. (2004). Nano Letters, 4, p. 419.
[112] Li. M.; Guo, Y.; Wei, Y; MacDiarmid, A.G.; Lelkes, P.Y. (2006). Biomaterials, 27: p. 2705.
[113] Wnek, G.E.; Carr, M.E.; Simpson, D.G.; Bowlin, G.I. (2003). Nano Letters, 3, p. 213.

[114] Schreuder-Gibson H., Gibson P., Senecal K., Sennett M., Walker J., Yeomans W., Ziegler D., Tsai P.P. (2002). Adv Mater; 34: p. 44.
[115] Gibson P., Gibson H.S., Rivin D. (2001). Colloids Surf A; 187-188: p. 469.
[116] Kim, C., Yang, K.S. (2003). Appl Phys Lett., 83, p. 1216.
[117] Choi S.W.; Jo, S.M.; Lee, W.S.; Kim, Y.R. (2003). Adv Mater., 15, p. 2027.
[118] Liu, H., Kameoka, J., Czaplewski, D.A., Craighead, H.G. (2004). Nano Letters, 4, p. 671.
[119] Aussawasathien, D., Dong, J.H.; Dai, L. (2005). Synth. Met., 154, p.37.
[120] Manesh, K.M.; Gopalan, A.I., Lee, K.P., Santhosh, P., Song, K.D.; Lee, D.D. (2007). IEEE Trans Nanotechnol, 6, p. 513.
[121] Shin, M. K; et al. (2008). Sensors and Actuators B, 134, p.122.
[122] Jeong, J.S.; Jeon, S.Y.; Lee, T.Y.; Park, J.H.; Shin, J.H.; Alegaonkar, P.S.; Berdinsky, A.S.; Yoo, J.B. (2006). Diamond & Related Materials, 15, p. 1839.
[123] Abreu,M.; Montanez, S.; Pinto, N. J. (2011). Journal of Applied Polymer Science, 119, p. 3640
[124] Laforgue, A. (2011). Journal of Power Sources, 196, p. 559.
[125] Lee, J. Y.; Bashur, C.A.; Goldstein, A. S.; Schmidt, C.E. (2009). Biomaterials, 30, p. 4325.
[126] McKeon,K. D.; Lewis,A.; Freeman, J. W. (2010). Journal of Applied Polymer Science. 115, p. 1566.
[127] Li, M., Guo, Y. Wei, Y. MacDiarmid, A.G.; Lelkes P.I. (2006). Biomaterials, 27, p. 2705.
[128] Borriello, A.; Guarino, V.; Schiavo, L.; Alvarez-Perez, M.; Ambrosio, A. L. (2011). J Mater Sci: Mater Med, 22, p. 1053.
[129] Aussawasathien, D.; Sahasithiwat, S.; Menbangpung, L. Teerawattananon, C. (2011). Sensors and Actuators B: Chemical, 151, p. 341.
[130] Kwon, O. S. et al. (2010). Talanta 82, p. 1338.
[131] Pinto, N. J. et al. (2008). Sensors and Actuators B: Chemical, 129, p. 621.

6

Electrospun Metallic Nanofibers Fabricated by Electrospinning and Metallization

Kai Wei, Hae-Rim Kim, Byoung-Suhk Kim and Ick-Soo Kim
Faculty of Textile Science and Technology, Shinshu University, Japan

1. Introduction

The development of high value added one-dimensional (1D) electrospun nanofibers and structures (such as wires, belts and tubes) have tremendously increased over the last decade. Today, a wide variety of nano-objects (such as metal nanoparticles,[1-3] clays,[4] carbon nanotubes,[5, 6] ceramics,[7] etc.) can be immobilized onto nanofibers, which bring new materials' properties and potential applications in a broad range of areas such as electronics, medicine, sensor, and controlled release technology. As a result, a tremendous amount of effort has been devoted to the synthesis, characterization and utilization of nanofibers composites with well-controlled dimensions and properties. Indeed, extensive research and development in the 21st century is directed towards the development of novel hybrid nanomaterials with tunable properties. The purpose of such formulations is to achieve unique and superior properties that can not be achieved using only a single component of the hybrids. In this way, one can combine the properties of an organic component (polymers, biomolecules) such as mechanical toughness and flexibility with the hardness and thermal stability of the inorganic component (carbon, metals, ceramics, and glass) into a single system.[8,9] The new nanofibrous products based on the immobilization of functional nanomaterials onto electrospun nanofibers have recently attracted significant attention from both the academic and industrial sectors.[10–12] The main strategies for developing new nanostructured nanofibers were mostly carried out by direct electrostatic spinning of various polymers with molecular building blocks or polymer solutions (or polymer melts) with well-dispersed nano-objects,[1–7] and by co-electrospinning core–shell polymer nanofiber using a single-nozzle technique,[13,14] and by mimicking natural self-assembled process to produce hierarchically ordered self-assembled nanofibers.[15]

This chapter concentrates on metallized nanofibers which can be serving as the necessary functional components in building nanoelectronics. Nanofibers with uniform diameters represent an ideal model system to fabricate metal nanofibers. Long polymer fibers with sub-micrometer diameters down to a few nanometers can be prepared by electrospinning, which is a process where polymer solutions or polymer melts are processed in an electrical field.[16, 17] Electrospun nanofiber has a history of more than 70 years. In 1934, Formhals patented his first invention relating to the process and the apparatus for producing artificial filaments using electric charge.[18] This technique involves the use of a high voltage to charge the surface of a polymer solution (or melt) droplet and thus to induce the ejection of a liquid jet through a capillary spinneret. From then, this technique had gradually received more and more attention to prepare nanosized polymer fibers. Nanofibers produced by electrospinning are of industrial and scientific interest due to their long lengths, small diameters and pores, and high surface area per unit volume which enable them have enormous applications in tissue scaffolds, protective clothing, filtration, and sensors, etc. Numerous polymer systems, including homopolymers, various kinds of copolymers, blends, and composites were successfully electrospun.[16, 19-21] Besides lots of efforts for developing these new nanofibers, tailoring the surface chemistry and function of electrospun nanofibers is also of utmost importance for the application of these materials: for example, modification of surface properties such as chemical composition, functionality, and charge as well as biocompatibility is essential for the utilization of the resulting nanofibers in specific applications. Recent studies by several groups have demonstrated the potential use of metal nanofibers as active components in fabricating nanoscale devices. For instance, Yang et al.[22] reported the electrospinning of ultrafine poly(acrylonitrile) (PAN) fibers containing Ag nanoparticles. Then many other metal nanoparticles such as gold, titanium, copper have been successfully dispersed on into a variety of polymer nanofibers via electrospinning. Kumar and co-workers[23] reported the fabrication of novel metal oxide-coated polymeric nanofibers using the electrospinning and solution dipping technique. Greiner and co-workers[24] reported that the coating of the template and the formation of the walls of the tubes could be achieved by chemical vapor deposition (CVD) with poly (p-xylylene) (PPX) via a template route. Parsons and co-workers[25] recently employed atomic layer deposition (ALD) on electrospun polymer fibers as a direct means by which to construct inorganic microtubes with well-defined nanoscale walls composed of Al_2O_3 after the removal of the templating polymer. In addition, noble metal nanofibers (such as Ag and Au) exhibit strong absorption in the visible and near infrared regions, an optical feature substantially different from those of planar metal surface or bulk materials. However, most of the methods are very time-consuming. To overcome these disadvantages, it is necessary to find a reliable and straightforward method to produce a variety of the nanostructured nanofibers with controlled morphologies and microstructures. The conductive surface is often required for applications such as anti-static, conductive shield, packing, and protective materials. The metallic nanostructured nanofibers have a great potential for a wide range of applications in many industries. In this chapter, we introduce a straight forward combined technology of electrostatic spinning and metallization for producing the conductive nanostructured hybrid nanofiber webs. Such metalizing technology can be widely used to deposit very thin films on various substrates for commercial and scientific purposes. The morphologies, microstructures and mechanical properties of the resulting metallized nanofibers were characterized by field emission scanning electron microscopy (FE-SEM), X-ray diffraction (XRD) and tensile tester, respectively. The ability to deposit well-controlled coatings onto the nanofibers would expand the application of nanofibers, for instance electromagnetic interference shielding effectiveness (EMI SE), based on changes to both the physical and chemical properties of the nanofibers.

2. Experimental procedure

2.1 Materials

The polyurethane (PU, ca. 14 wt%) stock solution, which is currently used for mass production electrospinning, was kindly provided by TECHNOS Co., Ltd., Japan and used for electrospinning.[26,27] Poly(vinyl alcohol) (PVA) (degree of hydrolysis = 88%, degree of polymerization (DP) ≈ 1,700) was obtained from Kuraray Co. Ltd., Japan. High purity aluminum (99.5%, grains 3-5 mm), tin (99.99%, grains 2-3 mm), copper (99.99%, wire-cut) and nickel (99.9% up, grains 2-5 mm) were purchased from Kojundo Chemical Laboratory Co., Ltd., Japan. All chemicals were of analytical grade and used without further purification.

2.2 Electrospinning

In order to produce electrospun nanofibers,[28-30] a high-voltage power supply (Har-100*12, Matsusada, Co., Japan) was used as the source of the electric field. The PVA was dissolved in distilled water, and the concentration of PVA solution was 12 wt%. Each polymer (PU and PVA) solution was supplied through a plastic syringe attached to a capillary tip with an inner diameter of 0.6 mm. The copper wire connected to a positive electrode (anode) was inserted into the polymer solution, and a negative electrode (cathode) was attached to a metallic collector (Fig. 1a). The voltage was controlled at the range of 8-12 kV. The distance between the capillary tip and the collector was fixed to be 15 cm. All solutions were electrospun onto a rotating metallic collector at room temperature under identical conditions.

2.3 Metallization

A system UEP-6000 (ULVAC, Inc., Japan) was used to deposit a metallic layer onto the organic nanofibers. A high-purity target metal was mounted on the vapor plate (Fig. 1b), and the nanofiber webs were fixed on the rotating holder with a side facing the target. The vacuum pressure was set at 2.0×10^{-3} Pa, the power was set at 10 kV, 270 mA. The deposition rate of the metallic layer was controlled to be about 1.0 Å/s (copper, thin deposition) and 0.2 Å/s (nickel deposition), respectively. The distance between the target and substrates was 400 mm. The thicknesses of metal layers were scanned by a coated thickness measurement with X-ray spectrometers (Fischerscope X-RAY XDL-B, Fischer Technology Inc., Germany).

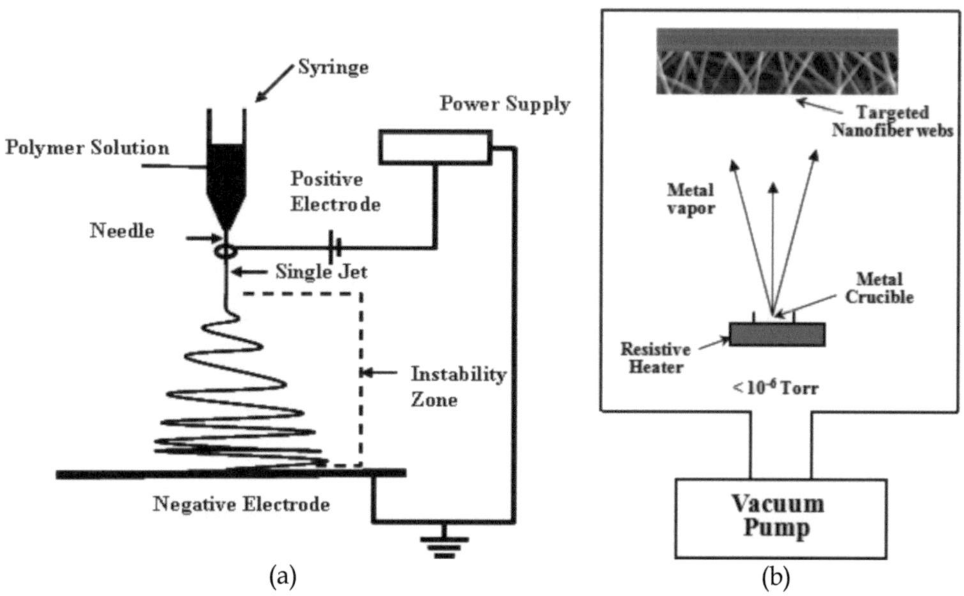

Fig. 1. Schematic of (a) electrospinning apparatus and (b) metallization.

2.4 Annealing process

After the Cu/Ni metal depositions, in order to produce the alloy nanofibers, the obtained Cu/Ni metal-deposited nanofibers were placed into the alundum crucible, and annealed in the electric furnace under N_2. Annealing process was conducted at various annealing temperatures and times, which were varied at the ranges of 250 ~ 600 °C and 6 ~ 24 hr, respectively. The heating rate was 5.0 °C/min.

2.5 Characterization

The morphologies of metallized nanofibers were characterized by field emission scanning electron microscope (FE-SEM) (Hitachi model S-5000). Wide angle X-ray diffraction (WAXD) experiments were carried out with a fixed anode X-ray generator operating at 40 kV and 150 mA (Rotaflex RTP300, Rigaku Co., Japan). The mechanical behaviors of metallized hybrid nanofiber webs were determined by a universal testing machine (TENSILON RTC1250A, A&D Company Ltd, Japan) under a crosshead speed 10mm/min at room temperature. In accordance with ASTM D-638, samples were prepared in the form of a dumbbell-shape and, then, at least five specimens were tested for tensile behavior and the averaged values were reported. For determining the tensile strength of metal deposited single nanofibers, we specially developed the test machine (FITRON NFR-1000, RHESCA Co., Japan). Three parameters were determined from each stress-strain curve: Young' modulus, tensile strength, and elongation at break. Elastic modulus or Young's modulus is the initial slope of the stress-strain curve. Tensile strength is the stress at failure and the strain corresponding to the tensile strength is the failure strain. The electrical volume resistivity ρ of the metallized hybrid nanofiber webs was measured using a standard four-point probe method at ambient conditions. The electrical volume resistivity ρ was calculated by Ohm's law, $\rho = (V/I)(wt/l)$, where V is voltage, I is current, l is distance between inside electrodes, w,t is width and thickness of the specimen, respectively. The shielding effect (*SE*) of the metalized nanofiber webs was analyzed using near-field antenna measurement systems, which consisted of a vector network analyzer (37169A Anritsu Co., Ltd.) and a near-field antenna measurement instrument (Tokai-techno Co., Ltd.) with a transmitting horn antenna and a receiving waveguide probe. The horn antenna and waveguide probe will be changed according to the frequency range to be measured.

3. Morphologies and microstructures of metallized nanofibers

3.1 Morphologies of metallized PU nanofibers

Fig. 2 shows typical FE-SEM micrographs of the pure PU and the metallized hybrid PU nanofiber webs with different thicknesses of the copper layer. As expected, electrospun nanofibers are deposited as a randomly oriented nanofiber web, forming a highly porous structure, which is held together by connecting sites such as crossing and bonding between the fibers. [31-33] The image at higher magnification (inset in Fig. 2a) shows that the surface morphology of pure PU nanofiber is smooth and uniform, with an average diameter of ca. 500 nm. On the other hand, after metallizing of pure PU nanofiber webs, the surface of metalized hybrid PU nanofibers becomes slightly coarse and looks rough (inset in Fig. 2b). When the thickness of the copper layer increases to 50 nm, it can be clearly seen that the tiny copper nanoparticles are formed through the surface of the nanofibers (inset in Fig. 2c); such nanoparticle becomes evident gradually with increase in the thickness of the copper layer (inset in Fig. 2d). The results suggest that the metallic coppers are well deposited on the surface of nanofibers. In addition, the fibrous morphologies were satisfactorily conserved even after removal of the nanofiber template (by heat treatment at 400 °C for 24 hr. The complete removal of the organic components was confirmed by FT-IR analysis) in the metallized hybrid PU nanofiber webs, indicating the successful deposition of metals on the surface of the nanofiber template. Furthermore, in the case of FE-SEM images in Fig. 2d (inset), nearly spherical copper nanoparticles were observed, with an average diameter of ca. 23 nm. This result is also well coincident with the results of WXAD, which will be explained below. The shape of copper nanoparticles is irregular and non-spherical, which may be attributed to the massive copper migration and aggregation. The migration and aggregation of copper nanoparticles are probably driven mostly by the instability of copper atoms due to their high surface free energy. [34] Therefore, the aggregation would produce thermodynamically stable particles with bigger sizes.

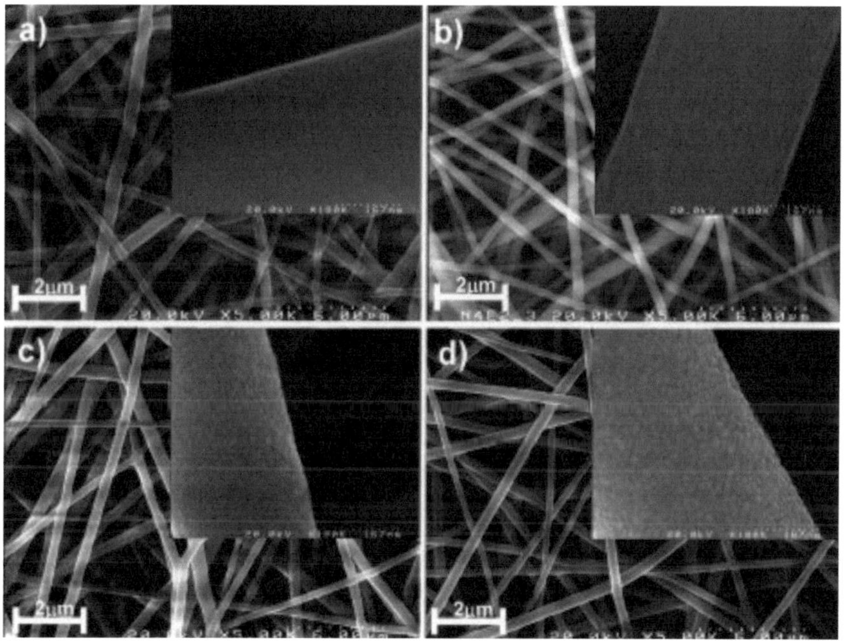

Fig. 2. FE-SEM photographs of pure PU (a) and metallized hybrid PU nanofiber webs with different copper layers of 10 nm (b), 50 nm (c), and 100 nm (d). Insets show FE-SEM images of each nanofibrous webs at higher magnification.

Fig. 3 shows the WAXD patterns of pure PU and metallized PU nanofibers. As seen in Fig. 3a, it was observed that the PU nanofibers have an amorphous structure. Moreover, the metallized PU nanofibers with the copper layer of 10 nm are also amorphous (Fig. 3b), suggesting that there is no formation of clear copper nanocrystals on the PU nanofibers, while the metallized PU nanofibers with the copper layer of 50 and 100 nm shows clearly typical crystalline peaks (Figs. 3c and 3d). The three peaks are observed at 2θ=43.5°, 50.0°and 73.5°, corresponding to the XRD peaks of crystalline copper and show that the copper crystals are in the form of (111), (200), and (220) reflections, [35] respectively. Furthermore, the full width at half-maximum (FWHM) of the strongest characteristic peak

(111) is used to estimate the average crystallite size by applying the following Debye - Scherrer equation, $D=\kappa\lambda/\beta\cos\theta$, where X-ray wavelength λ is 1.5402, κ is the shape factor which is often assigned a value of 0.89 if the shape is unknown, D is the average diameter of the crystals in angstroms, θ is the Bragg angle in degrees, and β is the full width at half-maximum of the strongest characteristic peak in radians. It was found that the averaged sizes of copper nanoparticles deposited on the PU nanofibers are ca. 12 and 23 nm for the metallized PU nanofibers with the copper layers of 50 and 100 nm, respectively. These results also agree well with FE-SEM data (FE-SEM images in Figs. 2c and 2d), indicating that spherical copper nanoparticles are observed.

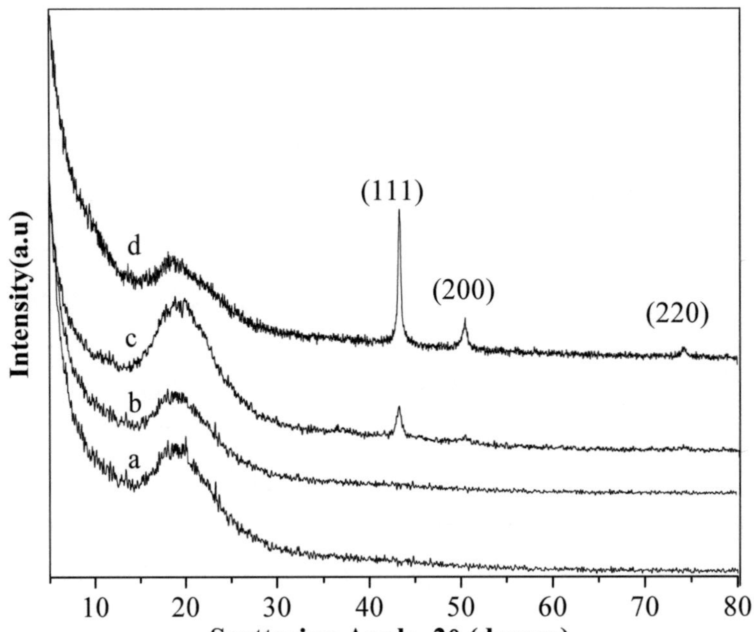

Fig. 3. X-ray diffraction patterns of pure PU nanofiber (a) and metalized hybrid PU nanofibers with different copper layers of 10 nm (b), 50 nm (c), and 100 nm (d).

3.2 Morphologies of metallic nanotubes

Fig. 4 shows typical FE-SEM micrographs of pure metallic nanotubes and nanofibers (a: 200 nm Cu-coated onto the nanofiber with the diameter of 200 - 250 nm, b: 200 nm Cu-coated onto the nanofiber with the diameter of 150 - 200 nm, c: 50 nm Cu-coated onto both side of the nanofiber with the diameter of 200 - 250 nm) after calcination at 400 °C for 24h, respectively. As seen in Fig. 4, pure metallic nanotubes exhibited a rough surface morphology compared to the metallized PVA nanofibers before calcination. The diameter of nanopores were found to be about 100 - 150 nm. On the other hand, the long metallic nanotubes were not obtained presumably due to incomplete coating of the metal layers onto the nanofiber template, but rather broken metallic nanotubes were observed (Fig. 4a). Therefore, in order to achieve complete metallic coverage of the nanofibers, the metallic coating onto both sides of the nanofibers were performed, and followed the removal of the nanofiber template to produce the long metallic nanotubes. The result is shown in Fig. 4c.

Clearly, it was observed that the long metallic nanotubes were successfully achieved without any breakages in pure metallic nanotubes. The formation of metallic nanotubes was also confirmed by transmission electron microscopy (TEM) analysis (data not shown). In addition, in order to investigate the diameter effect of the nanofibers on metallization, the electrospun PVA nanofibers with the smaller diameter ranging from 150 to 200 nm (PVA polymer solution: 8 - 10 wt%) were used and explored. Fig. 4b shows typical SEM images of pure metallic nanofibers. Interestingly, unusual pearl necklace-like morphology was observed, which is not clear yet and at present. As a result, we could successfully prepare the pure metallic nanotubes or nanofibers depending on the diameter of the nanofibers and the thickness of deposited metal layer by using the combined technology of electrospinning and metallization.

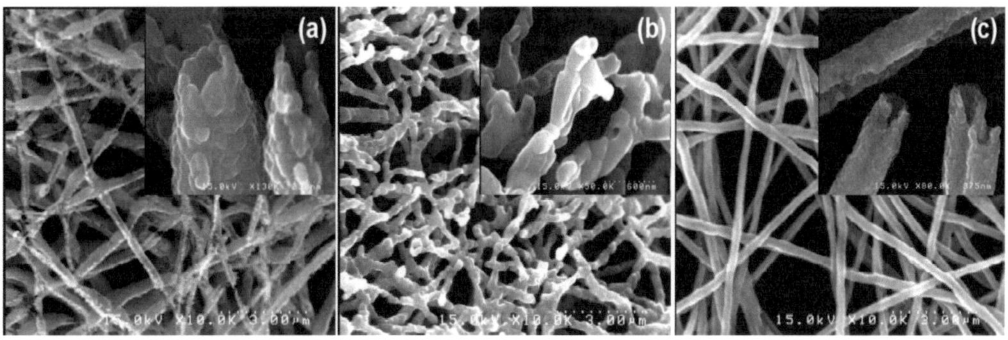

Fig. 4. FE-SEM micrographs of pure metallic nanotubes and nanofibers after calcination at 400 °C for 24h, respectively (a: 200 nm Cu-coated onto the nanofiber with the diameter of 200 - 250 nm, b: 200 nm Cu-coated onto the nanofiber with the diameter of 150 - 200 nm, c: 50 nm Cu-coated onto both side of the nanofiber with the diameter of 200 - 250 nm).

3.3 Metal alloy nanofibers

Metal alloy has been attracted considerable attention to metallurgists, materials engineers, and materials scientists in four major areas: (i) development of new alloys for specific applications; (ii) fabrication of these alloys into useful configurations; (iii) design and control of heat treatment procedures for specific alloys that will produce the required mechanical, physical, and chemical properties; and (iv) solving problems that arise with specific alloys in their performance in commercial applications, thus improving product predictability. In general, the use of phase diagrams is useful and allows research, development, and production to be done more efficiently and cost effectively. In the area of alloy development, phase diagrams have proved invaluable for tailoring existing alloys to avoid overdesign in current applications, designing improved alloys for existing and new applications, designing special alloys for special applications, and developing alternative alloys or alloys with substitute alloying elements to replace those containing scarce, expensive, hazardous, or "critical" alloying elements.[36] For instance, copper and nickel are completely soluble in all proportions in the solid state. Here, we attempt to study the annealing conditions for achieving the Cu/Ni metallized nanofibers from metal-deposited electrospun nanofibers prepared by electrospinning and metallization. Various annealing conditions, such as annealing temperature and time, and composition ratio of two metals (Cu and Ni) are investigated in order to find out optimum annealing process for the formation of alloy nanofibers. We chose two different conditions with composition ratio of Cu/Ni 90:10 and 40:60 in deposited thickness, and investigated the morphologies and crystal structures of metal-deposited nanofibers before and after annealing. The annealing conditions, such as annealing temperatures and times were varied to find out optimum annealing process for the formation of alloy nanofibers.

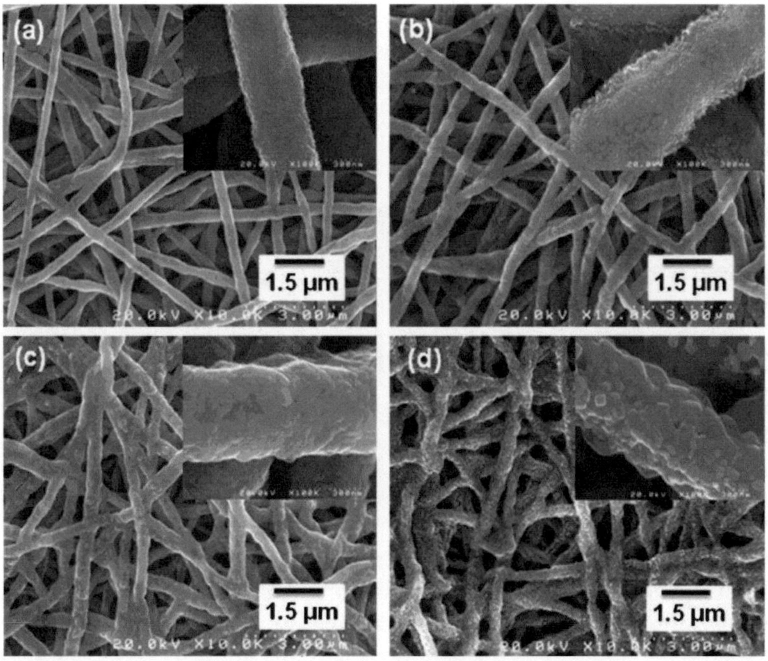

Fig. 5. FE-SEM images of Cu/Ni metallized (9:1, in deposition thickness) nanofibers before annealing (a) and after annealing at 250 °C (b), 400 °C (c), and 600 °C (d) for 12 h.

Fig. 5 shows FE-SEM images of Cu/Ni metallized (9:1, in deposition thickness) nanofibers before annealing (a) and after annealing at 250 °C (b), 400 °C (c), and 600 °C (d) for 12 h. It was found that the fibrous morphologies were satisfactorily conserved even after removal of the nanofiber template through an annealing process,[37] suggesting the successful deposition of metallic layers onto the surface of nanofiber template. As seen in Fig. 5, before annealing the surface of metal-deposited nanofibers exhibited a rough surface morphology due to the formation of the metallic granular nanoparticles (ca. 23 nm), whereas the surface roughness of the pure nanofibers was smooth (not shown). Moreover, the surface roughness of the metallized nanofibers annealed above 400 °C showed larger and irregular-shaped metal particles during annealing at higher temperature. On the other hand, the metallized nanofibers annealed at 250 °C exhibited the similar morphologies to the metal-deposited nanofibers before annealing, but irregular diameter with a node-structure, probably due to the partly flowing and deterioration of the viscoelastic polymer nanofiber around 250 °C.

Fig. 6 shows FE-SEM images of Cu/Ni metallized (4:6, in deposition thickness) nanofibers before annealing (a) and after annealing at 250 °C (b), 400 °C (c), and 600 °C (d) for 12 h. The similar annealing behavior was also observed at the composition ratio of Cu/Ni.40:60. That is, the surface roughness of Cu/Ni metallized nanofibers annealed at various temperatures for 12 h became coarse and rough as increasing the aging temperature.

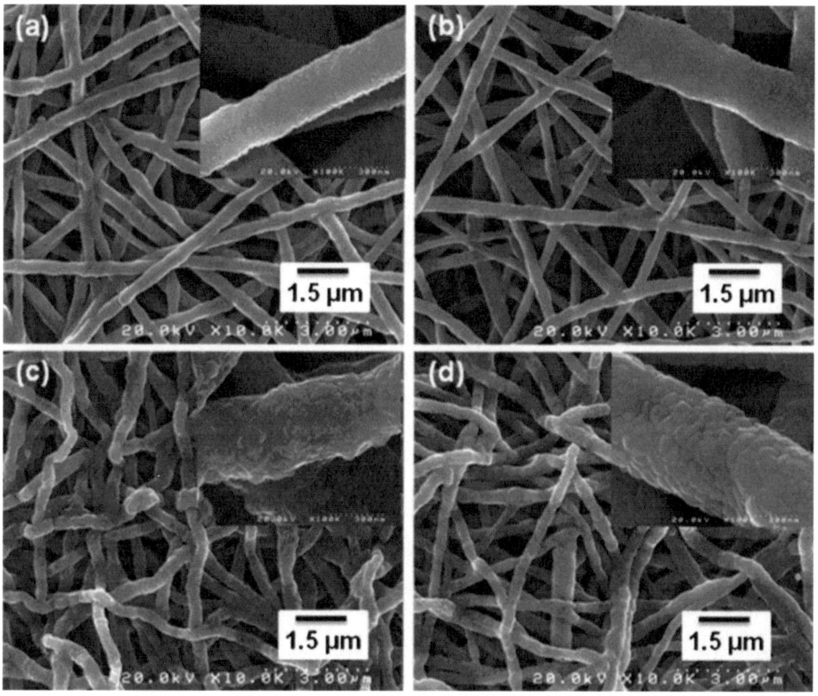

Fig. 6. FE-SEM images of Cu/Ni metallized (4:6, in deposition thickness) nanofibers before annealing (a) and after annealing at 250 °C (b), 400 °C (c), and 600 °C (d) for 12 h.

Fig. 7 shows FE-SEM images of Cu/Ni metallized (9:1, in deposition thickness) nanofibers annealed at 400 °C for (a, b) 6 h, (c, d) 12 h, and (e, f) 24 h, respectively. As seen in FE-SEM images at higher magnification (Figs. 7b, 7d and 7f), the irregular-shaped and different sized metal nanoparticles were observed. Moreover, it seemed that the irregular-shaped nanoparticles were converged gradually and its size became larger as increasing the annealing time, which may be attributed to the massive copper migration and aggregation during annealing at higher temperature. The migration and aggregation of metal nanoparticles are probably driven mostly by the instability of metal atoms due to their high surface free energy, and therefore would produce thermodynamically stable particles with bigger sizes.[20] As a result, these micrographs showed that sintering was occurred as increasing the annealing time at 400 °C, which was demonstrated by the variation in crystallite size during annealing, as confirmed by FE-SEM analysis. The result is well in accordance with the previously reported paper.[37]

3.4 Microstructures of Cu/Ni metallized nanofibers

Fig. 8 shows WAXD patterns of Cu/Ni metallized (9:1, in deposited thickness) nanofibers before annealing (a) and after annealing at 250 °C (b), 400 °C (c), and 600 °C (d) for 12 h. It was observed that the PU nanofibers had an amorphous structure, whereas the Cu/Ni deposited nanofibers before annealing clearly showed typical crystalline peaks of each pure metal, as seen in Fig. 8a. The typical crystalline peaks of both Cu and Ni metals were observed at 2θ = 43.6, 50.1, and 73.7, corresponding to the (111), (200), and (220) reflections, for pure Cu crystals[38] and at 2θ = 44.5, 51.9, and 76.6, corresponding to the (111), (200),

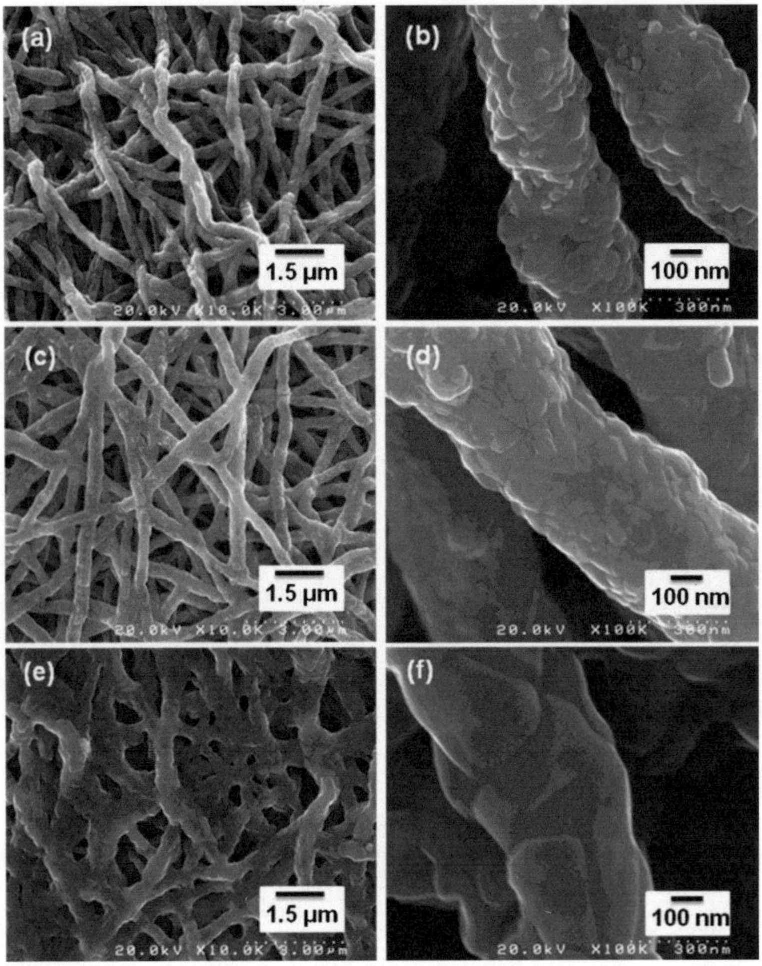

Fig. 7. FE-SEM images of Cu/Ni metallized (9:1, in deposition thickness) nanofibers annealed at 400 °C for (a, b) 6 h, (c, d) 12 h, and (e, f) 24 h.

and (220) reflections, for pure Ni crystals,[39] respectively. On the other hand, after annealing of Cu/Ni deposited nanofibers at higher temperature, such as 400 and 600 °C (Figs. 8c and 8d), new crystalline peaks were observed at 2θ = 32.5, 35.5, 38.7, 48.7, 53.4, and 58.5, corresponding to the (110), (002), (111), (202), (020), and (202) reflections, for CuO cubic crystals,[40] and 2θ = 37.3, 43.4, 63.0, 75.5, and 79.6, corresponding to the (111), (200), (220), (311), and (222) reflections, for NiO cubic crystals, respectively.[41, 42] The results suggested the transformation of pure metals to metallic oxides (CuO, NiO) during annealing processing. Specifically, in case of annealing temperature at 250 °C, different crystalline peaks were observed presumably due to the formation of intermediate metal oxide crystals (Cu_2O), which is however not clear at present. Fig. 9 shows WAXD patterns of Cu/Ni metallized (4:6, in deposited thickness) nanofibers before annealing (a) and after annealing at 250 °C (b), 400 °C (c), and 600 °C (d) for 12 h. The similar microstructures corresponding to metallic oxides (CuO, NiO) were also observed at the composition ratio of Cu/Ni=4:6 in deposited thickness.

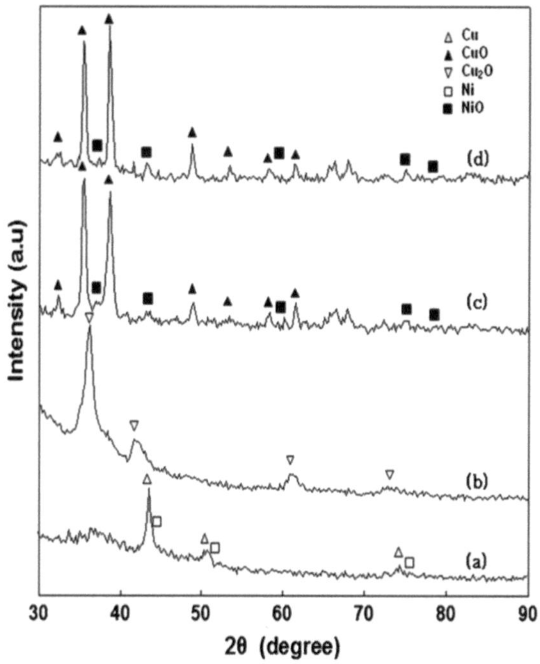

Fig. 8. WAXD patterns of Cu/Ni metallized (9:1, in deposited thickness) nanofibers before annealing and after annealing at 250 °C (b), 400 °C (c), and 600 °C (d) for 12 h.

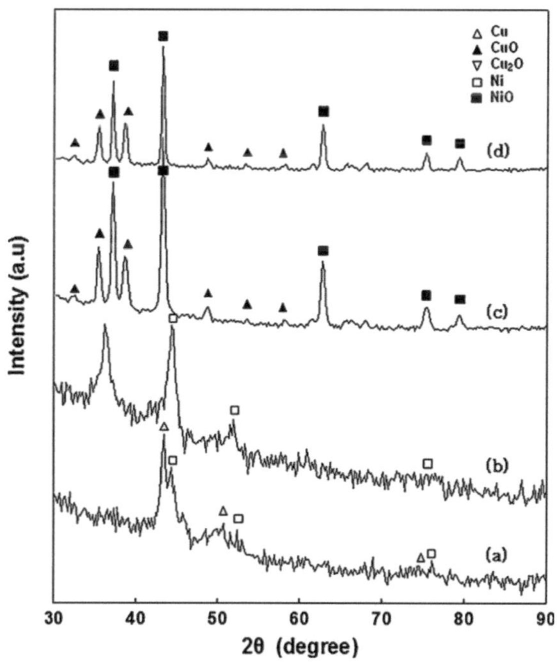

Fig. 9. WAXD patterns of Cu/Ni metallized (4:6, in deposited thickness) nanofibers before annealing (a) and after annealing at 250 °C (b), 400 °C (c), and 600 °C (d) for 12 h.

4. Mechanical properties of metallized nanofibers

4.1 Mechanical properties of metallized PU nanofiber webs

Stress-strain curves of pure PU nanofiber and metallized hybrid PU nanofiber webs are shown in Fig. 10. For all curves, stress-strain behavior shows a typical characteristic of thermoplastic elastomers, in that the stress-strain behaviors show a linear elasticity as their intrinsic materials' properties. Compared with pure PU nanofiber webs (Young's modulus 3.59 MPa), Young's modulus of metallized nanofiber webs was increased to 3.74 and 5.97 MPa for the metallized hybrid PU nanofiber webs with different copper layers of 10 and 100 nm, attributed to the metal hard-coating layers deposited on the nanofibers, which makes it a solid metal fiber resulting in an enhanced stiffness as well as conductivity. This result is crucial from an industrial point of view. Accordingly, the elongation of metallized nanofibers' webs decreases as the thickness of copper layer increases, suggesting that the elasticity of metallized hybrid nanofibers was reinforced by metallization: breaking elongation of metallized hybrid PU nanofiber webs decreased from 286% for pure PU nanofiber webs up to 131% for metallized hybrid PU nanofiber webs with the copper layer of 100 nm. On the other hand, the tensile strength of metallized nanofiber webs slightly decreased to 7.75 MPa for metallized hybrid PU nanofiber webs with the copper layer of 100 nm, compared with pure PU nanofiber webs (9.11 MPa).

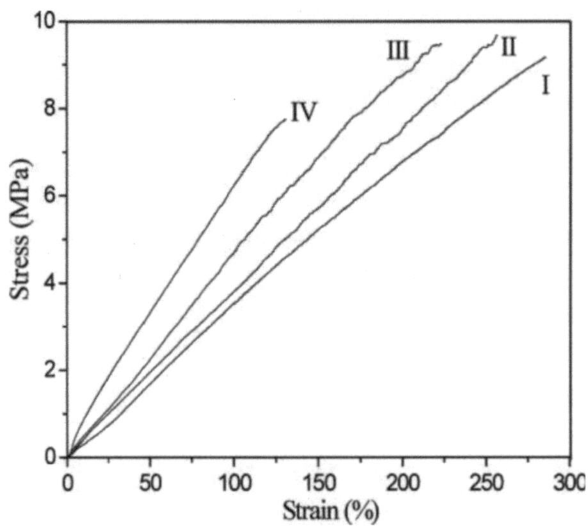

Fig. 10. Stress-strain curves of pure PU nanofiber web (I) and metalized PU nanofiber webs with different copper layers of 10 nm (II), 50 nm (III), and 100 nm (IV) at room temperature.

The detailed data are shown in Table 1. It can be, therefore, considered that the strength and ductility of electrospun nanofiber webs after metallization are greatly modified. The toughness, defined as the energy absorbed by the electrospun nanofiber webs until breaking, for metalized hybrid PU nanofiber webs with the copper layer of 100nm has the lower value (~0.010 J) than that of pure PU nanofiber webs (~0.026 J), suggesting that the higher amounts of the deposited copper layer onto PU nanofiber webs resulted in higher stiffness of metallized PU nanofiber webs. Although the decreased tensile strength behaviors of metallized hybrid nanofiber webs specifically with the copper layer of 100 nm are not fully understood, it could be considered that non-uniformly deposited thicker copper layer may result in an easier deformation via small cracks or other imperfects in the metallized hybrid PU nanofibers, and thereby give rise to a poor tensile strength.

	Tensile modulus (MPa)	Tensile strength (MPa)	Elongation at break (%)
Pure PU	3.59	9.11	286
10 nm	3.74	9.64	256
50 nm	4.35	9.49	226
100 nm	5.97	7.75	131

Table 1. Mechanical properties of pure and Cu-deposited PU nanofibers with different thickness of the metal layers webs.

4.2 Mechanical properties of single metallized nanofibers

The mechanical properties of single metallized nanofibers were investigated by using recently developed tensile test machine [37, 43-44]. Evaluating the mechanical properties for the single nanofiber had remained lots of problems to be figured out. Limited reports have been made so far with special methods. The development of a novel method is required to study the mechanical properties of single nanofibers. For determining the tensile properties of single nanofibers, we specially developed the test machine (FITRON NFR-1000, RHESCA Co., Japan, [43], Fig. 11). It was demonstrated that the test machine is well operated for measurement of mechanical properties of nano/micro-sized fibers. Sample frame for collecting a single nanofiber is shown in Fig. 12. Oriented nanofibers between two parallel metal templates were collected while several single fibers were also collected on paper cutting sheet (as show in Fig. 12a), the nanofibers which cross over each other were eliminated under microscope (Fig. 12b). The cutting sheet with chosen single nanofiber was fixed on the holder (Fig. 12c). After a suitable fiber is selected and pasted with adhesive on a sample holder, the fiber axis is precisely set to align the stress axis of the holder using an optical microscopy. The main specifications of the developed system are as follows: the maximum loading capacity; 500 mN, the stroke; 20 mm, the loading speed; 5 - 20 μm/s, the displacement sensitivity; 1.0 μm, loading sensitivity; 1.0 μN, etc. Three parameters were determined from each stress-strain curve: Young's modulus, tensile strength, and elongation at break.

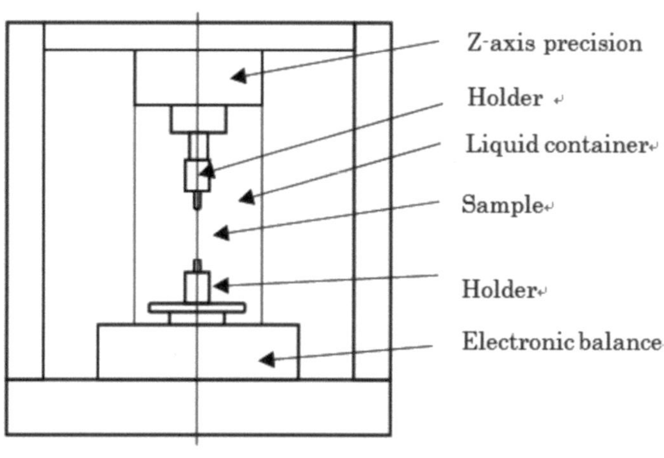

Fig. 11. Specially developed tensile test machine.

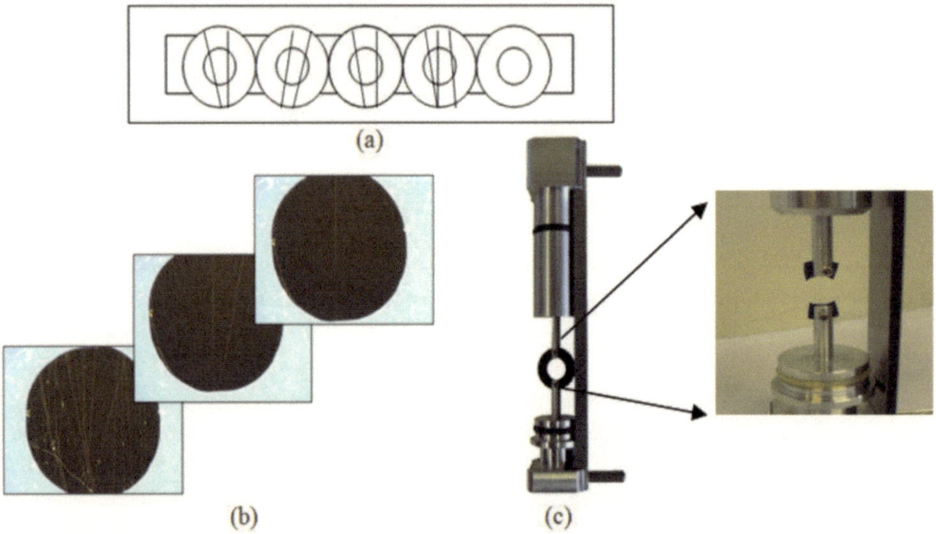

Fig. 12. Sample preparation, (a) Sample frame for collecting a single nanofibers, (b) single nanofibers preparation, (c) cutting sheet holder.

The tensile strength of pure PU, 30 and 50 nm Cu-deposited single PU nanofibers were also carried out at room temperature and the result was shown in Fig. 13. As seen in Fig. 13, the tensile strength of the Cu-deposited PU single nanofibers were increased with increasing the thickness of deposited copper layer. Compared to pure PU nanofibers, Young's modulus for the Cu-deposited PU nanofibers with the copper layers of 30 and 50 nm were increased to 610 and 750 MPa, respectively, due to the formation of metallic hard-coating layers onto the surface of PU nanofibers.[37] The detailed data were summarized in Table 2.

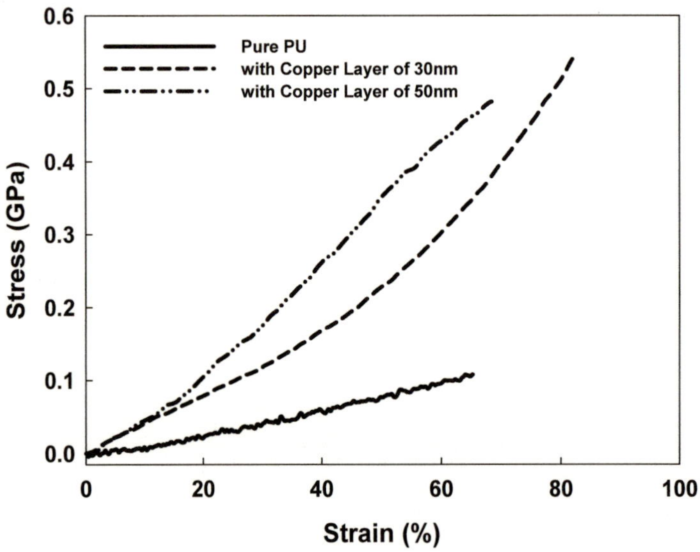

Fig. 13. Stress–strain curves of pure PU (solid line), 30 nm Cu-deposited (dotted line), and 50 nm deposited (dash-dot-dot line) single nanofibers (load: 10 mN, loading speed: 10 mm s^{-1}).

	Tensile modulus (MPa)	Tensile strength (MPa)	Elongation at break (%)
Pure PU	0.17	0.11	65.2
30 nm	0.61	0.54	82.0
50 nm	0.75	0.48	68.4

Table 2. Mechanical properties of pure PU single nanofiber and Cu-deposited PU single nanofiber with different thickness of the metal layers.

Here, we also report the tensile strength of various metals (for instance, Cu, Ni, Sn, and Al)-deposited PU nanofibers and various polymeric nanofibers measured by this recently developed tensile test machine. As previous said, we found that the tensile strength of metal-deposited PU single nanofibers was increased with increasing the thickness of deposited Cu layer.[37] Fig. 14 shows typical stress–strain curves of Cu-deposited (dash-dot line), Ni-deposited (medium dash line), Sn-deposited (solid line), Al-deposited (dotted line), and pure (dash-dot-dot line) single PU nanofibers (load: 10 mN, loading speed: 15 μm s^{-1}). In this work, the thickness of each metal layer was controlled and fixed to be about 50 nm. As seen in Fig. 14, compared with pure PU nanofibers (Young's modulus 170 MPa), mechanical properties of metal-deposited PU nanofibers were significantly increased and affected by the types of metals used for metallization. For instance, Young's modulus (8.81 GPa) of the Ni-deposited PU nanofibers was higher than those (Al-deposited, 1.0 GPa; Sn-deposited, 0.5 GPa; Cu-deposited, 1.9 GPa) of Al-, Sn-, and Cu-deposited PU nanofibers. On the other hand, among the metal-deposited PU nanofibers, Cu-deposited PU nanofibers showed the highest tensile strength and higher elongation at break, which resembles strong and tough thermoplastic properties. Although this is not fully understood, this can be considered due to the formation of polymer–metal nanocomposite layer in the interface areas of organic nanofibers and metallic layers.

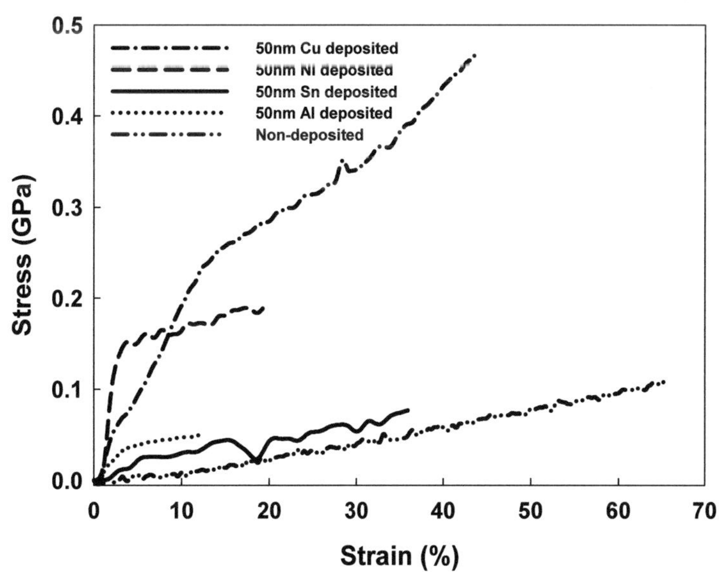

Fig. 14. Stress–strain curves of Cu-deposited (dash-dot line), Ni-deposited (medium dash line), Sn-deposited (solid line), Al-deposited (dotted line), and pure (dash-dot-dot line) single PU nanofibers (load: 10 mN, loading speed: 15 mm s^{-1}).

4.3 Electromagnetic interference (EMI) shield effects of metallized nanofiber webs

Shielding effectiveness (SE) is the ratio or impinging energy to the residual energy. When an electromagnetic wave pass through a shielding material, absorption and reflection takes place. Residual energy is part of the remaining energy that is neither absorbed nor reflected by the shielding material but it is emerged out from the shielding material. All electromagnetic waves consist of two essential components, a magnetic field (*H*) and electric field (*E*). These two fields are perpendicular to each other and the direction of wave propagation is at right angles to the plane containing the two components. The shielding should be in high conductance, thus metals, such as steel, copper, aluminum, etc., are the most common materials used for EMI shielding. However, metal shields have the inconvenience of poor mechanical flexibility, exceedingly high weight, propensity to corrosion, and limited tuning of the SE, while polymers offer lightness, low cost, easy shaping, etc [45]. Thus, polymer composites with discontinuous conducting fillers, such as metal particles, carbon particles, carbon fiber, are extensively employed in EMI shielding [46,47]. Among them, it is expected that the fiber with the high aspect ratio enables the formation of a conductive pathway through the resin matrix at low concentration. Fig. 15 shows the EMI SE (a) and the volume resistivity (b) of metal (Cu)-deposited nanofiber webs as a function of the thickness of metal layer. As can be seen in Fig. 15a, the EMI SE of metal-deposited nanofibers increased with increase in the thickness of deposited metal layer. The increase in EMI SE can be ascribed to the decrease in the electrical resistivity of metal-deposited nanofibers.[48]

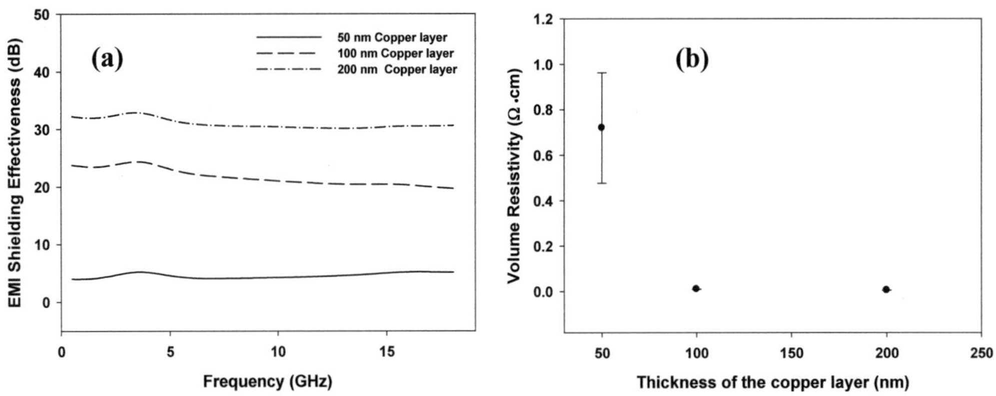

Fig. 15. EMI SE (a) and volume resistivity (b) of copper-deposited nanofiber webs as a function of the thickness of deposited Cu layer.

5. Summary

The design and development of electrospun fibers with sub-micrometer diameters from various kinds of materials has gained significant attention due to their large potential for manifold applications in electronics, optics, medicine, sensors, separation, storage, and so forth [49, 50]. The polymer fibers with sub-micrometer diameters down to a few nanometers can be prepared by electrospinning, which is a quick and facile technique to create high-surface-area fiber membranes where the fibers are several orders of magnitude smaller than those produced by conventional spinning techniques [51]. In this chapter, we have shown that the metallized nanofibers could be successfully prepared by using a combined technology of electrostatic spinning and metallization. The electrospun nanofibers were metallized with different thicknesses of metal layer via a metallization. The tensile strength of the metal-deposited single nanofibers was investigated by recently developed tensile test machine. As a result, it was found that the tensile strength of 50 nm metal-deposited single nanofibers was dramatically improved, and was higher than that of pure polymer single nanofibers, which is attributed to the formation of metallic hard-coating layers onto the surface of single nanofibers. Moreover, the tensile strength of the metal-deposited single nanofibers was also depended on the types of metals (Cu, Ni, Sn, and Al) used for metallization. We have furthermore investigated the annealing behavior of Cu/Ni deposited nanofibers to find out optimum conditions enabling to prepare the metal alloy nanofibers. Certainly, based on the combined technology of electrospinning and metallization, the broad organic and metallic components should be adjustable, which will accelerate new applications, such as EMI SE, of these interesting conductive nanofiber webs. Furthermore, the procedures and costs could be reduced, which is, therefore, a promising straightforward technique for a large-scale production of novel metallized hybrid nanofiber webs.

6. References

[1] A. C. Patel, S. Li, C. Wang, W. Zhang, Y. Wei, *Chem. Mater.* 19, 1231 (2007).
[2] M. K. Shin, S. S. Kim, S. J. Kim, S. K. Kim, H. Kee, *Appl. Phys. Lett.* 88, 193901 (2006).
[3] D. Christopher, L. Xin, Z. David,W. Xianyan, F. B. Ferdinando,W. James, A. S. Lynne, K. Jayant, *Nano Lett.* 3, 143 (2003).
[4] Y. Ji, B. Li, S. Ge, J. C. Sokolov, M. H. Rafailovich, *Langmuir* 22, 1321 (2006).
[5] S. D. McCullen, D. R. Stevens, W. A. Roberts, S. S. Ojha, L. I. Clarke, R. E. Gorga, *Macromolecules* 40, 997 (2007).
[6] K. Wei, J. H. Xia, B. S. Kim, I. S. Kim, *J. Polym. Res.* DOI: 10.1007/s10965-010-9451-z.
[7] C. K. Kim, B. S. Kim, F. A. Sheikh, U. S. Lee, M. S. Khil, H. Y. Kim, *Macromolecules* 40, 4823 (2007).
[8] S. R. Davis, A. R. Brough, A. J. Atkinson, *J. Non-Cryst. Solids* 315, 197 (2003).
[9] D. K. Chattopadhyay, K. V. SN. Raju, *Prog. Polym. Sci.* 32, 352 (2007).
[10] [10] R. A. Caruso, J. H. Schattka, A. Greiner, *Adv. Mater.* 13, 1577 (2001).
[11] [11] H. Hou, J. J. Ge, J. Zeng, Q. Li, D. H. Reneker, A. Greiner, S. Z. D. Cheng, *Chem. Mater.* 17, 967 (2005).
[12] A. C. Patel, S. Li, J. M. Yuan, Y. Wei, *Nano Lett.* 6, 1042 (2006).
[13] Z. Sun, E. Zussman, A. L. Yarin, J. H. Wendorff, A. Greiner, *Adv. Mater.* 15, 1929 (2003).
[14] A. V. Bazilevsky, A. L. Yarin, C. M. Megaridis, *Langmuir* 23, 2311 (2007).
[15] H. W. Jun, V. Yuwono, S. E. Paramonov, J. D. Hartgerink, *Adv. Mater.* 17, 2612 (2005).
[16] Z. M. Huang, Y. Z. Zhang, M. Kotaki, S. Ramakrishna, *Compos. Sci. Technol.* 63, 2223 (2003).
[17] D. Li, Y. Xia, *Adv. Mater.* 16, 1151 (2004).
[18] A. Formhals, US Patent 1975504, 1934.
[19] K. O. Kim, Y. A Seo, B. S. Kim, K. J. Yoon, M. S. Khil, H. Y. Kim, I. S. Kim, *Colloid & Polymer Science* 289, 863 (2011).
[20] N. Kimura, H. K. Kim, B. S. Kim, K. H. Lee, I. S. Kim, *Macromol. Mater. & Eng.* 295, 1090 (2010).

[21] H. Sato, K. O. Kim, H. K. Kim, B. S. Kim, Y. Enomoto, I. S. Kim, *Fibers and Polymers* 11, 1123 (2010).
[22] Q. B. Yang, D. M. Li, Y. L. Hong, Z. Y. Li, C. Wang, S. L. Qiu, Y. Wei, *Synth. Met.* 137, 973 (2003).
[23] M. F. Ottaviani, R. Valluzzi, L. Balogh, *Macromolecules* 35, 5105 - 5115 (2002).
[24] M. Bognitzki, H. Hou, M. Ishaque, T. Frese, M. Hellwig, C. Schwarte, A. Schaper, J. H. Wendorff, A. Greiner, *Adv. Mater.* 12, 637 (2000).
[25] Q. Peng, X. Y. Sun, J. C. Spagnola, G. K. Hyde, R. J. Spontak, G. N. Parsons, *Nano Lett.* 3, 719 (2003).
[26] K. Wei, T. Ohta, B. S. Kim, K. W. Kim, K. H. Lee, M. S. Khil, H. Y. Kim, I. S. Kim, *Polym. Adv. Technol.* 21, 746 (2010).
[27] H. R. Kim, T. Ito, B. S. Kim, Y. Watanabe, I. S. Kim, *Adv. Eng. Mater.* 13, 376 (2011).
[28] J. C. Park, T. Ito, K. O. Kim, K. W. Kim, B. S. Kim, M. S. Khil, H. Y. Kim, I. S. Kim, *Polym. J.* 42, 273 (2010).
[29] I. S. Kim, K. Wei, T. Ohta, B. S. Kim, Y. Watanabe, *Materials Science Forum 638-642*, 1719 (2010).
[30] O. Ohsawa, K. H. Lee, B. S. Kim, S. Lee, I. S. Kim, *Polymer* 51, 2007 (2010).
[31] K. Kim, M. Yu, X. Zong, J. Chiu, D. Fang, Y. S. Seo, B. S. Hsiao, B. Chu, M. Hadjiargyrou, *Biomaterials* 24, 4977 (2003).
[32] S. H. Tan, R. Inai, M. Kotaki, S. Ramakrishna, *Polymer* 46, 6128 (2005).
[33] J. H. Park, B. S. Kim, Y. C. Yoo, M. S. Khil, H. Y. Kim, *J. Appl. Polym. Sci.* 107, 2211 (2008).
[34] G. Carotenuto, *Appl. Organomet. Chem.* 15, 344 (2001).
[35] Z. Li, H. Huang, C. Wang, *Macromol. Rapid Commun.* 27, 152 (2001).
[36] ASM Handbook, ASM International, Metals Park, OH 1992, p. 5.
[37] H. R. Kim, N. Kimura, H. S. Bang, B. S. Kim, Y. Watanabe, I. S. Kim, *Mater. Sci. Forum* 654-656, 2463 (2010).
[38] Z. Li, H. Huang, C. Wang, *Macromol. Rapid Commun.* 27, 152 (2006).
[39] N. A. M. Barakat, B. S. Kim, H. Y. Kim, *J. Phys. Chem. C* 113, 531 (2009).
[40] S. Asbrink, A. Waskowska, *J. Phys. : Condens. Matter.* 3, 8173 (1991).
[41] I. Hotovy, J. Huran, L. Spiess, *J. Mater. Sci.* 39, 2609 (2004).
[42] H. Qiao, Z. Wei, H. Yang, L. Zhu, X. Yan, *J. Nanomater.* 2009, 1 (2009).
[43] I. S. Kim, Y. Enomoto, T. Takahashi, *Sen'i Gakkaishi* 65, 325 (2009).
[44] K. Wei, J. H. Xia, Z. J. Pan, G. Q. Chen, B. S. Kim, I. S. Kim, *Advanced Materials Research* 175-176, 294 (2011).
[45] J. M. Thomassin, C. Pagnoulle, L. Bednarz, I. Huynen, R. Jerome, C. Detrembleur, *J. Mater. Chem.* 18, 792 (2008).
[46] M. S. P. Shaffer, A. H. Windle, *Adv. Mater.* 11, 937 (1999).
[47] T. Makela, J. Sten, A. Hujanen, H. Isotalo, *Synth. Met.* 101, 707 (1999).
[48] M. H. Al-Saleh, U. Sundararaj, *Carbon* 47, 1738 (2009).
[49] M. Bognitzki, M. Becker, M. Graeser, W. Massa, J.H. Wendorff, A. Schaper, D. Weber, A. Beyer, A. Golzhauser, A. Greiner, *Adv. Mater.* 18, 2384 (2006).
[50] R. Dersch, M. Steinhart, U. Boudriot, A. Greiner, J.H. Wendorff, *Polym. Adv. Tech.* 15, 276 (2005).
[51] C. Drew, X. Liu, D. Ziegler, X. Wang, F.F. Bruno, J. Whitten, L.A. Samuelson, J. Kumar, *Nano Lett.* 3, 143 (2003).

7

Preparation and Characterization and Reducing Properties of MoO_3 Nano-Fibers

Zhao Peng
Applied Materials Research Lab of Chang'an University, China

1. Introduction

New methods of making new or existing inorganic materials are always interesting and challenging for materials scientists(Tenne 2006). Development of one-dimensional (1D) materials has become a focal area in nanostructured materials research, owing to their special characteristics which differ from those of respective bulk crystals(Romo-Herrera, 2007). These highly anisotropic 1D materials include elemental carbon, metals, semiconductor, alloys, sulfides, oxides, hydroxides, and so forth (Dmitruk, 2007). Among the important layered transition metal oxides and chalcogenides have been extensively investigated (Shabaev 2004).

In the course of the exploration of novel approaches for the preparation of metastable oxide materials, the main interest is focused on soft-chemical routes. Both, topochemical reactions involving ion exchange, intercalation, and pillaring as well as sol-gel, and hydrothermal reactions belong to these low-temperature methods. Especially host guest compounds realized by the intercalation of different guest species into layered inorganic frameworks represent a new and promising class of material that can be used for the controlled preparation of complex organized structures in the nanoscale regime(Markus 2001, Xiong 2003).

MoO_3 and its derivatives are widely used in industry as catalysts, display devices, sensors, smart windows, lubricants, battery electrodes(Zhao D.W.2008, Kim Youn-Su 2007, Ghorai T.K.2007, Dillon A.C.2008) . In particular, MoO_3 has been prepared into the forms of carbon-metal-oxide nanocomposites, nanotubes, and nanorods using carbon nanotubes (CNTs) as a host material or template. The MoO_3 nanofibers obtained with this novel method are up to 15μm long with their diameters ranging from 50 to 150 nm (Ajayan 1995, Satishkumar, 2000).

An efficient method to produce nanoscopic molybdenum oxide fibers was reported. The procedure is based on the intercalation of primary amines into the layered structure of molybdic acid and subsequent transformation of the lamellar molybdenum oxide–amine intermediate into the fibrous product. Standard synthesis: In a typical procedure, molybdic acid $MoO_3{\cdot}2H_2O$ (10 mmol) was mixed with the amine in 5 ml ethanol (molar ratio 2:1). After the addition of 15 ml distilled water, the yellow suspension was stirred at room temperature for 48 hours until a white precipitate was formed. The hydrothermal reaction of this composite was performed in an autoclave at 120°C for 3 to 5 days and, after filtering and washing with ethanol and diethyl ether, a white powder resulted. The molybdenum oxide-amine composite (2.5g) was stirred in 10 ml 33% nitric acid for 24 hours at room temperature and thereby the fibrous material was formed. The product was filtered, washed with ethanol and diethyl ether and then dried at 80°C under vacuum (10-3 mbar), (Markus,2001).

One-dimensional nanostructures of orthorhombic molybdenum trioxide (α-MoO_3) have been synthesized in the forms of ribbons or rods via acidification under hydrothermal conditions at 140-200°C. The reaction path has been revealed with kinetic investigations, which shows the following sequence: (i) from the starting compound $(NH_4)_6Mo_7O_{24}\cdot 4H_2O$ to (ii) formation of intermediate compound $((NH_4)_2O)_{0.0866}\cdot MoO_3\cdot 0.231H_2O$, and then to (iii) final α-MoO_3 nanoribbons or nanorods in 100% phase purity. The optimal growth temperature is in the range of 170-180°C under the current experimental settings. At higher reaction temperatures, this transformation can be accelerated, but with poorer crystal morphological homogeneity. It has been found that the dimensions of these rectangular nanorods are about 50 nm in thickness, 150-300 nm (mean value at 200 nm) in width, and a few tens of micrometers in length. The crystal morphology can be further altered with inorganic salts such as $NaNO_3$, KNO_3, $Mg(NO_3)_2$, and $Al(NO_3)_3$. Using an H_2S/H_2 stream, the above-prepared α-MoO_3 nanorods can be converted completely to 2H-MoS_2 nanorods at 600°C. The original rodlike morphology is well-retained, although the aspect ratio of the oxide template is reduced upon the sulfidation treatment,(Xiong Wen Lou 2002).

A new whisker, α-MoO_3, was fabricated via molybdenum thread oxidation at 973,1023, and 1173K. Various morphologies of the whiskers, such as flakelike, platelike, and needlelike, were observed under different fabrication conditions. The lateral surfaces of the whiskers were close-packed (010) plane, and the growth direction was [001], which related to the unique layered structure of the α-MoO_3 crystal. Growth of the MoO_3 whisker was attributed to a vapor-liquid–solid (VLS) mechanism at 1023 and 1173K, whereas, at 973K, growth was attributed to a vapor-liquid(VL) mechanism (Jianqiang Li 2004).

But, for nano Mo metal powders prepared from MoO_3 nanorods or nanofibers have not been reported in the literatures, the principle of preparation of MoO_3 nanofiber prepared by hydrothermal process is not clearly, and the cross-section shape of MoO_3 nanofiber never characterized by electro microscopy.

Here we report the process of MoO_3 nanofiber prepared by hydrothermal method, and, the morphology of MoO_3 nanofiber was studied under different reaction conditions, and nano Mo metal powder indeed can be reduced from the resultant MoO_3 nanofibers through reduction conversion by using H_2 stream.

2. Experimental

2.1 Materials preparation

The α-MoO_3 nanofibers were synthesized by a hydrothermal route. Ammonium heptamolybdate tetrahydrate (AHM; $(NH_4)_6Mo_7O_{24}\cdot 4H_2O$) and nitric acid were the two starting reagents. A saturated solution of precursor compound AHM was prepared at room temperature. For each run of experiment 30.0 ml of the saturated solution was diluted with deionized water (30.0 ml) and then further acidified using the 2.2M nitric acid to a total of 70-85.0 ml in volume. The resultant solution was transparent and transferred to a Teflon-lined stainless steel autoclave and heated at 170-180°C for 40 h. The product precipitate was

filtered and rinsed with deionized water, followed by drying at 62°C for 5h. The dried sample was fibrous and pale yellowish.
The above as-grown α-MoO_3 crystals were further used as metal oxide precursors for reduction investigation. The reactions were carried out in a tubular quartz reactor at 400-600°C using H_2 stream under normal atmospheric pressure.

2.2 Materials characterization

Crystallographic information of samples including reduced ones was investigated with X-ray diffraction (XRD; Shimadzu XRD-6000, Cu Kα,λ=1.5406 Å). Crystal sizes and the morphology of crystal samples was examined before and after the reduction reactions with scanning electron microscopy (SEM, JSM-5600LV, 15 kV).
High-resolution analytical transmission electron microscopy (TEM, JEM-2010, 200 kV) was also used to examine crystalline α-MoO_3 and its reduced products. The specimens for TEM imaging study were prepared by suspending solid samples in acetone.

3. Results and discussion

3.1 Preparation of polymorphous molybdenum trioxide by hydrothermal process

In order to get molybdenum trioxide powder with different morphologies, especially to get molybdenum trioxide nanofiber, using saturated solution of ammonium paramolybdate and nitric acid as raw materials, molybdenum trioxide powders with different micro-morphology were successfully synthesized through changing synthetic parameters of hydrothermal synthetic process in a hydrothermal reactor.
Synthesized molybdenum trioxide powders were characteristiced by XRD, SEM, and HRTEM. Results showed that molybdenum trioxide nanofibers could be got which were more than 20μm in length, 50~200 nm in diameter when the heat treatment temperature was more than 150°C, acidity coefficient (acidity coefficient is defined as ratio of number of moles of acid added in the solution with number of moles of Ammonium heptamolybdate tetrahydrate) of precursor solution was more than 15, and heat treatment time more than 20 h.
Crystal growth direction of molybdenum trioxide nanofibers was (001). Deviating of synthetic process parameters mentioned above,molybdenum trioxide powders with rodlike, platelike, and flakelike micro morphologies could be synthesized.

3.1.1 Synthesized at low acidity conditions

Low acidy conditions indicated that acidity coefficient in the reaction solution was less than 10, and the template often used in the synthesis process of nano fibers was not used in this research.
For the 25ml of AHM with 63% (wt) concentration, mixed with 2.2M HNO_3 solution 50ml, then, appropriate amount of milk like mixture got was put in the teflon-lined stainless steel autoclave and heated at 170°C for 20 h,the acidity coefficient is 8.6, the hexagonal short prismatic MoO_3 was obtained (Fig1.).
In the Fig2, the 25ml of AHM with 25% (wt) concentration, mixed with 2.2M HNO_3 solution 10ml, then, appropriate amount of milk like mixture got was put in the teflon-lined stainless steel autoclave and heated at 170°C for 20 h,the acidity coefficient is 4.3, the square flake MoO_3 was obtained.

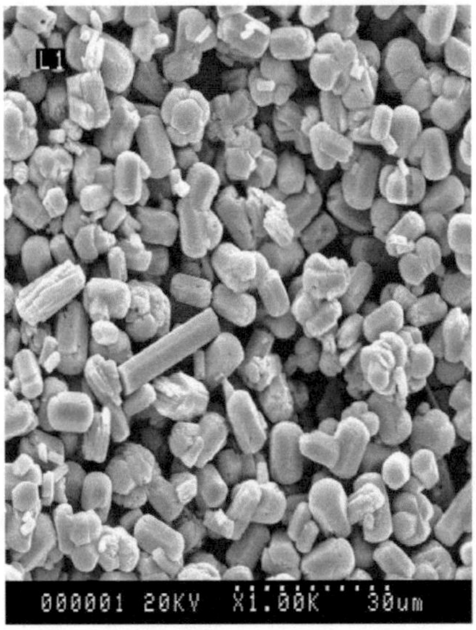

Fig. 1. MoO_3 powder morphology synthesized at low acidity conditions (acidity coefficient is 8.6).

From Fig2, the morphology of MoO_3 crystal is square flake crystal,the width and length are less than 10 micrometer, the thickness of crystal is less than 1 micrometer.

Fig. 2. MoO_3 powder morphology synthesized at low acidity conditions (acidity coefficient is 4.3).

In Fig 3, the 30ml of AHM with 27% (wt) concentration, mixed with 2.2M HNO_3 solution 10ml, then, appropriate amount of milk like mixture got was put in the teflon-lined stainless steel autoclave and heated at 170°C for 20 h,the acidity coefficient is 3.0, the hexagonal long prismatic MoO_3 was obtained.

Fig. 3. MoO_3 powder morphology synthesized at low acidity conditions (acidity coefficient is 3.0).

3.1.2 Synthesized with template at 170°C for 40h

To study the adding template on the morphology of molybdenum trioxide through hydrothermal process, the first use of cetyltrimethylammonium bromide(CTAB) as template agent. Because of CTAB is cationic surfactant, the particles in the AHM solution are negatively charged, so as different amount of CTAB was added, the emulsion solution easy to be formed,the heatreatment temperature is 170°C, the time is 40h, after hydrothermal reaction, the different morphologies of molybdenum trioxide crystal were obtaned (see in the Fig4-Fig5).

Fig. 4. MoO_3 morphology synthesized at different moles ratio of CTAB with AHM (CTAB / AHM =0.1).

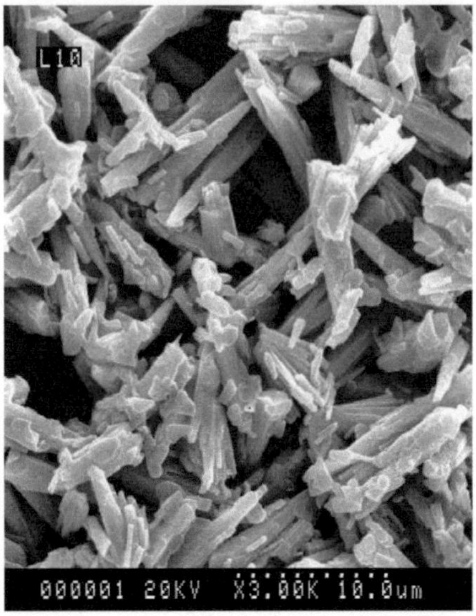

Fig. 5. MoO_3 morphology synthesized at different moles ratio of CTAB with AHM (CTAB / AHM =0.4).

3.1.3 Synthesized with template at 180°C for 40h

Increasing heattreatmeng temperature to 180°C, and increasing ratio of CTAB with AHM in the solution, a series of strip like MoO_3 crystals were obtained (in the Fig.6-Fig.8).
From Fig.6 to Fig.8, the results indicated that the size of MoO_3 decreaseed when moles ratio of CTAB with AHM increased, the shapes of samples were nearly strip fiber.

Fig. 6. MoO_3 morphology synthesized at 0.5 moles ratio of CTAB with AHM under 180°C for 40h.

Fig. 7. MoO_3 morphology synthesized at 0.7 moles ratio of CTAB with AHM under 180°C for 40h.

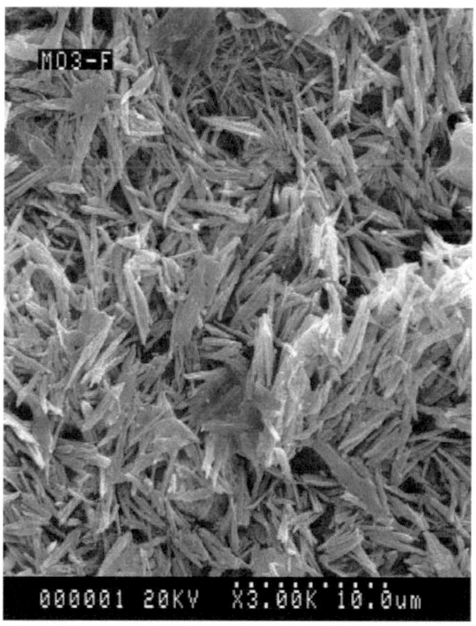

Fig. 8. MoO_3 morphology synthesized at 1.0 moles ratio of CTAB with AHM under 180°C for 40h.

3.1.4 Synthesized at high acidity without template at 170°C for 40h

High acidy conditions indicated that acidity coefficient in the reaction solution was more than 15, and the template CTAB was not used in this research.

For different concentration of AHM(wt%), 25 %, 12.5%, 26% respectively with different acidity coefficient 15, 20, 22, the precursor solutions were all transparent, after heattreatment at 170°C for 40h, fibers of MoO_3 were obtained (see in the Fig.9- Fig.11).

Fig. 9. MoO_3 morphology of precursor of 25% AHM at acidity coefficient 15(180°C for 40h).

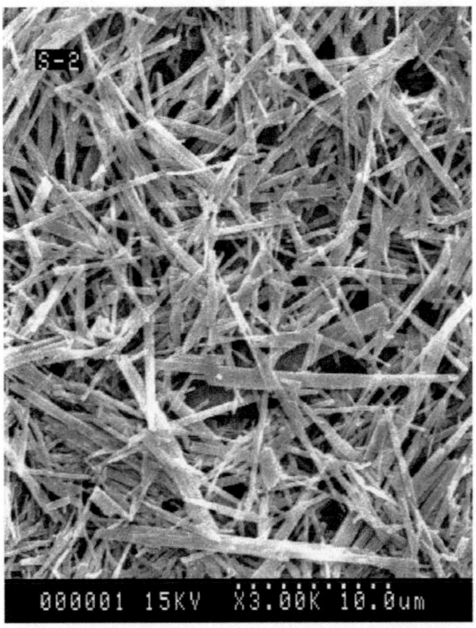

Fig. 10. MoO_3 morphology of precursor of 12.5% AHM at acidity coefficient 20(180°C for 40h).

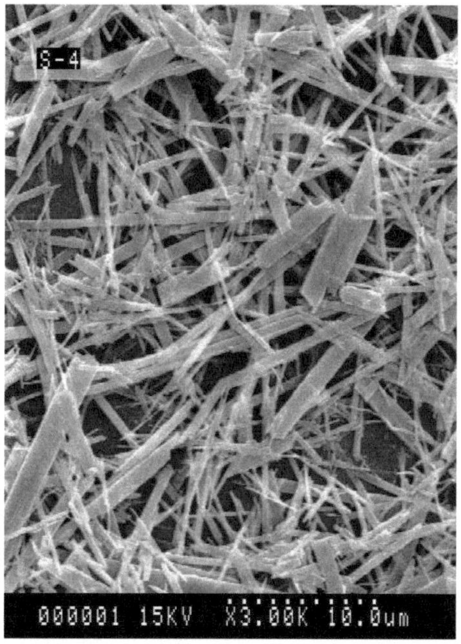

Fig. 11. MoO_3 morphology of precursor of 26% AHM at acidity coefficient 22 (180°C for 40h).

3.1.5 Optimization of synthesis process parameters

A SEM image of a typical product showed the sample as an entirely fiber-like structure after optimized synthesis process parameters(Fig.12),the nanofiber length is more than several tens of microns.

Fig. 12. Typical morphology of MoO_3 nano fibers synthesized under optimized conditions.

3.2 Characterization of MoO_3 nano-fibers

3.2.1 Characterized by HRTEM and FSEM

The result of typical MoO_3 nanofibers characterized by HRTEM is shown in fig.13, the nanofiber cross- section is not round, some nanofibers is not single fiber, combined by two single fiber. Most fibers are with width of about 50 to 200 nm, and length of about a few tens microns.

Nearly rectangle-like cross section of the nanofibers could be seen from the image of Fig.14. HRTEM image of an individual nanofiber (Fig.15) provided further insight into the structure of these products. HRTEM images recorded perpendicular to the growth axis of the single nanofibers could be attributed to the [010] of orthorhombic MoO_3, and suggested that the nanofibers grew along the [001] direction.

Fig. 13. Typical image of MoO_3 nano fibers characterized by HRTEM.

Fig. 14. Typical image of cross-section of MoO3 nano fibers characterized by FSEM.

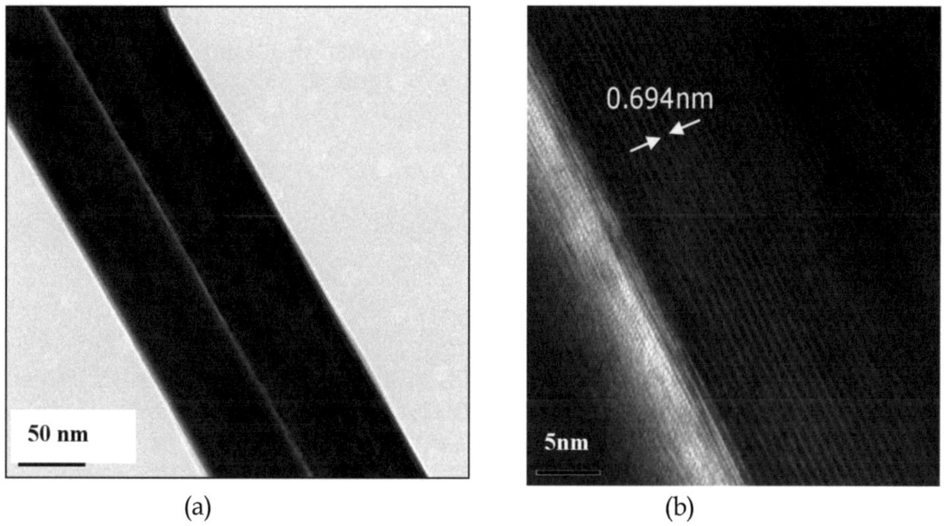

Fig. 15. HRTEM image of an individual nanofiber combined by two single fiber(a); and the lattice image of MoO_3 nanofiber crystal (b).

3.2.2 Characterized by XRD

X-ray diffraction showed that samples consisted only strongly oriented orthorhombic α-MoO_3 (a = 1.385 nm, b = 0.3696 nm, c = 0.3966 nm, JCPDS 89-7112). It can be observed in Fig.16., that(020)，(040)，(060) are strong peak respectivily, that means all the strong diffraction peaks correspond to (0h0) reflections. (110),(021),(002) are weak peaks, which shows that crystal growth is slow in these direction, causing the crystal diffraction line intensity decreased.

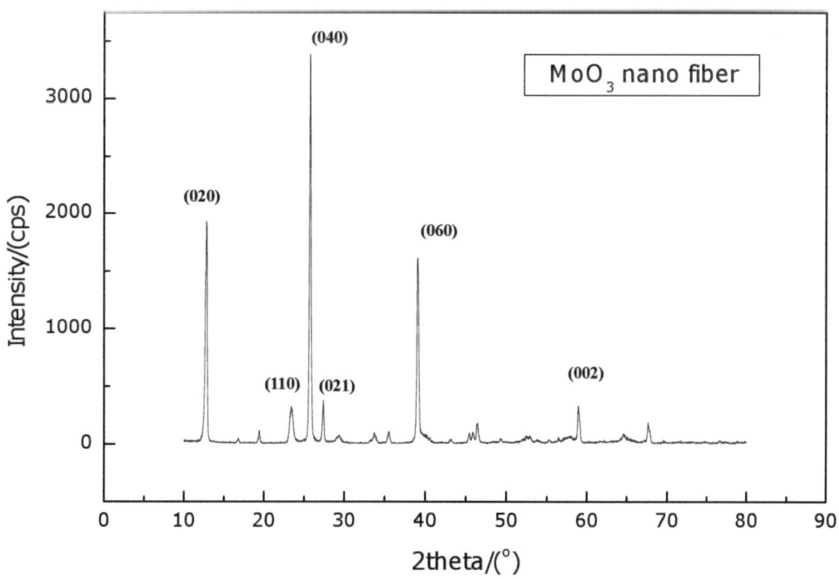

Fig. 16. The XRD patterns of MoO_3 nano fibers.

3.3 Principle of MoO_3 nano-fibers crystal growth

The crystal growth process of MoO_3 under hydrothermal conditions could be explained according literatures, and the MoO_3 nanofiber crystal growth modle is shown in the Fig.17. (Xiao Lin Li 2002).

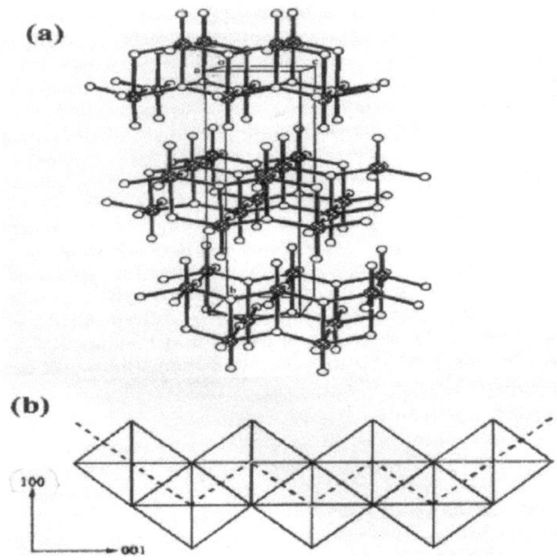

Fig. 17. The atom structure of MoO_3 (a), and schematic diagram of the crystal growth direction(b).

From Fig.17(a), It can be seen, that molybdenum trioxide is a typical orthogonal layered structure. Layered structure constitutes the smallest unit is MoO6 octahedron. Fig.17 (a) shows that the crystal chemistry cell of molybdenum trioxide is selected in three layers of MoO6 octahedron, where a, c values are relatively small, namely 3.966 and 3.693Å, while the b value is relatively large 13.858Å (JCPDS05-0508).

Each layer of molybdenum trioxide crystal is formed by MoO6 octahedrals twist to share angle connectivity in the a axis direction, and in the c-axis direction, MoO6 octahedrals connect by a sharing edge way, this sharing edge way twists and turns along acording the Z-shaped.

From Figure 17 (b) , It can be seen, along the [100] direction and [001] direction ,the same chemical environment is not same, which led to the [001] and [100] direction connection requires different chemical activation energy. If along the [001] direction is necessary to form two Mo-O bond, while along the [100] only the formation of one Mo-O bond, so MoO6 octahedron cenection along [001] will release more energy, so the crystal growth along [001] preferred orientation, especially in the presence of H^+, the occurrence of the following reaction:

$$Mo_7O_{24}^{6-} + 3H^+ \rightarrow (H\text{-}O)_3\text{-}Mo_7O_{23}^{3-}$$

$$(H\text{-}O)_3\text{-}Mo_7O_{21}^{3-} + 3H^+ \rightarrow (H\text{-}O)_6\text{-}Mo_7O_{18} \rightarrow 7MoO_3 + 3H_2O$$

As can be seen from the above reaction, increased the concentration of AHM and nitric acid, were very favorable reaction towards the right, which is the main reason to synthesis of nano-fibers with high concentration of molybdenum trioxide.

Secondly, in order to ensure that the oxygen ions prolapse from AHM to form water molecules, excessive acid concentration to be necessary, the minimum theoretical acidity coeffcent is of 7, the actual value in the this research increases to 15 or more,so synthesized the MoO_3 nano fibrous structures, that the high acidity is also needed.

Molybdenum trioxide crystal is the vertex connected along the [100] direction, the combination of atoms along [010] direction is van der Waals force, resulting in fiber diameter direction, the atoms combination is not close, so the double-fiber structure is formed.

3.4 Reducing of MoO_3 nano-fibers

Molybdenum and molybdenum alloys have good electrical conductivity, thermal conductivity, high temperature strength and high temperature hardness, corrosion resistance and good performance, therefore, widely used in chemical engineering, metallurgy and aerospace industry and other fields. With the rapid development of high-tech, material properties of molybdenum and its alloys in many aspects of the request beyond the traditional performance. For getting high performance molybdenum metal materials, ultra-fine powders of molybdenum is its first step (Tuominen 1980).

The methods of ultrafine Mo powder preparation are traditional MoO_3 reduction method, hydroxyl thermal decomposition method, molybdenum chloride vapor method, molybdenum chloride pyrolysis method, laser flash method, and microwave plasma method, etc.(Malikov 1997, Futaki Shoj 1992, Chow 1991, Liu Binghai 1999, Gonzalez 2001)).

However, all of these methods, to achieve large-scale industrial production of fine metal powders of molybdenum, the traditional reduction method still has a certain appeal. It is possible that ultrafine Mo powder could be produced by reducing industrialized nano molybdenum oxide.

3.4.1 MoO_3 nanofibers microstructure before H_2 reduction

The purpose of reduction test of nano-fibers of molybdenum trioxide synthesized by hydrothermal process, is to prepare ultrafine Mo powder. Reduction experiment of molybdenum trioxide nanofibers was performed in the hydrogen reduction furnace by using of the traditional hydrogen reduction method.

The XRD pattern of molybdenum oxide nanofibers befor reduction is shown in Fig.16,and nanofibers of molybdenum trioxide SEM photograph is shown in Fig.18.

Fig. 18. FSEM image of MoO_3 nanofiber before reduction.

In fig.18 FSEM photo, the nano-fiber length ranging is 10-30 microns, width of it is about 50-200 nm. The MoO_3 nano-fibers has better dispersion and smooth surface. The nano-fiber size is uniform befor hydrogen reduction.

3.4.2 Reduction result of MoO_3 nanofibers

The MoO_3 nanofibers were reduced at different temperature under H_2 stream atmosphere. From fig19, the XRD patterns indicated that MoO_3 nano fibers could be reduced as MoO_2 at 530°C for 1 hour, the materials had a pure phase of MoO_2. That means the MoO_3 nanofibers have been reduced as MoO_2 completely at 530°C. Corresponding to the fig 18, the morphology of the MoO_3 nanofibers have been changed, the fibers-like morphology was partial retained in the MoO_2 sample (Fig.20).

Fig.19 showed that the MoO_3 nanofibers had been reduced as Mo powders completely when the reduced temperature was increased to 600°C, corresponding to the fig21, the morphology of Mo powders is not fibers-like, most of metal Mo powders are irregular shape, neither the ball-like, nor fibers-like.

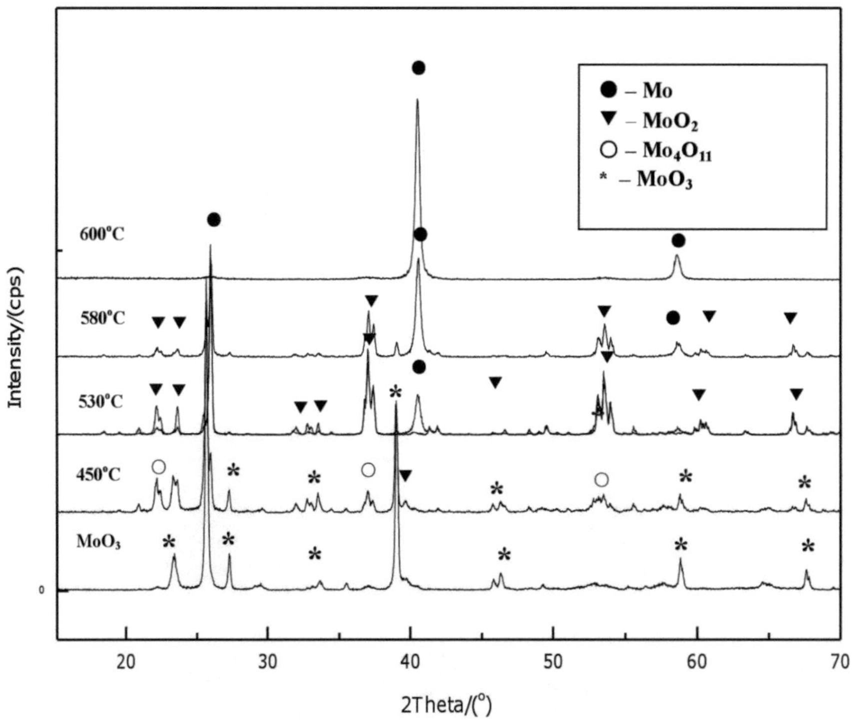

Fig. 19. The XRD patterns of MoO_3 nano fibers reduced at different temperatures by using H_2.

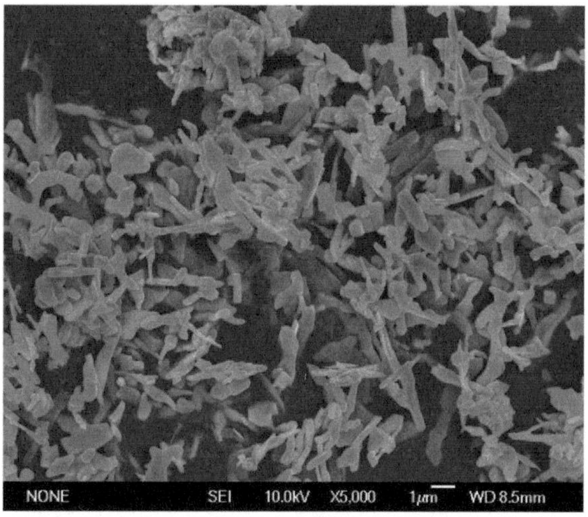

Fig. 20. The SEM image of nanofibers molybdenum trioxide reduction at 530°C for 1h.

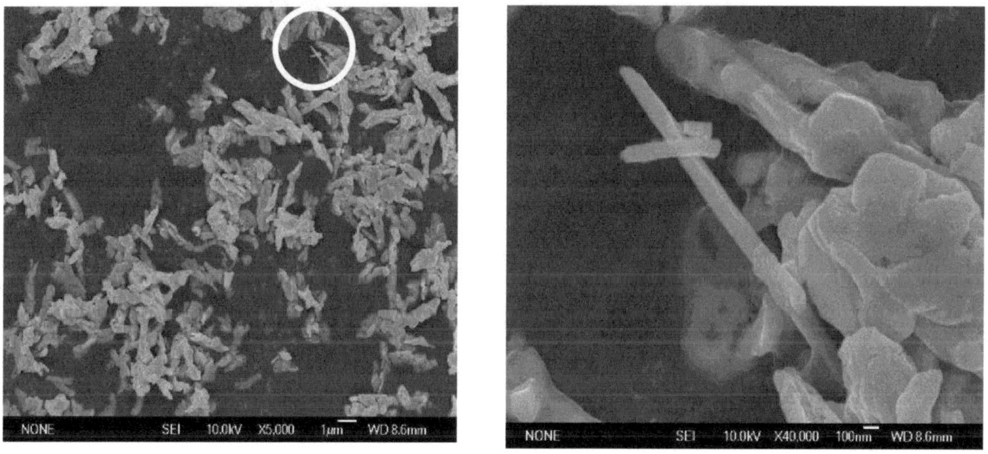

(a) (b)

Fig. 21. SEM images of Mo powder reduced at 600°C(a), and (b) is enlargement of the white circle in (a).

But in some case, the nanofibers-like shape of MoO_3 was retained, the result could be seen in the fig21. From fig.21(b), it could be found that needle-like metal Mo powder was formed. The reason of forming needle-like metal Mo powder could be explained that the MoO_3 nanofiber was restricted in a free space, no other MoO_3 nanofiber contact with each other, although orthorhombic crystal structure of MoO_3 changed as FCC crystal structure of metal Mo, no other metal Mo particles to combine with, so the nanofibers-like shape was retained.

3.4.3 Optimization of reduction result of MoO_3 nanofibers

According to the reason mentioned above, we using small bulk (about 2-3mm) MoO_3 nanofiber as raw materials instead of the MoO_3 nanofibers powders. The small bulk MoO_3

was reduced at 600°C for 1h, XRD indicated that MoO_3 had been completely changed as metal Mo powder. SEM result shown in fig22. Much more needle-like metal Mo powders were formed. It means nanofiber-like MoO_3 could be reduced as needle-like metal Mo powder in some special case.

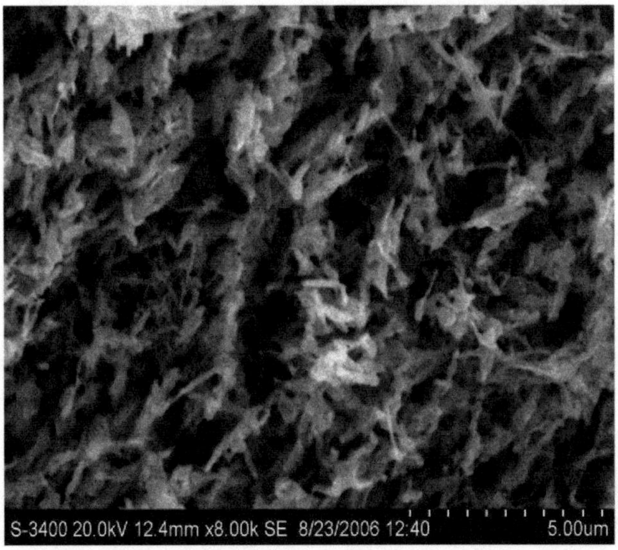

Fig. 22. SEM image of needle-like metal Mo powder.

3.5 Principle of MoO_3 nano-fibers reduced by using H_2

Through analysis the reduction process of nanofibers molybdenum trioxide, in general, the fibrous structure of molybdenum trioxide is difficult to maintain or inherited. Just because of molybdenum trioxide is orthogonal layered structure, and molybdenum dioxide is rutile structure.As we know, the reduction process of nanofibers molybdenum trioxide is a very complicated local chemical processes, including following steps (Tuominen 1980, Thorsten 2001):

1. adsorption of reduction gas
2. the chemical reaction between molybdenum oxide with the hydrogen molecules
3. changes in crystal chemistry and the desorption of gas in the reaction products

The first step of reduction process of molybdenum trioxide is mainly broken of crystal at 450°C or more, because of the molybdenum dioxide is gradually began to be formed on the surface of molybdenum trioxide and the layer of molybdenum dioxide could be peeled off, and the crystal of molybdenum trioxide could be broken or cracked.

As the reaction further, the layer thickness of molybdenum dioxide is more than half the thickness of the crystal of molybdenum trioxide, molybdenum dioxide peeling process could be stopped, the molybdenum trioxide completely transformed into molybdenum dioxide, the shape of crystal was changed generally.

Delamination of molybdenum dioxide should be occurred in the weak place of crystal surface of molybdenum trioxide, due to anisotropy of chemical bonding strength, so it is broken easily in the [001] direction of molybdenum trioxide. Therefore, when molybdenum trioxide nanofiber is reduced as molybdenum oxide, the crystal size is shorter (Fig.23).

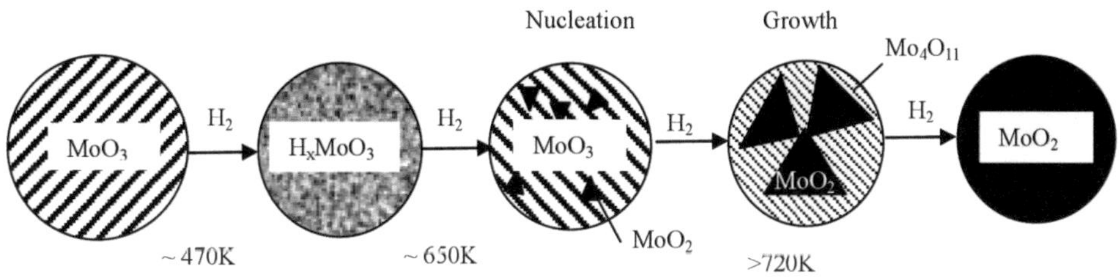

Fig. 23. Schematic diagram of the reduced process of MoO_3 to MoO_2.

4. Conclusion

One-dimensional nanostructures of orthorhombic molybdenum trioxide have been synthesized in the forms of fibers under hydrothermal conditions. The dimensions of the fibers are about 50nm in thickness, 100-200nm in width and a few tens of micrometers in length under our experimental settings.
The MoO_3 fibers prepared can be converted completely to Mo metal powders at 600°C by ussing H_2 stream. The original fiber-like morphology is not well-retained, but in some specials cases, fiber-like morphology can be partial retained, although the aspect ratio of the oxied fibers is reduced upon the reduction treatment.

5. Acknowledgment

The authors gratefully acknowledge research funding supported by "the Fundamental Research Funds for the Central Universities of China" and "National Natural Science Foundation of China 50472028".

6. References

R. Tenne(2006). Inorganic Nanotubes And Fullerene-Like Nanoparticles. *Nature Nanotechnology*, Vol.1(Jan.2006), pp103-111, ISSN 1748-3387

Romo-Herrera, J.M., Terrones, M., Terrones, H., Dag, S., & Meunier, V. .(2007), Covalent 2d And 3d Networks From 1d Nanostructures: Designing New Materials. *Nano Letters*, Vol.7,No.4,(Appil 2007), pp 570-576, ISSN 1530-6984

Dmitruk, N., Barlas, T., Kotova, N.,Romanyuk, V., & Dmytruk, A.(2007), Metal 1d Micro(Nano) Self-Organized Wires On Semiconductor Surface: Preparation, Topology, And Optical Properties. *Materials Science And Engineering C*, Vol.27 No.(5-8), (Sept. 2007),pp1141-1144, ISSN 0928-4931

Shabaev A., Efros & Al.L.(2004),1d Exciton Spectroscopy Of Semiconductor Nanorods. *Nano Letters*, Vol.4, No.10,(Oct. 2004), pp1821-1825, ISSN1530-6984

Markus N, Frank K, & Hans-Joachim M.(2001) Synthesis And Characterization Of Novel Nanoscopic Molybdenum Oxide Fibers. *J.Mater.Chem.* Vol.11, (Nov.2001), pp1941-1945, ISSN 0959-9428

Xiong W L, & Hua C Z.(2003), Complex A-MoO3 Nanostructures With External Bonding Capacity For Self-Assembly. *J.Am.Chem.Soc*, Vol.125, No9, (Sep. 2003), pp2697-2704, ISSN 0002-7863

Zhao D.W., Sun X.W., Jiang C.Y., Kyaw A.K.K., Lo G.Q., & Kwong D.L.(2008). Efficient Tandem Organic Solar Cells With An Al/MoO_3 Intermediate Layer. *Applied Physics Letters*, Vol.93,No.8, (Aug.2008), pp 3305-3 ISSN 1077-3118

Kim Youn-Su, Ahn Hyo-Jin, Shim Hee-Sang, Nam Sang Hoon, Seong Tae-Yeon, & Kim Won Bae. (2007), Pt-Embedded MoO3 Electrodes For Rechargeable Lithium Batteries. *Electrochemical And Solid-State Letters*, Vol.10, No. 8 (Oct.2008),pp180-183,ISSN 1099-0062

Ghorai T.K., Dhak D., Biswas S.K., Dalai S., & Pramanik P.(2007), Photocatalytic Oxidation Of Organic Dyes By Nano-Sized Metal Molybdate Incorporated Titanium Dioxide (Mxmoxti1-Xo6) (M = Ni, Cu, Zn) Photocatalysts. *Journal Of Molecular Catalysis A: Chemical*, Vol.273, No.(1-2),pp224-229,ISSN 1381-1169

Dillon A.C., Mahan A.H., Deshpande R., Parilla P.A., Jones K.M., & Lee S-H.(2008),Metal Oxide Nano-Particles For Improved Electrochromic And Lithium-Ion Battery Technologies. *Thin Solid Films*, Vol.516,No.5,(May.2008),Proceedings Of The Fourth International Conference On Hot-Wire Cat-Cvd Process,pp794-797, ISSN 0040-6090

Ajayan P. M., Stephan O., Redlich P., & Colliex C.(1995), Carbon Nanotubes As Removable Templates For Metal Oxide Nanocomposites And Nanostructures. *Nature* Vol.375, (June 1995) pp 564-567 ISSN 0028-0836.

B. C. Satishkumar, A. Govindaraj, M. Nath & C. N. R. Rao(2000), Synthesis of metal oxide nanorods using carbon nanotubes as templates, *J. Mater. Chem.*, Vol.10, (Oct. 2000) pp2115-2119. ISSN 0959-9428

Xiong Wen Lou, & Hua Chun Zeng, (2002). Hydrothermal Synthesis Of α-MoO3 Nanorods Via Acidification of Ammonium Heptamolybdate Tetrahydrate. *Chem. Mater.*, Vol.14, (Sept. 2002), pp 4781-4789, ISSN 0897-4756

Jianqiang Li, Pan Wei, Jian Chen, & Li Rongti (2004), Preparation and Growth Mechanism of Molybdenum Trioxide Whisker, Journal of the American Ceramic Society,Vol.85, No.8, (August 2002), pp2116–2118, ISSN 002-7820

Xiao Lin Li, Jun Feng Liu, Ya Dong Li (2002). Low-Temperature Synthesis Of Large-Scale Single-Crystal Molybdenum Trioxide MoO3 Nanobelts. *Appl. Phys. Lett.* Vol.81, No.25, (Dec.2002), pp4832-4834, ISSN 003-6951

Thorsten Ressler, Julia Wienold,& Rolf E. Jentoft (2001). Formation Of Bronzes During Temperature Programmed Reduction Of MoO3 With Hydrogen -An In Situ Xrd And Xafs Study. *Solid State Ionics* . Vol.141-142, (June,2001), pp243-251, ISSN 0167-2738

Tuominen S M, & Carpenter K H.(1980),Powder metallurgy molybdenum:influence of powder reduction processes on properties.*Journal of Metals*, Vol.32, No.1, (Jan.1980), pp23-27. ISSN 0148-6608

Malikov I V, & Mikhailov G M. (1997), Electrical resistivity of epitaxial molybdenum films grown by laser ablation deposition.*J.Appl.Phys.*, Vol.82,No.11,(Dec. 1997), pp5555-5559, ISSN 1089-7550

Futaki Shoj, Shiraishi Katsuzo, & Uemura Mikio(1992).Ultrafine refractory metal particles produced by hybrid plasma process.*Journal of the Japan Institute of Metals*, Vol.56,No.4,(Apr. 1992) pp4:64, ISSN 0021-4876

Chow.GM.(1991), Sputtering synthesis and properties of molybdenum nanocrystals and Al/Mo layered. *Journal of Material Research*, Vol.4, (Apr.1991) pp737-743, ISSN0884-2904

Liu Binghai, Gu Hongchen, & Chen Qiling(1999). Preparation of nanosized Mo powder by microwave plasma chemical vapor deposition method.*Materials Chemistry and Physice*,Vol. 59, No. 3, (Jun.1999), pp204-209, ISSN 0254-0584

Gonzalez.G(2001), Synthesis and characterization of nanophase Partieles abtained by D.C sputtering. *Scripta Material*, Vol.3,(Mar.2001), pp1883–1887, ISSN 1359-6462

8

Electrospinning of Continuous Nanofiber Bundles and Twisted Nanofiber Yarns

Usman Ali, Yaqiong Zhou, Xungai Wang and Tong Lin
Centre for Material and Fibre Innovation, Deakin University, Australia

1. Introduction

Spinning is a prehistoric technology in which endless filaments, shorter fibers or twisted fibers are put together to produce yarns that serve as key element to assemble multifarious structural designs for diverse functions. Electrospinning has been regarded as the most effective and versatile technology to produce nanofibers with controlled fiber morphology, dimension and functional components from various polymeric materials (Dersch et al., 2007, Frenot and Chronakis, 2003, Schreuder-Gibson et al., 2002). However, most electrospun fibers are produced in the form of randomly-oriented nonwoven fiber mats (Doshi and Reneker, 1995, Madhavamoorthi, 2005). The relatively low mechanical strength and difficulty in tailoring the fibrous structure have restricted their applications. With the rapid development in nanoscience and nanotechnology, yarns composed of nanofibers may uncover new opportunities for development of well-defined three dimensional nano fibrous architectures. This chapter focuses on recent research and advancement in electrospinning of nanofiber bundles and nanofiber yarns. The preparation, morphology, mechanical properties and potential applications of these fibrous materials are discussed in details.

2. Nanofibers and yarns

Nanofibers are defined as one dimensional nanomaterials with diameters less than 1 μm (1000 nm), and an aspect ratio (length/diameter) larger than 100:1. They are also named superfine or ultrathin fibers in some literatures (Supaphol et al., 2005, Alborzi et al.). When the fibers are in the range of 100–1000 nm, they are also referred to as submicron fibers (Bellan et al., 2006, Keun Kwon et al., 2005, Givens et al., 2007). This intrinsic feature offers them drastically increased surface to volume ratio and high aspect ratio. There are several methods to prepare nanofibers, encompassing top-down (melt-blown (Ellison et al., 2007), melt electrospinning (Dalton et al., 2007, Lyons et al., 2004), islands-in-the-sea (Nakata et al., 2007), and electrospinning (Bhardwaj and Kundu)), and bottom-up (interfacial polymerization (Xing et al., 2008), self-assembly (Viney, 2004), and phase separation (Zhao et al.)) approaches. Nanofibers produced from electrospinning have a naturally formed porous structure with excellent pore interconnectivity and the pores are in the range between tens of nanometers to a few micrometers. The open pore structure and high permeability to gas, along with the high surface area make them ideal porous membranes. Compared with other one-dimensional nanostructures (e.g. nanotubes or nano-rods), continuous nanofibers are advantageous in terms of fabrication cost and the possibility of being integrated into other desired assembly in one-step.

Ascribed to the high surface area, porous structure, and the ability to adopt functional molecules and nanomaterials (e.g. nanoparticles, nanotubes), nanofiber nonwoven webs have been used in areas as diverse as batteries (Norris et al., 2000), biomedical (Liang et al., 2007, Agarwal et al., 2008, Yoshimoto et al., 2003, Cao et al., 2009, Xu et al., 2004), filtrations (Gibson et al., 2001, F. DOTTI and MAZZUCHETTI*, October 2007), medical prostheses (Buchko et al., 1999, Buchko et al., 2001), sensors (Wang et al., 2002b, Wang et al., 2002a), fuel cells (Shabani et al., Li et al.), nanocomposites (Huang et al., 2003, Chronakis, 2005), and protective clothing (Lee and Obendorf, 2007, Ramakrishna et al., 2006).

Electrospun nanofibers are generally collected as nonwoven or randomly arranged structures, due to the "whipping instability" of the electrospinning jet. Early studies on fiber deposition and its assemblies were focused on controlling the fiber alignment. A few typical ways to electrospun-aligned nanofibers are shown in Figure 1. Aligned nanofibers can be collected using a dynamic mechanical collector such as cylindrical drum (Ding et al., 2002, Li and Xia, 2004), wired drum (Katta et al., 2004) or a tapered wheel (Xu et al., 2004) , either grounded or negatively charged (Theron et al., 2001). When the collector rotates at a high surface speed, such as 8 m/s, fibers start to align along the rotating axis. However, it is yet difficult to obtain perfect fiber alignment with this method, because the residual charge accumulation on the deposited fibers interferes with the incoming ones.

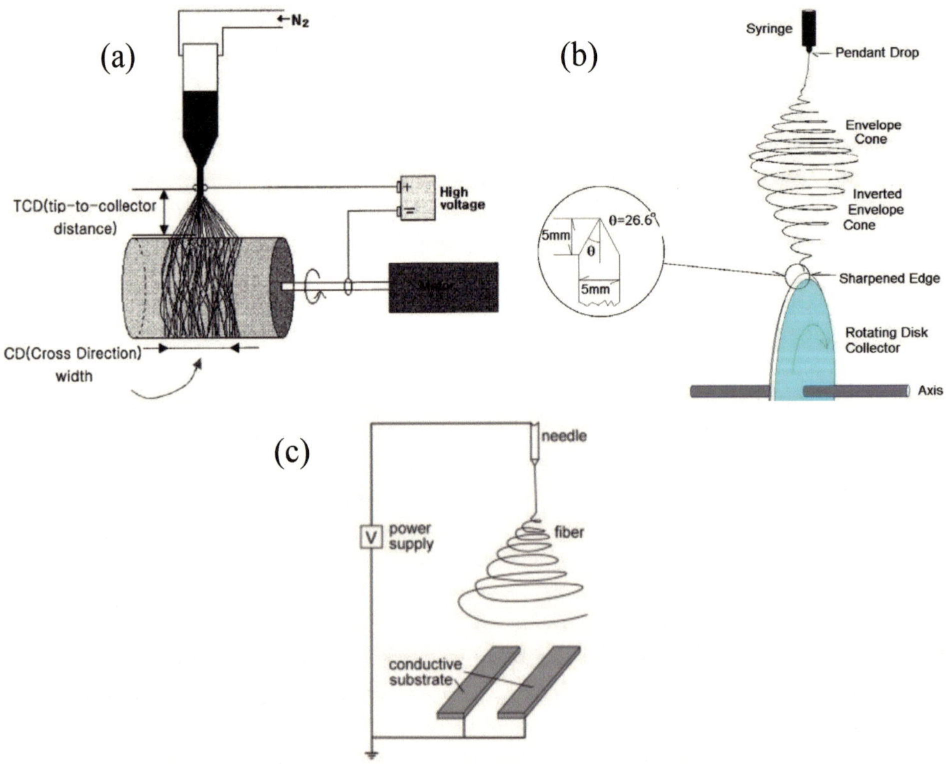

Fig. 1. Basic methods for controlling the alignment of electrospun nanofibers. (a) High speed rotating drum collector, b) a tapered wheel, c) stationary gap collector.

Since the electrostatic forces dominate the electrospinning process, it is also possible to obtain aligned nanofibers through the manipulation of the electric field. Electric field was manipulated by using two split parallel flat plates (Gap Collector) connected with grounded electrode (Li et al., 2004, Liu and Dzenis, 2008, Dan Li, *June 20, 2003*, Bazbouz and Stylios, 2008a, Kameoka et al., 2004, Ishii et al., 2008), or a frame-shaped collector (Dersch et al., 2003). The distance between two conductive electrodes is up to centimeters. A high degree of fiber alignment can be achieved with this method, and the produced fibers can be easily transferred to other substrates for device fabrication. However, the fibers thus produced are of limited length and thickness.

Yarns can be assembled or interlaced into many fibrous structures by the process of weaving, knitting or embroidery (P. J Denton and Daniels, 2002). Single nanofibers are too fragile and too thin for such processes, and it also requires sophisticated equipment for precise handling of these fibers (Tan et al., 2005). Attempts to produce nanofiber yarns from electrospun nanofibers were reported by Formhals as early as in 1934. In a series of patents (Anton, 1934, Anton, 1938b, Anton, 1944, Anton, 1938a), he described the experimental setups to produce polymer filaments by using electrostatic forces. Most of the early works on electrospinning of nanofiber yarns were actually focused on non-twisted nanofiber bundles. Although twist can be inserted by a post-electrospinning treatment, it is now highly preferred that twists can be added to continuous nanofiber yarns directly from an electrospinning process.

3. Discontinuous nanofiber bundles and twisted yarns

Generally, electrospinning produces random nanofiber mats, if a planar stationary plate is used as a collector. However, by manipulating the electric field or using moving collecting or spinneret systems, one can collect non-continuous nanofiber bundles or twisted yarns rather than mats.

3.1 Short nanofiber bundles

In 2001, Deitzel et al (Deitzel et al., Deitzel et al., 2001) demonstrated the deposition of nanofibers in a more targeted fashion by a series of charged rings in the electrospinning zone, as shown in Table 1A. The charged rings had the same polarity as the surface charge on the jet, which increased the downward force on the jet. This method resulted in nanofiber deposition in a narrow strip (0.6 cm wide) when the target was a rotating drum. The average diameter of fibers was 300 nm. The prepared short bundles of polyethylene oxide were analyzed using wide angle X-ray diffraction (WAXD) technique and some molecular orientation but poor crystalline microstructure was found in the fibers.

Theron et al (Theron et al., 2001) described an electrostatic field-assisted assembly technique in combination with the dynamic rotating collector to position and align individual nanofibers. Bundles of nanofibers were collected on a tapered wheel-like disc as shown in Table 1B. The disc was made of aluminum with a diameter of 200 mm, a thickness of 5 mm and a tapered angle of 26.6°. The tapered edge of the wheel collector had a strong converging electric field, thus enabling the jet emerged from the droplet to form an envelope cone at the start and then shrink and form an inverted cone, once it approached the sharp edge. As a result, nanofibers of polyethylene oxide (PEO) with a diameter varying within 100-300 nm would preferentially deposit on the wheel edges, which were assembled in a parallel array due to the rotation of the wheel.

Teo and Ramakrishna (Teo and Ramakrishna, 2005) demonstrated the fabrication of nanofiber bundles by collecting them across two negatively charged steel blades (3 cm apart) as shown in Table 1C, and then dipping them in water. During electrospinning, the fibers were preferentially deposited between the blades due to the concentration of electrostatic charges on the sharp edges. The fibers were first deposited on one blade, and then stretched to the other with the aid of electrostatic forces acting on the fibers. As a result, fibers were deposited across the gap. However, "stray" fibers could not be eliminated during the electrospinning process. A further step of dipping fibers in water was thus required to cluster the fibers together, by the force of water surface tension. The resulting fiber bundles had two fixed ends, which were of limited length, but they were free to be transferred to other substrates, and to be twisted or braided manually.

Year	Author	Setup	Morphology	Fiber Bundling/Polymer
A. 2001	Deitzel et al.	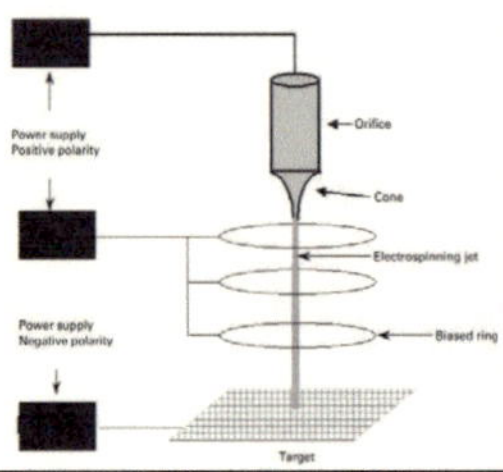	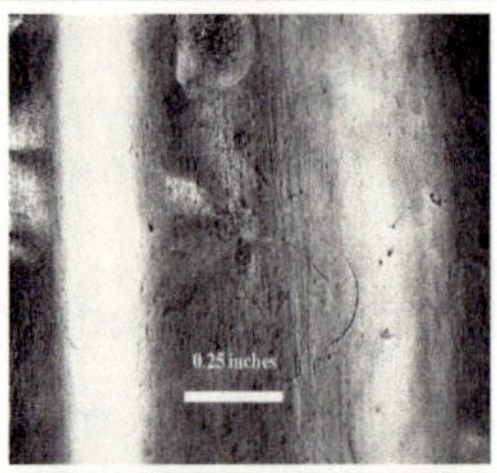	Jet was stabilized by placing a series of charged rings in the electrospinning zone and the rings were charged to have the same polarity to the spinneret; PEO
B. 2001	Theron et al.	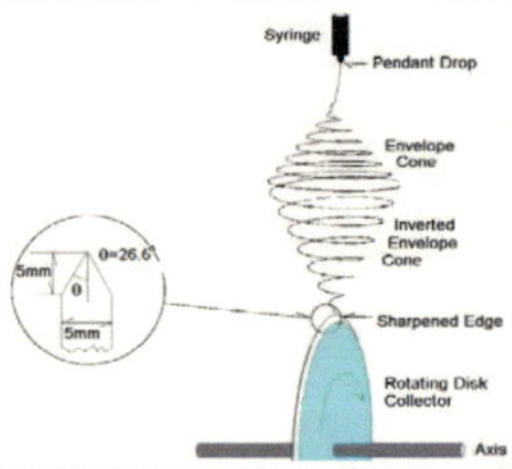	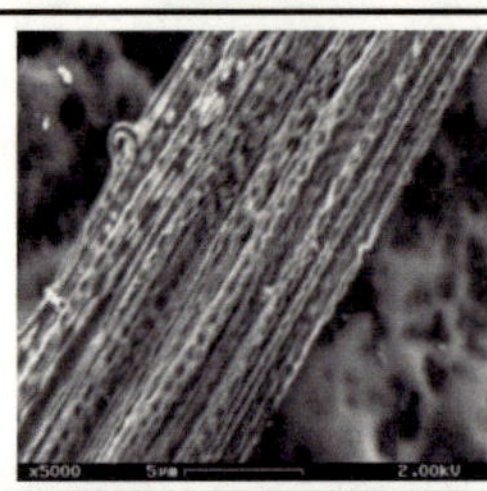	Electric field is concentrated on the tapered wheel which rotates for fiber alignment. The collector has a limited fiber deposition area; PEO
C. 2005	Teo et al.	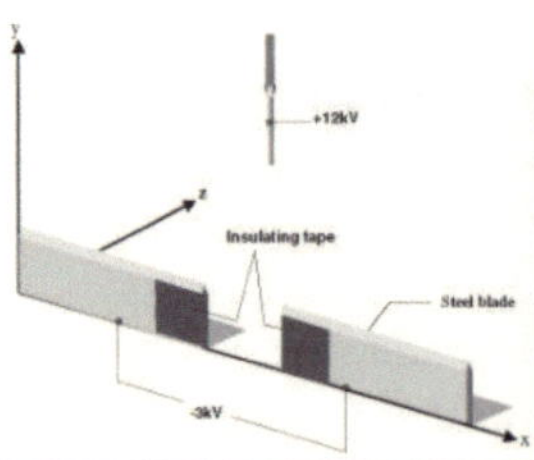	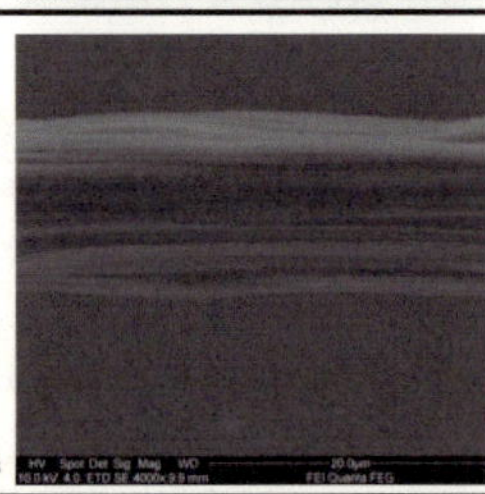	Gap collector using sharp blades for concentration of electric field; PCL

Table 1. Short Nanofiber Bundles.

Short nanofiber bundles with parallel orientated short fibers can also be prepared by cutting an aligned nanofiber membrane into narrow strips. In this case, preparing aligned nanofiber membranes is the focus. Different fiber alignment techniques have been used, such as using a rapidly oscillating frame (Fong et al., 2002, Dersch et al., 2003), and a copper comb collector in needleless electrospinning (Jiri Chvojka and Lukas, **2009**).

3.2 Short nanofiber yarns

High-speed rotating drum has been recognized as the easiest way to get highly aligned nanofibers. Fennessey et al (Fennessey and Farris, 2004) prepared short nanofiber yarns using aligned PAN nanofiber membranes. The nanofiber membrane was cut into short tows (approximately 32 cm × 2 m) and then linked together. An electric spinner was used to insert twist and the action of twist on yarn tensile strength and modulus was studied. They found that increasing the twist angles led to increased yarn ultimate strength and modulus. Liu et al (Liu et al., 2008) also used a high speed annual shaped collector to prepare aligned nanofiber membrane, and the membrane was cut and twisted in a similar way into short yarns. Later on, PAN nanofibers reinforced by carbon nanotubes (CNTs) were also prepared into short nanofiber yarn by Farouk et al (Farouk et al., 2009) who used a high speed rotating drum collector. They found that well dispersed CNTs in the polymer solution for electrospinning led to high yarn mechanical properties. Recently, Moon et al (Moon and Farris, 2009) prepared short PAN nanofiber yarns using a similar method for making carbon nanofiber yarns. They reported that the carbon nanofiber yarns had a tensile strength as high as 1 GPa.

Researchers have also paid attention to directly getting short twisted nanofiber yarns. For instance, Dalton et al (Dalton et al., 2005) used two parallel grounded rings to get aligned nanofibers (Table 2A). Nanofibers were first deposited at the lower part of the rings, which then proceeded upward to the ring top, aligning across the gap. By rotating one of the rings the suspended fibers were converted into a multi-filament yarn with a diameter of less than 5 μm and a length of 4 ~ 10 cm. The yarn length was limited by the distance between the two electrodes. Later on Liu et al (Liu et al., 2007) prepared nanofiber-based composite ropes of poly(methyl methacrylate) (PMMA)/MWCNT by using a gap collector (Table 2B). Nanofibers were electrospun vertically downward and bridged between the two electrodes. By rotating the point electrode, the as-spun PMMA/MWNT nanofibers were converted into yarns of 30 ~ 40 cm in length and 13 ~ 23 μm in diameter. This method was difficult to spin yarns longer than 40 cm, as the constant distance between two collectors had to be maintained so that nanofibers can be continuously deposited.

Gu et al (Gu et al., 2007) described a method of directly fabricating single twisted nanofibers by controlling the electric field around an auxiliary electrode (Table 2C). A polygon with several faces was placed in between the needle and the collecting plate. Each face of the polygon was subjected to an applied voltage. By switching the voltage rapidly, a rotated electric field was created, changing the direction of the electrostatic force which led to the twisted nanofibers.

Lotus et al (Lotus et al., 2008) also adopted the concept of rotating an electrode in the gap collector to produce twisted nanofiber yarns. The set-up consists of a rotating hollow hemisphere and a non-rotating but translating tapered metal rod. The tapered end of the rod and the edge of the hemisphere are conductive so fibers would preferentially deposit on their edges, and form a conical cone with its apex lying on the rod tip, as shown in Table 2D. by rotating the hollow the hollow hemisphere, and translating the metal rod away, 10cm long zinc oxide and nickel oxide yarns can be fabricated. The yarns are 5~30 μm in diameter and consist of nanofibers 60~100nm in diameter.

Year	Author	Setup	Morphology	Fiber Twisting/Polymer
A. 2005	Dalton et al.	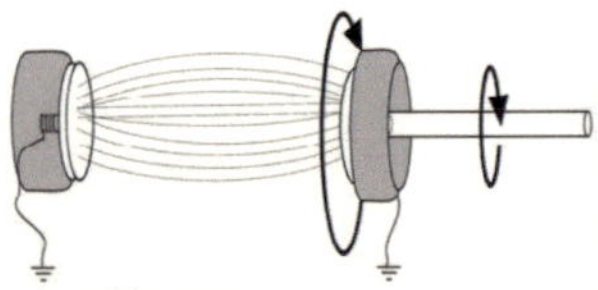		Aligned nanofibers were collected between the electrodes (gap collector) and twisted by rotating one electrodes; PCL
B. 2007	Liu et al.	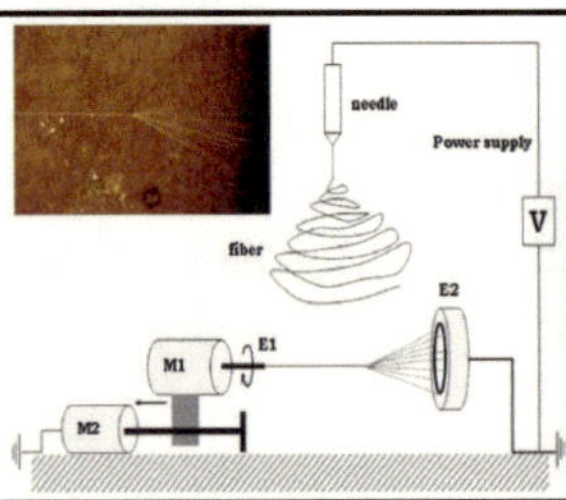	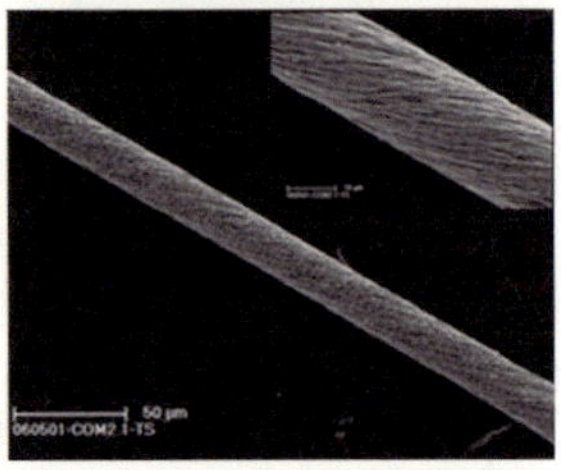	Aligned nanofibers collected by twisting needle electrode; PMMA/MWNTs
C. 2007	Gu et al.	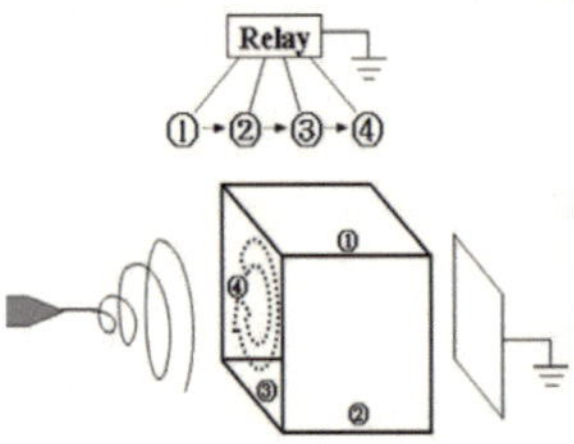	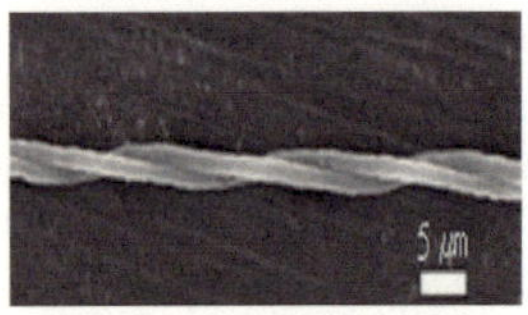	Nanofibers were twisted by electric field using an auxiliary polygon electrode; PEO
D. 2008	Lotus et al.	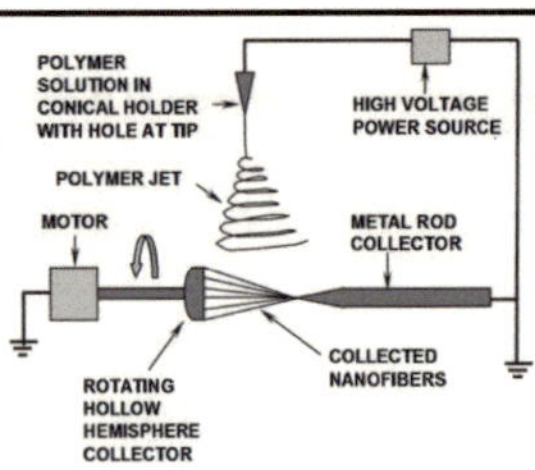	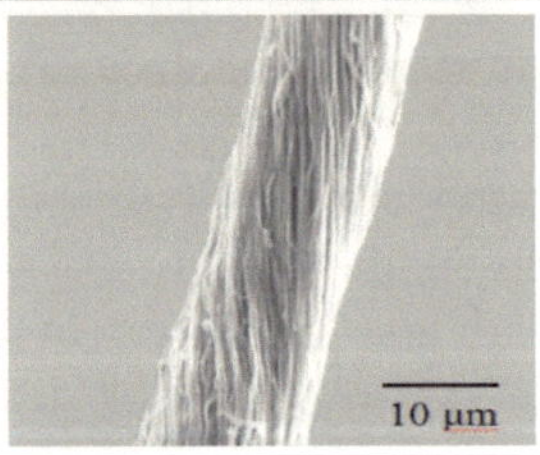	(Gap collector) aligned NFs twisted by rotating the round electrode; ZnO, NiO
E. 2010	Chang et al.	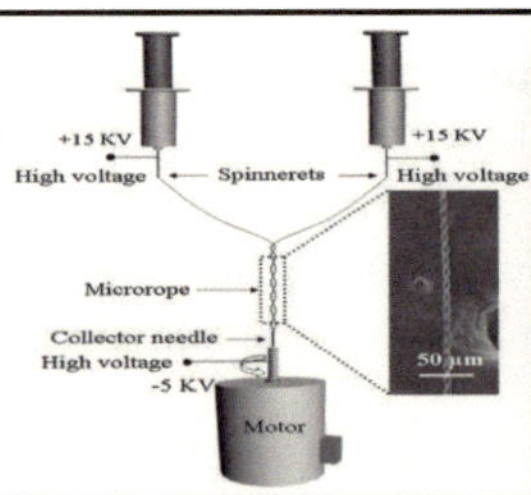	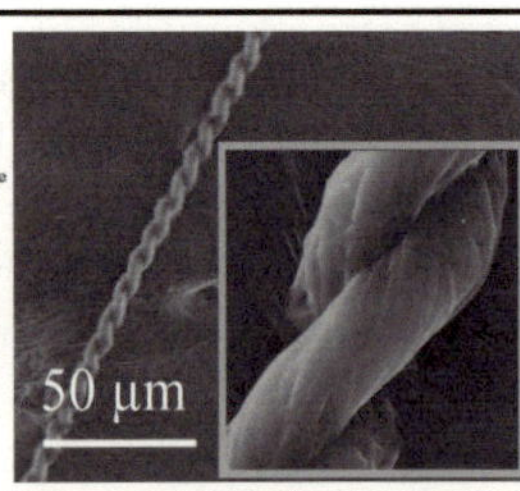	Helical ropes were prepared with two needle spinnerets and one rotating needle collector; PVP

Table 2. Short Nanofiber Yarns.

Recently Chang et al (Chang and Shen) prepared helical ropes of polyvinyl pyrrolidone (PVP) by using two needle electrospinning spinneret and a rotating needle collector (Table 2E). The diameter of the needle collector plays a key role in determining the formation of the ropes. A stable fiber bundle was formed when the diameter of the needle was smaller than 0.5 mm. The pitch of the ropes could be adjusted through the distance between the two spinnerets. When a needle collector of the diameter greater than 5 mm was used, fibers travelled randomly and deposited as a mat on the collector.

4. Continuous nanofiber bundles

Continuous nanofiber bundles are defined as a group of loosely assembled nanofibers, which have no twist. The main techniques to get continuous nanofiber bundles are described in Table 3.

A liquid medium was used separately by Khil et al (Khil et al., 2005) and Smit et al(Smit et al., 2005)to collect nanofibers and subsequently wind into continuous nanofiber bundles (Table 3A). Nanofibers initially formed a random web on the liquid surface, which were drawn and lifted off with a rotating mandrel to form a fiber bundle. Under the action of liquid surface tension, the bundle was elongated and collapsed into a round cross-section. By continuously depositing nanofibers into the liquid, while at the same time collecting them with the winder, a continuous nanofiber bundle was obtained. The fiber bundles could be drawn at a rate of 0.05 m/s ~ 0.5 m/s. To make the fiber suspend on the liquid surface could help the collection, and a liquid having a high density or high surface tension was normally selected as the collecting medium.

To obtain continuous nanofiber bundles, Pan et al (Pan et al., 2006) used a special electrospinning setup consisting of a pair of spinnerets with opposite polarities and a rotating shaft collector (Table 3B). In the process, the nanofibers from the two spinnerets carried opposite charges. They attracted to each other to form a fiber bundle and the fibers aligned within the bundle. The fiber bundle production rate was dependent on the rotating speed of the collector, the polymer material used and the fiber diameter produced. The yarn diameter can be controlled by the number of spinneret pairs. For PVP nanofibers with an average diameter of 670 nm, the maximum production speed was 14.9 m/s. Later, Li et al (Li et al., 2008) and Yao et al (Yao et al., 2009) respectively prepared nanocomposite yarns of poly(L-lactide)(PLLA)/nanotricalium phosphate and PLLA/Zein using this electrospinning setup. Li et al also studied the effect of electrospinning distance and found that nanofibers bonded tightly to each other at an electrospinning distance of 20 cm, however longer electrospinning distance resulted in loosely packed nanofiber bundles.

Okuzaki et al (Okuzaki et al., 2008) observed the formation of poly(p-phenylenevinylene) (PPV) precursor fiber bundles using just a conventional electrospinning setup (Table 3C). The spinneret was placed vertically on top of the grounded flat collector. The unusual bundle formation was explained by the conductive nature of the PPV precursor. The deposited fibers discharged through polyelectrolyte chains to the grounded target, which preferred the gravitated deposition so as to decrease the gap between the fibers and the needle. Centimeter-long nanofiber bundles were formed spontaneously between the spinneret and the grounded plate within a minute. Subsequent thermal conversion led to bundles of PPV nanofibers.

In 2008, Wang et al (Wang et al., 2008b) reported a method to self assemble continuous nanofiber bundles (Table 3D). A grounded needle tip was incorporated into the collecting system to induce self-bundling of nanofibers. It was observed that the liquid jet was a straight line initially, and then whipped around until it got close to the tip of the needle collector, splaying fibers to converge on the needle collector due to the large electrostatic attraction. By pulling back the bundle and winding it on a grounded rotating collector,

Year	Author	Setup	Morphology	Yarn Bundling/Polymer
A. 2005 2005	Khil et al. Smit et al.	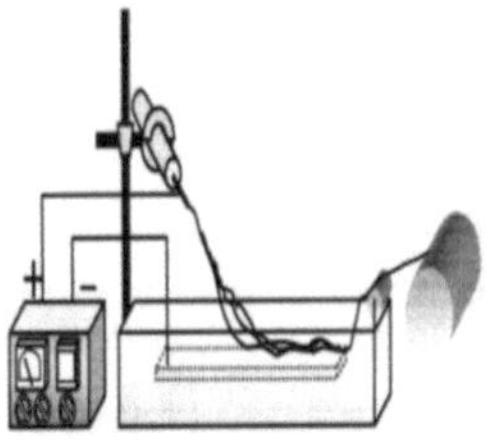	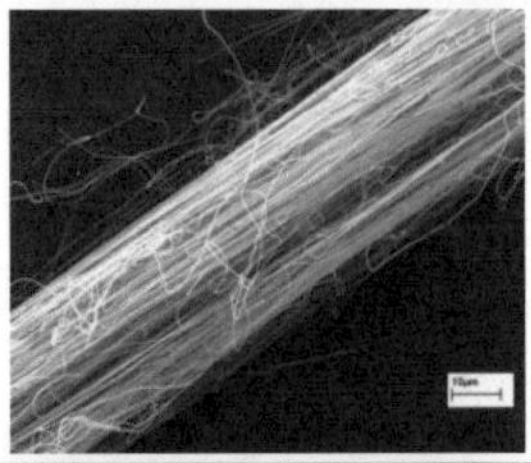	Flexible fiber web in liquid was withdrawn into fiber bundles; PCL, PVDF, PAN, PVAc
B. 2006 2007 2009	Pan et al. Li et al. Yao et al.	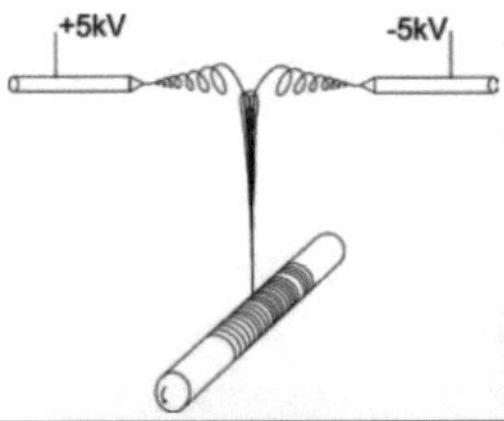	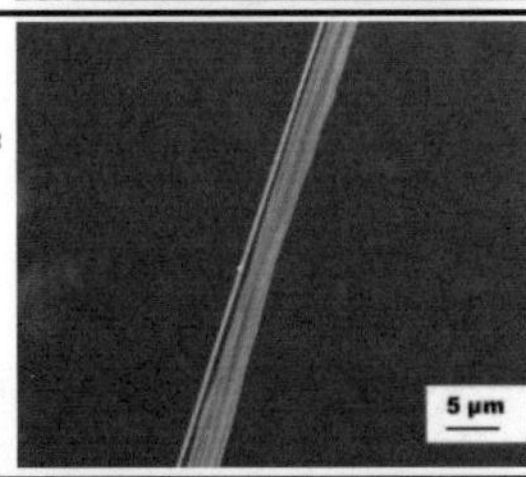	Nanofibers electrospun from opposite charged spinneret attract each other to form bundles; PVA; PVP
C. 2006	Okuzaki et al.	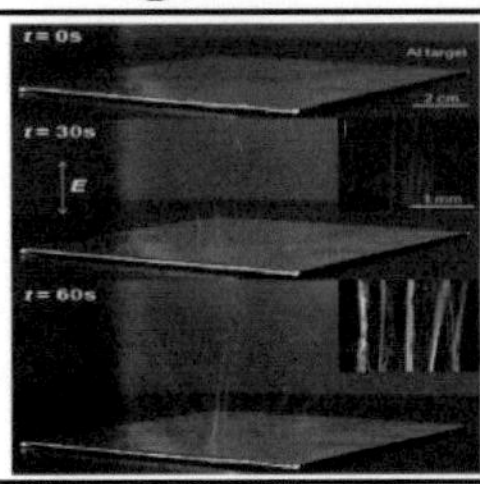		Conductive fiber induces self-bundling; PPV
D. 2008	Wang et al.	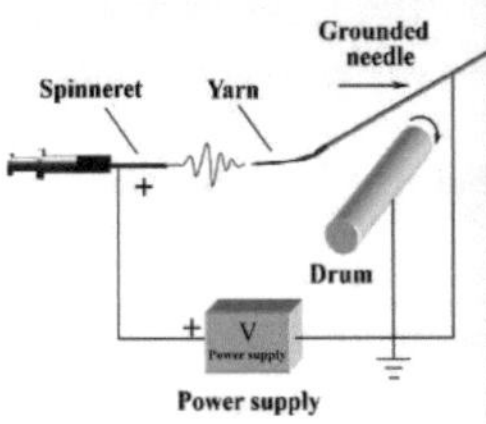	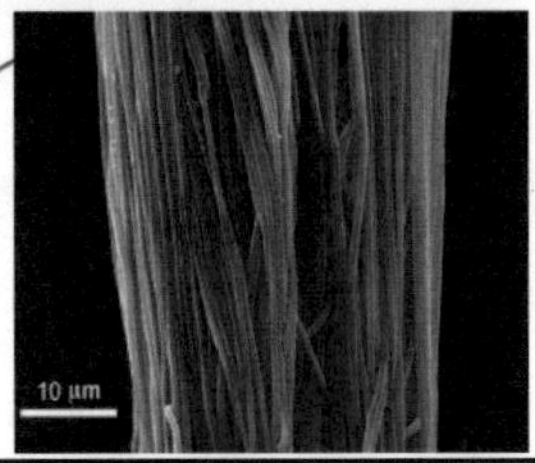	Needle collector induced self-bundling of nanofibers; PHBV
E. 2008	Mondal et al.	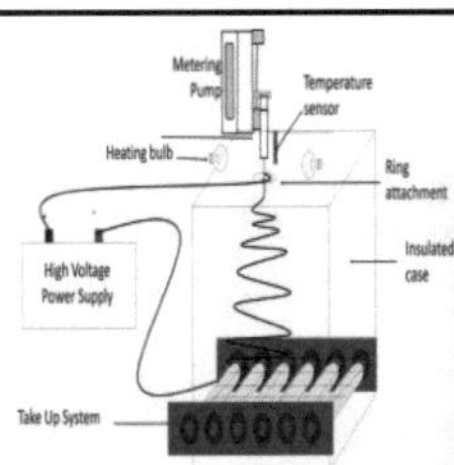		Ring induced self bundling of nanofibers; PAN
F. 2009	Maheshwari et al.	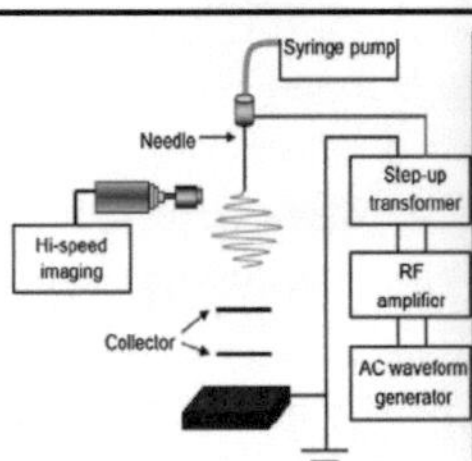	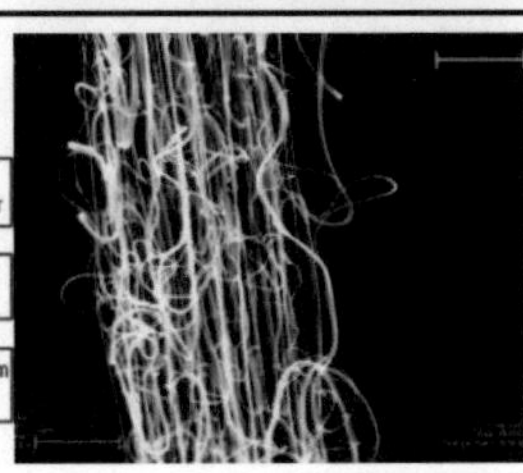	AC induced self-bundling of nanofibers; PVP

Table 3. Continuous Nanofiber Bundles.

continuous nanofiber bundle was generated, provided that the surface velocity of the rotating drum matched the velocity of the bundle growth. The self-bundling became easier when increasing the conductivity of polymer solution. When the solution conductivity was larger than 400 µS/cm, self-bundling occurred automatically without the induction of a needle collector. Greater fiber alignment was observed with the increase in the solution conductivity.
Another approach to prepare electrospun nanofiber yarn was introduced by Mondal et al (Mondal et al., 2008). In their design, the electric field was manipulated by attaching a ring to the spinning nozzle just above the orifice to self-assemble the fibers for yarn formation (Table 3E). To initiate the yarn formation, a glass rod was introduced to attract yarn on the glass rod end. By moving the glass rod away from the spinning zone, a continuous nanofiber yarn was prepared manually. The formation of nanofiber yarn was controlled by temperature, polymer concentration and applied voltage.
Maheshwari et al (Maheshwari and Chang, 2009) reported an assembly of multi-stranded nanofiber threads via AC electrospinning (Table 3F). Nanofibers with both positive and negative segments were electrospun because of polarity switching during electrospinning. The fibers experienced an alternation of attractive and repulsive forces, resulting in increased alignment. A visible thread was observed emerging downstream from the spinneret. The thread could be easily deflected away and collected because of the small attraction toward the counter electrode. The authors also showed how the AC voltage and frequency affected the fiber collection. Increasing the frequency suppressed the jet whipping instability, thus giving smaller thread width (thread with lower fiber density), but more beaded fibers. An opposite effect was seen with the increase in the voltage.

5. Continuous nanofiber yarns

A nanofiber yarn can be formed by twisting a thin strand of nanofibers around the axis. The fibers after twisting show a helical configuration. Twist functions to increase the fiber to fiber cohesion and strength of the yarn. In electrospinning research, considerable effort has gone into making continuous twisted nanofiber yarns in recent time. Table 4 summarizes the main techniques that have been developed to electrospin continuous nanofiber yarns.

Year	Author	Setup	Morphology	Yarn Twisting/Polymer
A. 2003	Ko et al.	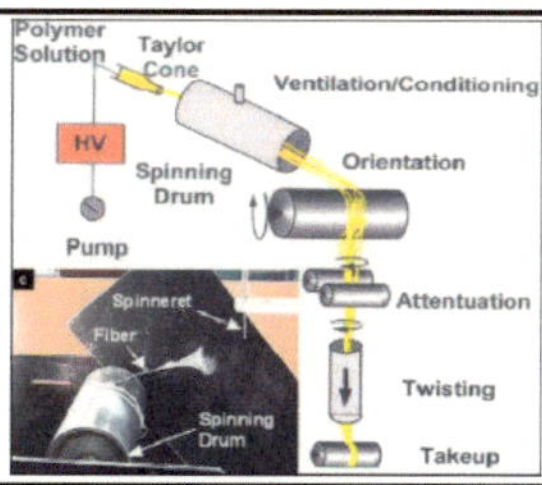	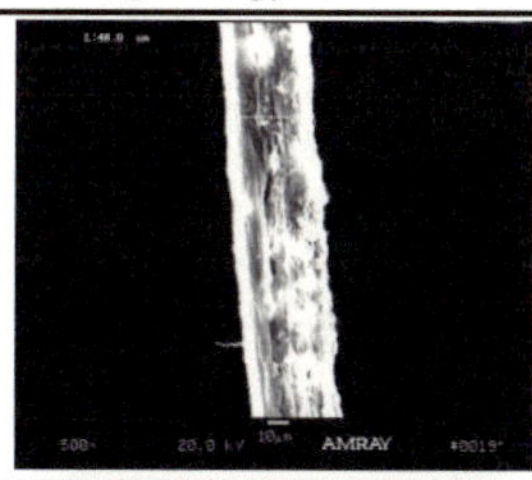	Twisting of nanofibers with a special setup; PAN,PLA
B. 2005	Kim et al.	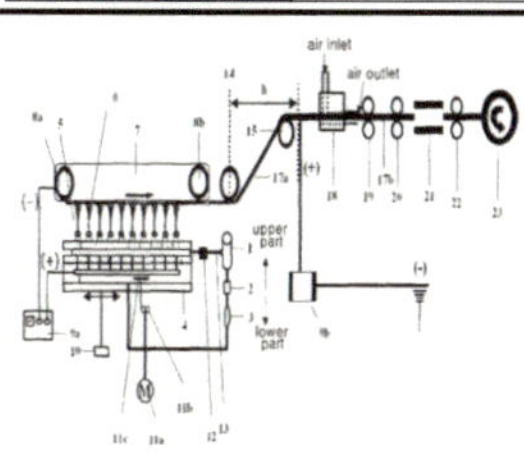		Manually drawing and twisting nanofibers electrospun from a multi-needle setup; Poly urethane

Year	Author	Setup	Morphology	Yarn Twisting/Polymer
C. 2006	Kim et al.	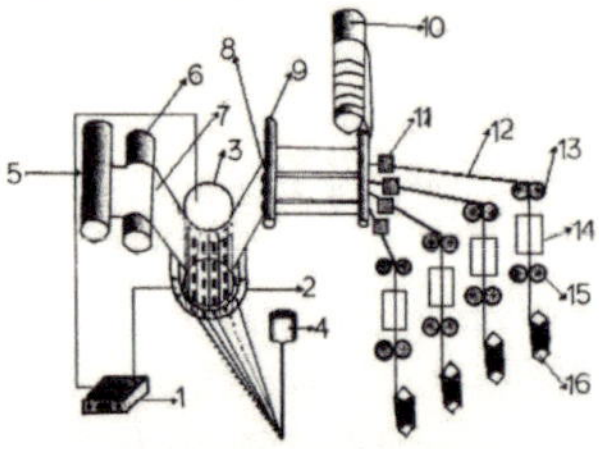		Manually drawing and twisting nanofibers electrospun from a needleless electrospinning setup;
D. 2007 2010	Teo et al. Yousef. et al.	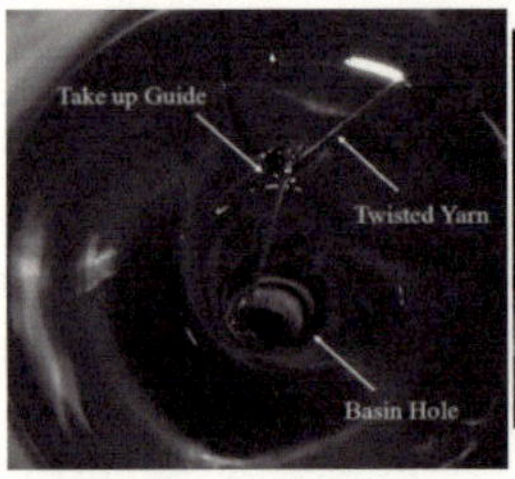	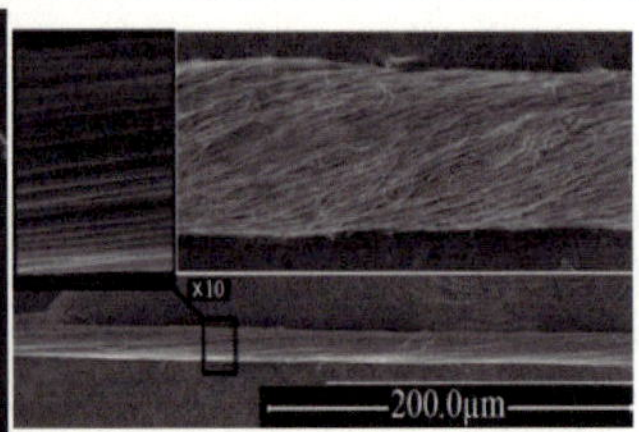	Nanofibers twisted due to rotation of liquid water; polymer (Helix angel 19.76°)
E. 2007	Dabirian et al.	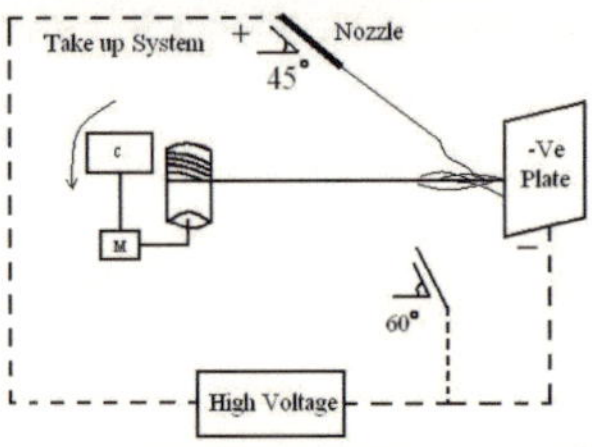		Nanofiber stream redirected and twisted by a twist winder; PAN
F. 2007	Bazbouz et al	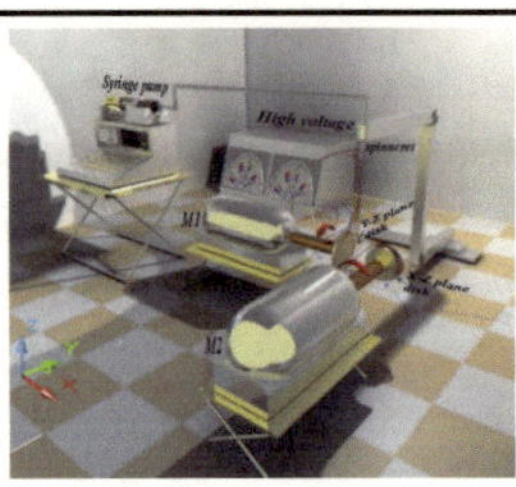	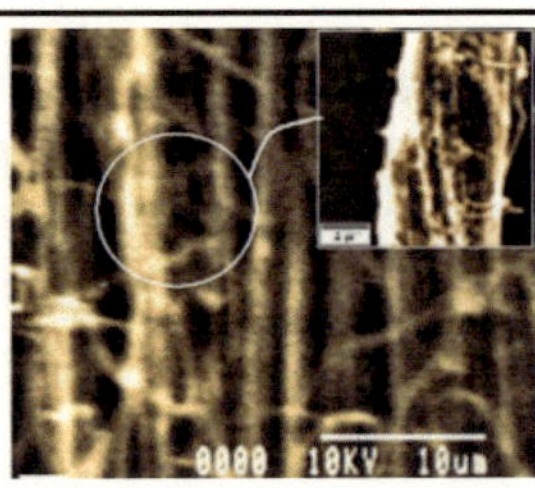	Nanofibers twisted by rotating the left disk and then taken up by the right disk; Nylon6/MWNT
G. 2008	Lee et al.	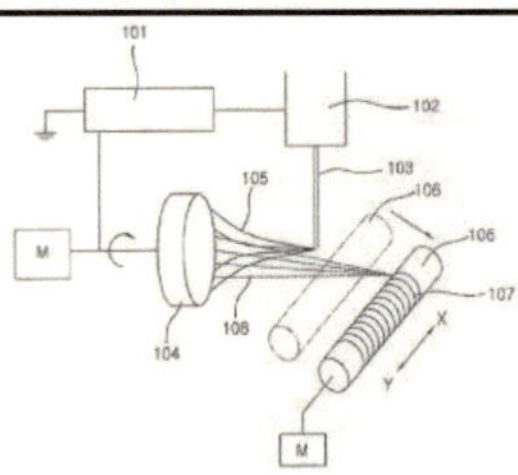	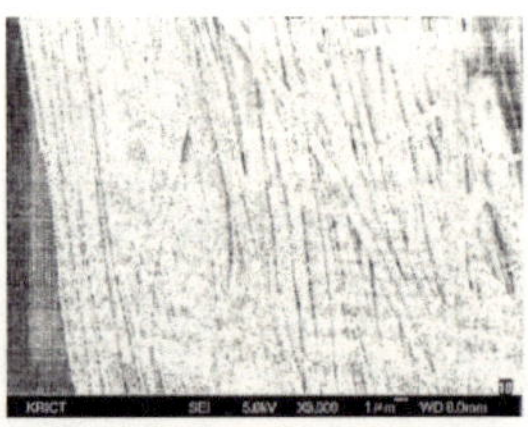	Nanofiber web directly twisted by rotating the disk collector; PAI
H. 2009	Dabirian et al.	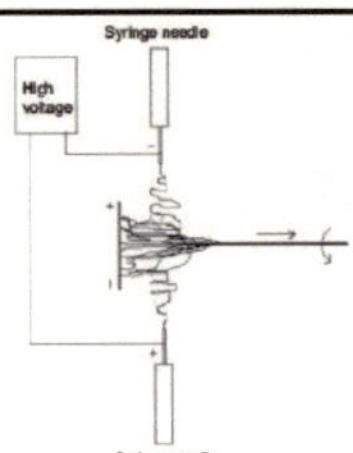	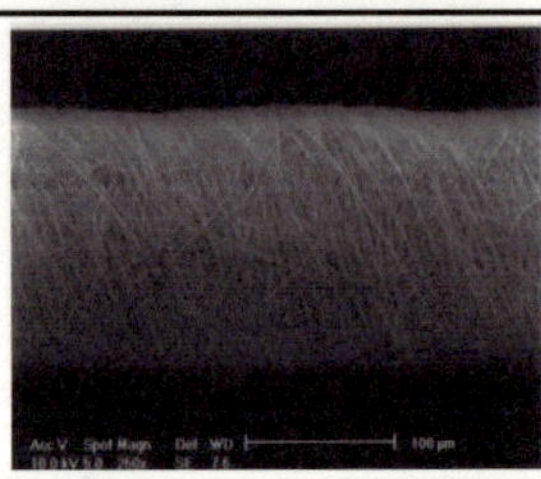	Nanofibers produced from two oppositely charged spinneret and twisted by using a electronic twister; PAN

Year	Author	Setup	Morphology	Yarn Twisting/Polymer
I.				
2009	Kim et al.	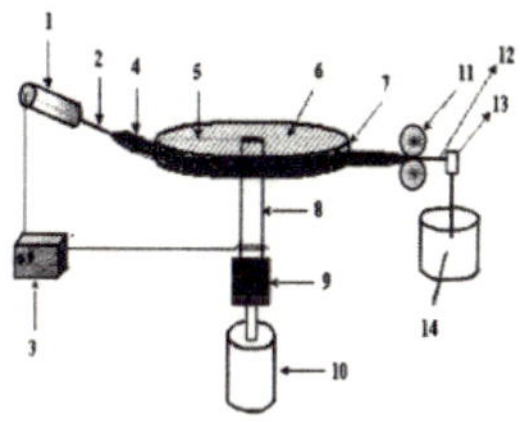	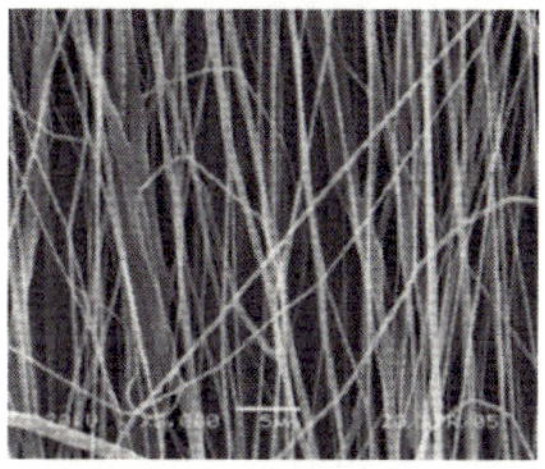	Single and multi nozzles are arranged along the circumferential direction. Nanofibers are drawn and twisted; Nylon6, Nylon66 (Helix angle up to15°)
J.				
2010	Afifi et al.	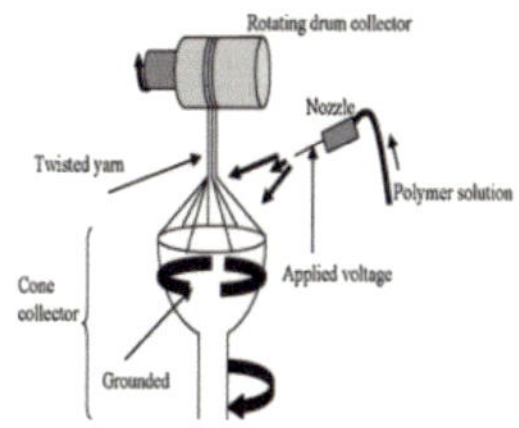	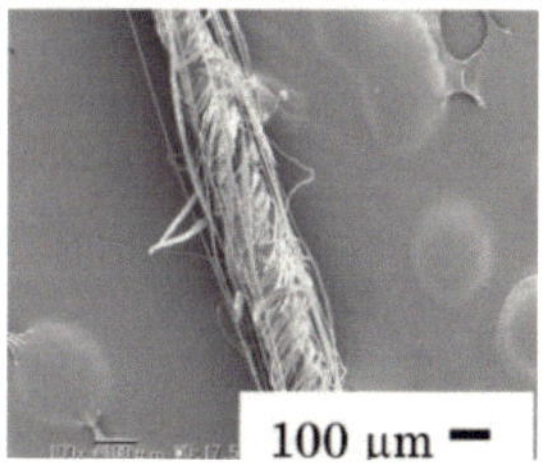	Fiber web was twisted by an earthed rotating funnel; PLLA
K.				
2011	Ali et al.	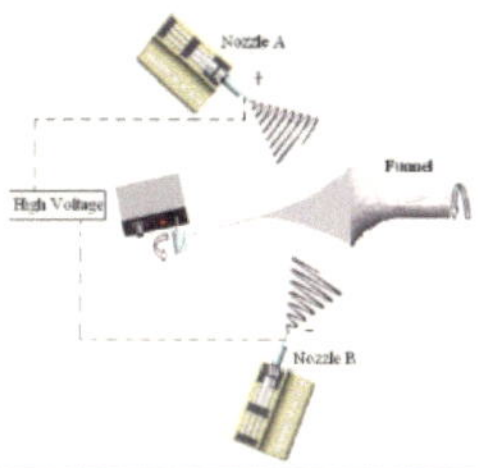	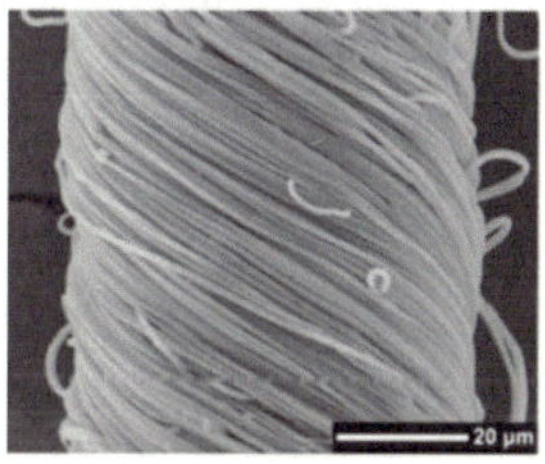	Nanofiber yarns were directly produced from an un-earthed fibrous cone formed on a rotating funnel; PVDF-HFP, PCL, PAN, PS (twist angle up to 0 - 61°)

Table 4. Continuous Nanofiber Yarns.

The concept of using nanofiber yarns to make complex fibrous structures for a new generation of composites was firstly presented by Frank Ko (Ko et al., 2003). Continuous poly(latic acid) (PLA) and PAN nanofiber yarns filled with carbon nanotubes in the fibers were produced by applying a yarn twisting and subsequent winding method (Table 4A). However, no details were presented to explain how the yarns were twisted and continuously wound, considering in particular that the twisting arrangement resembles a false twist texturing process rather than one that inserts any real twist into the yarn.

Kim et al (Kim, 2010a, Kim, 2009a, Kim, 2008b, Kim, 2009b, Kim, 2010b, Kim, 2005) claimed a series of patents to electrospin continuous filament composed of nanofibers. In one of these techniques, a multiple needle electrospinning system was used to produce nonwoven web on a rotating belt which was then stretched and collected on single winding roller (Table 4B) (Kim, 2008b). The nonwoven web was also prepared by a needleless electrospinning setup using a rotating cylinder spinneret before splitting it into several strips (Table 4C). Later on, the group also prepared continuous nanofiber bundles by collecting nanofibers on the circumference of a disc collector (Kim, 2008a) (Table 4I). By using 16,000 nozzles, the group has achieved a maximum speed of 490 m/min with an average fiber diameter of 480 nm.

Teo et al (Teo et al., 2007) demonstrated a technique for continuous formation of nanofiber bundles using the water vortex. Different from previous fiber bundling techniques (Table

3A), the nanofiber bundles were formed inside a liquid (e.g. water), as shown in Table 4D. A take-up rate of 1.05 m/s could be reached in this experiment. The nanofiber yarn had a maximum twist angle of 19.76° (Yousefzadeh et al.).

Dabirian et al (Dabirian et al., 2007) reported a method of continuous nanofiber yarn formation through the manipulation of the electric field. A negative bar and a negative surface were placed within the electric field to direct the electrospinning jet. A spinning triangle was formed as a result (Table 4E). Nanofibers were collected and twisted with a three-phase motor. The twist level and the take-up speed were inter-related. To increase the twist level, the take-up speed has to be compromised. The yarn production rate was about 14 m/hr, whereas the yarn diameter was in the range of 160~170 µm. Later on in 2009, Dabirian et al (Dabirian and Hosseini, 2009) also produced nanofiber yarns by using two oppositely charged nozzles placed on either side of an uncharged collector (Table 4H), and the yarn take-up velocity was 5.76 m/hr.

Bazbouz et al (Bazbouz and Stylios, 2008b) improved the gap collector by using two perpendicularly disks placed 4 cm apart. One disk was rotated to twist the nanofibers, and the other was for continuous yarn winding and both disks rotated at a constant rate (Table 4F). Different twist rates could be applied to provide fiber bundles with different degrees of lateral cohesion and friction. The yarn diameter was 5 ~ 10 µm and dependent on the number of fibers generated at the spinneret, as well as the twist and take-up speeds. The limitation of the method is the fiber sticking, due to the short spinning distance and the low evaporation rate of the solvent.

Lee's group (Lee, 2009, Lee, 2010b, Lee, 2010a, Lee, 2008) prepared a long filament consisting of nanofibers by using a multi-collector technique (Table 4G). They indicated that the first earthed collector should be made of a conductive metal plate or a metal mesh and the second collector a material capable of generating static electricity. The shape of the first collector was not fixed but the second collector must have a round shape like a tube. Different from previous methods based on gap collection, the glass rod is extendable, making the yarn collection continuous. However, due to its non-conductive nature, the fiber alignment could not be matched with the ones prepared by gap collector, where two electrodes are both conductive and grounded.

Afifi et al (Afifi et al.) used a slowly rotating grounded "funnel" to collect PLLA electrospun fiber yarns (Table 4J). The charged jet was ejected toward the grounded funnel target from a diagonal direction, forming a web on the mouth plane. The web was twisted as it was formed, and pulled upward to the winder. Continuous yarns were fabricated thereby, with diameters of around 164 µm. However, the authors only demonstrated that micro-sized fibers with diameters around 6 µm were electrospun for this yarn production.

Recently, Ali et al (Ali et al.) developed a method to directly electrospin highly twisted continuous nanofiber yarns. The setup combined a rotary funnel collector and two oppositely charged spinnerets, as shown in Table 4K. The funnel collector was not earthed. The deposition of nanofibers onto the funnel was largely due to the electrostatic attractions. Nanofibers mainly deposited to cover the funnel forming a fibrous membrane on its mouth. To initiate a twisted yarn, a plastic rod was first placed near the central area of the rotary membrane. When the rod was withdrawn from the funnel, the membrane formed a "fibrous cone" with its apex attaching to the rod, and its edges connecting to the funnel end. Further withdrawing of the cone apex induced the formation of continuous yarns as shown in Figure 2.

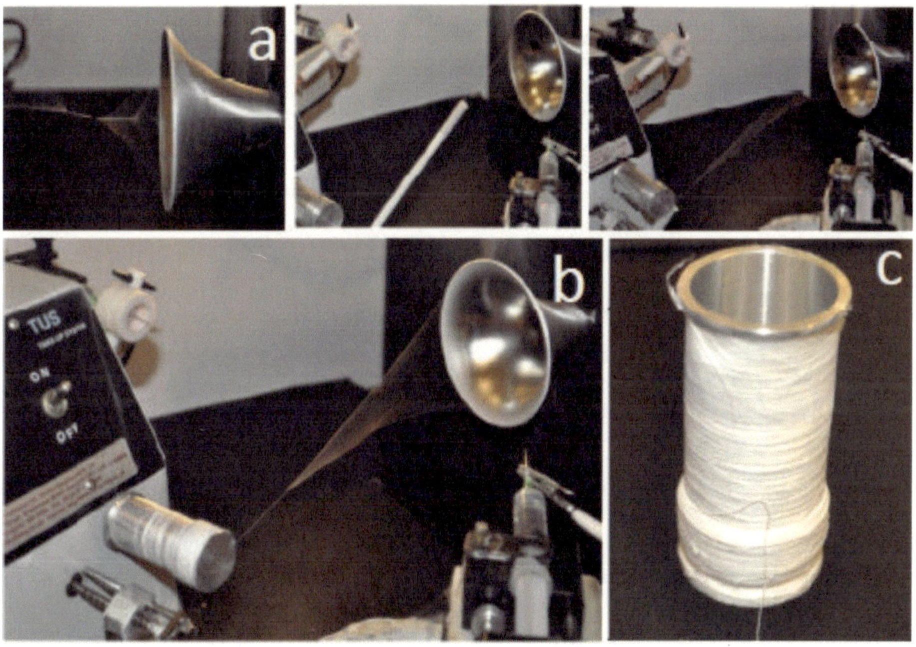

Fig. 2. a) Frames of a video showing the formation of nanofiber cone with the aid of a plastic rod, b) hollow nanofiber cone taken up by continuous drawing with a winder, c) a photograph of as-electrospun nanofiber yarn, d) SEM images of a nanofiber yarn section.

The yarn could be collected and wound incessantly as far as the fibrous cone was maintained dynamically. Kilometers of nanofiber yarns have been drawn from the fibrous cone, with up to 7400 tpm of twist inserted through the rotation of the funnel. The average yarn diameter can be adjusted in the range of 30 ~ 450 μm through the overall solution feeding rate and the twist rate, while the constituent fiber diameter can be changed over a wide range (between 480 - 1500 nm in the study).

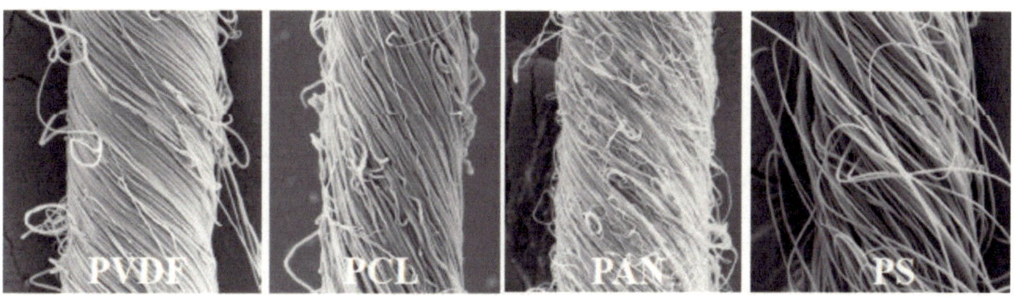

Fig. 3. SEM images of nanofiber yarns produced from different polymers.

Poly(vinylidene fluoride-co-hexafluoropropene) (PVDF-HFP), PAN, polycaprolactone (PCL) and polystyrene (PS) have been successfully electrospun into yarns using this setup. The SEM images of the yarns are shown in Figure 3. This electrospinning mode increased not only the stability of yarn formation but also the yarn spinning rate. In addition, the two spinnerets provide opportunities for fabrication of hybrid nanofiber yarns.

6. Morphology and mechanical properties

A lot of yarn manufacturing technologies have been explored (Zhou and Gong, 2008). It has been generally recognized that both the structure of fibers and their macroscopic arrangement play key roles in yarn mechanics.

Teo et al (Teo et al., 2007) reported the morphology of individual fibers in the yarns and the control through adjusting the electrospinning parameters, such as solution concentration, flow rate, spinning distance and applied voltage. They also reported that the average diameter of nanofiber yarn increased linearly with increasing the feed rate of polymer solution. Ali et al (Ali et al.) study the effect of applied voltage, solution concentration and flow rate on the morphology of nanofiber yarns in detail and indicated that with an increase in the applied voltage the individual fiber diameter decreased but the overall diameter of yarn increased linearly. Not only the individual fiber but also the yarn increased their diameters linearly with polymer concentration and flow rate. The morphology of nanofiber yarns also depended on the winding rate. It was demonstrated that with increasing the winding rate, the diameters of both the individual fiber and the nanofiber yarn decreased linearly (Ali et al.).

It is well known that the strength of conventional yarns increases with increasing the twist level in the yarn, up to a limit. Ali et al (Ali et al.) systematically studied the effect of twist on the mechanical properties of nanofiber yarns. They found that when the twist level was 420 tpm, the yarn tensile strength and elongation at break were 42.0 MPa and 250%, respectively. However the tensile strength started decreasing when the twist level reached 3500 tpm, as shown in Figure 4a. The maximum average tensile strength of PVDF-HFP nanofiber yarn produced from their method was 60.4 MPa. Figure 4b shows a comparison of stress-strain curves for the aligned twisted nanofiber strip, twisted nanofiber yarn and a yarn braid. It shows that when the nanofiber yarn is braided the tensile strength is improved considerably (Fig. 4b).

Electrospun fibers normally possess poorly developed crystalline structure owing to the quick solidification of the spinning jet. Post-treatment such as drawing and annealing used for improving crystalline structure and mechanical strength of conventional fibers has also been applied to increase the crystallinity and yarn tensile strength. By uniaxial drawing PAN nanofiber yarns in boiling water under tension and annealing, higher molecular orientation and increased degree of crystallinity were observed (Wang et al., 2008a). It was indicated that drawing plus annealing was an effective way to improve the strength of nanofiber yarns, approaching the value equivalent to conventional fibers.

Jalili et al (Jalili et al., 2006) also studied the mechanical properties after carrying out a post treatment in boiling water under tension. They found that fiber bundles became stronger but relatively low elongation after the post-treatment. The average values of E-Modulus and tensile strength before the treatment were 2786 ± 200 MPa, 99 ± 12% and after the post-treatments were 4575 ± 220 MPa and 178 ± 15%, respectively.

Apart from twisting, immersing nanofiber yarns in some solvents can also improve yarn strength. For example, when electrospun MWNT/PMMA rope was immersed in methanol overnight and then dried for 8 hours at 40 °C (Liu et al., 2007), the yarn strength was much improved, approaching that of the bulk material. This was attributed to the increased

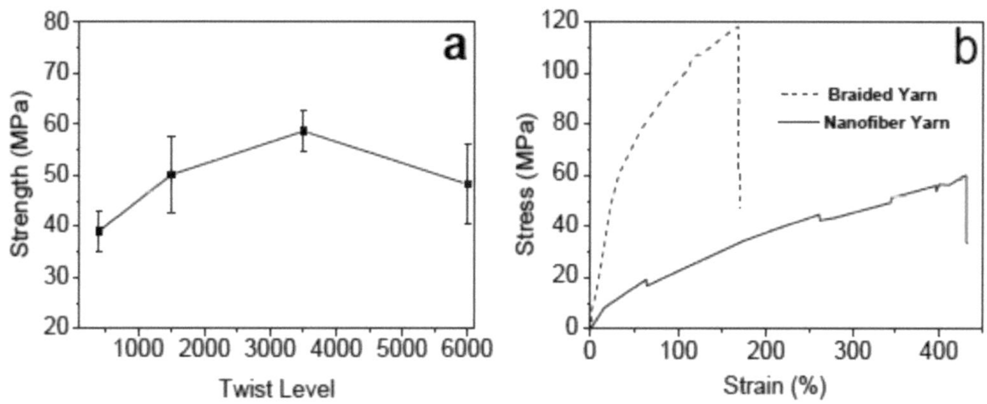

Fig. 4. (a)Influence of twist level on the tensile properties of nanofiber yarns, (b) Typical stress–strain curves of a nanofiber yarn (twist level 3500 tpm), a nanofiber yarn braid (four-ply).

packing density within yarns by reducing the repulsive interactions from the residual surface charge.

Carbon nanotube has been extensively used as a filler material to reinforce polymer nanofibers, as demonstrated and reviewed by many groups. Ko et al (Ko et al., 2003) used CNTs to improve the strength and toughness of nanofiber yarns. A two-fold increase in the modulus was observed in SWNT/PAN yarns. It was explained that polymer matrix was stiffened as a result of interaction with SWNTs. The structural changes of the polymer caused by nanotubes could also be seen from the improved thermal stability. Farouk et al (Farouk et al., 2009)reported that yarn tensile strength increased by about 35% due to MWCNT reinforcement in the PAN polymer with surfactant-aided CNT dispersion. Also the mechanical properties of the well dispersed sample were improved relative to those of the poorly dispersed samples. Similar effects were also observed with MWNT/PMMA composite ropes (Liu et al., 2007).

7. Potential applications

Through weaving, knitting, or other embroidery processes, nanofiber yarns can be assembled into many new fibrous structures, which may find uses in high performance clothing, filters and composites. Nanofiber yarns have been shown their potential for carbon fiber production, electronic devices, and tissue engineering. Many attempts have been devoted to making high quality PAN nanofiber yarns to produce super-strong carbon nanofibers (Moon and Farris, 2009). By twisting yarns composed of metal oxide nanofibers, which displays semi-conducting behavior, a novel p-n junction can be fabricated (Lotus et al., 2009). The typical rectifying current-voltage characteristics of the yarn have potential for constructing sensors and transistor devices. A plain woven fabric of PCL nanofibers has been shown successful for cell cultures, and used as a woven tissue scaffold (Khil et al., 2005).

Another application for yarns is composite reinforcement, to enhance the fracture toughness and damage tolerance. Yarns composed of nanofibers are also envisioned to have such potential, as nanofibers possess much greater surface area and stiffness. Besides these, nanofibers with diameters smaller than the wavelength of visible light should reinforce composite without changing the composite color. However, electrospun nanofibers haven't been used much in composite reinforcement so far.

8. References

Afifi, A. M., Nakano, S., Yamane, H. & Kimura, Y. Electrospinning of continuous aligning yarns with a 'funnel. Target. *Macromolecular Materials and Engineering,* 295, 660-665.

Agarwal, S., Wendorff, J. H. & Greiner, A. 2008. Use of electrospinning technique for biomedical applications. *Polymer,* 49, 5603-5621.

Alborzi, S., Lim, L.-T. & Kakuda, Y. Electrospinning of Sodium Alginate-Pectin Ultrafine Fibers. *Journal of Food Science,* 75, C100-C107.

Ali, U., Zhou, Y., Wang, X. & Lin, T. Direct electrospinning of highly twisted, continuous nanofiber yarns. *Journal of the Textile Institute.* 2011

Anton, F. 1934. *Process and apparatus for preparing artificial threads.* United States patent application 1975504.

Anton, F. 1938a. *Artificial fiber construction.* United States patent application 2109333.

Anton, F. 1938b. *Method and apparatus for the production of fibers.* United States patent application 2123992.

Anton, F. 1944. *Method and apparatus for spinning.* United States patent application 2349950.

Bazbouz, M. B. & Stylios, G. K. 2008a. Alignment and optimization of nylon 6 nanofibers by electrospinning. *Journal of Applied Polymer Science,* 107, 3023-3032.

Bazbouz, M. B. & Stylios, G. K. 2008b. Novel mechanism for spinning continuous twisted composite nanofiber yarns. *European Polymer Journal,* 44, 1-12.

Bellan, L. M., Coates, G. W. & Craighead, H. G. 2006. Poly(dicyclopentadiene) Submicron Fibers Produced by Electrospinning. *Macromolecular Rapid Communications,* 27, 511-515.

Bhardwaj, N. & Kundu, S. C. Electrospinning: A fascinating fiber fabrication technique. *Biotechnology Advances,* 28, 325-347.

Buchko, C. J., Chen, L. C., Shen, Y. & Martin, D. C. 1999. Processing and microstructural characterization of porous biocompatible protein polymer thin films. *Polymer,* 40, 7397-7407.

Buchko, C. J., Kozloff, K. M. & Martin, D. C. 2001. Surface characterization of porous, biocompatible protein polymer thin films. *Biomaterials,* 22, 1289-1300.

Cao, H., Liu, T. & Chew, S. Y. 2009. The application of nanofibrous scaffolds in neural tissue engineering. *Advanced Drug Delivery Reviews,* 61, 1055-1064.

Chang, G. & Shen, J. Fabrication of microropes via bi-electrospinning with a rotating needle collector. *Macromolecular Rapid Communications,* 31, 2151-2154.

Chronakis, I. S. 2005. Novel nanocomposites and nanoceramics based on polymer nanofibers using electrospinning process - A review. *Journal of Materials Processing Technology,* 167, 283-293.

Dabirian, F. & Hosseini, S. A. 2009. Novel method for nanofibre yarn production using two differently charged nozzles. *Fibres and Textiles in Eastern Europe,* 74, 45-47.

Dabirian, F., Hosseini, Y. & Ravandi, S. A. H. 2007. Manipulation of the electric field of electrospinning system to produce polyacrylonitrile nanofiber yarn. *Journal of the Textile Institute,* 98, 237 - 241.

Dalton, P. D., Grafahrend, D., Klinkhammer, K., Klee, D. & Mãller, M. 2007. Electrospinning of polymer melts: Phenomenological observations. *Polymer,* 48, 6823-6833.

Dalton, P. D., Klee, D. & Mãller, M. 2005. Electrospinning with dual collection rings. *Polymer,* 46, 611-614.

Dan Li, Y. W., And Younan Xia* *June 20, 2003*. Electrospinning of Polymeric and Ceramic Nanofibers as Uniaxially Aligned Arrays. *NANO LETTERS,* Vol.3, No. 8, 1167-1171.

Deitzel, J. M., Kleinmeyer, J. D., Hirvonen, J. K. & Beck Tan, N. C. Controlled deposition of electrospun poly(ethylene oxide) fibers. *Online URL: www.arl.army.mil/arlreports/2001/ARL-TR-2415.pdf, 2001*

Deitzel, J. M., Kleinmeyer, J. D., Hirvonen, J. K. & Beck Tan, N. C. 2001. Controlled deposition of electrospun poly(ethylene oxide) fibers. *Polymer,* 42, 8163-8170.

Dersch, R., Graeser, M., Greiner, A. & WendorfF, J. H. 2007. Electrospinning of Nanofibres: Towards New Techniques, Functions, and Applications. *Australian Journal of Chemistry,* 60, 719-728.

Dersch, R., Liu, T., Schaper, A. K., Greiner, A. & Wendorff, J. H. 2003. Electrospun nanofibers: Internal structure and intrinsic orientation. *Journal of Polymer Science, Part A: Polymer Chemistry,* 41, 545-553.

Ding, B., Kim, H. Y., Lee, S. C., Shao, C. L., Lee, D. R., Park, S. J., Kwag, G. B. & Choi, K. J. 2002. Preparation and characterization of a nanoscale poly(vinyl alcohol) fiber aggregate produced by an electrospinning method. *Journal of Polymer Science, Part B: Polymer Physics,* 40, 1261-1268.

Doshi, J. & Reneker, D. H. 1995. Electrospinning process and applications of electrospun fibers. *Journal of Electrostatics,* 35, 151-160.

Ellison, C. J., Phatak, A., Giles, D. W., Macosko, C. W. & Bates, F. S. 2007. Melt blown nanofibers: Fiber diameter distributions and onset of fiber breakup. *Polymer,* 48, 3306-3316.

F. Dotti, A. V., A. Montarsolo, A. Aluigi, & Mazzuchetti*, C. T. A. G. October 2007. Electrospun Porous Mats for High Efficiency Filtration. *Journal of Industrial Textiles,* Vol. 37, No. 2, 151-162.

Farouk, B., Uddin, N. M., Ko, F., Xiong, J. & Capaldi, F. 2009. Process, structure, and properties of electrospun carbon nanotube-reinforced nanocomposite yarns. *Research Letters in Materials Science,* 2009.

Fennessey, S. F. & Farris, R. J. 2004. Fabrication of aligned and molecularly oriented electrospun polyacrylonitrile nanofibers and the mechanical behavior of their twisted yarns. *Polymer,* 45, 4217-4225.

Fong, H., Liu, W., Wang, C. S. & Vaia, R. A. 2002. Generation of electrospun fibers of nylon 6 and nylon 6-montmorillonite nanocomposite. *Polymer,* 43, 775-780.

Frenot, A. & Chronakis, I. S. 2003. Polymer nanofibers assembled by electrospinning. *Current Opinion in Colloid & Interface Science,* 8, 64-75.

Gibson, P., Schreuder-Gibson, H. & Rivin, D. 2001. Transport properties of porous membranes based on electrospun nanofibers. *Colloids and Surfaces A: Physicochemical and Engineering Aspects,* 187-188, 469-481.

Givens, S. R., Gardner, K. H., Rabolt, J. F. & Chase, D. B. 2007. High-temperature electrospinning of polyethylene microfibers from solution. *Macromolecules,* 40, 608-610.

Gu, B. K., Shin, M. K., Sohn, K. W., Kim, S. I., Kim, S. J., Kim, S. K., Lee, H. & Park, J. S. 2007. Direct fabrication of twisted nanofibers by electrospinning. *Applied Physics Letters,* 90.

Huang, Z. M., Zhang, Y. Z., Kotaki, M. & Ramakrishna, S. 2003. A review on polymer nanofibers by electrospinning and their applications in nanocomposites. *Composites Science and Technology,* 63, 2223-2253.

Ishii, Y., Sakai, H. & Murata, H. 2008. A new electrospinning method to control the number and a diameter of uniaxially aligned polymer fibers. *Materials Letters,* 62, 3370-3372.

Jalili, R., Morshed, M. & Ravandi, S. A. H. 2006. Fundamental parameters affecting electrospinning of PAN nanofibers as uniaxially aligned fibers. *Journal of Applied Polymer Science,* 101, 4350-4357.

Jiri Chvojka & Lukas, D. 2009. Electrospun Nanoyarns Produced Using Special Collectors. *Online URL: http://www.nanocon.cz/data/nanocon2009/sbornik/Lists/Papers/127.pdf*

Kameoka, J., Czaplewski, D., Liu, H. & Craighead, H. G. 2004. Polymeric nanowire architecture. *Journal of Materials Chemistry,* 14, 1503-1505.

Katta, P., Alessandro, M., Ramsier, R. D. & Chase, G. G. 2004. Continuous electrospinning of aligned polymer nanofibers onto a wire drum collector. *Nano Letters,* 4, 2215-2218.

Keun Kwon, I., Kidoaki, S. & Matsuda, T. 2005. Electrospun nano- to microfiber fabrics made of biodegradable copolyesters: structural characteristics, mechanical properties and cell adhesion potential. *Biomaterials,* 26, 3929-3939.

Khil, M. S., Bhattarai, S. R., Kim, H. Y., Kim, S. Z. & Lee, K. H. 2005. Novel fabricated matrix via electrospinning for tissue engineering. *Journal of Biomedical Materials Research - Part B Applied Biomaterials,* 72, 117-124.

Kim, H.-Y.-., Pyunghwa-Dong 2 Ga, Wansan-Gu, Jeonju-Si, Jeollabuk-Do, 560-865, Kr), Gil, Myung-Seop (Jeollabuk-Do, Kr), Jung, Yoon-Ho (Jeollabuk-Do, Kr), Kim, Hyung-Jun (Jeollabuk-Do, Kr), Lee, Bong-Seok (Jeollabuk-Do, Kr). 2008a. *Process of preparing continuous filament composed of nano fiber*. United States patent application 7354546.

Kim, H.-Y.-., Pyunghwa-Dong 2ga, Wansan-Gu, Jeonju-Si 560-865, KR), Park, Jong-Cheol (109-1202, LG XI Apartment, 430, Ichon Dong, Yongsan-Gu, Seoul 140-030, KR). 2010a. *Process of preparing continuous filament composed of nanofibers.* United States patent application 7807094.

Kim, H.-Y. C.-D., KR). 2009a. *Process of preparing continuous filament composed of nanofibers.* United States patent application 20090189319.

Kim, H.-Y. C.-D., KR), Park, Jong-Cheol (Seoul, KR). 2008b. *Process of Preparing Continuous Filament Composed of Nanofibers.* United States patent application 20080122142.

Kim, H.-Y. J.-D., KR), GIL, MYUNG-SEOP (JEOLLABUK-DO, KR), JUNG, YOON-HO (Jeollabuk-Do, KR), Kim, Hyung-Jun (Jeollabuk-Do, KR), Lee, Bong-Seok

(Jeollabuk-Do, KR). 2005. *Process of preparing continuous filament composed of nano fiber*. United States patent application 20050253305.

Kim, H.-Y. J.-S., KR). 2009b. *Method Of Manufacturing A Continuous Filament By Electrospinning And Continuous Filament Manufactured Thereby*. United States patent application 20090324950.

Kim, H.-Y. J.-S., KR). 2010b. *Method of manufacturing a continuous filament by electrospinning*. United States patent application 7799262.

Ko, F., Gogotsi, Y., Ali, A., Naguib, N., Ye, H., Yang, G. L., Li, C. & Willis, P. 2003. Electrospinning of Continuous Carbon Nanotube-Filled Nanofiber Yarns. *Advanced Materials,* 15, 1161-1165.

Lee, J.-R. D., KR), Jee, Seung-Yong (Daejun, KR), Kim, Hyo-Jung (Daejun, KR), Hong, Young-Taik (Daejun, KR), Kim, Seok (Daejun, KR), Park, Soo-Jin (Daejun, KR). 2008. *Filament Bundle Type Nano Fiber and Manufacturing Method Thereof*. United States patent application 20080241538.

Lee, J.-R. D., KR), Jee, Seung-Yong (Daejun, KR), Kim, Hyo-Jung (Daejun, KR), Hong, Young-Taik (Daejun, KR), Kim, Seok (Daejun, KR), Park, Soo-Jin (Daejun, KR). 2010a. *Filament Bundle Type Nano Fiber And Manufacturing Method Thereof*. United States patent application 20100021732.

Lee, J.-R. Y.-G., KR), Jee, Seung-Yong (Daejun, KR), Kim, Hyo-Joong (Daejun, KR), Hong, Young-Taik (Daejun, KR), Kim, Seok (Daejun, KR), Park, Soo-Jin (Daejun, KR). 2009. *Filament Bundle Type Nano Fiber and Manufacturing Method Thereof*. United States patent application 20090117380.

Lee, J. R. D., KR), Jee, Seung Yong (Daejun, KR), Kim, Hyo Joong (Daejun, KR), Hong, Young Taik (Daejun, KR), Kim, Seok (Daejun, KR), Park, Soo Jin (Daejun, KR). 2010b. *Filament bundle type nano fiber and manufacturing method thereof*. United States patent application 7803460.

Lee, S. & Obendorf, S. K. 2007. Use of electrospun nanofiber web for protective textile materials as barriers to liquid penetration. *Textile Research Journal,* 77, 696-702.

Li, D., Wang, Y. & Xia, Y. 2004. Electrospinning Nanofibers as Uniaxially Aligned Arrays and Layer-by-Layer Stacked Films. *Advanced Materials,* 16, 361-366.

Li, D. & Xia, Y. 2004. Electrospinning of Nanofibers: Reinventing the Wheel? *Advanced Materials,* 16, 1151-1170.

Li, L., Zhang, P., Liu, R. & Guo, S. M. Preparation of fibrous Ni-coated-YSZ anodes for solid oxide fuel cells. *Journal of Power Sources,* 196, 1242-1247.

Li, X., Yao, C., Sun, F., Song, T., Li, Y. & Pu, Y. 2008. Conjugate electrospinning of continuous nanofiber yarn of poly(L-lactide)/nanotricalcium phosphate nanocomposite. *Journal of Applied Polymer Science,* 107, 3756-3764.

Liang, D., Hsiao, B. S. & Chu, B. 2007. Functional electrospun nanofibrous scaffolds for biomedical applications. *Advanced Drug Delivery Reviews,* 59, 1392-1412.

Liu, C. K., Sun, R. J., Lai, K., Sun, C. Q. & Wang, Y. W. 2008. Preparation of short submicron-fiber yarn by an annular collector through electrospinning. *Materials Letters,* 62, 4467-4469.

Liu, L.-Q., Eder, M., Burgert, I., Tasis, D., Prato, M. & Daniel Wagner, H. 2007. One-step electrospun nanofiber-based composite ropes. *Applied Physics Letters,* 90, 083108-083108-3.

Liu, L. & Dzenis, Y. A. 2008. Analysis of the effects of the residual charge and gap size on electrospun nanofiber alignment in a gap method. *Nanotechnology,* 19.

Lotus, A. F., Bender, E. T., Evans, E. A., Ramsier, R. D., Reneker, D. H. & Chase, G. G. 2008. Electrical, structural, and chemical properties of semiconducting metal oxide nanofiber yarns. *Journal of Applied Physics,* 103, 6.

Lotus, A. F., Bhargava, S., Bender, E. T., Evans, E. A., Ramsier, R. D., Reneker, D. H. & Chase, G. G. 2009. Electrospinning route for the fabrication of <equation>p-n</equation> junction using nanofiber yarns. *Journal of Applied Physics,* 106, 014303-014303-4.

Lyons, J., Li, C. & Ko, F. 2004. Melt-electrospinning part I: Processing parameters and geometric properties. *Polymer,* 45, 7597-7603.

Madhavamoorthi, P. 2005. "Nanofibers and nanofiber web": A new class of nonwovens. *Synthetic Fibres,* 34, 12-18.

Maheshwari, S. & Chang, H.-C. 2009. Assembly of Multi-Stranded Nanofiber Threads through AC Electrospinning. *Advanced Materials,* 21, 349-354.

Mondal, A., Borah, R., Mukherjee, A., Basu, S., Jassal, M. & Agrawal, A. K. 2008. Electrospun self-assembled nanofiber yarns. *Journal of Applied Polymer Science,* 110, 603-607.

Moon, S. & Farris, R. J. 2009. Strong electrospun nanometer-diameter polyacrylonitrile carbon fiber yarns. *Carbon,* 47, 2829-2839.

Nakata, K., Fujii, K., Ohkoshi, Y., Gotoh, Y., Nagura, M., Numata, M. & Kamiyama, M. 2007. Poly(ethylene terephthalate) nanofibers made by sea-island-type conjugated melt spinning and laser-heated flow drawing. *Macromolecular Rapid Communications,* 28, 792-795.

Norris, I. D., Shaker, M. M., Ko, F. K. & Macdiarmid, A. G. 2000. Electrostatic fabrication of ultrafine conducting fibers: Polyaniline/polyethylene oxide blends. *Synthetic Metals,* 114, 109-114.

Okuzaki, H., Takahashi, T., Hara, Y. & Yan, H. 2008. Uniaxially aligned carbon nanofibers derived from electrospun precursor yarns. *Journal of Polymer Science, Part B: Polymer Physics,* 46, 305-310.

P. J Denton & Daniels, P. N. 2002. Textile terms and definitions. , *A Textile Institute.*

Pan, H., Li, L., Hu, L. & Cui, X. 2006. Continuous aligned polymer fibers produced by a modified electrospinning method. *Polymer,* 47, 4901-4904.

Ramakrishna, S., Fujihara, K., Teo, W. E., Yong, T., Ma, Z. W. & Ramaseshan, R. 2006. Electrospun nanofibers: solving global issues. *Materials Today,* 9, 40-50.

Schreuder-Gibson, H., Gibson, P., Senecal, K., Sennett, M., Walker, J., Yeomans, W., Ziegler, D. & TSAI, P. P. 2002. Protective textile materials based on electrospun nanofibers. *Journal of Advanced Materials,* 34, 44-55.

Shabani, I., Hasani-Sadrabadi, M. M., Haddadi-Asl, V. & Soleimani, M. Nanofiber-based polyelectrolytes as novel membranes for fuel cell applications. *Journal of Membrane Science,* 368, 233-240.

Smit, E., Buttner, U. & Sanderson, R. D. 2005. Continuous yarns from electrospun fibers. *Polymer,* 46, 2419-2423.

Supaphol, P., Mit-Uppatham, C. & Nithitanakul, M. 2005. Ultrafine Electrospun Polyamide-6 Fibers: Effects of Solvent System and Emitting Electrode Polarity on Morphology and Average Fiber Diameter. *Macromolecular Materials and Engineering,* 290, 933-942.

Tan, E. P. S., Ng, S. Y. & Lim, C. T. 2005. Tensile testing of a single ultrafine polymeric fiber. *Biomaterials,* 26, 1453-1456.

Teo, W. E., Gopal, R., Ramaseshan, R., Fujihara, K. & Ramakrishna, S. 2007. A dynamic liquid support system for continuous electrospun yarn fabrication. *Polymer,* 48, 3400-3405.

Teo, W. E. & Ramakrishna, S. 2005. Electrospun fibre bundle made of aligned nanofibres over two fixed points. *Nanotechnology,* 16, 1878-1884.

Theron, A., Zussman, E. & Yarin, A. L. 2001. Electrostatic field-assisted alignment of electrospun nanofibres. *Nanotechnology,* 12, 384-390.

Viney, C. 2004. Self-assembly as a route to fibrous materials: concepts, opportunities and challenges. *Current Opinion in Solid State and Materials Science,* 8, 95-101.

Wang, X., Lee, S. H., Drew, C., Senecal, K. J., Kumar, J. & Samuelson, L. A. Year. Highly sensitive optical sensors using electrospun polymeric nanofibrous membranes. *In:* BAO, Z., BULOVIC, V., ERMER, S., JEN, A. & MAALIARAS, G., eds. Materials Research Society Symposium - Proceedings, 2002a Boston, MA. 397-402.

Wang, X., Zhang, K., Zhu, M., Hsiao, B. S. & Chu, B. 2008a. Enhanced Mechanical Performance of Self-Bundled Electrospun Fiber Yarns via Post-Treatments. *Macromolecular Rapid Communications,* 29, 826-831.

Wang, X., Zhang, K., Zhu, M., Yu, H., Zhou, Z., Chen, Y. & Hsiao, B. S. 2008b. Continuous polymer nanofiber yarns prepared by self-bundling electrospinning method. *Polymer,* 49, 2755-2761.

Wang, X. Y., Drew, C., Lee, S. H., Senecal, K. J., Kumar, J. & Samuelson, L. A. 2002b. Electrospinning technology: A novel approach to sensor application. *Journal of Macromolecular Science-Pure and Applied Chemistry,* A39, 1251-1258.

Xing, S., Zheng, H. & Zhao, G. 2008. Preparation of polyaniline nanofibers via a novel interfacial polymerization method. *Synthetic Metals,* 158, 59-63.

Xu, C. Y., Inai, R., Kotaki, M. & Ramakrishna, S. 2004. Aligned biodegradable nanotibrous structure: a potential scaffold for blood vessel engineering. *Biomaterials,* 25, 877-886.

Yao, C., Li, X. & Song, T. 2009. Preparation and characterization of zein and zein/poly-L-lactide nanofiber yarns. *Journal of Applied Polymer Science,* 114, 2079-2086.

Yoshimoto, H., Shin, Y. M., Terai, H. & Vacanti, J. P. 2003. A biodegradable nanofiber scaffold by electrospinning and its potential for bone tissue engineering. *Biomaterials,* 24, 2077-2082.

Yousefzadeh, M., Latifi, M., Teo, W.-E., Amani-Tehran, M. & Ramakrishna, S. Producing continuous twisted yarn from well-aligned nanofibers by water vortex. *Polymer Engineering & Science,* 51, 323-329.

Zhao, J., Han, W., Chen, H., Tu, M., Zeng, R., Shi, Y., Cha, Z. & Zhou, C. Preparation, structure and crystallinity of chitosan nano-fibers by a solid-liquid phase separation technique. *Carbohydrate Polymers,* 83, 1541-1546.

Zhou, F.-L. & Gong, R.-H. 2008. Manufacturing technologies of polymeric nanofibres and nanofibre yarns. *Polymer International,* 57, 837-845.

Part 2

Property and Characterization

9

The Microstructure Characterization and the Mechanical Properties of Electrospun Polyacrylonitrile-Based Nanofibers

Chen Zhang, Xuejia Ding and Sizhu Wu
Key Laboratory of Carbon Fiber and Functional Polymers, Ministry of Education, College of Materials Science & Engineering, Beijing University of Chemical Technology, Beijing, China

1. Introduction

Electrospinning provides a straightforward and cost-effective approach to produce fibers from polymer solutions or melts having the diameters ranging from sub-microns to nanometers (Reneker & Chun 1996; Dzenis 2004; Hammel et al. 2004; Greine & Wendorff 2007). The nanofiber fabrication by the electrospinning technique can also provides high orientation of the polymer chains by some special colleting accessories due to the high elongation forces that developed during the electrospinning process. This process is accompanied by massive solvent evaporation which leads to fibers with high packaging density (Doshi & Reneker 1995). The carbon fibers possess high mechanical strengths and moduli, superior stiffness, excellent thermal and electrical conductivities, as well as strong fatigue and corrosion resistance; therefore carbon nanofibers have recently received increased attention for their potential applications as composite reinforcing fillers, heat-management materials, high-temperature catalysts, membrane-based separation media, and components in nanoelectronics and photonics etc. (Chand 2000). There are several precursors for the production of carbon fiber, such as polyacrylonitrile (PAN), pitch and cellulose (Edie 1998; Basova et al. 2004) etc. But PAN is the most widely used precursor for manufacturing high-performance fibers due to its combination of tensile and compressive properties as well as its high carbon yield (Sutasinpromprae et al. 2006). Usually, the PAN or PAN-based carbon fiber is produced by the following steps: (1) first, spinning the precursor under different drawing ratio, (2) then, the precursor fiber is stabilized at 200 to ca. 300°C in an oxidation atmosphere; (3) finally, the stabilized fiber is heated at temperatures from 1200 to ca. 3000 °C in an inert atmosphere to wipe off nearly all of the non-carbon element to obtain the final carbon fibers. There are two critical factors in these manufacturing processes of PAN carbon fibers which one is to control the suitable orientation and crystallization of the precursors and the other is the post-treatment steps such as pre-oxidization, stabilization and carbonization. Conventional PAN-based carbon fibers typically have diameters ranging from 5 to 10 um (Morgan 2005). However, the electrospun PAN nanofibers are uniform with the diameters of approximately 300 nm (Chun et al. 1999; MacDiarmid et al. 2001), which is more than 30 times smaller than their conventional counterparts. The high specific surface area of electrospun polymer and carbon nanofibers leads to the enhanced properties in various applications such as electrodes in fuel cells and supercapacitors. The subsequent thermal treatments, including pre-oxidization, stabilization and carbonization, convert the PAN precursor nanofibers into carbon nanofibers that are very long (continuous) probably with desired microstructural, electrical, mechanical, and other properties. However, the mechanical properties of electrospun nanofibers need to be improved because of the limited crystallinity and orientation during the electrospinning (Li et al. 2003). The microstructures and the related mechanical and/or electrical properties of the electrospun carbon nanofibers are still largely unknown.

Carbon nanotubes (CNTs) have achieved a momentous research and application interest due to their unique properties, such as high tensile modulus, good heat and electrical conduction, unique optical and electronic properties, and so forth. It is found that reinforcement of polymers by CNTs may significantly improve their mechanical properties, thermal stability, electric conductivity, and other functional properties. For enhancing the properties of the nanofibers, the carbon nanotubes (CNTs) have also tried to introduce in the PAN-based composites due to their appealing mechanical (Wong et al. 1997; Yu et al. 2000), electrical (Javey et al. 2003), and thermal conductivity properties (Miaudet et al. 2007). Both single-walled carbon nanotubes (SWNTs) and multiwalled carbon nanotubes (MWNTs) are of scientific interest in nanotechnology and nanodevices because of their superior structural, mechanical, chemical, thermal, and electrical performance (Iijima 1991; Baughman et al. 2002; Collins et al. 2001). The SWNT reinforced pitch-based carbon microfibers(Andrews et al. 1999) and PAN/CNTs microfibers (Sreekumar et al. 2004) exhibited a significant improvement in mechanical properties and thermal stability. It has been shown that significant interactions exist between PAN chains and CNTs, which lead to higher orientation of PAN chains during the heating process (Rilutsky et al. 2010). It is also shown that CNTs embeds in electrospun PAN nanofiber serve as nucleation centers for graphitic structure formation during the carbonization process leading to a more ordered and oriented structure. For pre-oxidization, stabilization and carbonization, it is noteworthy that applying tension during thermal treatments, particularly during stabilization, is crucial for the development of carbon fibers with high mechanical strength; i.e., if the stabilization is carried out without tension, the resulting carbon fibers are mechanically weak (Donnet et al. 1998; Morgan 2005). However, before full realization of their reinforcing improvement, the following two crucial issues have to be solved: (i) dispersion and orientation of CNTs in the nanofiber (Chandrasekar et al. 2009; Hao et al. 2006), good interfacial bonding is required to achieve load transfer across the CNTs-matrix interface (Chen et al. 2006); (ii) the macroscopic alignment in the nanofibers (Na et al. 2009) and the orientation and crystallinity of polymer chains. Therefore, in this chapter, the manufacturing process and characterization methods for the microstructures and mechanical properties of PAN and PAN-based nanofibers has been reviewed.

2. Manufacturing and characterization of the PAN and CNTs/PAN composites nanofibers

2.1 Manufacturing of the PAN and PAN-based nanofibers

The PAN and PAN-based nanofibers can be made by electrospinning with the nominal electric field on the order of 1 kV/cm. Fig.1 is the schematic diagram of the electrospinning technique. In the process the PAN or CNTs/PAN solution is held by its surface tension at the end of a capillary, such as a stainless steel needle. As the intensity of the electric field is increased, usually by increasing the voltage, the hemispherical surface of the solution at the tip of the capillary tube elongates to form a structure named as a Taylor cone (Taylor 1969). At a certain voltage the electrical forces overcome the surface tension and a jet ejects from the Taylor cone. The jet travels some distance and then a whipping instability begins to further attenuate the jet into nanofibers. The fibers are collected onto a counter electrode, such as a screen, drum, plate, or the edge of a rotating disk. Coupled with the usual observations the fact that fibers can be electrospun using an Ac field indicates that a portion of the attenuation occurs in the stable jet region. The Dc electrospun fibers are smaller than Ac spun fibers indicates further drawing occurs in the instability region. The voltage between the electrode and the counter electrode could be controlled by the high voltage power supply such as setting at 14-16 kV. The collector rotated at some surface speed such as 6.6m/s, which the high speed rotating collector could align the nanofibers into the nanofiber sheets.

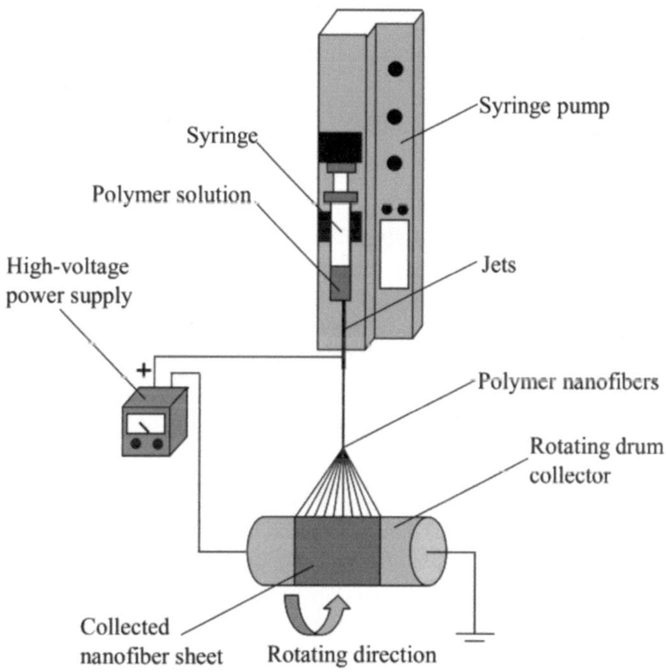

Fig. 1. Schematic diagram of the electrospinning technique.

Since N,N-dimethylformamide(DMF) is the common solvent of PAN-based carbon fibers manufacturing process which can easily evaporate during the electrospinning, the DMF can be selected as solvent and are through mild bath sonication to have a completely dissolution. Usually the suitable weight concentration of PAN/DMF can be selected higher than 10% with the PAN average molecular weight of 100,000 g/mol. Fig. 2 shows the SEM micrographs of different concentrations of electrospinning PAN/DMF solutions of 8 wt %, 10 wt % and 12 wt % respectively. At the 8 wt% PAN/DMF ratio (Fig. 2A), viscous forces within the jet are insufficient to stabilize disturbances, leading to capillary breakup. At 10 wt% and 12 wt% PAN/DMF ratios, irregular and regular beads formed, respectively (Fig. 2B,C). Therefore it can be said as the PAN/DMF ratio increases higher than 10wt%, viscous forces stabilize disturbances at the jet surface, leading to a beads-onstring configuration. However, if the concentration was too high, controlling and maintaining a stable flow rate becomes very difficult because the viscosity of the polymer solution is high.

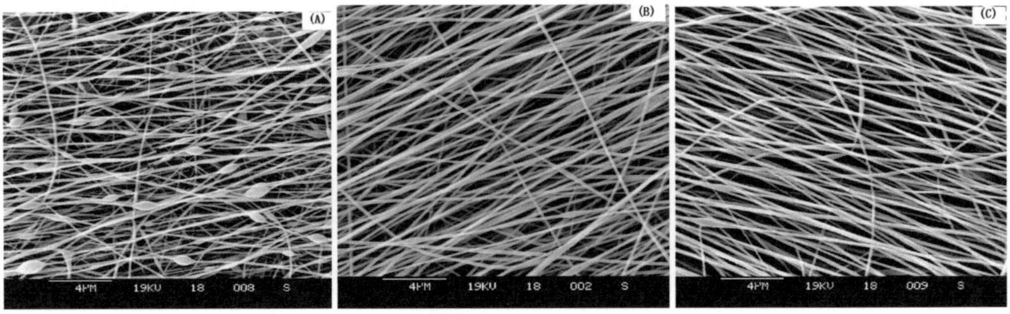

Fig. 2. SEM micrographs of PAN/DMF different concentrations of electrospinning nanofibers with the solutions of (A)8 wt %; (B)10 wt %; (C)12 wt %.

To uniformly disperse the CNTs in the organic polymer matrix, the CNTs are modified to form an individually polymer-wrapped structure (Waclawik et al. 2006). As is well known, polymer wrapped greatly inhibits the Van der Waals attraction between the polymer with solvent and the interactional polymer chains which normally observed between separate SWNTs with small ropes of SWNTs, such as the SWNTs can be wrapped by regioregular poly(3-hexyl thiophene(rrP3HT). These effects caused the wrapped nanotubes to be much more readily suspended in concentrated SWNTs solutions and suspensions, which in turn substantially enabled manipulation of SWNTs into various bulk materials, including films, fibers, solids, and composites of all kinds (Smalley et al. 2007).

For example, a 0.75 wt% SWNTs based PAN composite solution is prepared as follows; (1) first, a given weight of SWNTs is first dispersed for 2 h in DMF through mild bath sonication, which is followed by the addition of PAN (128.91 mg per milliliter of SWNTs/DMF solution); (2) then, the mixture is mechanically stirred overnight at 40 °C using a magnetic stirrer to yield a homogeneous solution. Fig.3 SEM shows the micrographs of PAN and PAN/SWNTs nanofibers by electrospinning with different SWNTs concentrations.

The relatively aligned PAN nanofibers and PAN/SWNTs composite nanofibers can be obtained by electrospinnin by the set-up of a rotating instrument. Thus, for collection of the large area aligned nanofibers, a parallel rotating drum can be adopted. Such as in our study, the 0.16 m perimeter collector rotated at a surface speed about 6.6 m/s, that the high speed rotating collector could align the nanofibers into the nanofiber sheets. Fig. 4 shows the schematic diagram of the frame to prepare the aligned nanofibers films with (a) the 16cm×12cm paper frame with the hollow of 16cm×4cm wrapping around (b) the rotating drum and (c) the sheet after electrospinning. And the SEM photographs of different speeds of rotation are shown in Fig. 5, which indicates the higher rotating speed leads to higher alignment of the nanofibers.

Fig. 3. SEM micrographs of PAN and PAN/SWNTs nanofibers with different SWNTs concentration (a)0wt% (b)0.25wt% (c)0.5wt% (d)0.75wt% (e)1wt%.

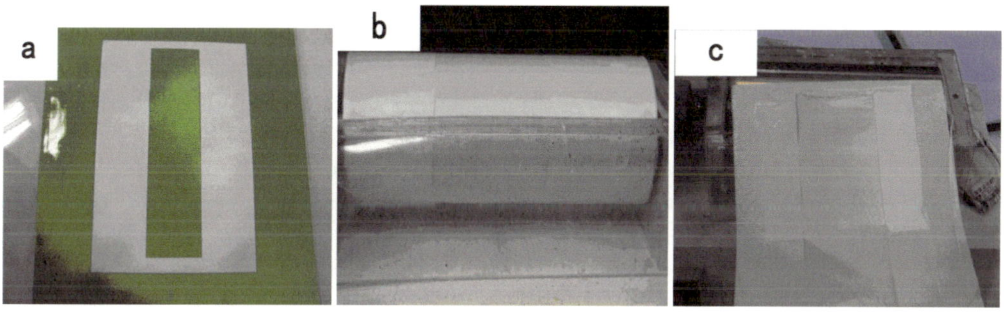

Fig. 4. Schematic diagram of the frame to prepare the aligned nanofibers films (a) the 16cm×12cm paper frame with the hollow of 16cm×4cm (b) on the rotating drum (c) after electrospinning.

During the electrospinning process, however, the whirlpool jet from the pinhead to the collector still made it difficult to get the unidirectional alignment in a large-area sheet (Huang et al. 2003), making the subsequent hot-stretched procedure particularly useful, which is also the key process during the manufacturing of carbon fibers. Therefore, the electrospun nanofibers needed a subsequent hot-stretch to improve the fiber alignment. The PAN nanofibers and PAN/SWNTs composite nanofibers can be hot-stretched according to the method proposed by Phillip and Johnson (Johnson 1965; 1966). Some methods have been tried, such as the both ends of the nanofiber sheet (size of 4 cm width × 10 cm length with 17 μm thickness) can be clamped with the pieces of graphite plates. Then, one end was fixed to the ceiling of the oven and the other end can be weighted by some of metal poise (75g) to give a desired tension and elongation in the temperature-controlled oven at 135 ± 2°C for 5 min. The schematic diagram of hot-stretching of nanofiber sheet is shown in Fig. 6.

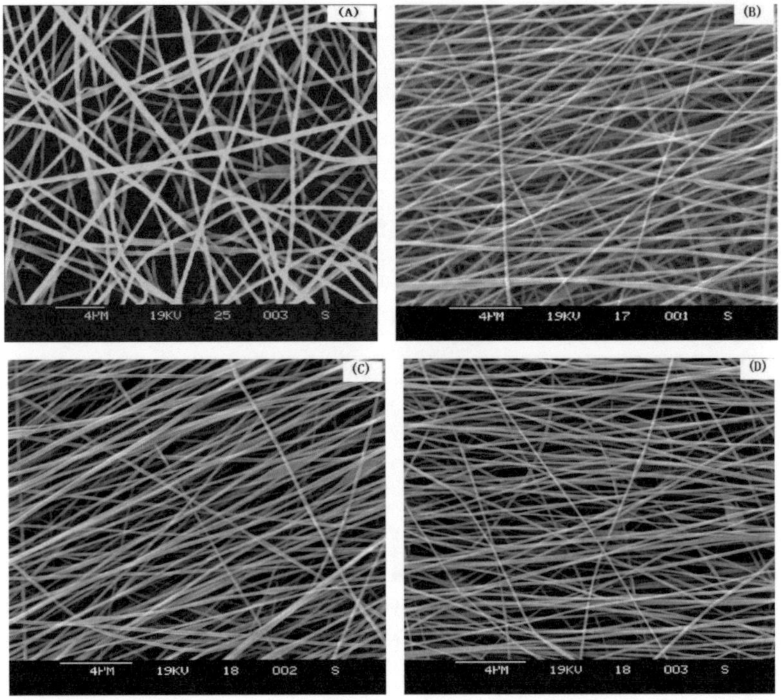

Fig. 5. SEM photographs of different speed of rotation (a) non-woven fabrics, (b) 6.28 m/s, (c) 10.47 m/s, and(d) 14.65m/s.

The stretching ratio, λ, can be calculated from $\lambda = L/L_0$, where L and L_0 are the lengths of nanofiber sheet after and before the hot-stretching, respectively. Fig. 7 is the SEM micrographs of as-spun and hot-stretched PAN nanofibers, and the cross sectional views of as-spun and hot-stretched one respectively. It can be seen that the high alignment and density can be achieved by hot-stretched process.

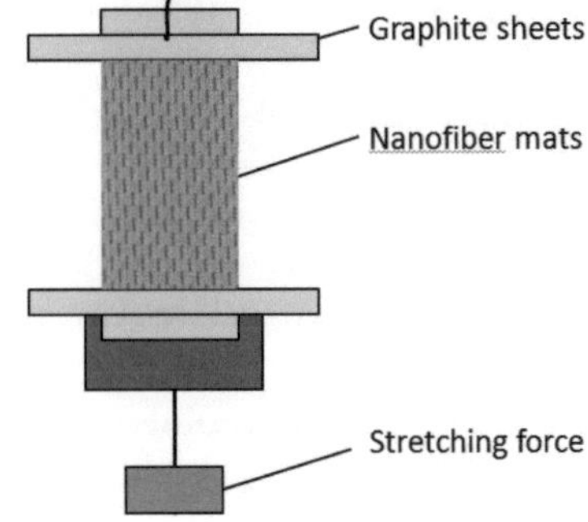

Fig. 6. The schematic representation of the experimental setup for hot-stretching process.

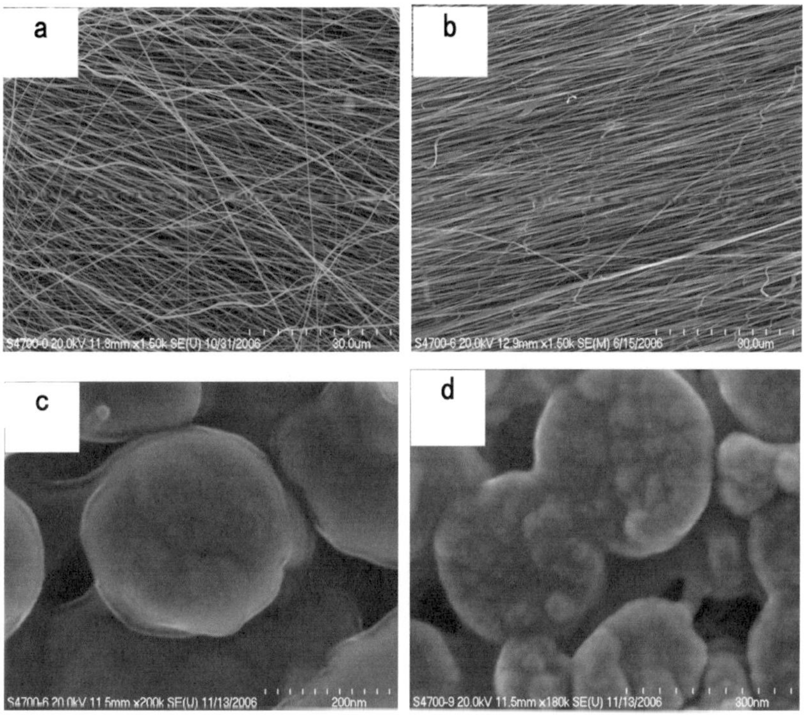

Fig. 7. SEM micrographs: (a) as-spun nanofibers, (b) hot-stretched nanofibers, (c) cross sectional views of as-spun nanofibers and (d) hot-stretched nanofibers.

2.2 Characterization of microstructures and mechanical properties

2.2.1 Morphology of PAN and PAN-based composite nanofibers

The optical micrograph (OM) images of electrospinning composites nanofibers mats are shown in Fig. 8 which the color becomes dark consequently. This is due to the different concentrations of the CNTs. Since the cross-sectional area of spinning nanofibers is very small, these nanofibers can be potentially used as optics transparency materials such as the optical filter etc.

Fig. 8. OM images of PAN and PAN/SWNTs nanofibers using (a)0wt% (b)0.25wt% (c)0.5wt% (d)0.75wt% (e)1wt% SWNTs concentrations.

The diameters of the PAN nanofibers which are directly electrospun on a TEM-copper-grid can be observed by scanning electron microscopy (SEM), which the diameters of electrospun nanofibers can be analyzed with image analyzer software (Image J). The results of the diameter distribution of PAN and PAN/SWNTs nanofibers with different SWNTs concentration 0wt%, 0.25wt%, 0.5wt%, 0.75wt% and 1wt% are shown in Fig. 9. The diameter locates between 100 to 500nm which is just the range of nano-scale.

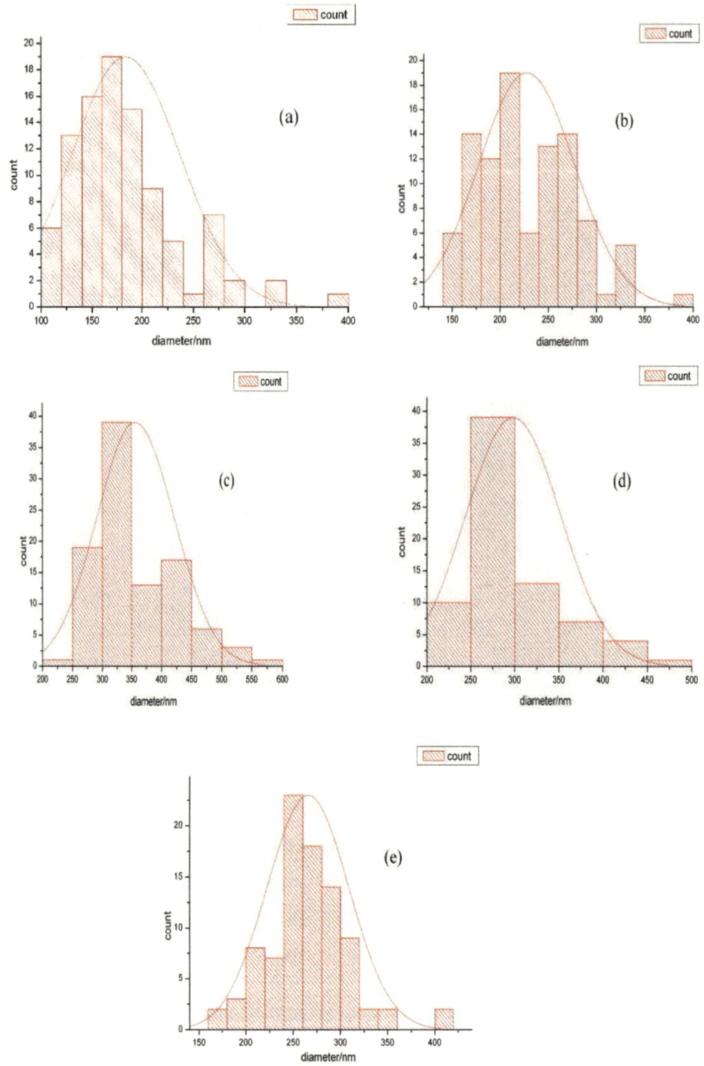

Fig. 9. The diameter distribution of PAN and PAN/SWNTs nanofibers with different SWNTs concentration (a)0wt% (b)0.25wt% (c)0.5wt% (d)0.75wt% (e)1wt%.

Fig. 10 compares the SEM micrographs of as-spun and hot-stretched PAN nanofibers and as-spun PAN/SWNTs composite nanofibers. It can be seen that there is no obvious conglutination in the nanofibers after the introduction of SWNTs. And Fig. 10(c) shows the hot-stretched PAN nanofibers, documenting the good alignment along the sheet axis after the hot-stretched process. It can further be found that the alignment of the fibers became closer to parallel after being hot-stretched. Also, the average diameters of the original as-spun fibers are significantly reduced from 200 nm to 120 nm after hot-stretching.

In order to demonstrate that the prepared nanofibers do contain some oriented SWNTs, transmission electron microscopy (TEM) can be utilized to view the alignment and orientation of SWNTs within the nanofibers manufacturing process. As seen by TEM in Fig. 11, the

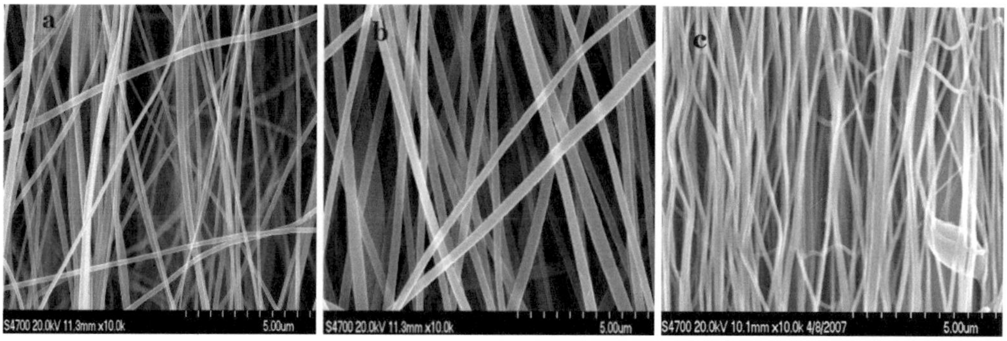

Fig. 10. SEM micrographs: (a) as-spun pure partially aligned PAN nanofibers; (b) PAN/SWNTs composite nanofibers with SWNTs concentration 1 wt%; (c) hot-stretched pure PAN nanofibers.

surface morphology of PAN nanofibers is smooth (Fig. 11(a)), but that of the PAN/SWNTs nanofibers is much rougher (Fig. 11(b)). Since the SWNTs possess a high electron density compared with the PAN polymer matrix, SWNTs appear as darker tubular structures embedded in the PAN/SWNTs composite nanofibers. It can be seen that the SWNTs are completely wrapped by the PAN matrix. TEM images reveals that in some regions nanotubes oriented well along the fiber axis but the nanotube distribution (number and orientation of the tubes) within a fiber may vary quite significantly (Fig. 11(b) and 11(c)). The nanotube distribution within a given fiber is usually quite different from the distribution in others. Topological defects such as entanglements, twisted sections, and knots can at times be observed (Fig. 11(d)). It is plausible that the extremely fast electrospinning process, by which carbon nanotubes cannot fully stretch within a millisecond range, leads to such defects.

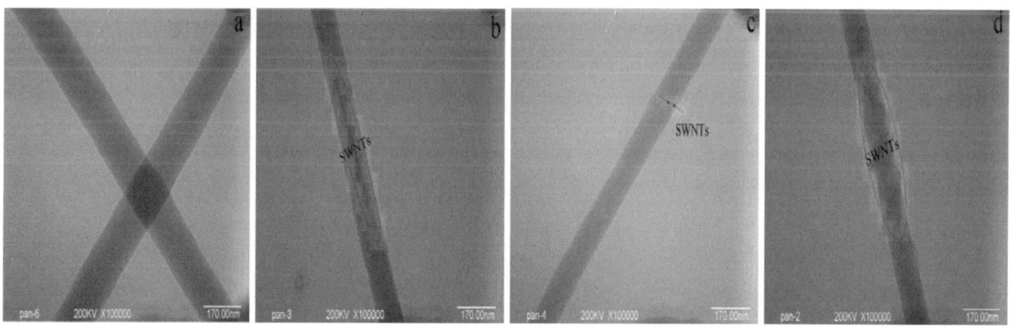

Fig. 11. TEM images: (a) PAN nanofibers; (b–d) PAN/SWNTs nanofibers with SWNTs concentration 1 wt%.

2.2.2 The alignment or the orientation of the as-spun, hot-stretched PAN and PAN-based nanofibers

The key determining the properties of the carbon fiber is the orientation of the polymer chain in the precursors. Usually the high degree of preferential orientation along the fiber axis of the layer planes is mainly responsible for the extraordinarily high Young's modulus of the fibers. Researchers now realize that understanding and controlling the orientation structure during the precursor formation step is critical if the properties of the carbon fibers are to be optimized (Ruland 1990). Therefore, the research on the microstructure of the precursors under different processing conditions is important in the study of the final mechanical properties of the carbon fibers (Ko et al. 1989). However, due to the complicated process in the manufacture of carbon fibers, it is not easy and fast to determine the orientation factor for PAN precursors, especially the less theoretical studies on PAN precursors.

There are several methods to determine the alignment or orientation of the fibers, which are polarizing microscope, infrared dichroism, sound velocity method, X-ray diffraction (XRD) and pole figures ect.(Xu et al. 2005). For electrospinning nanofibers, since these are not really fibers from macroscopic point of view, usually only infrared dichroism, X-ray diffraction (XRD) and pole figures are suitable. Fig. 12(A), (B) and (C) show the XRD patterns from the as-spun and hot-stretched PAN nanofibers, and the pole figures of the electrospun PAN nanofibers as-spun and hot-stretched nanofibers respectively. The operation conditions are the 40 kV and 200 mA with produce CuKα radiation (λ = 1.54 Å). There are two diffraction peaks at 17 and 29° which are corresponding to the d≈5.3Å from the (100) and the d≈3.03Å from the (110) reflections(Ra et al. 2005). The ratio of the d-spacing calculated by Bragg equation of these two peaks (1.74) is very close to √3:1, indicating hexagonal packing of the rod-like PAN chains (Zussman et al. 2005). The diffraction pattern of the as-spun nanofiber show one weak peak with a value of 2θ at 17°. And it also can be found that the peak at 2θ = 29.5° became much bigger after the process of hot-stretched. This indicates that electrospinning of the nanofibers onto a rotating drum generates limited crystallinity. In contrast, the oriented nanofibers after hot-stretching also show two diffraction peaks indexed with values of 2θ of 17 and 29°. From Fig. 13(A), noted that the X-ray beam is directed perpendicular to normal of the nanofibers, and thus the beam is also parallel to the winding direction of the nanofibers. The pole figures also show the same results with the elliptical shape after hot-stretching. It means the nanofibers are highly oriented along the winding direction of the rotating drum collector. The orientation coefficient after hot-stretching increased about 21.5% in comparison with those of as-spun nanofiber.

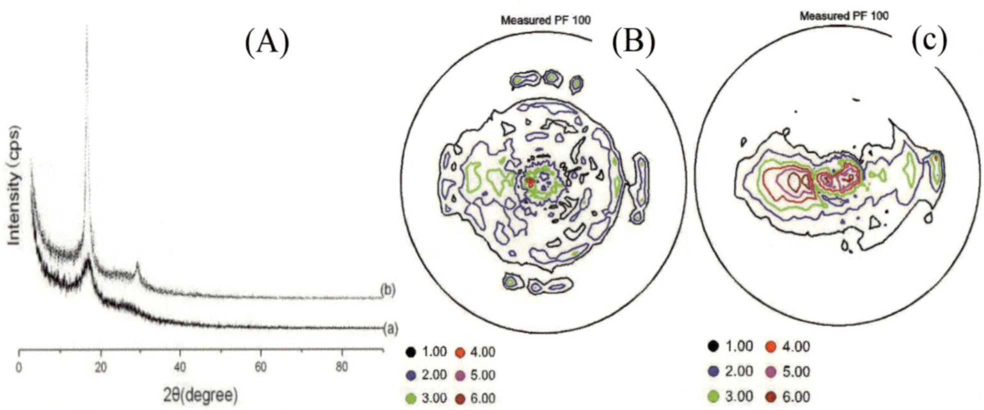

Fig. 12. (A) X-ray diffraction patterns for the nanofibers: (a) as-spun; (b) hot-stretched and the pole figures of the electrospun PAN nanofibers (B) as-spun nanofibers and (C) hot-stretched nanofibers.

The degree of orientation is usually expressed by orientation coefficient, which can be calculated from the half-width of the corresponding peak in the XRD curve. And the

Herman's orientation factor, f, can be determined from the fully corrected azimuthal intensity distribution diffracted from the (*100*) reflection at $d \approx 5.30$ Å by equation (1).

$$f = \frac{\int_0^{90} \frac{(3\cos^2\varphi - 1)}{2} I \sin\varphi d\varphi}{\int_0^{90} I \sin\varphi d\varphi} \tag{1}$$

Where, φ is the azimuthal angle between the axis of the molecular segment and the fiber alignment, and I is the scattering intensity of the (100) reflection at that angle(Fennessey & Farris 2004). The orientation factor f, which can be calculated from equation (1), increased from 0.22 to 0.76 after the hot-stretched process correspondingly.

The electrospun fiber bundles can be examined using the infrared spectrometer (FTIR) with a polarized wire-grid to measure the dichroism of the nitrile-stretching (-CN) group vibration at 2240 cm^{-1} (Bashir et al. 1994). Spectra are acquired with the draw direction of the electrospun fibers positioned both parallel and perpendicular to the electric vector direction of the polarizer, which the results before and after hot-stretching show in Fig. 13. The spectra are recorded over the range of 700–4000 cm^{-1} with typically 64 scans. The dichroic ratio, D, and the chain orientation factor, f, can be calculated from Eqs. (2), (3) and (4) using the transition moment angle α of 70° (or 73°), which α is the angle between the direction of the nitrile group's dipole moment change and the chain axis.

$$D = A_{//} / A_{\perp} \tag{2}$$

$$D_0 = 2\cot^2\alpha \tag{3}$$

$$f = \frac{(D-1)(D_0+1)}{(D_0-1)(D+2)} \tag{4}$$

where $A_{//}$ is the absorbance when the electric vector direction of the polarizer is oriented parallel to the fiber draw direction and $A_{\perp}$ is the absorbance when the electric vector is oriented perpendicular to the fiber draw direction, D_0 is the dichroic ratio of an ideally oriented polymer, α is the transition moment angle.

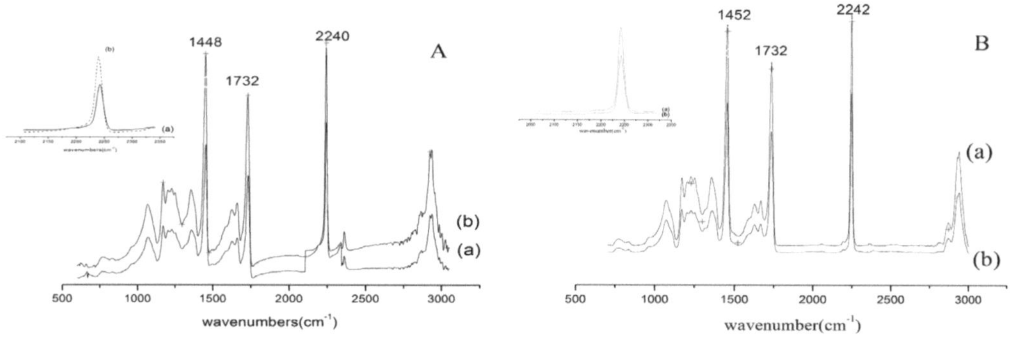

Fig. 13. Polarized FTIR spectra of PAN nanofiber: (A) before and (B) after hot-stretching, absorbent spectra at the perpendicular (a) and parallel (b) directions respectively.

In Fig. 13, the 2240 or 2242cm^{-1} is the characteristic peak of PAN corresponding with the stretching vibration of -C≡N group, and the 1732cm^{-1}, 1448 or 1452cm^{-1} are the stretching vibration of C=O and bending vibration of C-H. It shows that no much change of the microstructure by hot-stretching process. The chain orientation factor, *f*, calculated from Eq. (4) is 0.46 and 0.65 before and after hot-stretching respectively, which shows the hot-stretching can enhance the chain orientation by the applied stretching force.

2.2.3 Crystallinities of the as-spun and hot-stretched PAN and PAN-based nanofibers

The crystallinities of the as-spun and hot-stretched PAN nanofiber can be also investigated with *X*-ray diffractometer (XRD). The percent crystallinity can be obtained by extrapolation of the crystalline and amorphous parts of the diffraction pattern. The crystallite size is calculated by using the formula as following equation:

$$L_c = k\lambda / (\beta \cos\theta) \quad (5)$$

where $k = 0.89$, $\lambda = 1.54$ Å, β is the full width Half maximum (FWHM), and θ is Bragg angle (Norman et al. 1998).

Table 1 presents values of the percent crystallinity and the orientation coefficient for these two types of nanofibers. The percent crystallinity of the hot-stretched nanofiber increased about 3-times in comparison with those of as-spun nanofiber. The crystallite size also increased about 162%, indicating highly oriented PAN nanofibers, which was in agreement with the results of the orientation factor. Fibers with larger PAN crystals and higher polymer molecular orientation are expected to lead to a more perfect and higher orientation carbon fiber with improved mechanical properties (Chae et al. 2005).

Nanofiber	Crystallinity (%)	Crystallite size(nm)
As-spun	11.27	4.14
Hot-stretched	38.34	10.83

Table 1. Percent crystallinity and crystallite size obtained from X-ray diffraction curves.

The T_g of the PAN nanofiber and PAN/SWNTs nanofiber can be examined using Differential scanning calorimetry (DSC) method. To obtain the DSC curve in Fig. 14, the samples can be heated at a scanning rate of 20 °C/min under nitrogen atmosphere in order to diminish oxidation. The value of T_g is found by differentiating the heat flow curve with the temperature.

Fig. 14(A) shows the DSC curves to determine the glass transition temperature (T_g) of PAN and PAN/SWNTs nanofibers. It can be seen that, compares with PAN nanofiber (T_g = 102.3°C), the T_g is increased by about 3°C by incorporating only 0.75 wt% SWNTs into the PAN matrix (T_g = 105.4°C). Increase in the glass transition temperature as compared to the PAN fiber provides the evidence of interaction between PAN and SWNTs. The improvement in the T_g stemmed from a stronger interfacial interaction and possible covalent bonding between PAN and the SWNTs. Fig. 14(B) shows the DSC thermograms which the peak of PAN/SWNTs is higher than the one of PAN nanofibers. All these results suggests that the mobility of PAN chains is reduced due to the constraint effect of SWNTs (Chou et al. 2008).

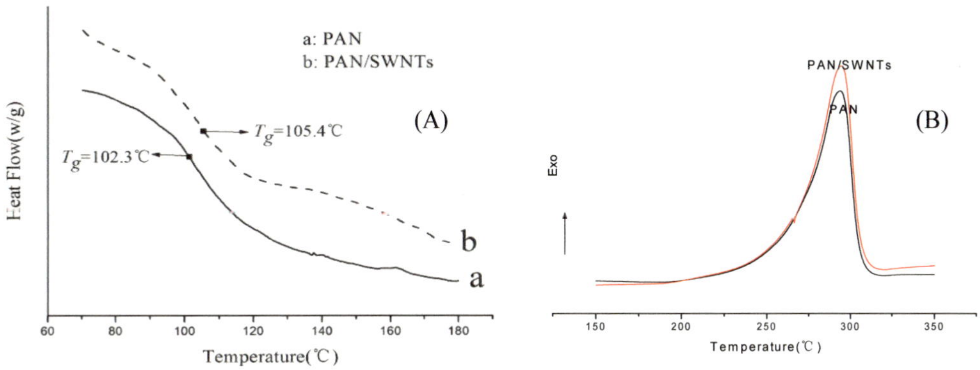

Fig. 14. (A) Tg from DSC of electrospun nanofibers: (a) PAN nanofibers; (b) PAN/SWNTs composite nanofibers; (B) DSC thermograms of electrospun PAN and PAN/SWNTs nanofibers.

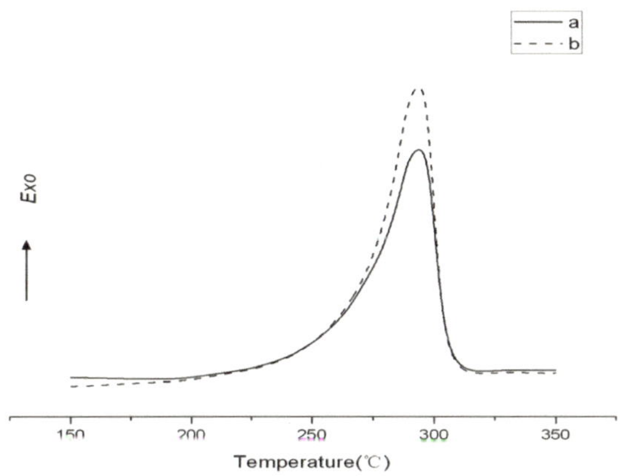

Fig. 15. DSC thermograms of PAN nanofiber sheets (a)before and (b)after hot-stretched.

Fig. 15 showed the DSC thermograms of PAN nanofiber sheets before and after hot-stretched process respectively. It can be seen that after hot-stretching, the peak and the peak area all increased. It means the cyclization heat of reaction enhances (from 510.53 J/g to 530.42 J/g) and the degree of cyclization increases correspondingly which is better for the manufacturing of the final carbon fibers.

2.2.4 Pre-oxidation treatment

The material, the process and the conditions used to form the precursor fiber, are the critical factors that define the final properties of the carbon fibers. Post-treatment steps such as pre-oxidation, stabilization and carbonization merely refine the as-spun structure (Paiva et al. 2001). The fundamental fiber structure needs to develop high properties must be created during the initial fiber formation step. However, this is not to say that fiber properties cannot be dramatically altered during the post-treatment (Edie 1998).

As infrared spectroscopy in Fig. 16(A) shows, with the pre-oxidation temperature increasing, -C≡N characteristic absorption band at 2,243 cm^{-1} is decreasing. And the absorption peaks of -C = N- band characteristics at 1,591 cm^{-1} and C - H features in the band at 1,371 cm^{-1} increase significantly with the increasing pre-oxidation temperature, which indicates that cyclization and dehydrogenation reaction has occurred in the pre-oxidation process of PAN nanofibers [9]. A new peak appears in the band at 810 cm^{-1}. It indicates that -C = C- has come into being in the molecules during pre-oxidation process, which shows that aromatic structure and heat-resistant structure have formed. Fig. 16(B) shows the XRD curves under different pre-oxidization temperatures. It has been known that two diffraction peaks indexed with values of 2 θ of 17 and 29° are corresponding with the (100) and (101) crystal planes. With the pre-oxidation temperature increased to 230°C, there is a new diffraction peak at 25.2° which is corresponding to the characteristic cyclization and aromatization structures of the (002) planes in graphite. It presented the dehydrogenation process in manufacturing of carbon fibers and the structure of linear PAN chain began to form the ladder structures which are corresponding to the decrease of the -C≡N characteristic peaks at 2 θ of 17 and 29° and increase of the aromatization diffraction peak at 25.2°.

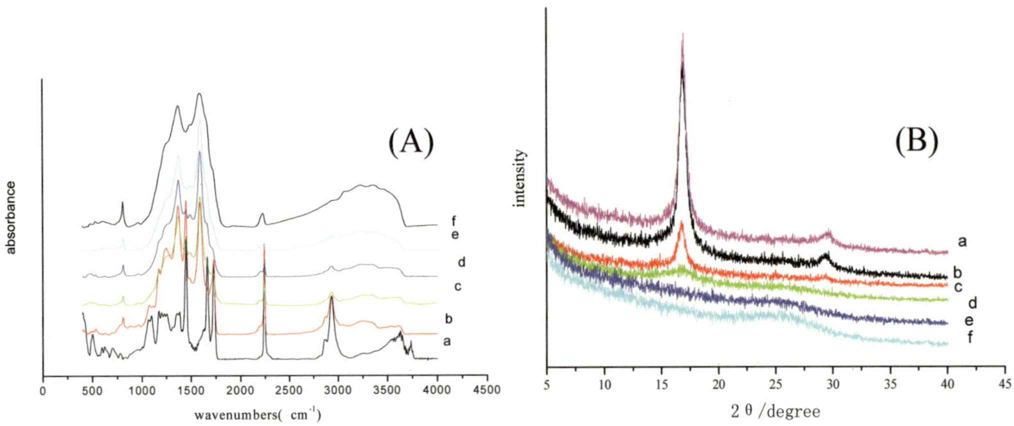

Fig. 16. (A) Infrared spectroscopy and (B) XRD curves under different pre-oxidization temperatures (a) 135 °C(b) 210 °C(c) 230 °C (d) 250 ℃(e) 270 °C (f) 290 °C.

The aromatization index (*AI*) can be calculated with the following equation (Koganemaru et al. 2004):

$$AI=I_c / (I_c+I_p) \tag{6}$$

where *Ip* and *Ic* are the intensity of PAN at 17° and aromatization structure at 25.2° respectively. It can be calculated that the extent of aromatization increasing from 21.86% to 41.98% at pre-oxidization temperatures of 230 and 270°C respectively. And if the temperature increased to 270 °C, the characteristic diffraction peak of -C≡N at 17° disappeared completely, which means that the higher pre-oxidation temperature is good for aromatization reaction.

2.2.5 Mechanical properties and electrical conductivities

A better representation of the nanocomposites characteristics is attempted by measuring the macroscopic nanofiber sheets. Stretching a piece of the nanofiber sheet gives an assessment of the average mechanical properties of the nanofibers rather than measuring an individual segment of a nanofiber composite (Moniruzzaman & Winey 2006). Mechanical property testing can be performed by the normal Instron tensile tests such as a LR30K Electromechanical Universal Testing Machine. The aligned nanofibers films taken down from the 16cm×12cm paper frame with the hollow of 16cm×4cm in Fig. 3 can be used for further tests which usually there are eight specimens used for each nanofiber sample in the tensile test. The samples can be cut into the size of 5mm width and 20mm length. The tensile speed in the mechanical test is selected as normal one such as 20 mm/min. The cross section areas of the samples are calculated via the weights of the samples and the densities of PAN and SWNTs.

The stress-strain curves of PAN nanofibers, before and after hot-stretched respectively, are presented in Fig. 17(A). Hot-stretched improved tensile strength and the modulus of PAN nanofibers. The tensile strength and tensile modulus increased by 55.32% and 156.48% respectively. Table 2 lists the tensile strength, tensile modulus, and elongation at break can be obtained from the stress-strain curves.

Sample	Tensile strength(MPa)	Impro.(%)	Tensile modulus(GPa)	Impro.(%)	Elongation at break(%)
As-spun	51.84		1.08		22.05
hot-stretched	80.52	55.32	2.77	156.48	11.61

Table 2. Mechanical properties of PAN nanofiber sheets.

It can be concluded that the hot-stretched method can improve the mechanical properties of PAN nanofibers. During the process of hot-stretched, the PAN molecular chain moves and arranges again along the fiber axis, the orientation and crystallinity are also improved. Therefore the mechanical properties of PAN nanofibers are improved due to the improvement of orientation and crystallinity. The improvement of crystallite size results in the elongation at break decreased obviously. The increased polymer orientation and crystal size point to the potential of PAN/SWNTs composite nanofibers as the precursor for the next generation carbon fiber. The stress-strain curves of PAN/SWNTs nanofibers, before and after being hot-stretched, are presented in Fig. 17(B). Hot-stretched also improves the tensile strength and the modulus of PAN/SWNTs nanofibers, which the tensile strength and tensile modulus increased by 54.70% and 125.40% respectively. It can be concluded that the hot-stretching can notably improve the mechanical properties of PAN/SWNTs nanofibers.

Fig. 18 shows the stress–strain curves of the PAN nanofibers and PAN/SWNTs nanofiber composites with different concentrations (hot-stretching). It shows that the SWNTs improves the modulus and tensile strength of the nanofiber. The tensile strength 128.76MPa of the nanocomposites at about 0.75% SWNTs by weight is increased with 58.9%. Also the tensile modulus shows a peak value of 4.62GPa with 66.8% improvement. The (e) curve in Fig. 18 deviates from the trend, which might be the non-uniform dispersion of SWNTs in high concentration. The significant improvement in strength and modulus is likely related to the good dispersion and orientation of SWNTs within the polymer matrix, and the strong interfacial adhesion due to the SWNTs surface modification (Chou et al. 2008). It can be concluded that both hot-stretching and the introduction of SWNTs can improve the mechanical properties of PAN-based nanofibers significantly.

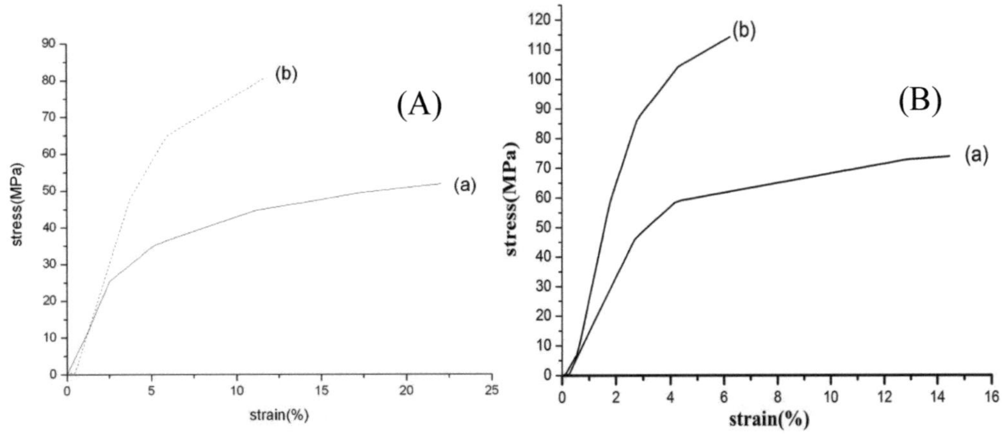

Fig. 17. Stress-strain curves of (A) PAN nanofiber sheets (a) before and (b) after hot-stretched and (B) PAN/SWNTs nanofiber sheets with SWNTs concentration 1 wt% (a) before and (b) after being hot-stretched.

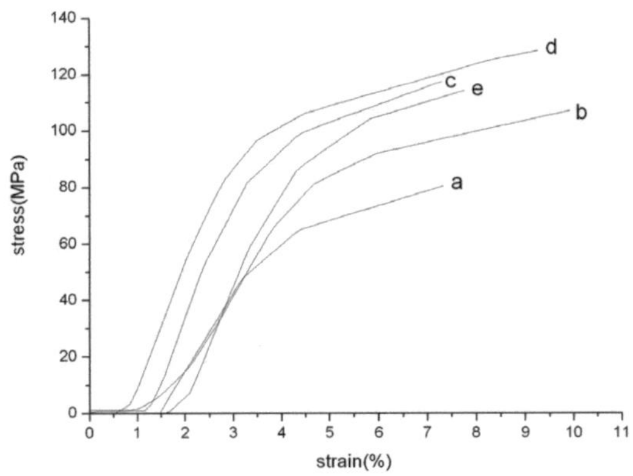

Fig. 18. Stress-strain curves for PAN and PAN/SWNTs Nanofiber: (a) pure PAN; (b) 0.25% SWNTs; (c) 0.5% SWNTs; (d) 0.75% SWNTs; (e) 1% SWNTs.

The electrical conductivities of electrospun PAN/SWNTs nanofiber composites can also be measured using the ultra-high resistance measuring machine at room temperature and ambient condition. The electrical conductivities can be obtained according to the formula as following:

$$\rho_v = R_v \times \frac{21.23}{t} \quad (7)$$

where ρ_v is volume resistivity, R_v is resistance, and t is the thickness of the nanofiber films. The electrical conductivity of the pure PAN nanofiber is 0.2-0.5 S/cm (Ge et al. 2004). Due to the superb electrical properties of SWNTs, a better electrical conductivity in PAN/SWNTs nanofiber composites is expected. Since electrical conductivity requires a percolating network be formed by the SWNTs, it can be concluded that the composite nanofibers, at a concentration of 0.75wt.% SWNTs, has formed the percolating network so that the

PAN/SWNTs nanofiber composites could possess electrical conductivity of up to 2.5 S/cm. Therefore these conductive PAN/SWNTs nanofiber sheets have potential applications in conductive nanoelectrodes, supercapacitors and nanosensors etc.

3. Conclusions

PAN nanofibers and PAN/SWNTs nanofiber composites can be prepared by electrospinning from PAN-based solution. An additional centrifugal field applied to an electrostatic field can produce a strong stretching force in electrospinning processes. Hot-stretched method and the incorporated CNTs are the key processes used to increase the degree of crystallinity and molecular orientation of PAN nanofibers and PAN/SWNTs nanofiber composites. The strong hot-stretching force can change the molecular orientation and an addition of only 0.75 wt% SWNTs to PAN increases the polymer mechanical properties significantly. SEM and TEM results show that SWNTs has a high orientation in PAN/SWNTs composite nanofibers. The crystallinity of the stretched sheet confirmed by X-ray diffraction has enhanced about 3 times in comparison with those of as-spun sheets and the chain orientation factor, *f*, is 0.46 and 0.65 before and after hot-stretching respective. Compared to pure PAN nanofibers, tensile strength and Young's modulus of the hot-stretched nanofibers exhibit considerable improvement. Thus, the improvement of orientation and crystallinity, the better PAN nanofibers alignment, are all contributed to the obvious increases of mechanical properties of the nanofibers. Incorporation of SWNTs into the nanofibers also increases the electrical conductivity to 2.5 S/cm for PAN/0.75% SWNTs nanofiber composites. Such PAN-based composite nanofiber sheets represent an important step toward utilizing carbon nanotubes in materials to achieve remarkably enhanced physical properties. Thus, the composite nanofibers with the component of SWNTs and the hot-stretched process can be used as the potential precursor to produce high-performance carbon nanofibers. The mechanical properties of the PAN nanofibers and PAN/SWNTs composite nanofibers can be improved more by extensive studies of electrospinning and the hot-stretching conditions. Thus, this novel electrospinning technique creates aligned and molecularly oriented PAN and PAN-based nanofibers that can be used to prepare carbon nanofibers with superior mechanical properties.

4. Acknowledgment

The authors gratefully acknowledge the financial support sponsored by the NSF of China(50973007) and the Program for Changjiang Scholars and Innovative Research Team in University (PCSIRT IRT0807).

5. References

Andrews, R.; Jacques, D.; Rao, A.M.; Rantell, T.; Derbyshire F.; Chen, Y.; Chen, J.; Haddon, R.C. (1999). Nanotube composite carbon fibers, *Appl. Phys. Lett.*, Vol.75, pp.1329-1331

Bashir, Z.; Church, S.; and Waldron, D. (1994). Interaction of water and hydrated crystallization in water-plasticized polyacrylonitrile films, *Polymer*, Vol.35, pp.967-97

Basova, Y.V.; Edie, D.D.; Lee, Y.S.; Reid, L.K. (2004). Effect of precursor composition on the activation of pitch-based carbon fibers, *Carbon,* Vol.42, pp.485–495

Baughman, R.H.; Zakhidov, A.A.; de Heer, W.A. (2002). Carbon nanotubes-the route toward applications, *Science,* Vol.297, pp.787-792

Chae, H.G.; Sreekumar, T.V.; Uchida, T; Kumar, S. (2005). A comparison of reinforcement efficiency of various types of carbon nanotubes in polyacrylonitrile fiber. *Polymer,* Vol.46, pp.10925-10935

Chand, S. (2000). Carbon fibers for composites, *J. Mater. Sci.* Vol.35, pp.1303-1313

Chandrasekar, R.; Zhang, L.F.; Howe, J.Y.; Hedin, N.E.; Zhang, Y.; Fong, H. (2009). Fabrication and characterization of electrospun titania nanofibers. *J. Mater. Sci.* Vol.44, pp.1198-1205

Chen, J.; Ramasubramaniam, R.; Xue, C.; Liu, H. (2006). A versatile molecular engineering approach to simultaneously enhanced, multifunctional carbon-nanotube-polymer composites. *Adv. Func.t Mater., V*ol.16, pp.114-119

Chou, W.J.; Wang, C.C.; Chen, C.Y. (2008). Characteristics of polyimide-based nanocomposites containing plasma-modified multi-walled carbon nanotubes. *Compos. Sci. Technol.,* Vol.68, pp.2208-2213

Chun, I.; Reneker, D.H.; Fong, H.; Fang, X. ; Deitzel, J.; Beck-Tan, N. (1999). Carbon nanofibers from polyacrylonitrile and mesophase pitch. *J. Adv. Mater.,* Vol.31, No.1, pp.36–41

Collins, P.G.; Arnold, M.S.; Avouris, Ph. (2001). Engineering carbon nanotubes and nanotube circuits using electrical breakdown, *Science,* Vol.292, pp.706.-709

Donnet, J.B.; Wang, T.K.; Peng, J.C.; Rebouillat, S. (1998). Carbon fibers. New York, NY: Marcel Dekker; pp.231-309

Doshi, J.; Reneker, D. H. (1995). Electrospinning process and applications of electrospun fibers, *J. Electrostat.,* Vol.35, pp.151–160

Dzenis, Y. (2004). Spinning continuous fibers for nanotechnology, *Science,* Vol.304(5679), pp.1917–9

Edie, D. D. (1998). The effect of processing on the structure and properties of carbon fibers, *Carbon,* Vol.36, pp.345–362.

Fennessey, S.F.; Farris, R.J. (2004). Fabrication of aligned and molecularly oriented electrospun polyacrylonitrile nanofibers and the mechanical behavior of their twisted yarns. *Polymer,* Vol.45, pp.4217-4225

Ge, J.J.; Hou, H.Q.; Li, Q,; Graham, M.J.; Greiner, A.; Reneker, D.H.; Harris, F.W.; and Cheng, S.Z.D. (2004).Assembly of Well-Aligned Multiwalled Carbon Nanotubes in Confined Polyacrylonitrile Environments: Electrospun Composite Nanofiber Sheets, *J. AM. CHEM. SOC.,* Vol.126, pp.15754-15761

Hammel, E.; Tang, X.; Trampert, M.; Schmitt T.; Mauthner, K.; Eder, A. and Pötschke, P. (2004). Carbon nanofibers for composite applications, *Carbon,* Vol.42, pp.1153–1158

Hao, R.; Yuan, J.Y.; Peng, Q. (2006). Fabrication and sensing behavior of Cr_2O_3 nanofibers via in situ gelation and electrospinning. *Chem. Lett.,* Vol.35, pp.1248-1249

Huang, Z.M.; Zhang, Y.Z.; Kotaki, M.; Ramakrishna, S. (2003). A review on polymer nanofibers by electrospinning and their applications in nanocomposites. *Compos. Sci. Technol.,* Vol.63, pp.2223-2253

Iijima, S. (1991). Helical microtubules of graphitic carbon, *Nature,* Vol.354, pp.56-58

Javey, A.; Guo, J.; Wang, Q.; Lundstrom, M.; Dai, H.J. (2003). Ballistic carbon nanotube field-effect transistors. *Nature,* Vol.424, pp.654-657

Johnson, J.; Phillips L.N. and Watt, W. (1965). The production of carbon fibers. *British Patent,* 1,110,790

Johnson, J.; Watt,W.; Phillips, L.N.; and Moreton, R. (1966). Improvements in or relating to carbonisable fibre and carbon fibre and their production. *British Patent,* 1,166,251

Ko, T.H.; Lin, C.H.; Ting, H.Y. (1989). Structure changes and molecular motion of polyacrylonitrile fibers during pyrolysis. *J. Appl. Sci.,* Vol.37, pp.553–556

Koganemaru, A.; Bin Y.Z.; Matsuo, M. (2004). Composites of Polyacrylonitrile and Mutiwalled Carbon Nanotubes Prepared by Gelation/Crystallization from solution. *Adv. Funct. Mater.,* Vol.14, pp. 842-850

Li, D.; Wang, Y.; and Xia, Y. (2003). Electrospinning of polymeric and ceramic nanofibers as uniaxially aligned arrays, *Nano Letts,* Vol. 3, pp.1167-1171

MacDiarmid, A., Jones, W., Norris, I., Gao, J., Johson, A., (2001). Electrostatically-generated nanofibers of electronic polymers, *Synthetic Metals,* Vol.119, 27-30

Miaudet, P.; Bartholome, C.; Derre, A.; Maugey, M.; Sigaud, G.; Zakri, C.; Poulin, P. (2007). Thermo-electrical properties of PVA-nanotube composite fibers. *Polymer,* Vol.48, pp.4068-4074

Moniruzzaman, M.; Winey, K.I. (2006). Polymer nanocomposites containing carbon nanotubes. *Macromolecules,* Vol.39, pp.5194-5205

Morgan, P.E. (2005). Carbon fibers and their composites. Boca Raton, FL: CRC Press; pp. 185–267

Na, H.; Li, Q.Y.; Sun, H.; Zhao, C.; Yuan, X.Y. (2009). Anisotropic mechanical properties of hot-pressed pvdf membranes with higher fiber alignments via electrospinning. *Polym. Eng. Sci.,* Vol.49, pp.1291-1298

Norman, W.H.; Cheetham, W.H.; Tao, L.P. (1998). Variation in crystalline type with amylose content in maize starch granules: An X-ray powder diffraction study. *Carbohyd. Polym.,* Vol.36, pp.277-284

Paiva, M.C.; Bernardo, C.A.; Edie, D.D. (2001). A comparative analysis of alternative models to predict the tensile strength of untreated and surface oxidized carbon fibers. *Carbon,* Vol.39, pp.1091–1101

Ra, E.J.; An, K.H.; Kim, K.K.; Jeong, S.Y.; Lee, Y.H. (2005). Anisotropic electrical conductivity of MWCNT/PAN nanofiber paper. *Chem. Phys. Lett.,* Vol.413, pp.188-19

Reneker D.H.; Chun, I. (1996). Nanometer diameter fibers of polymer, produced by electrospinning. *Nanotechnology,* Vol.7, No.3, pp.216–23

Rilutsky, S.; Zussman, E.; Cohen Y. (2010). Carbonization of Electrospun Poly(acrylonitrile) Nanofibers Containing Multiwalled Carbon Nanotubes Observed by Transmission Electron Microscope with In Situ Heating, *Journal of Polymer Science: Part B: Polymer Physics,* Vol.48, pp.2121–2128

Ruland W. (1990). Carbon-fibers. *Adv. Mater.;* Vol.2, pp.528–536

Smalley, R.E.; Colbert, D.T.; Smith, K.A.; Michael, O. (2007). Polymer-wrapped single wall carbon nanotubes. *US Patent,* 201,001,437,18A

Sreekumar, T.V.; Liu, T.; Min, B.G.; Guo, H.; Kumar S.; Hauge, R.H.; Smalley, R.E. (2004). Polyacrylonitrile single-walled carbon nanotube composite fibers, *Adv. Mater,* Vol.16, pp.58-61

Sutasinpromprae, J.; Jitjaicham, S.; Nithitanakul, M.; Meechaisue, C.; (2006). Supaphol, *P. Polym. Int.,* Vol.55, pp.825-833

Taylor, G. (1969). Electrically Driven Jets,, *Proc. R. Soc. London, Ser. A,* Vol.313, pp.453-475

Yu, M.F.; Lourie, O.; Dyer, M.J.; Moloni, K.; Kelly, T.F.; Ruoff, R.S. (2000). Strength and breaking mechanism of multi-walled carbon nanotubes under tensile load. *Science,* Vol.287, pp.637-640

Waclawik, E.R.; John, B.J.; Goh, SGR; Anthony, M.; Nunzio, M. (2006). Self-organization in composites of poly(3-hexylthiophene) and single-walled carbon nanotubes designed for use in photovoltaic applications. *Proc. of SPIE;* Vol.6036, pp.603607-4.

Wong, E.W.; Sheehan, P.E.; Lieber, C.M. (1997). Nanobeam mechanics: Elasticity, strength, and toughness of nanorods and nanotubes. *Science,* Vol.277, pp.1971-1975

Xu, Q.; Xu, L.H.; Cao, W.Y.; and Wu S.Z. (2005). A study on the orientation structure and mechanical properties of polyacrylonitrile precursors. *Polym. Adv. Technol.,* Vol.16, pp.642–645

Zussman, E.; Chen, X.; Ding, W.; Calabri, L.; Dikin, D.A.; Quintana, J.P.; Ruoff, R.S. (2005). Mechanical and structural characterization of electrospun PAN-derived carbon nanofibers. *Carbon,* Vol.43, pp.2175-2185

10

Photophysics and Photonics of Heteroepitaxial Organic Nanofibers

Francesco Quochi, Michele Saba, Andrea Mura and Giovanni Bongiovanni
Dipartimento di Fisica, Università di Cagliari, Italy

1. Introduction

Organic epitaxy is a powerful technique to grow highly ordered organic aggregates at crystalline surfaces. Long-range epitaxial order could be successfully exploited to enhance optoelectronic properties of organic molecular thin films for device applications. A relevant example of organic epitaxy on inorganic substrates is represented by high-vacuum deposition of para-sexiphenyl (*p*-6P) on (001)-oriented muscovite mica (Simbrunner et al., 2011), which yields large domains of parallel and crystalline nanofibers with extremely high surface-to-volume ratios, sub-wavelength cross-sectional dimensions, and lengths up to the millimeter scale (Yanagi et al., 1999; Andreev et al., 2000; Balzer & Rubahn, 2001). These nanofibers exhibit excellent photonic and optoelectronic properties: high blue photoluminescence efficiency (Stampfl et al., 1995), high in-fiber carrier mobility (Birendra-Singh et al., 2006; Kjelstrup-Hansen et al., 2006), luminescence guidance (Balzer et al., 2003), in-fiber amplification of spontaneous emission with large gain factors (Quochi et al., 2006; Cordella et al., 2007), and optically-induced laser action (Quochi et al., 2004, 2005, 2008). Very recently, organic-organic heteroepitaxy of mixed *p*-6P/alpha-sexithiophene (6T) nanofibers with highly polarized blue, green and red emission has been successfully demonstrated (Simbrunner et al., 2010), further broadening the range of potential applications of organic epitaxial nanofibers in photonics and optoelectronics.
In this paper, we will review the optical and photonic properties of *p*-6P and mixed 6T/ *p*-6P nanofibers epitaxially grown on oriented muscovite mica by hot-wall epitaxy and organic molecular beam epitaxy. We will first present ensemble studies of the amplified spontaneous emission and lasing action properties of nanofibers. The results of combined optical and morphological studies on individual nanofibers will be reported and the waveguide optical gain and lasing threshold figures will be given. We will then discuss the excited-state dynamics of nanofibers, as well as the intrinsic stimulated-emission perfomance of nanofibers in terms of gain lifetime and bandwidth. Potential applications of organic epitaxial nanofibers in photonics and sensing technologies will also be touched upon. Last, we will present recent results obtained in mixed 6T/*p*-6P nanofibers displaying high degree of epitaxial alignment, and highly polarized and broadband emission extending from the blue to the red.

2. Growth and characterization of *p*-6P epitaxial nanofibers

Para-sexiphenyl nanofiber films are grown on freshly-cleaved, (001)-oriented muscovite mica by Hot-Wall Epitaxy (Andreev et al., 2000) and Organic Molecular Beam Epitaxy

(Balzer & Rubahn, 2011). In the hot-wall epitaxy technique, *p*-6P is purified by threefold sublimation under dynamic vacuum. The system is operated with a nominal base pressure during growth of about 9×10^{-6} mbar and the *p*-6P source temperature is set to the optimized value of 240 °C. The substrate temperature is varied between 90 and 180°C, while the growth time is varied between 10 s and 120 min. Further details can be found in (Plank et al., 2001; Andreev et al., 2001). In the organic molecular beam epitaxy technique, sheets of muscovite mica are cleaved in air and are transferred immediately after cleavage into a high-vacuum apparatus (base pressure of 5×10^{-8} mbar). Before organic material is deposited, the samples are outgassed at a temperature of around 130 °C such that low energy electron diffraction shows the well-known hexagonal surface structure of clean mica with electric surface dipoles present. *Para*-sexiphenyl is deposited from a home-built Knudsen cell by vacuum sublimation; during the deposition the pressure inside the vacuum system rises to 2×10^{-7} mbar. Long *p*-6P needles grow for deposition rates of 0.1 Å/s and at substrate temperatures around 150 °C.
Morphological characterization of the nanofibers is performed by scanning-probe atomic force microscopy using Si-tip probes in tapping mode in air. Complementary optical characterization is carried out by cw epifluorescence measurements using an inverted microscope with a Hg high-pressure lamp as the excitation source.

2.1 Dependence of nanofiber morphology on deposition time

Topography studies of the surface morphology of films realized by hot-wall epitaxy at 130°C substrate temperature with different growth times demonstrate that oriented and mutually parallel nanofibers are formed by progressive regrouping of individual crystallites originating at the early growth stages for deposition times <10–25 s (Andreev et al., 2006). For growth times longer than 5 min., only linear fibers are observed. Surface topographic images of films obtained with various growth times are shown in Fig. 1. For long (>40 min.) deposition times, close-packed and interconnected nanofibers are realized.
X-ray diffraction studies demonstrated that *p*-6P nanocrystals packed in nanofibers are co-oriented and characterized by well-defined epitaxial relationships to the mica substrate: the long molecular axis of *p*-6P is nearly parallel to the mica surface plane and nearly pependicular to the fiber axes (Plank et al., 2003). Long-range azimuthal order is realized, which results in high optical anisotropy and high carrier mobility (Andreev et al., 2000; Birendra Singh et al., 2006).
Statistical analysis of topographic images reveal that the morphological parameters of nanofibers are strongly correlated to the growth time. Raising growth time from 40 to 120 min. results in an increase of fiber mean height from ~110 to ~290 nm, whereas the mean base width increases from ~210 to ~350 nm. Deposition time turns out to be a knob to tune cross-sectional dimensions of nanofibers; cross-sectional size is crucial to the attainment of optical waveguiding and amplificaton in nanofibers.
Closer inspection to the topographic images shows that nanofibers are characterized by the presence of breaks, resulting in fiber segmentation (clearly visible in the images shown in Fig. 2). Such breaks, ~50 to ~300 nm in width, occur at the end of the material growth process as a possible result of a surface thermal gradient during substrate cooling (Balzer & Rubahn, 2001). Breaks play a crucial role for the active photonic properties of nanofibers since they are responsible for the establishment of optical feedback along the nanofiber's axes and for light scattering into out-of-plane directions (see Sec. 3.2).

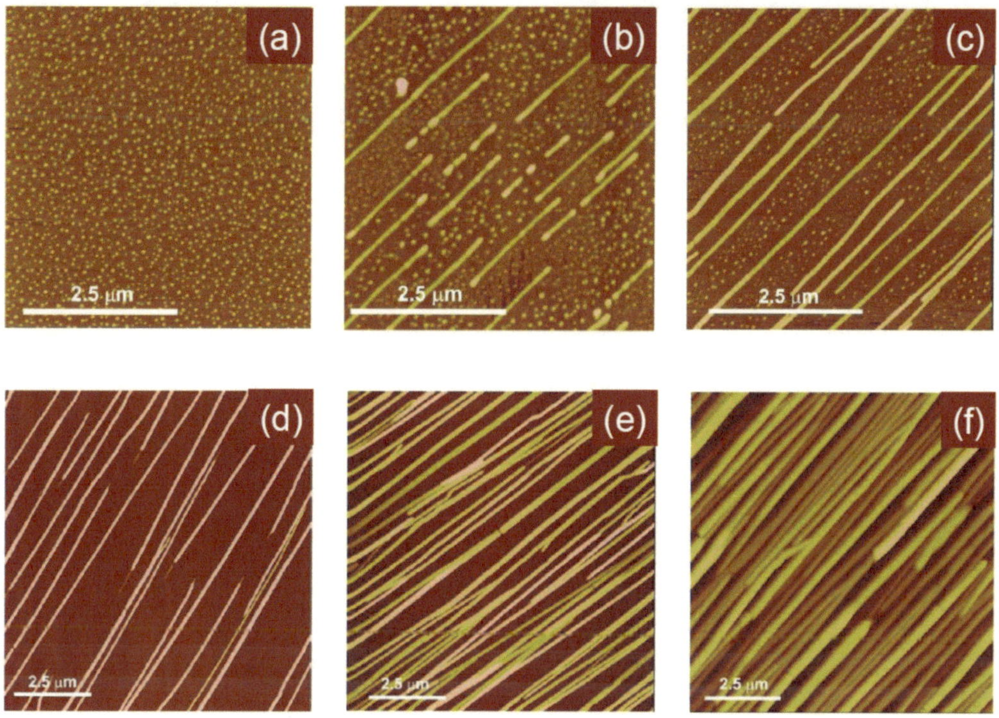

Fig. 1. AFM topography images of *p*-6P nanofiber films grown by hot-wall epitaxy on (001) oriented muscovite mica with (a) 10 s, (b) 25 s, (c) 90 s, (d) 5 min., (e) 40 min., and (f) 120 min. deposition time. The growth temperature is 130 °C. The height scale is 0–50 nm in (a)-(c), 0-100 nm in (d), 0–220 nm in (e), and 0–700 nm in (f).

2.2 Dependence of nanofiber morphology on substrate temperature

Substrate temperature is also an important growth parameter for tuning the morphological properties of nanofibers. In Fig. 2 are displayed topographic images of nanofiber films grown for 60 min. by hot-wall epitaxy at different substrate temperatures. As temperature is raised from 90 to 170°C, mean fiber height and cross-sectional area increase from ~80 to ~300 nm and from ~0.05 to ~0.3 µm^2, respectively, whereas the surface coverage (percentage of film surface covered by nanofibers) decreases from 70 to 20%, approximately. These behaviors are consistent with Arrhenius dependences, that is, nanofiber nucleation is a thermally activated process (Kankate et al., 2008). Activation energies range from a few tens to hundreds of meV, depending on whether hot-wall or beam deposition is used and on the morphological parameter being analyzed.

Strong correlation is found between fiber morphology and substrate temperature during growth. Hence, substrate temperature can also be used to control the photonic properties of nanofibers.

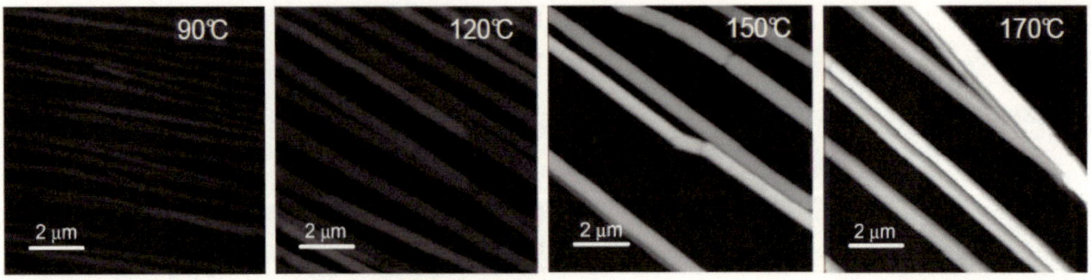

Fig. 2. Topographic images of the surface morphology of *p*-6P nanofiber films grown by hot-wall epitaxy on (001)-oriented muscovite mica at different substrate temperatures and with 60 min. deposition time. The height scale is 0–500 nm.

2.3 Epifluorescence microspectroscopy of nanofibers

Epifluorescence microscopy and microspectroscopy characterization techniques provide valuable information on the emission properties of nanofiber films. The inset of Fig. 3 shows an epifluorescence micrograph of an ensemble of nanofibers under UV illumination.

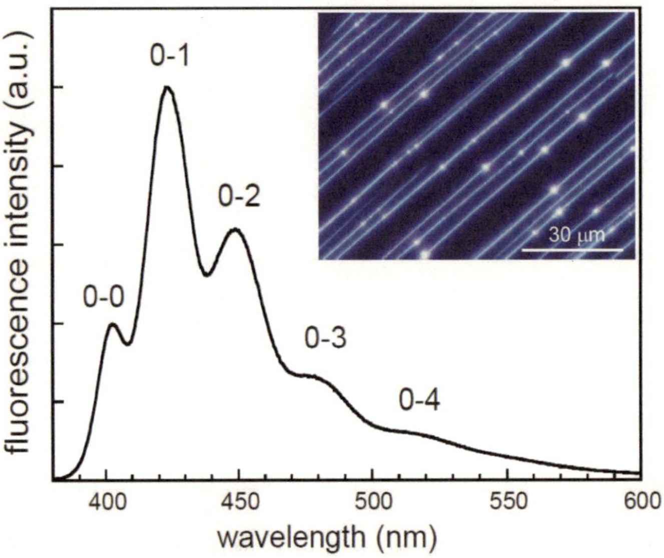

Fig. 3. Epifluorescence spectrum (main panel) and micrograph (inset) of an ensemble of *p*-6P nanofibers photoexcited by a Hg high-pressure lamp spectrally narrowed in the 330–360 nm band. Emission bands originating from vibronic coupling with C-C stretching mode are labeled for clarity.

Large enhancement of fluorescence scattering into out-of-plane directions is found to occur in correspondence of fiber breaks. These bright spots provide high sensitivity detection of *p*-6P epifluorescence. The main panel of Fig. 3 displays the room-temperature epifluorescence spectrum of photoexcited *p*-6P epitaxial nanofibers. The spectrum shows the typical features of an H-type molecular aggregate, that is, a weak electronic (0–0) transition and a more intense vibronic progression involving the C–C stretching mode.

3. Optical amplification and lasing action in *p*-6P epitaxial nanofibers

Nonlinear emission properties of *p*-6P epitaxial nanofibers are investigated by transient fluorescence spectroscopy using ultrashort (~150 fs) laser pulses delivered by a frequency-doubled (~390 nm wavelength) Ti:sapphire amplifield laser running at a repetition frequency of 1 kHz. The pump laser beam is focused to circular spots ranging from 120 to 180 μm in diameter on the nanofiber films, allowing to excite simultaneously tens of nanofibers with pulsed fluences up to 1 mJ/cm². For nonlinear emission experiments, the pump field polarization is set perpendicular to the axis of the nanofibers (thus parallel to the long axis of the *p*-6P molecules) for maximum optical absorption and the excitation fluence is varied using a variable density filter. In polarization-resolved experiments, the polarization of the optical emission is rotated using a retardation plate. The optical emission is dispersed in a single grating spectrometer equipped with a liquid-N_2-cooled charge-coupled device allowing for high-sensitivity, time-integrated measurements.

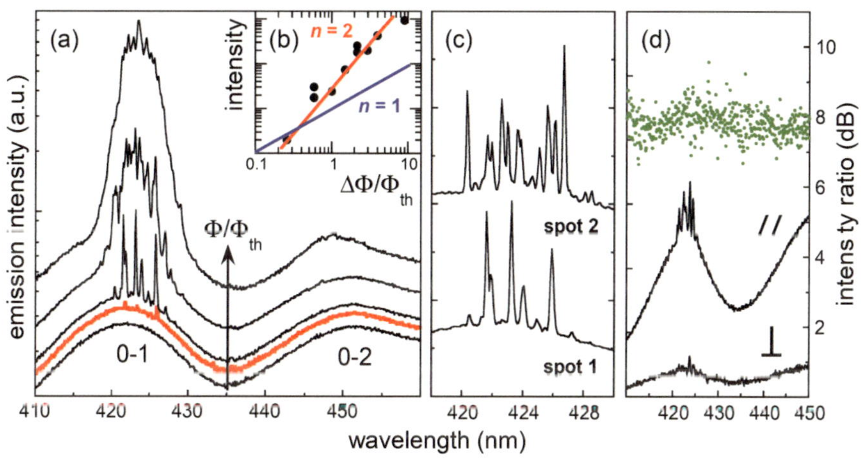

Fig. 4. (a) Time-integrated optical emission spectra of *p*-6P epitaxial nanofibers excited by subpicosecond pulses as a function of pump fluence: 0.8, 1.0, 1.6, 3.0, 10. Relative units refer to the Φ/Φ_{th} ratio, where Φ_{th} is lasing threshold fluence. (b) Time- and spectrally-integrated intensity of nonlinear emission above lasing threshold. Linear and quadratic dependences are shown by the red and blue line, respectively. (c) Position sensitivity of lasing modes: emission spectra are taken in different positions of the pump laser spot on the film surface. (d) Polarization-resolved emission spectra, acquired upon polarization filtering of the emission along the direction parallel (//) and perpendicular (⊥) to the long molecular axis of *p*-6P. Spectral dependence of intensity ratio is displayed as green dots.

The evolution from spontaneous emission to coherent (lasing) emission for increasing pump pulsed fluence is shown in Fig. 4(a). The emission spectrum relates to tens of interconnected nanofibers and thus represents an ensemble-averaged response of nanofibers. The films are kept at room temperature. When the pump fluence exceeds a threshold value Φ_{th}, resolution-limited peaks emerge from the spontaneous emission spectrum on the 0-1 vibronic band. Threshold fluences are position dependent and can vary from a few μJ/cm² to hundreds of μJ/cm². As pump fluence is much increased above threshold, the visibility of the narrow lines decreases until spectral narrowing of the vibronic peak dominates the system response. The narrow peaks denote the presence of coherent optical feedback within the *p*-6P nanofiber films, whereas spectral narrowing of the 0-1 and 0-2 emission peaks indicates that amplified spontaneaous emission occurs at the high pump levels. In Fig. 4(b), the emission intensity is plotted as a function of the normalized pump excess fluence, defined as $\Delta\Phi/\Phi_{th} = (\Phi-\Phi_{th})/\Phi_{th}$. The signal intensity is spectrally integrated over the 0-1 band upon subtraction of the spontaneous emission contribution. The nonlinear dependence of the emission intensity on $\Delta\Phi/\Phi_{th}$ (slope efficiency that increases with increasing pump fluence) is partially attributed to the fact that the number of coherent modes reaching oscillation theshold increases with raising pump excess fluence (Quochi et al., 2004).

The spatial sensitivity of nolinear lasing in *p*-6P crystalline nanofibers are reported in Fig. 4(c). Different distributions of unequally spaced modes are realized in different positions of the excitation spot on the film surface, suggesting the occurrence of coherent random lasing (Cao, 2003, 2005). Moving the pump spots across the fim surface results in the excitaton coherent modes having different resonance wavelengths and optical losses; as a consequence, strong variation in lasing threshold fluence is registered as the excited area is changed. As expected from the film structural anisotropy and epitaxial alignment, both linear and nonlinear optical emissions exhibit strong polarization anisotropy with intensity ratios as large as 10 dB (Fig. 4(d)).

Detailed studies on optical amplification and random lasing action are conducted on singly selected, isolated *p*-6P epitaxial nanofibers.

3.1 Guided amplification of spontaneous emission

Determination of net optical gain in isolated fibers yields the optical amplification performance of *p*-6P epitaxial nanofiber waveguides. The experiments are performed on nanofibers not showing the presence of strong scattering centers and displaying fairly uniform epifluorescence efficiency (Quochi et al., 2006).

Singly selected fibers grown by organic molecular beam epitaxy and excited by 150 fs-long pulses with 390 nm central wavelength are investigated by optical microspectroscopy. The optical emission, excited through the back (substrate) surface, is collected from the front surface using a microscope objective and focused onto the input slit of the spectrometer. Setting the spectrometer to zeroth-order diffraction and fully opening the input slit, fluorescing fibers covering the photoexcited area can be imaged with ~2 μm spatial resolution. Tuning the spectrometer to first-order diffraction and narrowing the input slit, the emission of individual nanofibers aligned parallel to the input slit can be spectrally resolved. Results obtained on a ~40 μm-long nanofiber are shown in Fig. 5. At low excitation levels, the emission micrograph of a selected nanofiber shows uniform emission intensity between the fiber tips (Fig. 5(a)). Increasing the pump fluence above a threshold value of ~100 $\mu J/cm^2$ per pulse, enhancement of the emission intensity is detected near the fiber tips (Fig. 5(b)). The emission intensity increases continuously as the position approaches the tips, where the guided light is efficiently outcoupled (Fig. 5(c)).

Assuming linear amplification of spontaneous emission and uniform scattering efficiency across the fiber, the emission intensity profile can be fitted by the function $I_T(z) = I(z)+I(L-z)$, where $I(z)\sim[\exp(gz)-1]/g$; z is the distance from a fiber tip, g the net modal gain coefficient, and L the fiber length.

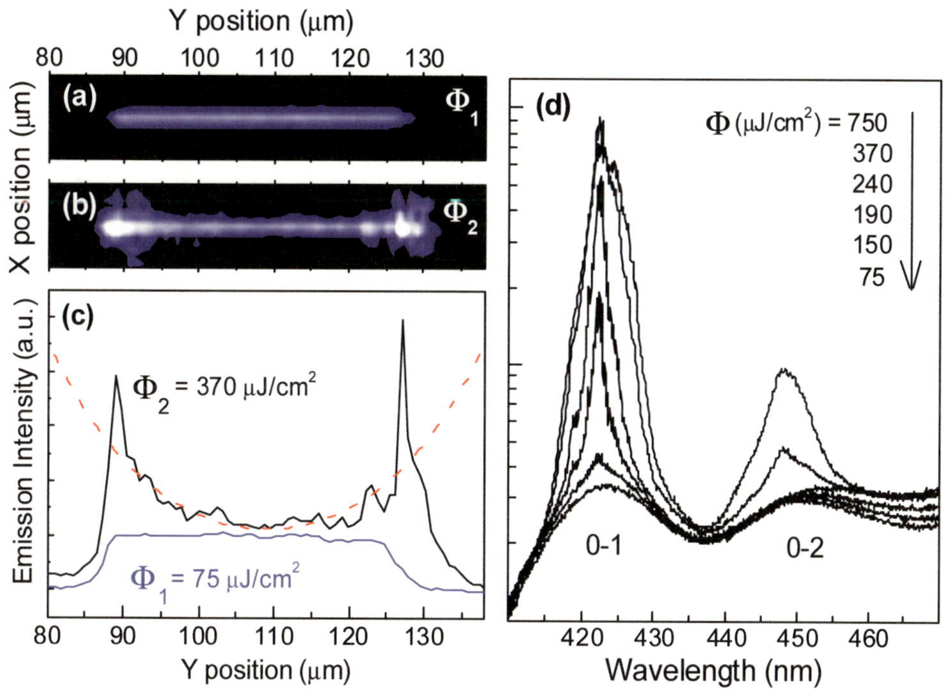

Fig. 5. (a,b) Time-integrated emission micrographs of an individual nanofiber excited by ultrafast laser pulses, for a pump pulsed fluence $\Phi_1 = 75$ (a) and $\Phi_2 = 370$ μJ/cm^2. (c) Spatial profiles of the emission intensity. The dashed line is the fit of a linear amplification function (see text) to the measured profile. (d) Emission spectra for different values of the excitation fluence. The spectra are spatially integrated over the nanofiber region.

The occurrence of amplified spontaneous emission is confirmed by the spectral analysis of the emission intensity above amplification threshold, which shows line narrowing of the emission towards the center of the vibronic bands (Fig 5(d)), whereas spectral fringes relating to ~40 μm-long Fabry-Perot resonator are not visible. Curve fitting of intensity profiles yields g_1 ~1200 cm^{-1} for the 0-1 emission band and g_2 ~700 cm^{-1} for the 0-2 emission band at the highest excitation fluence of 750 μJ/cm^2. Overall, the results demonstrate that *p*-6P epitaxial nanofibers are able to guide and amplify their spontaneous emission with large amplification factors, and thus hold potential as active nanoamplifiers at surfaces.

3.2 Random lasing action

Optical studies on individual *p*-6P nanofibers with strong scattering centers allow us to understand the origin of coherent optical feedback, which is responsible for random lasing action. Isolated nanofibers give the opportunity to study one-dimensional random lasing in self-assembled epitaxial nanostructures. To this aim, we apply the same microspectrographic technique used to characterize amplified spontaneous emission in homogeneous, break-free fibers. Emission spectra of isolated nanofibers grown by organic molecular beam epitaxy show that random lasing threshold is reached at both the 0-1 and 0-2 vibronic peaks. As expected for a single laser emitter, spontaneous emission saturates at the value reached at threshold (Quochi et al., 2006). On the contrary, fluorescence clamping is not observed in ensemble-averaged measurements, since not all the excited fibers reach lasing threshold (Fig. 4).

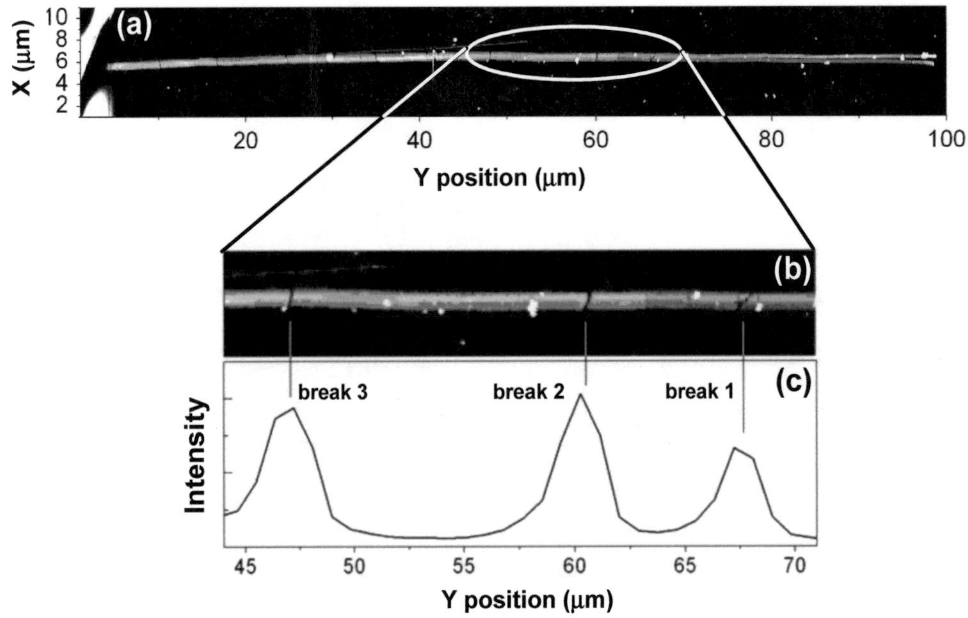

Fig. 6. (a) Topographic image of a segmented *p*-6P epitaxial nanofiber. (b) Image of the indicated subsection of the nanofiber. (c) Lasing intensity profile relating to the same fiber segment as in (b). Vertical marks highlight the correspondence between the positions of breaks and those of optical scattering centers responsible for coherent feedback and laser light outcoupling towards collecting optics.

Correlated lasing measurements and atomic-force topographic measurements give insight into the origin of one-dimensional coherent feedback in individual *p*-6P nanofibers. Fig. 6(a) shows the topographic image of a ~100-μm-long, selected nanofiber. When the fiber morphology is compared to the lasing intensity profile on the same region (Figs. 6(b) and 6(c)), it clearly turns out that scattering of the guided lasing emission occurs at the fiber breaks. In fact, excellent correspondence is found between the positions of the bright lasing spots and those of the fiber breaks. These findings strongly suggest that back-reflections of the fiber waveguide modes at the fiber break interfaces are the main source of optical feedback along the nanofibers' axes. Our interpretation of the experimental data in terms of one-dimensional coherent random lasing is in fact supported by calculations of the resonant optical modes of one-dimensionally disordered systems (Quochi et al., 2005).

Substrate temperature used during growth has a big influence on the lasing threshold performance of *p*-6P epitaxial nanofibers. Increasing substrate temperature from 90 to 170°C, mean lasing threshold, as measured in films grown by hot-wall epitaxy with 60 min. deposition time, decreases from ~300 to ~50 μJ/cm^2. These findings are traced back to the thermally activated growth of *p*-6P epitaxial nanofibers; in fact, thermal activation of the cross-sectional size of nanofibers should result in strong temperature increase of the optical confinement factor of the guided optical emission. Therefore, substrate temperature can also be used to tune the active photonic properties of *p*-6P epitaxial nanofibers.

4. Excited-state dynamics and photophysics of *p*-6P epitaxial nanofibers

The potential of organic films for laser device technologies stems from the ability to achieve lasing thresholds compatible with indirect electrical pumping by unexpensive and convenient light sources. Lasing performance results from both extrinsic factors (*i.e.*, optical confinement of lasing modes in the gain medium, propagation and feedback losses), and intrinsic gain performance of the active medium. The latter is quantified by the figure of merit for optical amplification, which can be defined as $FOM = \sigma_G B_G \tau$, where σ_G is the net-gain cross-section, B_G its bandwidth, and τ the excited-state lifetime at operating excitation intensity. These parameters can be retrieved from the analysis of the excited-state dynamics of the system.

We resort to time-resolved fluorescence and transient differential absorption measurements with ultrafast excitation to study the excited-state dynamics of *p*-6P epitaxial nanofiber films. We investigate ensembles of close-placked nanofibers displaying high amplification and lasing thresholds (>100 μJ/cm^2) to ensure that stimulated emission affects population dynamics only weakly. Ultrafast excitation experiments are performed using 150-fs-long pulses delivered by an optical parametric amplifier allowing to tune the excitation wavelength across the 350–400 nm interval. The sample emission is dispersed in a single grating spectrometer and temporally resolved by a visible streak camera with ~20 ps time resolution. A cold-finger cryostat fed with liquid air is used to vary the sample temperature in the 80–300 K range. Complementary differential transmission ($\Delta T/T$) measurements are performed using broadband pulses obtained by supercontinuum generation in a sapphire plate as the optical probe. Time delay betwen the pump and probe pulses is controlled by a motorized optical delay stage.

4.1 Ultrafast time-resolved photoluminescence spectroscopy

Main results of time-resolved fluorescence studies are reported in Fig. 7, where fluorescence decay rate constant is plotted as a function of the pump pulsed fluence. Fluorescence decay traces taken at 80 K are drawn in the inset of Fig. 7 (Quochi et al., 2008).

At high pump fluences, the excited-state decay dynamics is dominated by nonradiative, density-dependent processes ascribed to singlet-singlet (bimolecular) annihilations. Bimolecular recombination strongly depends on lattice temperature. Upon fitting the initial $1/e$ decay times with the results of model simulations of the decay dynamics, we estimate that the bimolecular coefficient (κ_{SS}) increases from 3×10^{-9} to 4×10^{-8} cm^3/s when lattice temperature is increased from 80 to 300 K. Linear (monomolecular) recombination rate constant (k_0) also exhibits a temperature dependence, concomitanly increasing from 1×10^9 to 2.3×10^9 s^{-1}. At 80 K, bimolecular processes kick in for pump fluences of ~2–4 μJ/cm^2, whereas the threshold goes down to 0.1 μJ/cm^2 at room temperature. Lattice temperature dependence of bimolecular processes suggests that exciton migration is a thermally activated process (Wiesenhofer et al., 2006). Activation energy is possibly related to the presence of energetic disorder in *p*-6P nanostructured films (Kadashchuk et al., 2004).

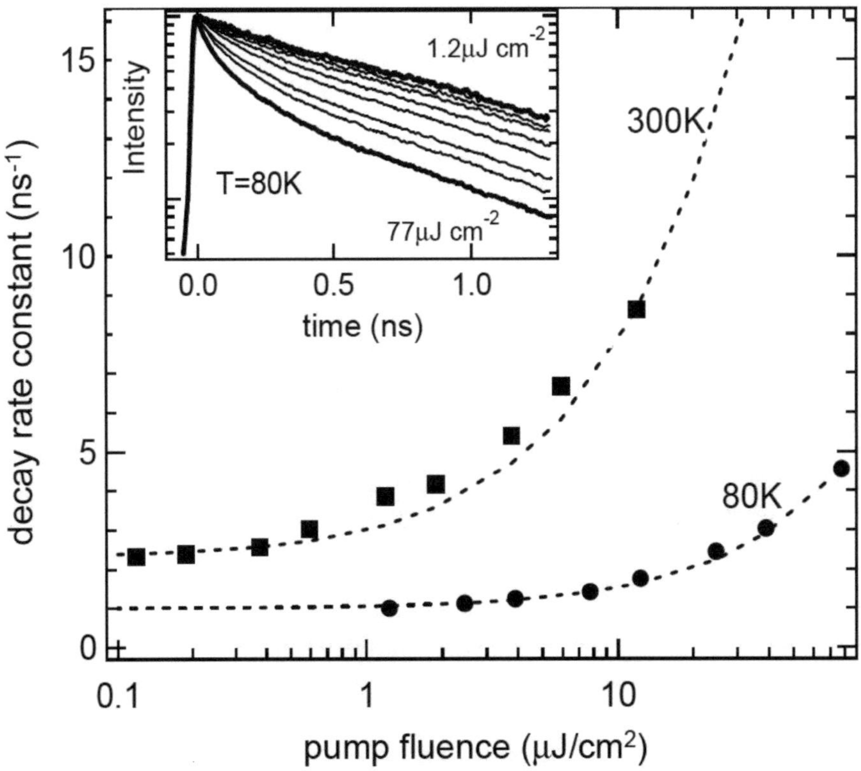

Fig. 7. Fluorescence decay rate constant of *p*-6P nanofibers excited by ultrafast pulses at 380 nm, at 300 K (squares) and 80 K (circles). Dashed lines: Calculated decay rate constant; the calculation includes both monomolecular and bimolecular recombination processes. Inset: Fluorescence decay traces measured at 80 K with pump pulsed fluence of 1.2, 2.5, 3.9, 7.8, 12, 25, 39, and 77 μJ/cm^2.

4.2 Ultrafast transient absorption spectroscopy

Excited-state photophysical processes such as stimulated emission and photoinduced absorption are studied by differential transmission measurements as a function of pump-probe time delay. Transmission spectra taken for various pump-probe delays are shown in Fig. 8. At short delays ($\leq$1 ps), the system response is characterized by broadband stimulated emission extending from the deep blue to the orange (B_G ~1 eV) and exhibiting the vibronic progression of singlet excitons in *p*-6P crystalline nanofibers. The stimulated-emission cross-section is found to be ~2×10^{-16} cm^2, as estimated from $\Delta T/T$ signal amplitude at the 0-2 vibronic peak at zero time delay, assuming that primary photoexcitations are singlet excitons. On the long wavelength side, the spectrum is dominated by photoinduced absorption of triplet excitons and polarons promptly excited by the pump pulses. At room temperature and high pump fluences (~90 μJ/cm^2), broadband stimulated emission decays rapidly due to singlet-singlet annihilations and a photoinduced absorption band arises near the optical gap of the material at ~400 nm.

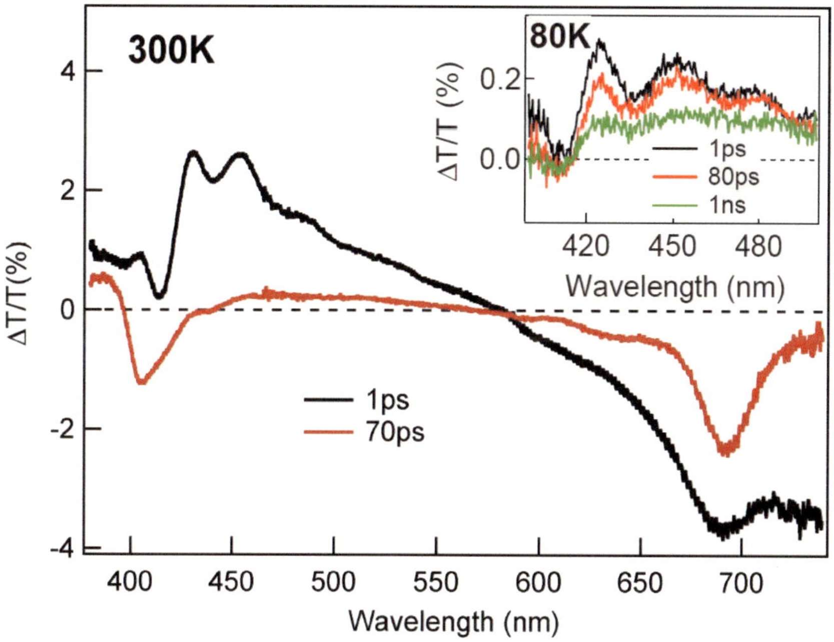

Fig. 8. Differential transmission spectra of *p*-6P epitaxial nanofibers excited by ultrafast pulses centered at 360 nm, for different lattice temperatures and pump-probe time delays. Pump pulsed fluence: 90 μJ/cm^2 (main panel); 10 μJ/cm^2 (inset).

This long-lived photoinduced absorption band (decay time > 1 ns) is attributed to intermolecular excitons with charge-transfer character and generated via singlet-singlet annihilations (Cordella et al., 2007). At cryogenic temperatures (80 K) and lower excitation fluences (~10 μJ/cm^2), bimolecular recombination is strongly suppressed: excited-state lifetime (τ) increases up to ~1 ns and the photoinduced absorption band associated to secondary charge-transfer excitons disappears (inset of Fig. 8).
Excited-state lifetime shortening is the primary effect of singlet-singlet annihilations. As a secondary effect, a (nonlinear) population of intermolecular excitons is created, whose absorption spectrum overlaps with gain spectrum of emissive (singlet) excitons. Combined effects of lifetime shortening, net-gain reduction and bandwidth shrinking by photoinduced absorption are detrimental to lasing action in nanofibers excited by long (nanosecond) pulses.

4.3 Monomolecular lasing

Sample cooling to cryogenic temperatures makes it possible to circumvent fast bimolecular decay and photoinduced absoption by intermolecular excitons. Direct demonstration of occurrence of *monomolecular lasing*, that is, lasing in the linear recombination regime at cryogenic temperatures is provided in Fig. 9, which reports the time-wavelength spectrogram of the emission intensity of nanofibers excited by subpicosecond pulses.

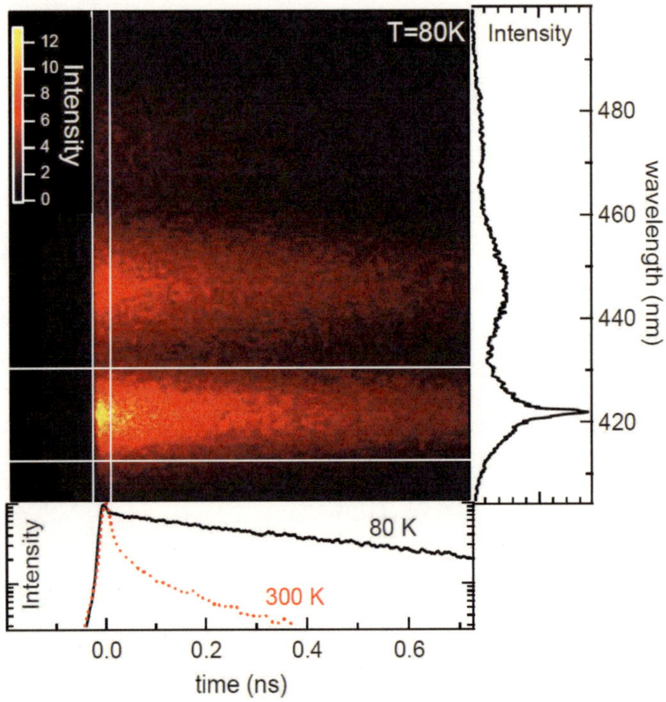

Fig. 9. Time-wavelength spectrogram of the fluorescence intensity of cryogenically cooled nanofibers, excited by subpicosecond pulses with 360 nm central wavelength. Colors code for intensity levels on an arbitrary unit scale. Pump pulsed fluence is 7 μJ/cm^2. Vertical (horizontal) white lines delimit the spectrogram region used for temporal (spectral) integration. Right panel: emission spectrum integrated in the first 30 ps after excitation pulse arrival. Bottom panel: Emission decay traces integrated over the 0-1 vibronic band of the *p*-6P emission. The solid (dashed) line depicts the time profile at 80(300) K.

The spectral profile (right panel) shows evidence of lasing emission on top of the 0-1 spontaneous emission band. The time profile analysis (bottom panel) demonstrates that at 80 K prompt laser emission decays rapidly, leaving the system with a population of singlet excitons undergoing monomolecular recombination with ~1 ns decay time (Quochi et al., 2008).

5. *Para*-sexiphenyl epitaxial nanofibers for active photonic sensing

Low-threshold random lasing action could be exploited to achieve high photonic sensitivity to various agents; miniaturized random laser sources are thus be envisaged to enable new functionalities for next-generation technologies (Wiersma, 2000; Wiersma & Cavalieri, 2001). Based on model simulations of the coherent optical response of a random medium (Quochi et al., 2005), we estimate that a typical *p*-6P epitaxial nanofiber with a length of ~100 μm and a dozen of thin (~200 nm) breaks would display *attoliter* sensitivity to contamination by index matching fluids (Fig. 10(a)). Also, strain sensors with very large (>10^3) gauge factor and high dynamic range could be obtained by optical interrogation of single nanofibers aligned parallel to the strain axis (Figs. 10(b) and 10(c)).

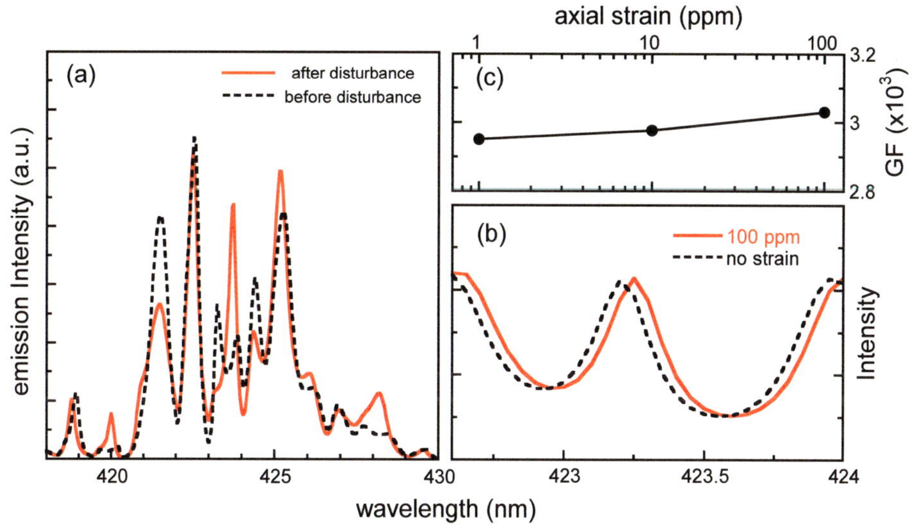

Fig. 10. Model calculations of the photonic sensitivity of a *p*-6P epitaxial nanofiber with one-dimensional random optical feedback. (a) Coherent emission spectrum before and after optical neutralization of a fiber break upon air-gap filling with index matching fluid. (b) Resonance wavelength shift in response to 100 ppm axial strain. (c) Strain gauge factor as a function of strain strength. Gauge factor (*GF*) is defined from the relation. $\Delta I/I$ (ppm) – $GF \cdot$ strain (ppm), where I is the optical emission intensity.

Surface adsorption of molecular species in nanofibers assembled from suitably functionalized oligomers (Schiek et al., 2005) could generate photonic chemosensing, e.g., by modulation of the effective refractive index of the nanofiber resonance modes. Photonic sensitivity could be further enhanced near lasing threshold, as demonstrated in recent reports (Rose et al., 2005; Chen et al., 2011).

6. Heteroepitaxial nanofibers based on the *p*-6P/6T system

Tuning of lasing wavelength of *p*-6P epitaxial nanofibers can possibly be achieved by crystal doping with long-wavelength emitting molecules. With this aim we study the growth and optical properties of *p*-6P/sexithiophene (6T) heteroepitaxial films by hot-wall epitaxy on (001)-oriented muscovite mica. Both *p*-6P and 6T are good model systems for heteroepitaxy and important oligomers for optoelectronic device applications. It has been demonstrated that both *p*-6P and 6T crystalline thin films can be used as organic templates to obtain highly crystalline organic-organic heterostructures in which *p*-6P and 6T molecules are both standing or parallel to the substrate (Oehzelt et al., 2006; Koller et al., 2006). We show by hot-wall heteroepitaxy that *p*-6P nanofiber templates grown on muscovite mica can be used to obtain oriented and mutually parallel 6T/*p*-6P nanofibers whose emission color depends on the relative concentrations of the two components. We resort to a growth chamber equipped with two hot-wall epitaxy reactors for serial deposition of *p*-6P and 6T. The substrate is first exposed to the reactor with the *p*-6P source and than transferred to the 6T reactor; details can be found in (Simbrunner, 2010).

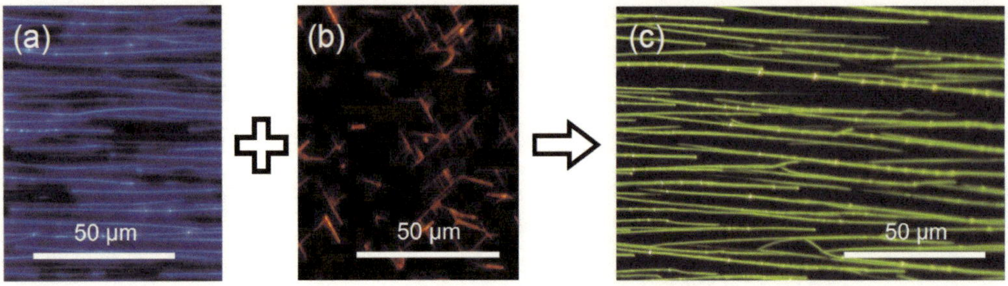

Fig. 11. Fluorescence micrographs of heteroepitaxial nanofibers grown on (001)-oriented muscovite mica by hot-wall epitaxy at 120°C substrate temperature: (a) *p*-6P; (b) 6T; (c) bilayer of 6T (deposition time 90 min.) on *p*-6P (deposition time 40 min.). Epifluorescence is excited by UV lamp irradiation.

When 6T is directly grown on muscovite, short nanofibers with red-orange fluorescence are yielded with several orientations, whereas 6T deposition on a template of *p*-6T nanofibers results in 6T/*p*-6P bilayer fibers which maintain the fiber template's orientation, with no 6T material being deposited in between adjacent p-6P fibers. This is clearly demonstrated by the epifluorescence micrographs shown in Fig. 11. Green emitting fibers are obtained with a 6T deposition time of 5 s, corresponding to submonolayer coverage. In this limit, the 6T fluorescence spectrum is similar to that of a solvated 6T phase (green line spectrum labeled as 6T(i) in Fig. 12).

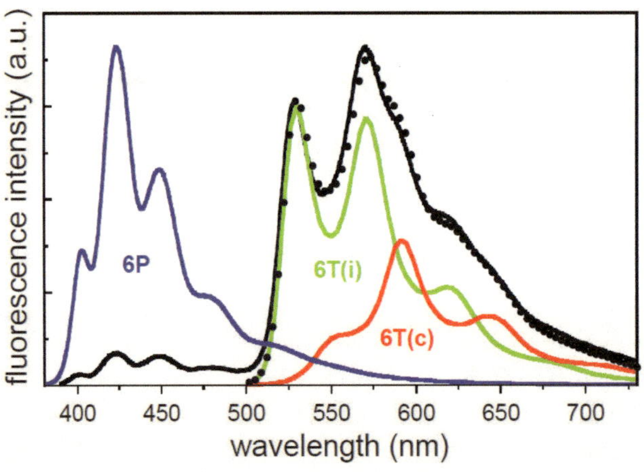

Fig. 12. Fluorescence spectra of pure *p*-6P nanofibers (blue line) and 6T/*p*-6P bilayer nanofibers (black line). Fluorescence is excited by 370 nm laser pulses at 80 MHz repetition rate. Green and red spectra belong to interfacial (i) and bulk crystal (c) 6T in the bilayer sample. Black dots are the sum of 6T(i) and 6T(c) spectra.

Increasing 6T deposition time (and so the 6T-to-6P concentration ratio), oriented 6T crystalline fibers are deposited on top of *p*-6P templating fibers. Crystal phase emission, as obtained in pure 6T films, is shown as the red line labeled as 6T(c) in Fig. 12. The overall emission spectrum of the 6T/*p*-6P bilayer nanofiber film grown with 40 min. (*p*-6P) and 90

min. (6T) deposition times (black line in Fig. 12) shows evidence of strong quenching of the *p*-6P emission. The interfacial, green emitting 6T sheet is highly emissive, its integrated intensity being comparable to that of the ~400-nm-thick 6T crystalline overlayer. The emission strengths of the interfacial and crystal components are determined by fitting a linear combination of the spectra of the two 6T phases to the total emission spectrum.
Quenching of *p*-6P emission is ascribed to efficient sensitization of 6T via resonance energy transfer (Wieb Van Der Meer et al., 1994). The *p*-6P/6T material couple is thus inferred to display type-I alignment of electronic levels in the bilayer structure, which could allow for extending operation of *p*-6P epitaxial nanofiber lasers to 6T emission wavelengths in the green and in the red via resonance energy transfer from *p*-6P to 6T.

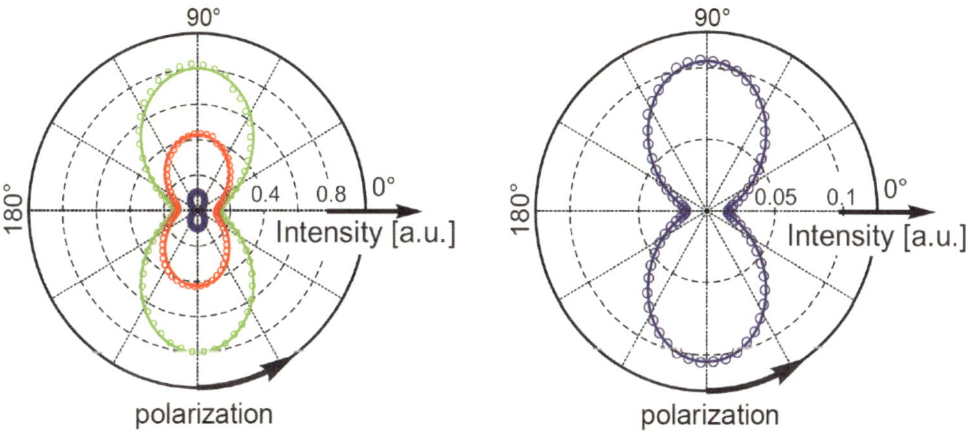

Fig. 13. Polar plots of the emission intensity of 6T/*p*-6P bilayer film versus analyzer polarizaton angle. Dots: experimental points; lines: fits of Malus' law function to the experimental data. Blue, green and red colors code for spectrally-integrated emission intensity of *p*-6P, 6T(i) and 6T(c) material phases, respectively. A magnified view of *p*-6P intensity data is shown in the right panel for better visibility.

Long-range azimuthal order is demonstrated in 6T/*p*-6P bilayer nanofibers by polarization-resolved fluorescence intensity measurements. While the pump laser polarization is set parallel to the *p*-6P long molecular axis for maximum absorption, the collected fluorescence signal is analyzed by a rotating linear polarized. For each angular position of the polarizer, the emission spectrum is acquired and decomposed in its *p*-6P (blue), interfacial 6T (green) and crystalline 6T (red) components. The integrated intensities of the three materials' components are then plotted versus polarizer rotation angle. The results fit quite well to Malus' ($\sin^2$) law dependences for all the three materials' phases. In particular, optical emission dipoles of *p*-6P and 6T are found to be parallel over domains as large in area as ~1 cm^2. These optical properties support the results X-ray diffraction studies, which determine the exact epitaxial relationships between the 6T/*p*-6P crystal phases and muscovite mica substrate (Simbrunner et al., 2010). Despite the potential exhibited by 6T/*p*-6P heteroepitaxial fibers on oriented muscovite mica for broadband laser operation, at the time this review is being written no evidence of 6T lasing action in such nanostructures or in other epitaxial thin films has been reported yet.

7. Conclusion

We reviewed the photonic and photophysical properties of *p*-6P epitaxial nanofibers grown by hot-wall epitaxy and organic molecular beam epitaxy on (001)-oriented muscovite mica. We presented experimental data on amplified spontaneous emission and coherent random lasing in ensembles of nanofibers, as well as in singly selected fibers, where waveguide amplification performance and coherent feedback mechanisms were found out. We also reported the excited-state nonlinear dynamics of nanofibers, which was studied by transient photoluminescence and absorption spectroscopy, and presented a quantitative account of the main photophysical processes occurring in photoexcited nanofibers, namely, stimulated emission and excited-state absorption. After a brief discussion on potential applications of organic epitaxial nanofibers in photonic sensing, we concluded by showing recent results obtained on mixed 6T/*p*-6P heteroepitaxial nanofibers. The 6T/p-6P material system, combined with the potential of organic epitaxy techniques, looks promising for the realization of nanoscale laser devices operating in the blue, green and red spectral regions.

8. Acknowledgment

We acknowledge the scientific collaboration of F. Cordella, F. Floris, M. Marceddu, R. Orrù, A. Andreev, G. Hernandez-Sosa, N. S. Sariciftci, G. Schwabegger, C. Simbrunner, H. Sitter, M. Oehzelt, R. Resel, F. Balzer, V.G. Bordo and H.-G. Rubahn.

9. References

Andreev, A.; Matt, G.; Brabec, C. J.; Sitter, H.; Badt, D.; Seyringer, H. & Sariciftci, N. S. (2000). Highly Anisotropically Self-Assembled Structures of *para*-Sexiphenyl Grown by Hot-Wall Epitaxy. *Adanced Materials*, Vol. 12, No. 9, (May 2000), pp. 629-633, ISSN 0935-9648

Andreev, A., Sitter, H., Brabec, C. J.; Hinterdorfer, P.; Springholz, G. & Sariciftci, N. S. (2001). Self-assembled growth of highly oriented para-sexiphenyl thin films. *Synthetic Metals*, Vol. 121, No. 1-3, (March 2001), pp. 1379-1380, ISSN 0379-6779

Andreev, A.; Quochi, F.; Cordella, F.; Mura, A.; Bongiovanni, G.; Sitter, H.; Hlawacek, G.; Teichert, C. & Sariciftci, N. S. (2006). Coherent random lasing in the deep blue from self-assembled organic nanofibers. *Journal of Applied Physics*, Vol. 99, No. 3, (February 2006), pp. 034305(1-6), ISSN 0021-8979

Balzer, F. & Rubahn, H.-G. (2001). Dipole-assisted self-assembly of light-emitting *p*-nP needles on mica. *Applied Physics Letters*, Vol. 79, No. 23, (December 2001), pp. 3860-3862, ISSN 0003-6951

Balzer, F.; Bordo, V. G.; Simonsen, A. C. & Rubahn, H.-G. (2003). Optical waveguiding in individual nanometer-scale organic fibers. *Physical Review B*, Vol. 67, No. 11, (March 2003), pp. 115408(1-8), ISSN 1098-0121

Birendra-Singh, T.; Hernandez-Sosa, G.; Neugebauer, H.; Andreev, A.; Sitter, H. & Sariciftci, N. S. (2006). Electrical transport properties of hot wall epitaxially grown *para*-sexiphenyl nano-needles. *physica status solidi (b)*, Vol. 243, No. 13, (November 2006), pp. 3329-3332, ISSN 0370-1972

Cao, H. (2003). Lasing in random media. *Waves in Random and Complex Media*, Vol. 13, No. 3, (? 2003), pp. R1-R39, ISSN 1745-5030

Cao, H. (2005). Review on latest developments in random lasers with coherent feedback. *Journal of Physics A: Mathematical and General*, Vol. 38, No. 49, (December 2005), pp. 10497-10535, ISSN 0305-4470

Chen, R.; Ling, B.; Sun, X. W. & Sun, H. D. (2011). Room Temperature Excitonic Whispering Gallery Mode Lasing from High-Quality Hexagonal ZnO Microdisks. *Advanced Materials*, Vol. xx, No. Xx, (published online April 2011), DOI: 10.1002/adma.201100423, ISSN 0935-9648

Cordella, F.; Quochi, F.; Saba, M.; Andreev, A.; Sitter, H., Sariciftci, N. S.; Mura, A. & Bongiovanni, B. (2007). Optical Gain Performance of Epitaxially Grown *para*-Sexiphenyl Films. *Advanced Materials*, Vol. 19, No. 17, (September 2007), pp. 2252-2256, ISSN 0935-9648

Kadashchuk, A.; Andreev, A.; Sitter, H.; Sariciftci, N. S.; Skryshevsky, Y.; Piryatinski, Y.; Blonsky, I. & Meissner, D. (2004). Aggregate States and Energetic Disorder in Highly Ordered Nanostructures of *para*-Sexiphenyl Grown by Hot Wall Epitaxy. *Advanced Functional Materials*, Vol. 14, No. 10, (October 2004), pp. 970-978, ISSN 1616-301X

Kankate, L.; Balzer, F.; Niehus, H. & Rubahn, H.-G. (2008). From clusters to fibers: Parameters for discontinuous *para*-hexaphenylene thin film growth. *Journal of Chemical Physics*, Vol. 128, No. 8, (February 2008), pp. 084709(1-12), ISSN 0021-9606

Kjelstrup-Hansen, J.; Henrichsen, H. H.; Bøggild, P. & Rubahn, H.-G. (2006). Electrical properties of a single *p*-hexaphenylene nanofiber. *Thin Solid Films*, Vol. 515, No. 2, (October 2006), pp. 827-830, ISSN 0040-6090

Koller, G.; Berkebile, S.; Krenn, J. R., Netzer, F. P.; Oehzelt, M.; Haber, T.; Resel, R. & Ramsey, M. G. (2006). Heteroepitaxy of Organic-Organic Nanostructures. *Nano Letters*, Vol. 6, No. 6, (June 2006), pp. 1207–1212, ISSN 1530-6984

Oehzelt, M.; Koller, G.; Ivanco, J.; Berkebile, S.; Haber, T.; Resel, R.; Netzer, F. P. & Ramsey, M. G. Organic (2006). Heteroepitaxy: *p*-Sexiphenyl on Uniaxially Oriented α-Sexithiophene. *Advanced Materials*, Vol. 18, No. 18, (September 2006), pp. 2466–2470, ISSN 0935-9648

Plank, H.; Resel, R.; Purger, S.; Keckes, J.; Thierry, A.; Lotz, B.; Andreev, A.; Sariciftci, N. S. & Sitter, H. (2001). Heteroepitaxial growth of self-assembled highly ordered *para*-sexiphenyl films: A crystallographic study. *Physical Review* B, Vol. 64, No. 23, (December 2001), pp. 235423(1-5), ISSN 1098-0121

Plank, H.; Resel, R.; Sitter, H.; Andreev, A.; Sariciftci, N. S.; Hlawacek, G.; Teichert, C.; Thierry, A. & Lotz, B. (2003). Molecular alignments in sexiphenyl thin films epitaxially grown on muscovite. *Thin Solid Films*, Vol. 443, No. 1-2, (October 2003), pp. 108-114, ISSN 0040-6090

Quochi, F.; Cordella, F.; Orrù, R.; Communal, J. E.; Verzeroli, P.; Mura, A.; Bongiovanni, G.; Andreev, A.; Sitter, H. & Sariciftci, N. S. (2004). Random laser action in self-organized para-sexiphenyl nanofibers grown by hot-wall epitaxy. *Applied Physics Letters*, Vol. 84, No. 22, (May 2004), pp. 4454-4456, ISSN 0003-6951

Quochi, F.; Cordella, F.; Mura, A.; Bongiovanni, G.; Balzer, F. & Rubahn, H.-G. (2005). One-Dimensional Random Lasing in a Single Organic Nanofiber. *The Journal of Physical Chemistry B*, Vol. 109, No. 46, (November 2005), pp. 21690-21693, ISSN 1520-6106

Quochi, F.; Cordella, F.; Mura, A.; Bongiovanni, G.; Balzer, F. & Rubahn, H.-G. (2006). Gain amplification and lasing properties of individual organic nanofibers. *Applied Physics Letters,* Vol. 88, No. 4, (January 2006), pp. 041106(1-3), ISSN 0003-6951

Quochi, F.; Saba, M.; Cordella, F.; Gocalinska, A.; Corpino, R.; Marceddu, M.; Anedda, A.; Andreev, A.; Sitter, H.; Sariciftci, N. S.; Mura, A. & Bongiovanni, G. (2008). Temperature Tuning of Nonlinear Exciton Processes in Self-Assembled Oligophenyl Nanofibers under Laser Action. *Advanced Materials,* Vol. 20, No. 16, (August 2008), pp. 3017-3021, ISSN 0935-9648

Rose, A.; Zhu, Z.; Madigan, C. F.; Swager, T. M. And Bulović, V. (2005). Sensitivity gains in chemosensing by lasing action in organic polymers. *Nature,* Vol. 434, No. 7035, (April 2005), pp. 876-879, ISSN 0028-0836

Schiek, M.; Lützen, A.; Koch, R.; Al-Shamery, K.; Balzer, F.; Frese, R. & Rubahn, H.-G. (2005). Nanofibers from functionalized para-phenylene molecules. *Applied Physics Letters,* Vol. 86, No. 15, (Dec 2005), pp. 153107(1- 3), ISSN 0003-6951

Simbrunner, C.; Quochi, F.; Hernandez-Sosa, G.; Oehzelt, M.; Resel, R.; Hesser, G.; Arndt, M.; Saba, M.; Mura, A.; Bongiovanni, G. & Sitter, H. (2010). Organic−Organic Heteroepitaxy of Red-, Green-, and Blue-Emitting Nanofibers, *ACSNano,* Vol. 4, No. 10, (October 2010), pp. 6244-6250, ISSN 1936-0851

Simbrunner, C.; Nabok, D.; Hernandez-Sosa, G.; Oehzelt, M.; Djuric, T.; Resel, R.; Romaner, L.; Puschnig, P.; Ambrosch-Draxl, C.; Salzmann, I.; Schwabegger, G.; Watzinger, I. & Sitter, H. (2011). Epitaxy of Rodlike Organic Molecules on Sheet Silicates—A Growth Model Based on Experiments and Simulations, *Journal of The American Chemical Society,* Vol. 133, No. 9, (March 2011), pp. 3056-3062, ISSN 0002-7863

Stampfl, J.; Tasch, S.; Leising, G. & Sherf, U. (1995). Quantum Efficiencies of Electroluminescent Poly(*para*-phenylenes). *Synthetic Metals,* Vol. 71, No. 1-3, (April 1995), pp. 2125-2128, ISSN 0379-6779

Wieb Van Der Meer, B.; Coker, G. III & Simon Chen, S.-Y. (August 9, 1994). *Resonance Energy Transfer: heory and Data* (1st edition), John Wiley & Sons, ISBN 0471185892, New York

Wiersma, D. S. (2000). The smallest random laser. *Nature,* Vol. 406, No. 6792, (July 2000), pp. 132-133, ISSN 0028-0836

Wiersma, D. S. & Cavalieri, S. (2001). Light emission: A temperature-tunable random laser. *Nature,* Vol. 414, No. 6865, (December 2001), pp. 708-709, ISSN 0028-0836

Wiesenhofer, H.; Zojer, E.; List, E. J. W.; Scherf, U.; Brédas, J.-L. & Beljonne, D. (2006). Molecular Origin of the Temperature-Dependent Energy Migration in a Rigid-Rod Ladder-Phenylene Molecular Host. *Advanced Materials,* Vol. 18, No. 3, (February 2006), pp. 310-314, ISSN 0935-9648

Yanagi, H. & Morikawa, T. (1999). Self-waveguided blue light emission in *p*-sexiphenyl crystals epitaxially grown by mask-shadowing vapor deposition. *Applied Physics Letters,* Vol. 75, No. 2, (July 1999), pp. 187-189, ISSN 0003-6951

11

Nano-Scale Reinforcing and Toughening Thermoplastics: Processing, Structure and Mechanical Properties

Suchart Siengchin
The Sirindhorn International Thai-German Graduate School of Engineering (TGGS), King Mongkut's University of Technology North Bangkok, Thailand

1. Introduction

Albeit the terms "nanocomposite" were introduced recently, such materials (e.g. carbon black filled rubbers) have been produced in industrial scale for more than five decades. Nowadays great effects are undertaken to improve the mechanical, thermal and other properties (e.g. flame resistance, barrier properties, electric conductivity) of polymers. To achieve this property upgrade nanocomposites are produced using fillers of various shape factors and dispersing them on nano-scale resulting in the formation of nanocomposites. The development of thermoplastics nanocomposites is rapidly emerging as a multidisciplinary research activity whose results could widen the applications of thermoplastics to the benefit of many different industries. Nanocomposites are a new class of composites that are particle-filled thermoplastics for which at least one dimension of the dispersed particle is in the nanometer range. In the related area thermoplastic/nanofiber composites have attracted considerable interest because they often exhibit remarkable property improvements when compared to virgin thermoplastic or conventional micro- and macro- composites. Accordingly, major goal of this book chapter is to explore a systematic study of enhancing the stiffness, strength and toughness characteristics of nanocomposites at the same time. In order to produce toughened nano-reinforced thermoplastics by water-assisted melt compounding, the toughener must be available also in aqueous dispersion. Moreover, most rubbers are available in latex form. Latex is a more or less stable aqueous dispersion of fine rubber particles being in submicron-micron range. The related particle size distribution strongly depends on the manufacturing of the latex. The mean size of the rubber particles in the latexes is exactly that what we need for toughening, impact modification of thermoplastics, in the range from 0.1-0.8 μm. The mean particle size is controlled by the viscosity ratio of the components and their interfacial tension in the melt.

In this book chapter different thermoplastic matrices are used (e.g. polyoxymethylene (POM) and polyamide-6 (PA-6)). Polyurethane (PU) and hydrogenated nitrile rubber (HNBR) were selected as the toughening agent for thermoplastics and used in its latex form. It is noteworthy that the mean size of rubber latices is closely matched with that of conventional toughening agents, impact modifiers. Synthetic nanofillers (e.g. carbon nanofiber (CNF), boehmite alumina (BA) and sodium fluorohectorite (FH)) were used as reinforcements. One of the criteria for selecting these fillers was that they are water swellable/dispersible and thus their nanoscale dispersion can be achieved also in aqueous polymer latex.

2. Concepts of nano-scale reinforcement and toughness in thermoplastics

The major problem with nanoparticles reinforcement is that it causes severe material embitterment which is in close analogy with discontinuous fiber-reinforced thermoplastics. On the other hand, exactly this amount of nano -scale particles is needed to get optimum stiffness and strength properties. Blending the thermoplastic with elastomeric modifiers is a successful way to overcome this problem and to improve the toughness characteristics. Many works have documented the effect of various parameters of the modifiers on the toughness response including rubber particle size (Borggreve, 1987; Margolina & Wu, 1988), rubber concentration (Baldi et al., 2006) and interparticle distance (Wu, 1988; Jiang et al., 2002). These studies revealed that the rubber concentration should be above a critical level in function of its particle size. To improve the toughness of a polymer, the interparticle distance should fall below a critical value which depends on the volume fraction and the particle size of the rubbers. In this case, namely, the stress fields of neighboring particles overlap and a larger volume fraction of the matrix supports an average load higher than the applied load. Multiple localized plastic deformations are hereby initiated in the plastic zone ahead of the crack tip. Typical rubber particle sizes for effective toughening should be in nano-scale range. Rubber particles with micro size have been found to be relative inefficient, although they might be active in crack bridging.
In additional, the search for high performance nanofiber polymer composites produced successful results in related with toughness enhancements. Few research studies addressed the property improvements of polymers by adding carbon- nanofibers (CNFs) and nanotubes (CNTs). Most of CNF/CNT reinforced polymer composites show improvements in strength and stiffness with loss of elongation at break (Kumar et al., 2002; Chung, 2001; Sandler et al., 2002). There have been a number of experimental studies reported in the literature about the influence of CNT/CNF on the toughness enhancements of various types of polymers (Puglia et al., 2003; Song & Youn, 2005; Kinloch et al., 2002; Gryschuck et al., 2006). (Ruan et al., 2003) showed for enhancements in ductility of MWCNT/ reinforced ultrahigh molecular weight polyethylene (PE). The increased ductility in PE is most likely due to the chain mobility enhancement and attributed to the secondary crystal formation due to the presence of MWCNT. (Lozano et al., 2004) published extremely high extensibility; high toughness of CNF reinforced polyethylene bulk composites. The total tensile elongation of more than 1600% is higher than any published value for nanoreinforced composites. The large ductility increase in PE has been attributed to the excellent dispersion of CNF and interfacial bonding produced by the high shear history. In addition, CNF has been applied as the toughening filler of the structural material for resin based composites. (Seyhan et al., 2009) reported that the addition of low content of silanized vapor grown carbon nanofiber (VGCNF) improved the fracture toughness of the epoxy resin by about 12%. They concluded that attachment of epoxide end groups containing silane molecules to surfaces of CNF enhanced the compatibility between CNF and epoxy. Considerable interest is devoted to nanocomposites due to the attractive properties. It has been shown that the dielectric and mechanical properties of polymer based nanocomposites were markedly

improved by the addition of CNF (Sui et al., 2008, 2009). Similar to CNT, CNF has also a high reinforcing efficiency. Its presence in a small amount in the corresponding polymer, typically less than 5 wt. %, can result in significant improvement in mechanical properties. It has been reported that adequate interfacial adhesion of CNF to the related matrix and its uniform dispersion are the key factors in respect to the reinforcing effect (Prolongo et al., 2009). The length of CNF can very between 50 and 100 μm, therefore high aspect ratios can be obtained. As previously reported, CNF were purified and functionalized to remove amorphous carbon and to open the highly tangled fiber (Lozano et al., 1999). Being much longer than CNT, the disentanglement, dispersion of CNF is very problematic. The goal of improving the carbon nanofiber matrix interfacial adhesion issue and complete dispersion must be solved before achieving the full potential of CNF nanocomposites.

3. Preparation of nanocomposites

It was early noticed that the preparation technique of the nanocomposites has a strong impact on the dispersion of the nanoparticles (Park et al., 2007). Accordingly, various methods, like in-situ polymerization, melt blending and solution/dispersion techniques have been tried to disperse nanoparticles/nanofibers in polymers. In situ polymerization is a challenging one, as it allows us to adjust the chemistry to optimize the affinity between filler and the resulting polymer. Different types of polymerization methods have been used to prepare polymer nanocomposites, such as solution, suspension or emulsion, and free radical polymerizations (Akelah & Moet, 1996; Wang et al., 2002; Chen et al., 2000; Zhou et al., 2001). This variety to prepare polymer nanocomposites by in situ polymerization was mostly explored with clays. However, this process is cumbersome and also costly. If the organophilic modification of inorganic fillers could be eliminated, the procedure might be simplified and the production cost substantially lowered compared to state-of-art processes.
The use of conventional melt compounding techniques to prepare nanocomposites is usually more practical and economical than in situ polymerization. For that purpose common polymer processing equipments, such as extruders and internal mixers are suited. The shear, accommodated in the melt during processing may also be helpful to support the nanoparticles dispersion. However, it is not always enough to break up big particles agglomerates as the resulting nanoparticles dispersion may remain further poor. A homogeneous dispersion of nanoparticles in a polymer by using conventional compounding techniques is very difficult due to the strong tendency of fine particles to agglomerate (Jana & Jain, 2001). An alternative way to prepare composite materials is the latex compounding/latex coagulation technique. Latex compounding is a promising technique compared for example to in-situ polymerization and solution techniques which are used to produce nanocomposites. It is becoming more and more important because of the following benefits: simple many polymers are available in latex form, as typically Figures 1a-b shown in TEM image of PS latex (Siengchin et al., 2007) and AFM image of HNBR latex, respectively. The mean particle size is at ca. 200 nm and the particles are present in a very narrow distribution (cf. Figure1a). Latex can be introduced in polymer melt during compounding in line.
In an aqueous polymer latex, microscopic solid polymer particles are suspended in water. During drying they form a film through coalescence. When combined with nano-sized filler, the polymer particles and the filler create a segregated network. This may show excellent

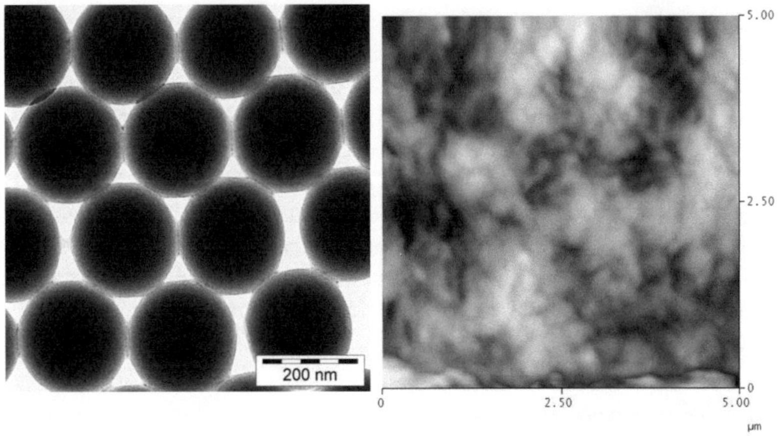

Fig. 1. (a) TEM image of PS latex (Siengchin et al., 2007) and (b) AFM image of HNBR latex.

properties at very low filler concentration (Grunlan et al., 2004), compared to a melt blended version containing the same polymer and filler. The advantage of this latex compounding process is that reduced health risk, good dispersibility, the expensive chemical modification of the fillers, giving the necessary affinity to the polymer, can be avoided. Moreover, when the nanofiller is well dispersed in the polymer latex, the latter can be used as a masterbatch (dry and liquid forms) for subsequent melt compounding. In addition, nanofiller containing latices or slurries can be injected in the melt during continuous melt compounding. Exactly, these are those options which have been explored in this chapter.

3.1 Latex melt compounding (LMC)

One very promising technique being developed for nanocomposite is latex melt compounding. Polymer blend, binary and ternary composite systems can be prepared by melt compounding using a masterbatch produced from polymer latex containing nanofillers (masterbatch technique, MB). A scheme of the masterbatch technique of polymer composite systems is given in Figure 2. First, an aqueous nanofillers slurry was produced at ambient temperature through mechanical stirring. Then the rubber latex was introduced in this slurry and stirred. The resulting slurry was poured in a framed glass plate and dried for few days at room temperature (RT). This resulted in a rubber film as the glass transition temperature (T_g) of rubber (PU or HNBR) is much lower than RT. The nanofiller-containing rubber masterbatch (MB) was introduced in the polymer system after melt mastication of the latter for few min. The overall duration of the melt mixing for blend, binary and ternary nanocomposites was prepared by laboratory kneader at the same condition.

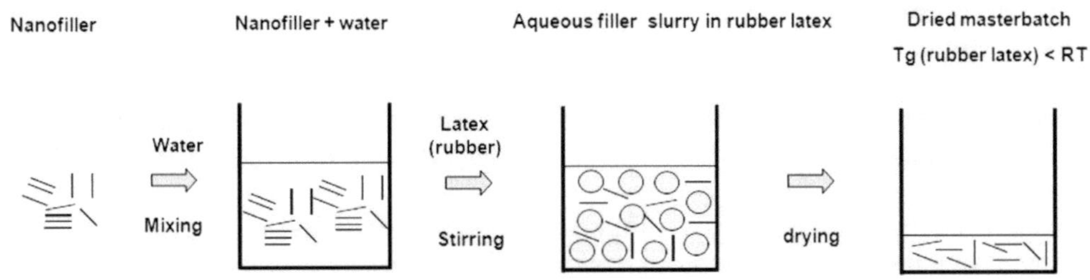

Fig. 2. Scheme of polymer nanocomposites prepared by the masterbatch technique.

3.2 Water-assisted melt compounding (WMC)

In this water-assisted melt compounding method, different nanoreinforced and/or toughened thermoplastics -based composite systems can be produced continuously in a twin-screw extruder whereby nanofillers in water slurry and rubber latices are introduced in the polymer melt. For compounding of the composites a conventional mixing screw was used the design of which is given in Figure 3. First, nanofillers were dispersed in water at ambient temperature under continuous mechanical stirring to obtain the aqueous nanofillers slurry. Thermoplastics–based composite systems were supplied so that sum totals were set to 10 kg/h. Rubber latex and/or aqueous nanofiller slurry were injected into the extruder, using pump. The extrusion temperature was selected between 150 °C and 190 °C systems from hopper to the die for all composite. Polymer pellets were charged in feeder in the first zone of the extruder. As shown in Figure 3, the nanoreinforced and/or toughened thermoplastics -based composite were prepared by using a pump: the resulting nanofiller slurry/rubber latex was injected in the melting zone into the extruder. A second pump was used to produce the ternary composite, whereby the rubber latex was introduced also in the barrels section 4. The water was eventually degassed in the transport zone (barrels section 7-9) of the screw and the evaporation was completed using a vacuum pump.

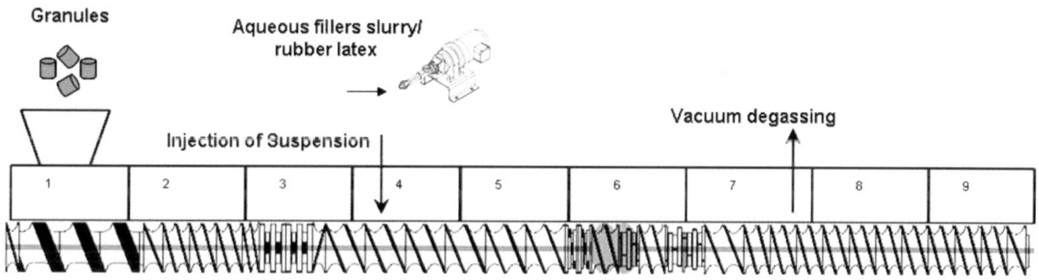

Fig. 3. Screw configuration and barrel sections of the co-rotating twin-screw extruder.

4. Structure – property relationships

From an effective process development point of view, it is interesting to be able to understand the relationships between process setting, physical molecular structure and (thermo) mechanical properties of nanoreinforcing and toughening thermoplastics. To obtain these relationships, a systematic study was performed on different characterization techniques for the related nanocomposites. The dispersion of nanofillers was characterized by transmission, scanning electron and atom force microcopies (TEM, SEM and AFM, respectively) techniques. The mechanical and thermomechanical properties of the composites were determined in Charpy impact, dynamic-mechanical thermal analysis (DMA), creep, stress relaxation tests, and thermogravimetric analysis (TGA). In additional, the dielectric response of nanocomposites was investigated by means of broadband dielectric spectroscopy.

4.1 Nanocomposites produced by LMC

4.1.1 Creep response

Nanocomposite materials used for structural applications of practical interest may exhibit viscoelastic behavior which has a profound influence on their performance. Viscoelasticity is of interest in materials science and engineering since it is causally linked to a variety of microphysical processes and can be used as an experimental probe of those processes. The causal links between viscoelasticity and microstructure are explored in various viscoelastic tests. The viscoelastic behavior of polymers and related composites is usually characterized in creep test (whereby the creep strain is measured in time under a constant load) or dynamic mechanical test, where the variation in storage and loss moduli is observed as a function of the temperature, (Xia et al., 2007; Eckstein et al., 1998).

To describe the creep results, phenomenological models, composed of spring and dashpot elements, are frequently used. A spring element (elastic) behaves exactly like a metal spring, stretching instantly when stress is applied, maintaining the stress indefinitely, and returning to its original dimension instantly when stress is removed. In a dashpot (viscous) under stress, the plunger moves through the fluid at a rate that is proportional to the stress. In a creep experiment, the stress is kept constant and the change in the deformation of the polymer as function of time is recorded.

In the linear viscoelastic range, the parameters of this series do not depend on the level of the applied load. For creep under applied constant stress σ_0, the material response is:

$$D(t) = \frac{\varepsilon(t)}{\sigma_0} \tag{1}$$

where D(t) is the viscoelastic creep compliance and $\varepsilon(t)$ is creep strain at time, t.

In the nonlinear range the dependence upon the level of the applied load can be expressed by multiplying the linear parameters by so-called nonlinearity factors, which, of course, are load, time and temperature dependent (Schapery, 1969; Brueller, 1987). The nonlinear creep compliance is given by:

$$D(t,\sigma(t),T) = \frac{\varepsilon(t,\sigma(t),T)}{\sigma_0} \tag{2}$$

where $\sigma(t)$ is the real stress at time, t and T is temperature. This equation can be simplified with:

$$\varepsilon(t,\sigma(t),T) = \varepsilon(t,\sigma_0,T) = \varepsilon(t,T)c\sigma_0 \tag{3}$$

Thus, the following equation can be also expressed in terms of creep compliance

$$D(t,\sigma(t),T) = D(t,\sigma_0,T) = c\varepsilon(t,T) \tag{4}$$

where c is a constant. In the above equation creep compliance is only a function of the time and temperature.

Creep test is an important method to study the reinforcing ability of nanofiber in the thermoplastic matrix. As shown in Figure 4, the effects of CNF and PU incorporations on the creep response of POM (Siengchin et al., 2010). Introducing CNF into POM resulted in a considerable reduction in the creep as shown by the plots of creep compliance vs. time. Accordingly, CNF acts as reinforcing phase in POM. However, the presence of PU rubber particles increases the creep compliance data. SEM picture also confirms that the CNF is well dispersed in POM matrix (cf. Figure 5). Additional incorporation of CNF into the

POM/PU blend improves the resistance to creep confirming the reinforcing action of CNF also in this blend.

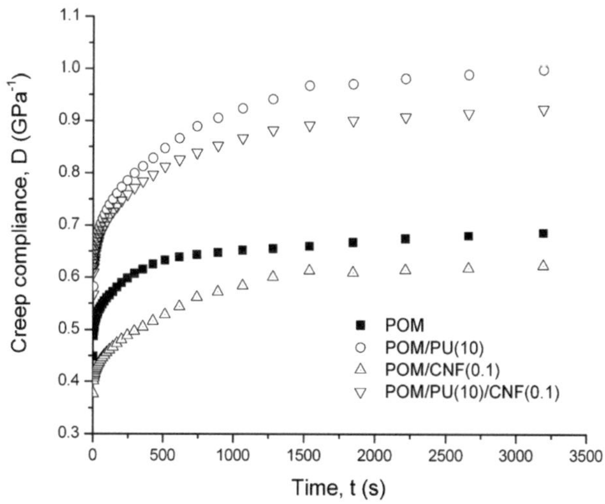

Fig. 4. Creep compliance for CNF based on POM composites (Siengchin et al., 2010).

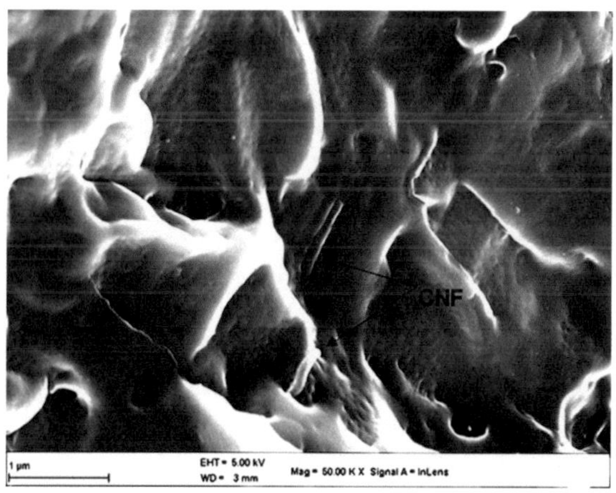

Fig. 5. SEM image of POM/CNF composite (Siengchin et al., 2010).

The considering outcomes are schematically depicted in Figure 6. The external mechanical loading included changes on lamellar level in a semicrystalline polymer, first lamellae rotation and separation take place after the amorphous molecules are highly stretched. These rotation and separation mechanisms, the occurrence of which depends on the relative orientation of the lamellae in respect to the loading, are at work also in the linear viscoelastic range. The nanofibers, being restricted by polymer chains and adhering well to the matrix strengthen the amorphous phase. So, the stretching of the amorphous and entangled chains occurs with considerable time delay under the same load in presence than in absence of nanofiber. On the other hand, the rubber particles act as stress concentrators due to the high difference in the elastic modulus between thermoplastic and rubber. This stress concentration effect supports the stretching of the amorphous chains. The lamellae separation and rotation are also favored via the tie-molecules. The outcome is increased creep and creep rate.

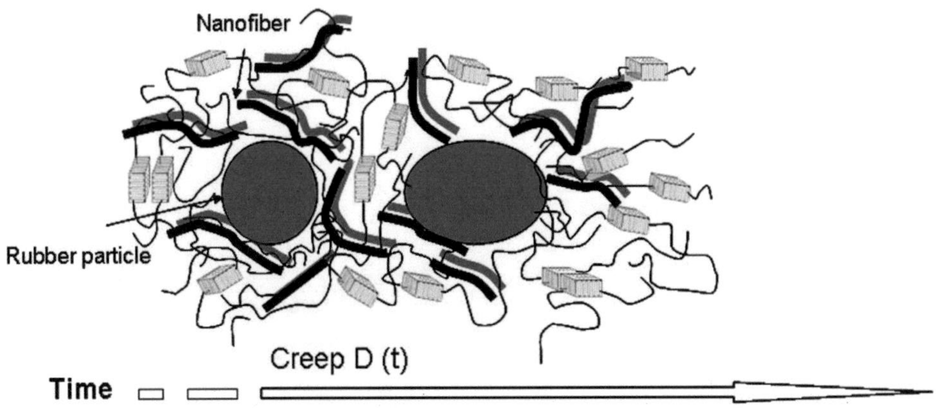

Fig. 6. Schema of creep development in reinforcing and toughening semicrystalline polymer.

(Siengchin & Karger-Kocsis, 2009) investigated by considering the morphological differences between the PA-6/HNBR/FH and PA-6/HNBR/BA and found that the FH stacks are dispersed in the PA-6 and located in the HNBR/PA-6 interphase. By contrast, BA is mostly embedded in the HNBR phase (cf. Figure 9). This creep effect is strongly suppressed in the FH-containing system due to the location of the FH in the interphase region. Embedding of BA in the HNBR reduced also the creep and creep rate as the stress concentration effect is diminished compared to the PA-6/HNBR binary blend. This reduction is due to the fact that the difference in the modulus between PA-6 and HNBR became smaller owing to the BA reinforcement of HNBR.

Further hints for the enhancement in the creep result were caused by the dispersed state (Siengchin et al., 2008). The authors found that the addition of PU into POM matrix resulted in a considerable increase in the creep compliance. The creep compliance values of the composites are smaller compared to the POM/PU blend showing the reinforcing effect of the alumina. The most striking finding is that the primary particle size of the alumina likely affects the creep behavior. More exactly the dispersion state of the alumina, are well reflected in the creep response. The change from micro- to nanocomposite reduces further the creep compliance. This may be associated with pronounced changes in the creep rate. A more interesting observation is that the creep response of all POM systems is sensitive to the temperature. This resulted in an increase in the mobility of the amorphous segments.

4.1.2 Dynamic mechanical thermal analysis

In addition of creep experiments, dynamic mechanical thermal analysis (DMA) is quite common method understand better some of the added complications that arise from the time and temperature dependence of the modulus of viscoelastic materials. Usually, the viscoelastic behaviour can be determined by measuring the storage modulus and the loss modulus over a wide range of frequencies and temperatures. The storage modulus is proportional to the degree of elasticity of the system. The loss modulus is proportional to the dissipation or loss energy as heat in a cycle deformation, reflecting a certain degree of the viscosity of the system (Shaw & MacKnight, 2005). Figure 7 illustrates a schematic of dynamic experiment which shows graphically the nature of the two sinusoidal signals against time.

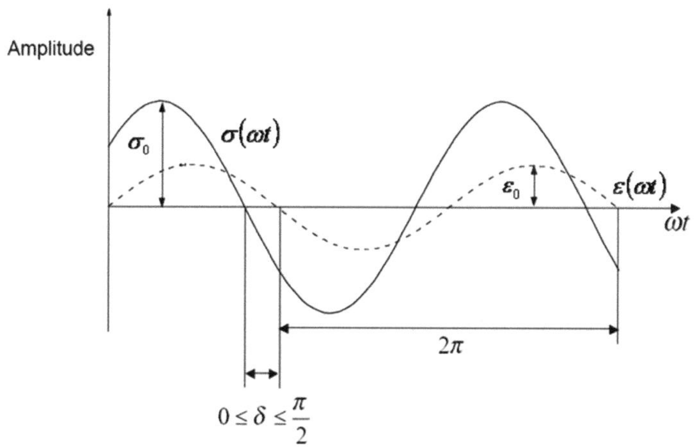

Fig. 7. Schematic of the dynamic mechanical experiment.

Consider the application of a sinusoidal strain, which can be presented by:

$$\varepsilon(t) = \varepsilon_0 \sin \omega t \tag{5}$$

where ε_0 is the maximum amplitude of the strain, ω is the angular frequency and t is time. The phase of the strain is arbitrarily set at zero, which can be done without loss of generality (perfectly elastic). The resulting stress $\sigma(t)$ is given by:

$$\sigma(t) = \sigma_0 \sin(\omega t + \delta) \tag{6}$$

A complex Young's modulus (E*) reflects the contribution of both storage (E′) and loss (E″) components to the stiffness of material, as follows:

$$E^* = \frac{\sigma_0}{\varepsilon_0} = E' + iE'' \tag{7}$$

According to schematic of DMA, the modulus at these two points can be calculated from equations 6 and 7. The storage modulus (E′) gives directly:

$$E' = E^* \sin \delta \tag{8}$$

And in term of loss modulus (E″):

$$E'' = E^* \sin(\pi / 2 + \delta) = E^* \cos \delta \tag{9}$$

The ratio of the loss modulus to the storage modulus is the tangent of the phase angle shift δ between the stress and strain vectors, thus:

$$\tan\delta = \frac{E''}{E'} = \frac{\sigma''}{\sigma'} \tag{10}$$

(Siengchin & Karger-Kocsis, 2009) examined the effect nanofillers, viz. FH and BA, in a HNBR toughened polyamide-6 in dynamic mechanical thermal analysis. DMA spectra in form of storage modulus (E′) and loss factor (tan δ) as function of temperature were plotted and are demonstrated in Figures 8, respectively. It can be seen in Figure 8a that the blend with 9 wt% of HNBR exhibits markedly lower stiffness than the PA-6 at least below the glass transition temperature (T_g) of PA-6 (ca. 50 °C). This is due to the rubbery character of the incorporated HNBR. However, the incorporation of sodium fluorohectorite (FH) and Boehmite alumina (BA) particles resulted in increase in the storage modulus in the whole temperature range, compared to that of the PA-6/HNBR blend. This can well be explained by the reinforcing effect of the nanoparticles leading to increased stiffness. One can also recognize that the storage modulus of FH is higher than BA nanocomposites. This stiffness increase suggests that FH stacks are well dispersed in the PA-6 matrix and have much higher aspect ratio in comparison with BA. This, along with the fact that BA was encapsulated in the HNBR phase, was confirmed by SEM image (cf. Figure 9).
The plots in Figure 8b reveal the T_g of PA-6 (at ca. 50 °C) and a strong secondary transition at ca. -50 °C. The latter peak in PA-6 (β-relaxation) is usually attributed to the mobility of non hydrogen bonded amide groups on adjacent chains (Araújo et al., 2004). Incorporation of HNBR yielded an additional maximum at -18 °C (T_g of HNBR) in the tan δ vs. T curves. Filling with BA and FH particles was accompanied with a shift in the T_g of PA-6 towards higher temperature (ca. 70 °C). The increment in the tan δ by adding HNBR is due to chain flexibilization, likely in the interphase regions. However, the tan δ values were reduced after additional incorporation of the nanofillers, as expected. The T_g of the HNBR may also be affected by the fillers. The related change may be traced to the preferred location of the corresponding fillers.

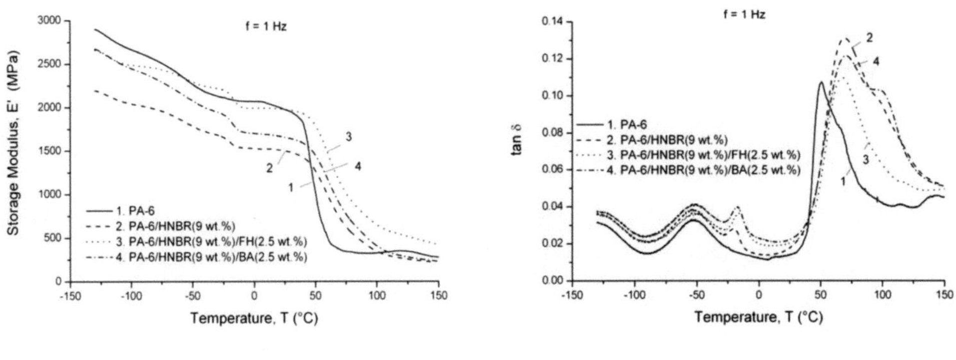

Fig. 8. (a) Storage modulus (E′) and (b) mechanical loss factor tan (δ) as a function of temperature for FH and BA based on PA-6 composites (Siengchin & Karger-Kocsis, 2009).

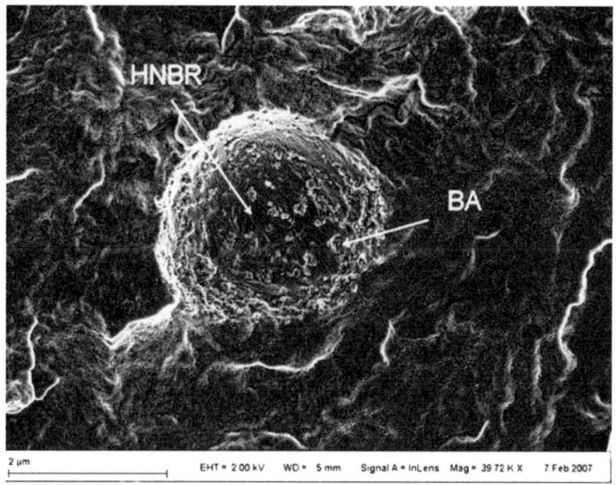

Fig. 9. SEM image of PA-6/HNBRBA ternary composite (Siengchin & Karger-Kocsis, 2009).

In another work, the effects of CNF and PU into POM on the DMA data are shown in Figure 10 (Siengchin et al., 2010). One can notice that the stiffness values of the POM/CNF composite were slightly higher than that of the neat POM in the whole temperature range. This indicates the reinforcing effect of CNF. As expected, adding PU rubber was associated with a drop in the storage modulus of the POM/PU blend compared to the neat POM. The storage modulus vs. temperature curves of the POM/PU and POM/PU/CNF systems were practically the same. The plots in Figure 10b reveal three relaxation transitions in POM. The glass transition temperature (T_g) is located at around -60 °C (γ-relaxation transition). Another two peaks, observed at ca. -2 °C and ca. 130 °C correspond to the β- and α-relaxations, respectively. These relaxation processes are assigned to the motions of long molecular segments in disordered (β) and well ordered crystalline phases (α), respectively (Hojfors et al., 1977). Incorporation of PU yielded an additional maximum peak at ca. -50 °C in the tan δ vs. T trace, which represents the T_g of the PU component. Adding CNF and/or PU was accompanied with a shift in both the α- and β-relaxations to lower temperatures (ca. 5 °C). One possible explanation is that small agglomerates of CNF and matrix voids within this network support the movement of macromolecular chains yielding a small reduction in the T_g (β-relaxation) of the matrix (Shen et al., 2007).

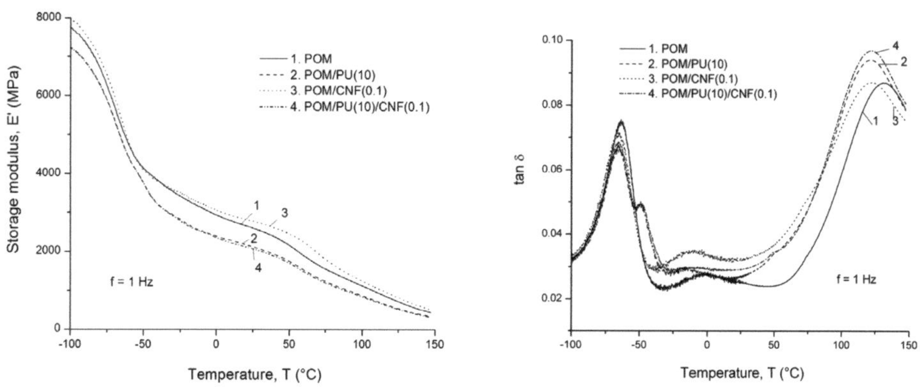

Fig. 10. (a) Storage modulus (E′) and (b) mechanical loss factor tan (δ) as a function of temperature for CNF based on POM composites (Siengchin et al., 2010).

4.1.3 Tensile test

The main advantage of reinforcing thermoplastic composites is the enhancement in the mechanical properties of the thermoplastic matrix at relatively low inorganic filler content. To monitor the reinforcing ability of nanofillers in the thermoplastics matrix of the compound, tensile mechanical test appears as an excellent tool. The effects of CNF and PU on the tensile mechanical data are shown in Figure 11. The enhancement in tensile strength of nanocomposites containing 0.1 wt. % CNF was ca. 18 % compared to the neat POM. The primary reason for that is that due to the uniform dispersion of CNF in POM, an efficient load transfer occurs from the matrix towards the CNF. SEM picture also confirms that the CNF is well dispersed in POM matrix (cf. Figure 5). A fine CNF dispersion is expected already based on the low concentration of CNF in the aqueous suspension used. SEM pictures taken from the fracture surfaces of the specimens indicated that the residual mean length of CNF is ca. 6 μm. Note that a reinforcing effect is generally accompanied with reduced ductility in all composite materials. This is the case also for the POM/CNF nanocomposite the elongation at break value of which is below that of the neat POM. Introducing PU has an adverse effect: the tensile strength decreases, whereas the elongation at break increases markedly. Note that this is the usual response of thermoplastics with impact modifier under static loading conditions. Additional incorporation of CNF increases the tensile strength and reduces the elongation at break of the POM/PU blend confirming that CNF acts as reinforcement.

For the incorporation of FH and BA particles based on HNBR/PA-6 systems (Siengchin & Karger-Kocsis, 2011), the ductility was strongly reduced compared to the PA-6 matrix. Due to the rubbery character of the incorporated HNBR, the PA-6/HNBR blend exhibits markedly lower stiffness than the parent PA-6. However, the elongation at break increases markedly by adding HNBR to PA-6. It is usually accepted that rubber domains first cavitate provoking locally a plane strain/plane stress transition in fracture mechanical terms (Paul & Bucknall, 2000). This supports the development of crazing (with further elongation of the craze fibrils) and superimposed shear yielding, which all enhance the ductility. As expected, the nanoparticle reinforcement leads to increased stiffness and reduced ductility (Siengchin & Karger-Kocsis, 2009; Karger-Kocsis, 2009). The latter is mainly due the stress concentration effects of filler agglomerates which cannot be released by matrix-related events (crazing, shear deformation) owing to inhomogeneous filler dispersion. Stress concentration induces filler/matrix debonding, and the related voids coalescence causing final, premature fracture. The nanocomposite containing FH exhibits higher E-modulus than the companion composite containing BA, though the related value remains under that of the neat PA-6. This relative stiffness increment should be traced to the difference in the dispersion characteristics of FH and BA. Recall that FH was intercalated and mostly dispersed in the PA-6 matrix by contrast to BA being located in the HNBR phase.

To order to investigate the effect of PU and alumina in POM systems,the tensile mechanical data in form of tensile strength, modulus and elongation at break were also studied. Addition of PU to POM or the combined use of PU and alumina improves the elongation at break, however, at cost of the tensile strength. The POM/PU/alumina ternary composites prepared by the masterbatch (MB) method exhibited higher tensile strength accompanied by much higher elongation at break when compared with the composite produced by direct melt compounding (DM) or to the POM/PU blend. Recall that the alumina particles are by far better dispersed in the matrix when prepared via the MB instead of the DM technique. This is in line with recent reports claiming that the tensile mechanical response is strongly affected by the dispersion state of nanofillers in polystyrene (PS) (Siengchin & Karger-Kocsis, 2006). Moreover, it is worth noting that increased ductility (elongation at break) generally manifests in improved toughness.

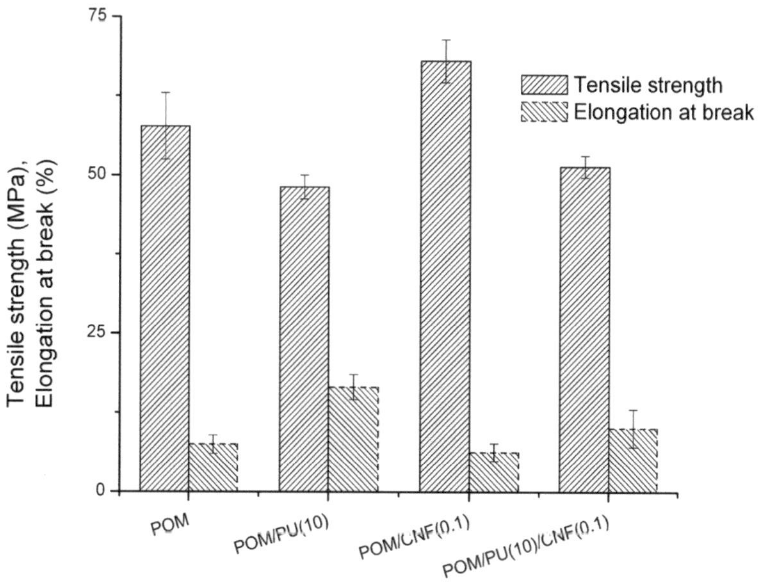

Fig. 11. Tensile strength and elongation at break for CNF based on POM composites (Siengchin et al., 2010).

4.1.4 Dielectric response

Dielectric spectroscopy is a powerful experimental tool for the investigation of polarization and conductivity mechanisms, molecular mobility, phase changes and interfacial phenomena in polymers and polymer matrix composites. Experimental data can be analyzed by means of different formalisms, such as dielectric permittivity, ac conductivity and electric modulus formalism. Although the recorded electrical effects can be described and analyzed in terms of any of the aforementioned formalisms, under certain conditions a specific formalism could be proved more helpful in extracting information with respect to the occurring physical processes. Figure 12 presents the dependence of the dielectric permittivity (ε') and dissipation factor (tan δ) on the frequency, measured at room temperature for all the examined specimens. The dielectric permittivity and dissipation factor attain high values at the low frequency edge, which are decreasing steeply with increasing frequency. Considering the insulating nature of the examined systems, the enhanced values in the low frequency range indicate the co-existence of electrode polarization and interfacial relaxation phenomena. Electrode polarization is an undesirable effect related to the charged electrode-specimen contacts. Interfacial polarization (IP) or Maxwell-Wagner-Sillars (MWS) effects are present in heterogeneous systems because of the accumulation of mobile charges at the interfaces of the composite (Tsangaris & Psarras,

1999). Separating the contribution of each effect is not an easy procedure in the dielectric permittivity mode and thus the electric modulus formalism will be employed for the interpretation of dielectric data.

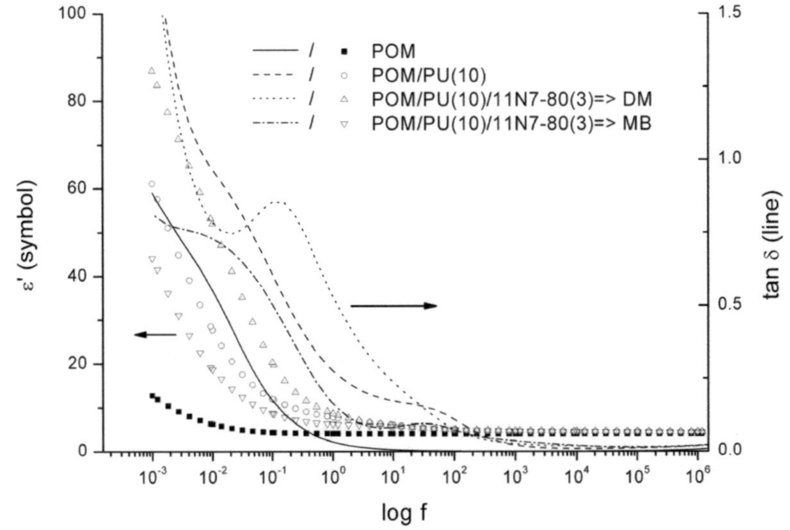

Fig. 12. Dielectric permittivity (ε') and dissipation factor (tan δ) as a function of frequency (f) for the composites produced by masterbatch (MB) and direct melt compounding (DM) (Siengchin et al., 2008).

Electric modulus presentation exhibits some advantages in the interpretation of slow relaxation phenomena (processes with relatively enhanced relaxation time) in complex systems, mainly because of the elimination of the undesirable effect of electrode polarization (Tsangaris et al., 1998; Psarras et al., 2007).

Electric modulus is defined as the inverse quantity of complex permittivity by the following equation:

$$M^* = \frac{1}{\varepsilon^*} = \frac{1}{\varepsilon' - j\varepsilon''} = \frac{\varepsilon'}{\varepsilon'^2 + \varepsilon''^2} + j\frac{\varepsilon''}{\varepsilon'^2 + \varepsilon''^2} = M' + jM'' \tag{11}$$

where ε', M′ and ε'', M″ are the real the imaginary part of dielectric permittivity and electric modulus respectively.

Figure 13a depicts the real part of electric modulus (M′) as a function of frequency for POM, POM/PU blend and the POM/CNF binary- and POM/PU/CNF ternary composites (Siengchin et al., 2010). The real part of electric modulus (M′) as a function of frequency undergoes a step-like transition from low to high values for POM and POM/CNF binary composite. On other hand the POM/PU blend and the POM/PU/CNF ternary composite exhibit, at least, two step-like transitions. These transitions correspond to relaxation processes, which become evident in the loss modulus index vs. frequency plots.

Figure 13b depicts the dielectric spectra of the imaginary part of electric modulus (M'') as a function of the frequency of the applied field. In the case of pure POM a single loss peak is recorded in the low frequency region, which is assigned as α-relaxation. However, at room temperature, and in the examined frequency range only the slower transition is recorded. The origin of this process cannot be ascribed undisputedly. It has been attributed to rearrangements in the crystalline parts of POM, although contributions from the amorphous phase cannot be excluded (Siengchin et al., 2008; Psarras et al., 2007). Since this is the slower occurring relaxation process in POM, it is reasonable to suggest that is related to interfacial polarization phenomena between crystalline and amorphous phases. Interfacial polarization (IP) occurs in heterogeneous and complex systems because of the accumulation of unbounded charges at the interfaces of the phases. The corresponding peak of POM/CNF (0.1) system is not recorded in the examined frequency range, and it seems that is shifted to lower frequencies (cf. Figure 13b). This is a typical behavior for composite systems, where the presence of the reinforcing phase enhances heterogeneity, interfacial polarization and the relative relaxation time (Tsangaris & Psarras, 1999). The situation differs in the case of POM/PU (10) blend and the ternary system POM/PU(10)/CNF(0.1). In their dielectric spectra two relaxation processes are recorded in the intermediate frequency region, while the tendency for the formation of a third peak is present in the high frequency edge. From previous studies (Siengchin et al., 2008; Psarras et al., 2007) it is known that PU, at ambient, has a peak in the vicinity of 100Hz, related to its glass to rubber transition, and tends to form a second one at high frequencies. The latter is attributed to reorientation of polar side groups of the main chain (Psarras et al., 2007). Systems containing PU exhibit the aforementioned loss peaks. However, the third peak (the slower process) recorded close to 0.1Hz cannot be connected with PU, and thus it should be assigned to POM. The peak locus of POM's α-mode appears to shift to higher frequencies by almost an order of magnitude in the POM/PU(10) and POM/PU(10)/CNF(0.1) systems, indicating that the processes became faster.

The dielectric response of the examined systems is influenced by their variations in morphology. As it is already stated the presence of PU affects the morphology of POM by disturbing the spherulitic structure of the latter. The extent of rigid crystalline regions is diminishing, rearrangement of crystallites and/or relaxation of isolated amorphous segments is facilitated and thus the peak of α-mode shifts to higher frequencies. Moreover the incorporation of CNF reduces further the size of spherulites (cf. Figure 14) and consequently increases the interfacial area between the constituents and phases of the nanocomposite. The presence of CNF in the ternary system slightly influences the position of the peak by shifting to lower frequency, and at the same time reduces the intensity of electric modulus loss index (M''). In the dielectric permittivity formalism, enhanced heterogeneity and extensive interfacial area yield to increased values of the real and imaginary part of permittivity (ε') and (ε'') respectively, at low frequencies. Recalling electric modulus equation, it should be noted that the same data expressed in the electric modulus formalism would provide reducing values of the real and imaginary part of electric modulus (M') and (M'') respectively, with the increase of heterogeneity and interfacial area. Under this point of view, it is reasonable to suggest that the specific relaxation mode, recorded in the vicinity of 0.1Hz, is influenced by interfacial polarization phenomena, which are strengthen in the case of the ternary system.

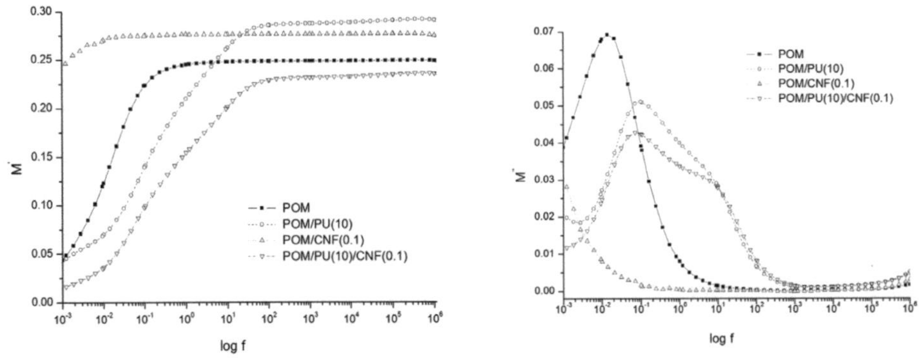

Fig. 13. (a) real (M') and (b) imaginary part (M'') versus frequency traces for CNF based on POM composites (Siengchin et al., 2010).

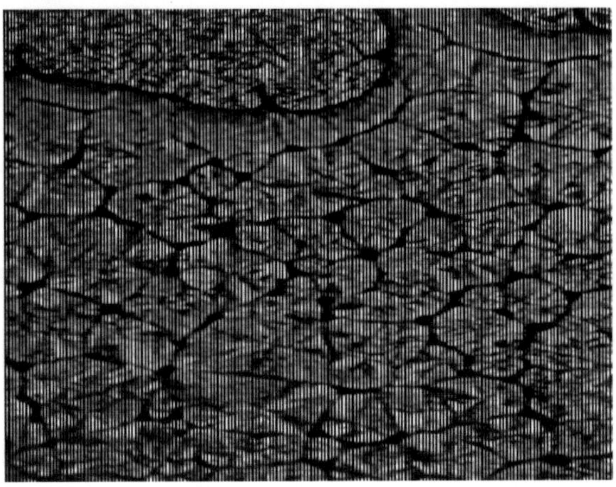

Fig. 14. Polarized light microscopy image for POM/PU/CNF ternary composite (Siengchin et al., 2010).

4.1.5 Stress relaxation

In order to describe the stress relaxation results phenomenological models, composed of spring and dashpot elements, are frequently used. Their parameters do not depend on the level of the applied deformation in the linear viscoelastic range according to the theory of viscoelasticity. For stress relaxation under applied constant strain (ε_0) the material response is given by:

$$E_r(t) = \frac{\sigma(t)}{\varepsilon_0} \tag{12}$$

where $E_r(t)$ is the stress relaxation modulus, and $\sigma(t)$ is stress, both as a function of time (t).

In the nonlinear range the dependence upon the level of the applied deformation can be expressed by multiplying the linear parameters by so-called nonlinearity factors, which are deformation-, time- and temperature-dependent. The nonlinear stress relaxation modulus is given by:

$$E_r\left(t,\varepsilon(t),T\right)=\frac{\sigma\left(t,\varepsilon(t),T\right)}{\varepsilon_0} \tag{13}$$

where $\varepsilon(t)$ is the real strain at time (t) and T is the temperature.
The stress relaxation data, measured in function of time for the POM, POM/PU blend and the CNF-containing binary-, and ternary composites is shown in Figure 15. The shape of the relaxation curves of the systems is similar to that of the neat POM. However, their relaxation modulus decreased remarkably with incorporation of PU compared to the neat POM. One can see that the relaxation modulus of the POM/CNF binary composite is slightly higher in the whole relaxation time range than those of the POM, POM/PU blend and POM/PU/CNF ternary composite. The increase of the relaxation modulus with CNF content is due to its reinforcing effects. Note that the long CNFs may relieve very efficiently such stress concentration effects which are induced by material heterogeneities. This finding is coherent with those deduced from creep (cf. Figures 4), DMA (cf. Figure 10) and tensile mechanical tests (cf. Figure 11).

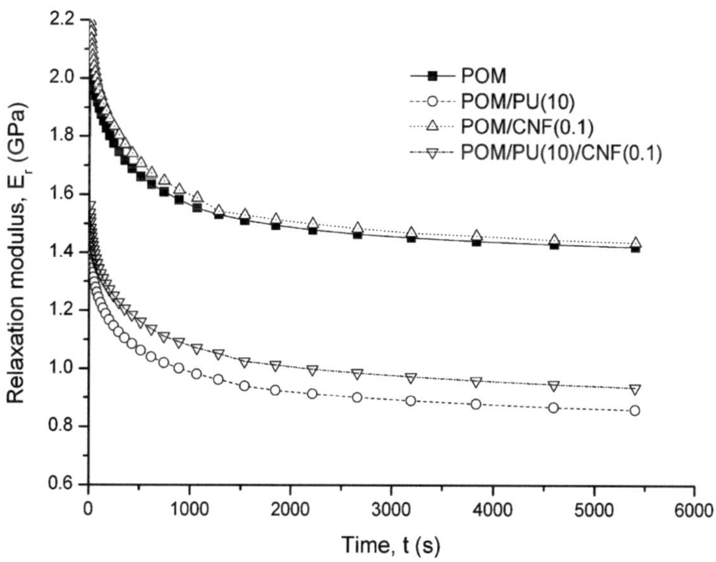

Fig. 15. Relaxation modulus for CNF based on POM composites (Siengchin et al., 2010).

4.1.6 Thermo-oxidative properties

It has been reported that by adding inorganic fillers (Sun et al., 2007), may increase the thermal stability of thermoplastics. This is at odds with our TGA observation. Figure 16 shows the weight loss vs. temperature traces for the POM, POM/PU blend, POM/CNF binary, and POM/PU/CNF ternary composites (Siengchin et al., 2010). The thermal degradation of the neat POM is a one-step procedure representing depolymerization. Incorporation of PU reduces slightly, whereas CNF markedly the onset of the thermal degradation. The follow-up degradation of the POM/CNF runs, however, at higher temperature than both POM and POM/PU blend. The POM/PU/CNF ternary composite has the highest onset decomposition temperature. Moreover, the related mass loss vs. temperature trace is parallel shifted to that of the POM towards higher temperatures. The common use of PU and CNF resulted in a synergistic effect in the thermo-oxidative stability of POM.

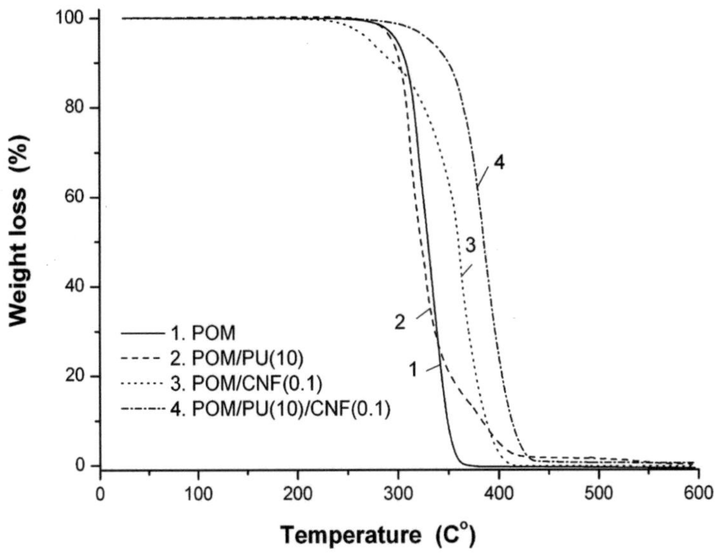

Fig. 16. Weight loss versus temperature for CNF based on POM composites (Siengchin et al., 2010).

4.2 Nanocomposites produced by WMC

Latex melt compounding techniques to disperse various rubber and nanofillers was already tried for various thermoplastic systems. This part of book chapter addresses the toughening and reinforcement of polyoxymethylene (POM) via a water-assisted melt compounding technique (above process) which is allow us to produce nanocomposites with additional toughening. Polyurethane (PU) was selected as toughening agent for POM according to the state-of-art. It was used in its latex form. It is noteworthy that the mean size of rubber latices is closely matched with that of conventional toughening agents, impact modifiers. Synthetic boehmite alumina was used as nanofiller. One of the criteria for selecting this filler was that it is water dispersible and thus its nanoscale dispersion can be achieved also in aqueous polymer latex.

Toughened or nanofilled POM composites have been successfully produced by water-assisted melt compounding technique. TEM results have shown that the alumina and rubber dispersions became fine and homogeneous when using the water-assisted extrusion melt compounding. The alumina particles are nanoscaled and homogeneously dispersed in the matrix (cf. Figure 17a). Figure 17b demonstrates that the PU rubber particles are also well dispersed in the POM matrix. The mean diameter of the PU is 700-900 nm, which is closely matched with the initial size of the PU particles in the corresponding latex. In the POM/PU/alumina ternary composites the alumina particles are mainly in the PU phase. On the other hand, the alumina and rubber particles are also separately dispersed in the POM matrix.

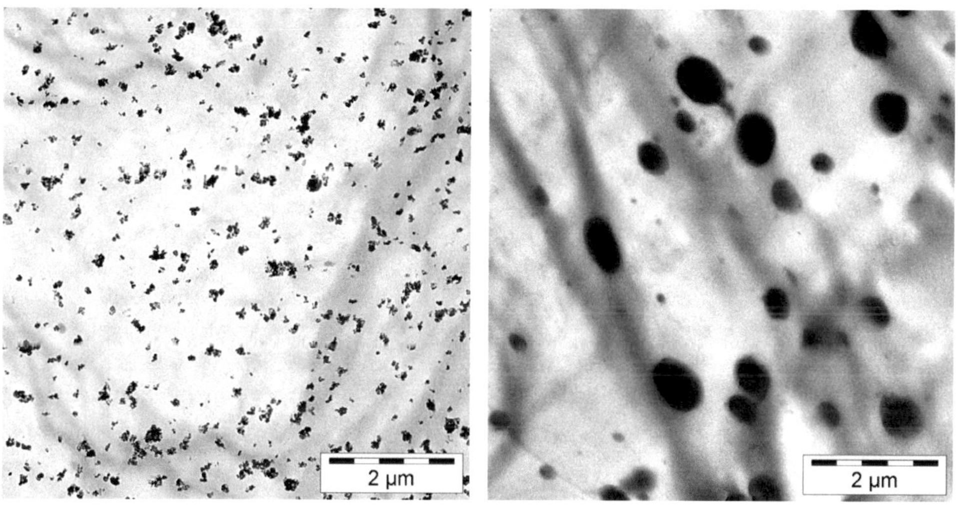

Fig. 17. TEM images of (a) POM/alumina composite and (b) POM/PU blend.

Reinforcement is defined as the ability of nanofillers to improve a variety of thermoplastic properties, e.g. tensile strength and stiffness. The reinforcement effect can be explained using DMA and creep diagrams. The tensile storage- and tan δ curve of nanoreinforced thermoplastic composites is shows up to a higher stiffness compared to an unfilled matrix. From this observation a reinforcing effect can be seen in Figure 18 (Siengchin et al., 2008). The modulus E′ of POM systems decreases with increasing temperature and the stiffness of the POM/PU blend was lower than the POM. However, the incorporation of alumina particles resulted in some increase in the storage modulus in the whole temperature range, compared to that of the pure POM. This can be well explained by reinforcing effect of alumina particles leading to an increased stiffness. In addition, the blend containing 10 wt% of PU exhibits markedly higher loss modulus than the other systems. This is due to the rubbery character of the incorporated PU.

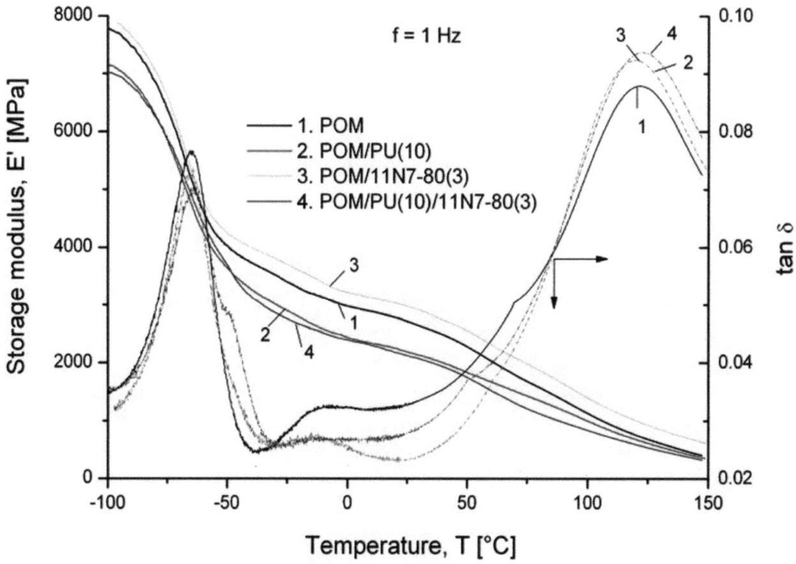

Fig. 18. Storage modulus (E′) and mechanical loss factor tan (δ) as a function of temperature for the composites produced by water-assisted extrusion melt compounding (Siengchin et al., 2008).

Generally, the creep behaviour of material is predicted by Burgers model (Findley et al., 1989). This creep model proved to be successfully applicable for many polymers. Therefore, this model is used to evaluate the effect of rubber particles and fillers in composites prepared by water-assisted melt compounding method. The creep experiment can be calculated from a discrete retardation spectrum of the Burgers model consisting of a Maxwell and a Kevin units connected in series by the equation (Kaschta & Schwarzl, 1994):

$$D(t)=D_0+\Psi(t)+\frac{t}{\mu_0} \tag{14}$$

$$\Psi(t)=\sum_{k=1}^{N} D_k * \left[1-e^{-t/\tau_k}\right] \tag{15}$$

where t is the time, D_0 is the instantaneous compliance, $\Psi(t)$ is equilibrium compliance part of creep compliance, μ_0 is viscosity and τ is different retardation times of the Kelvin unit D_k.

Figure 19 displays the traces of the creep compliance and its simulated value using the Burgers model, as a function of time at T=30 °C for the POM, POM/alumina composite, POM/PU blend and the alumina-containing ternary composites produced by water-assisted melt compounding. The incorporation of alumina particles into POM matrix resulted in a considerable decrease of the creep compliance, as expected owing to their reinforcing effect. Conversely, the addition of PU increased the creep of POM. The compliance values were reduced by approx. 12 % compared to the POM/PU blend when alumina was additionally incorporated. This creep response suggests that alumina and rubber particles are fine dispersed in the POM matrix. This was supported by the SEM inspection (cf. Figure 17). This phenomenological Burgers model was used to evaluate the effect of rubber and/or alumina particles on the instantaneous and equilibrium parts (approximated by six Voigt-Kelvin elements) of the creep compliance. The highest instantaneous compliance (D_0) was observed when PU was added into the POM. The instantaneous compliance of the POM/alumina composite is lower than those of the POM and POM/PU blend. The enhancement in the instantaneous part of compliance is caused by the alumina particles and their dispersed state. This resulted in a decrease in the mobility of the amorphous segments.

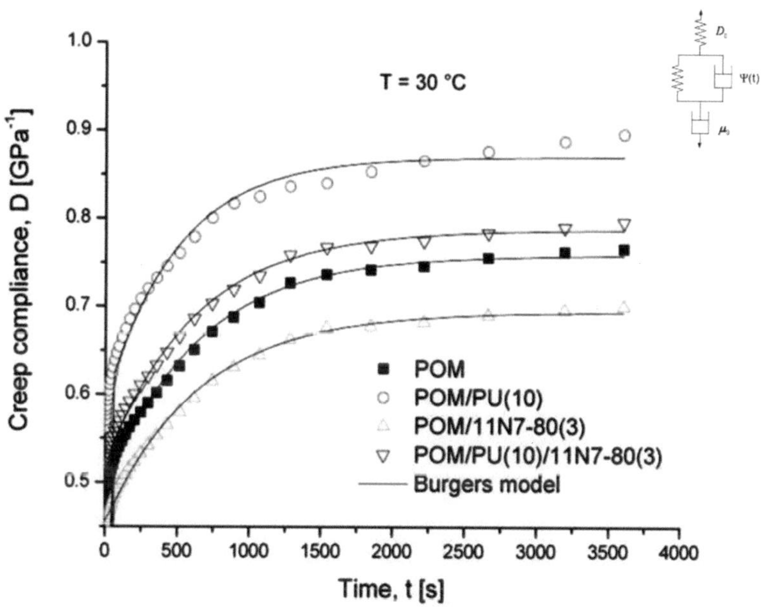

Fig. 19. Creep compliance for the composites produced by water-assisted extrusion melt compounding, fitted by Burgers model (Siengchin et al., 2008).

Results of the tensile mechanical tests of composites prepared by WMC are given in Table 1 model (Siengchin et al., 2008). It is clear by seen that the incorporation of alumina and PU particles strongly affected the mechanical behavior of POM. Adding alumina increased the stiffness slightly reduced the strength, the elongation at break. Conversely, the ductility of the POM/PU blend increased and the stiffness and strength decreased compared to pure POM. The reduction in the tensile strength may be attributed to the presence of rubber particles acting as stress concentrators and forcing the yielding at lower stresses. The decrease in the tensile modulus in the POM/PU blend may be accounted for the softening effect of PU, since the tensile modulus of rubber is considerably lower than that of the pure POM. The POM/PU/alumina ternary composites exhibited significant decrease in modulus accompanied by much higher elongation at break as compared to the POM, POM/alumina composite and POM/PU blend.

Sample Designation	Tensile strength [MPa]	Tensile modulus [MPa]	Elongation at break [%]
POM	63± 0.2	2646±71	17.8±2.8
POM/PU(10)	55± 0.2	2392±101	31.9±4.5
POM/alumina	61± 0.1	2847±143	9.1±2.8
POM/PU(10)/alumina	47± 0.5	2323±62	57.1±8

Table 1. Tensile mechanical characteristics of the composites produced by water-assisted extrusion melt compounding model (Siengchin et al., 2008).

Toughness is the ability of a material to absorb energy. The toughness can be done by using a pendulum and basic physics to measure how much energy it will be hold. Figure 20 shows the impact energy of POM, POM/PU blend and the composites with alumina particles produced by WM-CT model (Siengchin et al., 2008). Even though the tensile modulus of POM is increased by the incorporation of rigid alumina particles, the toughness was reduced. However, POM modified with 10 wt.% PU exhibit not only higher elongation at break, but similarly also higher impact energy (ca. 22 %) compared to the pure POM. This is due to the submicron dispersion of the rubber particles. So, the incorporation of rubbery particles in a suitable size range increases the impact resistance of POM. Furthermore, the impact energy of POM/PU/alumina ternary composites was marginally decreased compared to the POM/PU blend but remained still higher than POM.

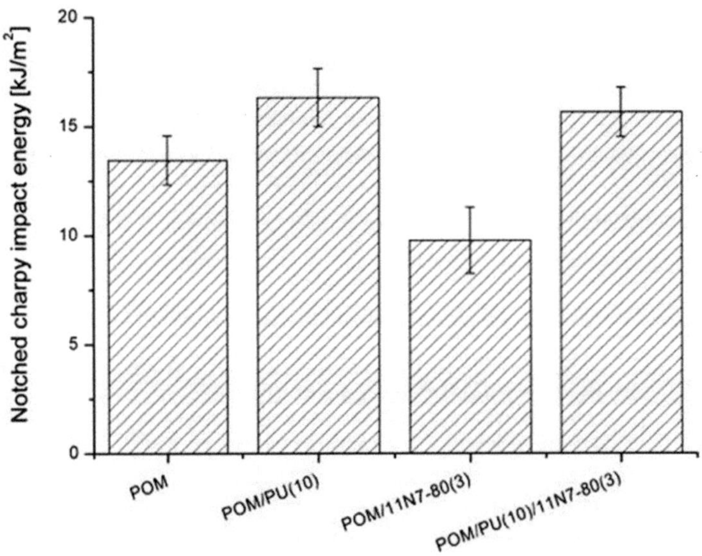

Fig. 20. the impact energy for the composites produced by water-assisted extrusion melt compounding model (Siengchin et al., 2008).

5. Summary

This chapter was devoted to develop the melt-compounding methods to manufacture tough, nanofiller-reinforced thermoplastics by adding an aqueous filler slurry and/or rubber dispersion to the molten polymer followed by evaporation of the water carrier. The work also compared the morphological, mechanical and thermal properties of nanoreinforced/toughened thermoplastic composites produced by latex melt compounding (LMC) and water-assisted melt compounding (WMC). Toughened or reinforced thermoplastic composites have been successfully produced by WMC. The mechanical properties (stiffness and creep resistance) of the composites were superior to those of the POM and POM/PU blend. The impact resistance of POM was highly enhanced by the addition of PU rubber. The dispersed rubber particle size was closely matched with that one required from usual toughening agents, impact modifiers. Besides, this WMC is a very effective process being simple and cost efficient for producing composite materials, as the expensive chemical modification of the nanofillers can be avoided. The water-assisted melt compounding to produce nanocomposites containing fillers- and rubber particles as nanoscale reinforcements and toughness offers a general way to modify the properties, such as stiffness and toughness. However, the toughened properties of and reinforced nanocomposites should be optimized via various rubber/filler ratios and concentrations.

Greek Symbols

$\varepsilon, \varepsilon_0$	Strain-, amplitude of the strain
$\varepsilon', \varepsilon''$	Real-, imaginary permittivity
ψ	Equilibrium compliance part
σ	Stress
σ_0	Applied constant stress
ω	Angular frequency
μ_0	Viscosity (Strain)
τ	Retardation time
δ	Phase angle shift

6. Acknowledgement

I am grateful to King Mongkut's University of Technology North Bangkok for providing me the opportunity of working in an outstanding environment from the scientific and technical point of view. Especially I would like to express my special thanks to Ms. Pawinee Boonyasopon who constantly support and encourage me.

7. References

Borggreve, R.J.M., Gaymans, R.J. & Schuijer, J. (1987). Impact behaviour of nylon-rubber blends: 5. Influence of the mechanical properties of the elastomer. *Polymer,* Vol. 30, No.1, pp. 71-77, ISSN 0032-3861

Margolina, A. & Wu, S. (1988). Percolation model for brittle-tough transition in nylon/rubber blends. *Polymer,* Vol. 29, No. 12, pp. 2170-2173, ISSN 0032-3861

Baldi, F., Bignotti, F., Tieghi, G. & Riccò T. (2006). Rubber toughening of polyamide 6/organoclay nanocomposites obtained by melt blending. *Journal of Applied Polymer Science,* Vol. 99, No. 6, pp. 3406-3416, ISSN 0021-8995

Wu, S. (1988). A generalized criterion for rubber toughening: The critical matrix ligament thickness. *Journal of Applied Polymer Science,* Vol. 35, No. 2, pp. 549-561, ISSN 0021-8995

Jiang, W., Yuan, Q., An, L. & Jiang, B. (2002). Effect of cavitations on brittle-ductile transition of particle toughened thermoplastic. *Polymer Communication,* Vol. 43, No. 4, pp. 1555-1558, ISSN 0032-3861

Kumar, S., Doshi, H., Srinivasarao, M., Park, J.O. & Schiraldi, D.A. (2002). Fibers from polypropylene/nano carbon fiber composites. *Polymer,* Vol. 43, No.5, pp. 1701-1703, ISSN 0032-3861

Chung D.D.L. (2001). Comparison of submicron-diameter carbon filaments and conventional carbon fibers as fillers in composite materials. *Carbon,* Vol. 39, No. 8, pp. 1119-1125, ISSN 0008-6223

Sandler, J. Werner, P., Shaffer, M.S.P., Demchuk, V., Altstaedt, V. & Windle, A.H. (2002). Carbon nanofibre reinforced poly(ether ether ketone) composites. *Composites Part A,* Vol. 33, No. 8, pp. 1033-1039, ISSN 1359-835x

Puglia, D., Valentine, L. & Kenny, J.M. (2003). Analysis of the cure reaction of carbon nanotubes/epoxy resin composites through thermal analysis and Raman spectroscopy. *Journal of Applied Polymer Science,* Vol. 88, No. 2, pp. 452-458, ISSN 0021-8995

Song, Y.S. & Youn, J.R. (2005). Influence of dispersion states of carbon nanotubes on physical properties of epoxy nanocomposites. *Carbon,* Vol. 43, No. 7, pp. 1378-1385, ISSN 0008-6223

Kinloch, I.A., Roberts, S.A. & Windle, A.H. (2002). A rheological study of concentrated aqueous nanotube dispersions. *Polymer,* Vol. 43, No. 26, pp. 7483-7491, ISSN 0032-3861

Gryschuck, O., Karger-Kocsis, J., Thomann, R., Kanya, Z. & Kiricsi, I. (2006). Multiwall carbon nanotube modified vinylester and vinylester - based hybrid resins. *Composites Part A,* Vol. 37, No. 9, pp. 1252-1259, ISSN 1359-835x

Ruan, S.L., Gao, P., Yang, X.G. & Yu, T.X. (2003). Toughening high performance unltrahigh molecular weight polyethylene using multiwalled carbon nanotubes. *Polymer,* Vol. 44, No.19, pp. 5643-5654, ISSN 0032-3861

Lozano, K., Yang, S. & Jones, R.E. (2004). Nanofiber toughened polyethylene composites. *Carbon,* Vol. 42, No. 11, pp. 2329-2331, ISSN 0008-6223

Seyhan A.T., Sun, Z., Deitzel. J., Tanoglu, M. & Heider, D. (2009). Cure kinetics of vapor grown carbon nanofiber (VGCNF) modified epoxy resin suspensions and fracture toughness of their resulting nanocomposites. *Materials Chemistry and Physics.* Vol. 118, No. 1, pp. 234-242, ISSN 0254-0584

Sui, G., Jana, S., Zhong, W.H., Fuqua, M.A. & Ulven, C.A. (2008). Dielectric properties and conductivity of carbon nanofiber/semi-crystalline polymer composites. Acta Materialia, Vol. 56. No. 10, pp. 2381-2388, ISSN 1359-6454

Sui, G., Zhong, W.H., Ren, X., Wang, X.Q. & Yang, X.P. (2009). Structure, mechanical properties and friction behavior of UHMWPE/HDPE/carbon nanofibers. *Materials Chemistry and Physics.* Vol. 115, No. 1, pp. 404-412, ISSN 0254-0584

Prolongo, S.G., Campo, M., Gude, M.R., Chaos-Morán, R. & Ureña, A. (2009). Thermo-physical characterisation of epoxy resin reinforced by amino-functionalized carbon nanofibers. *Composites Science and Technology,* Vol. 69, No. 3-4. pp. 349-357, ISSN 0266-3538

Lozano, K., Files, B., Rodriguez-Macias, F. & Barrera, E.V. (1999). Purification and functionalization of vapor grown carbon fibers and single wall nanotubes.

Symposium Powder Materials: Current Research and Industrial Practices (Cincinnati OH, USA), *TMS fall meeting*, pp. 333-340

Park, E.J., Heo, H. & Lim, K. T. (2007). Synthesis and characterization of $Al(OH)_3$/ polystyrene nanocomposite latex particles by emulsion polymerization. *Macromolecular Symposia*, Vol. 249-250, No. 1, pp. 247-250, ISSN 1521-3900

Akelah, A. & Moet, A. (1996). Polymer-clay nanocomposites: free-radical grafting of polystyrene on to organophilic montmorillonite interlayers. *Journal of Materials Science*, Vol. 31, No. 13, pp. 3589-3596, ISSN 0022-2461

Wang, D., Zhu, J., Yao, Q. & Wilkie, C.A. (2002). A comparison of various methods for the preparation of polystyrene and poly(methylmethacrylate) clay nanocomposites. *Chemistry of Materials*, Vol. 14, No. 9, pp. 3837-3843, ISSN 0897-4756

Chen, G., Qi, Z. & Shen, D. (2000). Shear-induced ordered structure in polystyrene/clay nanocomposite. *Journal of Materials Research*, Vol. 15, No. 2, pp. 351-356, ISSN 0884-2914

Zhou, Q., Fan, X., Xia, C., Mays, J. & Advincula, R. (2001). Living anionic surface initiated polymerization (SIP) of styrene from clay surfaces. *Chemistry of Materials*, Vol. 13, No. 8, pp. 2465-2467, ISSN 0897-4756

Jana, S.C. & Jain, S. (2001). Dispersion of nanofillers in high performance polymers using reactive solvents as processing aids, *Polymer*, Vol. 42, No. 16, pp. 6897–6905, ISSN 0032-3861

Siengchin, S., Karger-Kocsis, J., Apostolov, A.A. & Thomann, R. (2007). Polystyrene-fluorohectorite nanocomposites prepared by melt mixing with latex precompounding: Structure and mechanical properties, *Journal of Applied Polymer Science*, Vol. 106, No. 1, pp. 248-254, ISSN 0021-8995

Grunlan, J.C., Mehrabi, A.R. & Bannon, M.V. (2004). Water-based single-walled-nanotube-filled polymer composite with an exceptionally low percolation thershold. *Advanced Materials*, Vol. 16, No. 2, pp. 150-153, ISSN 0935-9648

Xia, H., song, M., Zhang, Z. & Richardson, M. (2007). Microphase separation, stress relaxation, and creep behavior of polyurethane nanocomposites. *Journal of Applied Polymer Science*, Vol. 103, No. 5, pp. 2992-3002, ISSN 0021-8995

Eckstein, A., Suhm, J., Friedrich, C., Maier, R.D., Sassmannshausen, J., Bochmann, M. & Mülhaupt, R. (1998). Determination of plateau moduli and entanglement molecular weights of isotatic, sydiotactic, and atactic polypropylenes synthesized with metallocene catalysts. *Macromolecules*, Vol. 31, No. 4, pp. 1335-1340, ISSN 0024-9297

Schapery, R.A. (1969). On the characterization of nonlinear viscoelastic materials. *Polymer Engineering and Science*, Vol. 9, No. 4, pp. 295-310, ISSN 0032-3888

Brueller, O.S. (1987). On the nonlinear characterization of the long term behavior of polymeric materials. *Polymer Engineering and Science*, Vol. 27, No. 2, pp. 144-148, ISSN 0032-3888

Siengchin, S., Psarras, G.C. & Karger-Kocsis, J. (2010). POM/PU/carbon nanofiber composites produced by water-mediated melt compounding: structure, thermomechanical and dielectrical properties. *Journal of Applied Polymer Science*, Vol. 117, No. 3, pp. 1804-1812, ISSN 0021-8995

Siengchin, S. & Karger-Kocsis, J. (2009). Structure and creep response of toughened and nanoreinforced polyamides produced via the latex route: Effect of nanofiller type. *Composites Science and Technology*, Vol. 69, No. 5. pp. 677-683, ISSN 0266-3538

Siengchin, S., Karger-Kocsis, J., Psarras, G.C. & Thomann, R. (2008). Polyoxymethylene/polyurethane/alumina ternary composites: structure, mechanical, thermal and dielectrical properties. *Journal of Applied Polymer Science,* Vol. 110, No. 3, pp. 1613-1623, ISSN 0021-8995

Shaw, M.T. & MacKnight, W.J (2005) Introduction to polymer viscoelasticity, John Wiley & Sons, ISBN 9780471740452, New York

Araújo, E.M., Hage, E. & Carvalho, A.J.F. (2004). Thermal properties of nylon6/ABS polymer blends: Compatibilizer effect. J Materials Science, Vol. 39, No. 4, pp. 1173-1178, ISSN 0022-2461

Hojfors, R.J., Baer, E. & Geil, P.H. (1977). Dynamic-mechanical study of molecular motions in solid polyoxymethylene copolymer from 4 to 315 °K. *Journal of Macromolecular Science, Part B Physics,* Vol. B13, No. 3, pp. 323-348, ISSN: 0022-2348

Shen, J., Huang, W., Wu, L., Hu, Y. & Ye, M. (2007). The reinforcement role of different amino-functionalized multi-walled carbon nanotubes in epoxy nanocomposites. *Composites Science and Technology,* Vol. 67, No. 15-16. pp. 3041, ISSN 0266-3538

Siengchin, S. & Karger-Kocsis, J. (2011). Structure, mechanical and fracture properties of nanoreinforced and HNBR-toughened polyamide-6. *Journal of Applied Polymer Science,* in press, ISSN 0021-8995

Paul, D.R. & Bucknall, C.B. (2000). *Polymer blends: Formulation and Performance,* Vol. 2, John Wiley & Sons, ISBN 0-471-35280-2, New York

Karger-Kocsis, J. (2009). On the toughness of "nanomodified" polymers and their traditional polymer composites, In: *Nano- and Micromechanics of Polymer Blends and Composites,* Chapter 12, Karger-Kocsis, J., Fakirov, S., (Eds.), 425-470, Hanser, ISBN 978-3-446-41323-8, Munich, Germany

Siengchin S. & Karger-Kocsis, J. (2006). Creep behaviour of polystyrene/fluorohectorite micro- and nanocomposites. *Macromolecular Rapid Communications,* Vol. 27, No. 24, pp. 2090-2094, ISSN: 1521-3927

Tsangaris, G. M. & Psarras, G. C. (1999). The dielectric response of a polymeric three-component composite. *Journal of Materials Science,* Vol. 34, No. 9, pp. 2151-2157, ISSN 0022-2461

Tsangaris, G.M., Psarras, G.C. & Kouloumbi, N. (1998). Electric modulus and interfacial polarization in composite polymeric systems. *Journal of Materials Science,* Vol. 33, No. 8, pp. 2027- 2037, ISSN 0022-2461

Psarras, G.C., Gatos, K.G., Karahaliou, P.K., Georga, S.N., Krontiras, C.A. & Karger-Kocsis, J. (2007). Relaxation phenomena in rubber/layered silicate nanocomposites. *Express Polymer letters,* Vol. 1, No. 12, pp. 837-845, ISSN 1788-618X

Psarras, G.C.; Gatos, K.G. & Karger-Kocsis, J. (2007). Dielectric properties of layered silicate-reinforced natural and polyurethane rubber nanocomposites. *Journal of Applied Polymer Science,* Vol. 106, No. 1, pp. 1405-1411, ISSN 0021-8995

Sun, T. J., Ye, L. & Zhao, X. W. (2007). Thermostabilising and nucleating effect of montmorillonite on polyoxymethylene. *Plastics Rubber Composites,* Vol. 36, No. 7-8, pp. 350, ISSN 1465-8011

Siengchin, S., Karger-Kocsis, J. & Thomann, R. (2008). Nanofilled and/or toughened POM composites produced by water-mediated melt compounding: Structure and mechanical properties. *Express Polymer letters,* Vol. 2, No. 10, pp. 746-756, ISSN 1788-618X

Findley, W.N., Lai, J.S. & Onaran, K. (1989). *Creep and relaxation of nonlinear viscoelastic materials,* Dover Publications, ISBN 978-0-486-66016-5, New York, USA

Kaschta, J. & Schwarzl, F. R. (1994). Calculation of discrete retardation spectra from creep data-I. Method. *Rheologica Acta,* Vol. 33, No. 6, pp. 517-529, ISSN 0035-4511

12

Carbon Nanofibers Reinforced Ceramic Matrix Composites

Pavol Hvizdoš, Viktor Puchý, Annamária Duszová and Ján Dusza
Institute of Materials Research, Slovak Academy of Sciences, Slovakia

1. Introduction

Modern ceramic materials have, thanks to their crystallographic structure and strong atomic bonds, many excellent properties, such as extremely high hardness, strength, high thermal and chemical stability, high corrosion resistance, and wear resistance. Their weakness is low fracture toughness and crack growth resistance and hence high brittleness and lower reliability. One of the ways how to overcome these drawbacks is preparation of composite materials, where the base ceramic matrix is reinforced by secondary phases in forms of particles/whiskers and in recent years increasingly in a form of fibrous structures. In advanced fine grained ceramics these usually take form of nanofibers and/or nanotubes.

Among the most promising candidates are carbon-based filamentous nanomaterials such as carbon nanotubes (CNTs) and also carbon nanofibers (CNFs), which attracted a lot of attention due to their outstanding mechanical properties, excellent thermal performance and useful electrical characteristics (high electrical conductivity).

Nowadays, new ceramic/carbon nanotube composites are being developed mostly with two aims: to improve the mechanical properties of the ceramic materials by reinforcing with carbon nanofibers and to develop functionalized ceramics with improved magnetic and electric properties. Studies show that CNTs (both single-wall and multi-wall) should be ideal reinforcing/functionalizing elements for composites due to their small size, low density and good electrical and thermal conductivity.

This work focuses on investigations of ceramic matrix composites based on alumina, zirconia and silicon nitride reinforced by carbon nanofibers and nanotubes. The basic characteristics of commercially available nanofibers/nanotubes are studied by various techniques.

The chapter then focuses on mechanical properties of reference monolithic and experimental composite materials. The effect of volume fraction of carbon nanofibers on hardness and fracture toughness is illustrated. Further, the possibilities of improving the tribological and wear properties are discussed. The chapter concludes with the section that explores important aspect of functionalization of ceramics composites by improving their electrical properties, namely electrical conductivity.

2. Description of constituents

Recently a lot of experimental work and development has been going on in preparation of ceramic matrix composites with carbon nanofibers as great improvements of material

properties are expected. Up–to-now only modest enhancement of properties have been reported in CNT reinforced silicon carbides (Ma et al., 1998) and silicon nitride (Balázsi et al., 2003; Belmonte et al., 2010). Some other works reported improved fracture toughness achieved in alumina matrix systems (Zhan et al., 2003a) but rather often the performance of new composite materials was disappointing due to difficulties in dispersing the CNTs in ceramic microstructures and rather weak bond with them. More success has been achieved in increasing the electrical conductivity in many originally insulating ceramic materials (Zhan et al., 2003b, Balázsi et al., 2006) and in optimizing tribological behaviour of brittle systems (González-Julián et al., 2010; Hvizdoš et al., 2010).
This section describes basic characteristics of alumina, zirconia and silicon nitride as materials of ceramic matrices and provides the information about the ongoing effort in combining these materials with carbon nanofibers and nanotubes.

2.1 Aluminium oxide

Aluminium oxide (Al_2O_3), commonly referred to as alumina, is the most cost effective and widely used material in the family of engineering ceramics. The raw materials from which this high performance technical grade ceramic is made are readily available and reasonably priced, resulting in good value for the cost in fabricated alumina shapes. With an excellent combination of properties and an attractive price, it is no surprise that fine grain technical grade alumina has a very wide range of applications.
Alumina possesses strong ionic interatomic bonding giving rise to its desirable material characteristics. It can exist in several crystalline phases which all revert to the most stable hexagonal alpha phase at elevated temperatures. This is the phase of particular interest for structural applications. Alpha phase alumina is the strongest and stiffest of the oxide ceramics.
Among its key properties one can emphasize high hardness and wear resistance, excellent dielectric properties from DC to GHz frequencies, resistance to strong acid and alkali attack at elevated temperatures, good thermal conductivity, excellent size and shape capability, high strength and stiffness. The material is industrially available in purity ranges from 94% (an easily metallizable composition), to 99.5% for the most demanding high temperature applications.
Typical uses of alumina include gas laser tubes, wear pads, seal rings, high temperature electrical insulators, high voltage insulators, furnace liner tubes, thread and wire guides, electronic substrates, ballistic armor, abrasion resistant tube and elbow liners, thermometry sensors, laboratory instrument tubes and sample holders, instrumentation parts for thermal property test machines, grinding media, etc.

2.2 Zirconium oxide

For polycrystalline zirconia ceramics is typical combination of high strength, fracture toughness and chemical stability, which enables its use in extreme conditions (Garvie et al., 1975). It is an extremely refractory material which can be used at temperatures up to 2400°C. It offers chemical and corrosion inertness to temperatures well above the melting point of alumina. It has relatively high density and low thermal conductivity (20% that of alumina). It resists molten metals and has high wear resistance. It is electrically conductive above 600°C and is used in oxygen sensor cells and as the susceptor (heater) in high temperature induction furnaces, where it can be used at over 2000°C in oxidizing atmospheres. Other typical uses include cutting tools for cutting problematic materials such as Kevlar, polymer films and paper. It is an attractive material for nuclear industry, where it is used for inert matrices for plutonium storage and as immobilizer of nuclear waste. Another applications include seals and pistons in pumps for chemically aggressive and abrasive suspensions, bearings, biomedical implants, e.g. dental, hip, and knee joints (Garvie et al., 1984), and thermal barriers coatings (using plasma sprayed TBCs the working temperature of thermally exposed parts, such as gas turbine blades, can be increased of about 200°C).

2.3 Silicon nitride

Silicon nitride (Si_3N_4) is a man made compound synthesized through several different chemical reaction methods. Parts are pressed and sintered by well developed methods to produce a ceramic with a unique set of outstanding properties. Silicon nitride has relatively high strength and fracture toughness, low density (3.2 $g.cm^{-3}$), good corrosion and oxidation resistance at room and elevated temperatures thanks to presence of surface layer of SiO_2. It has quite good thermal shock resistance thanks to low thermal expansion coefficient (3.1 ÷ 3.6 x 10^{-6} K^{-1} from room temperature up to 1000°C).

Silicon nitride can not be sintered directly, because above 1850°C at 0.1 MPa it dissociates. One of the common methods of preparation Si_3N_4 bulk ceramics is reaction sintering, its product is called reaction-bonded silicon nitride (RBSN).This methods is based on heating of silicate powder compact in nitrogen atmosphere. The resulting material tends to have higher open porosity. Hot pressing and hot isostatic pressing sinter the silicon nitride powders at temperatures 1700 to 1800 °C with oxides, which form liquid phase, as sintering additives and produce fully dense materials. Microstructure of resulting materials consists of fine grains of α-Si_3N_4 and larger β-Si_3N_4 with higher aspect ratio. Such microstructures are desired for high strength and better fracture toughness (Morrell, 1985).

Silicon nitride is a rather expensive material, but its performance to cost benefit ratio is excellent in the applications where it can outperform the normally utilized materials with long life and very reliable low maintenance operation. It has outstanding wear resistance in both impingement and friction modes.

High performance silicon nitride materials were developed for automotive engine wear parts, such as valves and cam followers and proven effective. The cost of the ceramic parts never dropped enough to make the ceramics feasible in engines and turbochargers. The very high quality bodies developed for these demanding high reliability applications are available today and can be used in many severe mechanical, thermal and wear applications.

2.4 Carbon nanotubes (CNT) and carbon nanofibers (CNF)

Carbon nanofibers CNFs are cylindrical or conical structures that have diameters varying from a few to hundreds of nanometers and lengths ranging from less than a micron to millimeters. The internal structure of carbon nanofibers varies and is comprised of different arrangements of modified graphene sheets. A graphene layer can be defined as a hexagonal network of covalently bonded carbon atoms or a single two-dimensional 2D layer of a three-dimensional 3D graphite Fig. 1a. In general, a nanofiber consists of stacked curved graphite layers that form cones Fig. 1b or "cups." (Krishnan et al., 1997; Endo et al., 2002) The stacked cone structure is often referred to as herringbone or fishbone while the stacked cups structure is most often referred to as a bamboo type, resembling the compartmentalized structure of a bamboo stem. Currently there is no strict classification of nanofiber structures. The main distinguishing characteristic of nanofibers from nanotubes is the stacking of graphene sheets of varying shapes.

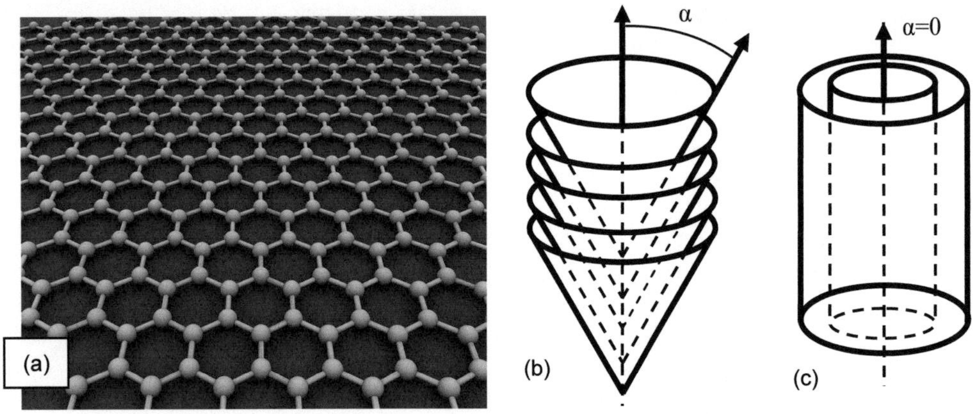

Fig. 1. Schematic structure of carbon nanofibers and nanotubes. (a) Graphene layer, (b) stacked cone (herringbone) nanofiber, and (c) nanotube. After (Melechko et al., 2005).

We can define α as an angle between the fiber axis and the graphene sheet near the sidewall surface Fig. 1b. The nanofiber with α=0 is a special case in which one or more graphene layers form cylinders that run the full length of the nanostructure, Fig. 1c. This arrangement, with its closed and semi-infinite surface, discovered and endorsed by Iijima (Iijima, 1991), results in extraordinary properties that made this type of a nanofiber known to the world as a carbon nanotube CNT.

The chemical properties of nanofibers and nanotubes are quite different since defect-free nanotube walls do not contain the exposed edges and unsaturated bonds of graphene planes (compare Fig.1 b and c), and as a result nanotubes are far less reactive than nanofibers. The scientific community and the popular press have focused more on the special case of the CNT structure than the more general case that includes nanofibers. The reason for the attention given to nanotubes is clear: the beauty and perfection of their mathematical description and the resulting extraordinary mechanical and electron-transport properties.

2.4.1 Mechanical properties of CNTs and CNFs

Important characteristic of most materials is that a small diameter (nanoscale) fiber is much stronger than the bulk material, (Callister, 2003) due to the lower probability of critical surface flaws with decreasing specimen volume. The majority of engineering forms of carbon have more or less disordered graphite microstructures. The microstructure of the crystalline layers within the fiber influences properties such as strength, stiffness, deformation modes, fracture behavior, and toughness, (McEnaney, 2001). The literature shows great variance in the mechanical properties of carbon forms such as carbon nanotubes and carbon fibers. (Melechko et al., 2005) gives some general ranges for the strength and modulus of engineering carbons. The high volumetric density of short, strong sp3 bonds gives diamond the highest stiffness of any known materials (Young's modulus ~ 1 TPa). Single crystal graphite has a Young's modulus over 28 times higher in the direction parallel to the basal planes than in the perpendicular direction. Thus, while there is a preferred orientation of basal planes parallel to the fiber axis, this creates poor transverse properties. Carbon fibers can also have as high as 100 times more stiffness along the fiber axis than perpendicular to it, (McEnaney, 2001). Single-crystal carbon nanotubes have extremely large length-to-diameter aspect ratios, with diameters as small as only a few nanometers. They are

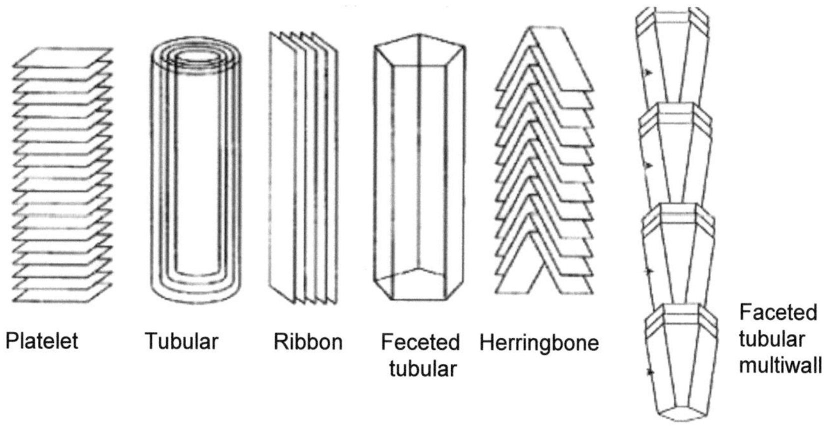

Fig. 2. Typical structures of CNFs, after (Al-Saleh, 2009).

the strongest known material, with a specific tensile strength as high as 100 times that of steel, (Ruoff et al., 2003). For SWCNTs, tensile strengths range between 50 and 200 GPa, Young's modulus is on the order of 1 TPa and fracture strains are between 5% and 20%. As a consequence of their size and high degree of crystalline perfection, CNTs are virtually flaw-free, which contributes to their exceptional strength. However, in spite of their incredible mechanical properties, SWCNTs are not utilized extensively as a reinforcement medium because they are expensive to produce and purify. On the other hand, MWCNTs are easier to produce since they do not require as stringent catalyst particle preparation. MWCNTs are composed of concentric graphene sheets and have diameters on the order of tens of nanometers depending on the number of graphite layers. One drawback is that these layers can slide past each other easily, often failing by the "telescoping" mechanism, (Yu et al., 2000). This failure and the increased probability of defects with greater mass make MWCNTs less desirable than SWCNTs, but their strength still surpasses that of steel and they have a high modulus.

Carbon fibers, composed of polycrystalline and amorphous carbon, have diameters ranging from hundreds to thousands of nanometers and can be grown thousands of micrometers long. Crystallographically, they can be found in many forms (Fig. 2). Their tensile strengths are about an order of magnitude lower than for CNTs, ranging from 1.5 to 4.8 GPa and the Young's modulus is between 228 and 724 GPa, (Callister, 2003). Carbon fibers retain their high tensile modulus and strength at elevated temperatures and are not affected by water, solvents, acids, or bases at room temperature. Even though their mechanical properties are not as astounding as those of carbon nanotubes, carbon fibers are relatively easy to produce as well as economical.

2.5 Preparation of ceramic matrix composites with carbon nanofibers

The processing steps of advanced ceramics include powder synthesis, preparation of the powders mixture (mixing, milling, etc.), forming (pressing, injection moulding, slip casting, etc.) and sintering (pressureless sintering, hot pressing, hot isostatic pressing, etc.). In the following we will summarize the processing steps used for the preparation of ceramic-CNTs/CNFs composites.

The inclusion of CNTs in a ceramic matrix is expected to create composites exhibiting high toughness and improved mechanical properties compared to the single-phase ceramic

material (Mukhopadhyay et al., 2010). Ideally, the CNTs dispersed in a ceramic matrix should serve as sites for fracture energy dissipation through mechanisms such as crack deflection and nanotube debonding/breakages, (Dusza & Šajgalík, 2009). Hot-pressing of ceramic powder mixed with CNT is the most common method to prepare such composites.
Hot-pressing is mostly applied in the pressure range of 20–40 MPa and at temperatures of 1300–2000 °C, but slip casting and injection molding followed by pressureless sintering is also used. There are only a limited number of authors dealing with the hot-pressing processing of CNT/ ZrO_2 composites.
Spark plasma sintering is a new process where ceramic powder can be sintered very fast to full density. It is similar to hot-pressing which is carried out in a graphite die, but the heating is accomplished by spark discharges in void between particles generated by an instantaneous pulsed direct current applied through electrodes at the top and bottom punches of the graphite die. Due to these discharges, the particle surface is activated and purified, and a self-heat phenomenon is generated between the particles, thus the heat-transfer and mass-transfer can be completed instantaneously, (Perera et al., 1998). SPS use heating and sintering temperature at 1000-1800 °C, at a heating rate of 500-600 °C/min under a pressure of 40-60 MPa and than fast cooling to 600 °C/2-3min.

2.5.1 Processing of ZrO_2-based ceramic-CNT composites

Fully-dense 3 mol. % yttria stabilized tetragonal polycrystalline zirconia (3Y-TZP) matrix composites containing 0.1–1 wt. % of MWCNTs and SWCNTs were fabricated by SPS at 1300 °C for 5 min, under an applied pressure of 20 MPa, (Sun et al., 2005).
MWCNT/3Y-TZP composites with different MWCNT contents were prepared by SPS by several groups (Shi & Liang, 2006; Mazaheri et al., 2011). The DC conductivity of the composites was found to follow a typical percolation behavior with a very low percolation threshold between 1.0 and 2.0 wt. % MWCNT content. The dielectric constant was greatly increased when the MWCNT concentration was close to the percolation threshold, which was attributed to dielectric relaxation, the space charge polarization effect, and the percolation effect. The electrical conductivity of a composite prepared by dispersing multiwall carbon nanotubes in yttria-stabilized tetragonal zirconia matrix, and subsequent spark plasma sintering, was investigated by Shi and Liang in 2007 (Shi & Liang, 2007). The experimental analysis of the effect of temperature on the conductivity suggested that for temperatures higher than 35 K, conduction can be attributed to thermal fluctuation-induced tunneling of the charge carriers through the insulating zirconia separating by the multiwall carbon nanotube clusters.
Sun et al., 2005, reported the CNTs-3Y-TZP composite powder sintered by SPS. The prepared composite powder was carefully placed into a 20 mm diameter graphite die, and heated to the sintering temperatures at 1300 °C for 5 min. A pressure of approximately 20 MPa was applied from the beginning of sintering and relaxed after the temperature below 800 °C. The composites with the content of carbon nanotubes ranging from 0.1 to 1 wt. % were prepared. It was found that the hardness values decreased prominently with the addition of CNTs. The fracture toughness kept almost the same value as that of the matrix when the amount of MWCNTs was kept 0.5 wt. %, however it decreased to 4.47 MPa $m^{1/2}$ when the MWCNTs content was raised to 1.0 wt. %. 0.5 wt. % SWCNTs-3Y-TZP composites showed no improvement on the mechanical properties of 3Y-TZP either. The existence of the agglomerated CNTs in the grain boundary and the weak bonding between carbon nanotube and zirconia matrix were reasons that led to the failure in reinforcement.

An alternative route for the synthesis of ceramic-CNT composites is a process, where the CNTs act as nucleation sites for the ceramic. Lupo et al., 2004, reported the successful application of such a route to prepare ZrO_2, which is grown on carbon nanotubes by a hydrothermal process. The hydrothermal process employs high temperatures and high pressures for the crystallization of hydroxide into the oxide [$Zr(OH)_4 \rightarrow ZrO_2 + H_2O$]. The end product consisted of CNTs coated with nano-sized ZrO_2 particles. Other interesting approaches include the process involving direct in-situ growth of MWNTs on zirconia particles (Datye et al., 2010) or colloidal processing (Garmendia et al., 2010).

2.5.2 Processing of Al_2O_3-based ceramic-CNT composites

Alumina based materials are usually prepared by conventional methods, as they are generally easy to sinter. When using the pure carbon phases, special care is necessary, so that they are not destroyed in the process. Inam (Inam et al., 2010) developed alumina-carbon black and alumina-CNTs composites with the aim to improve the electrical conductivity of the alumina. Multi-wall CNTs with average outer diameter 9.5 nm; lengths of up to 1.5 μm; and density 1.7 $g.cm^{-3}$) were dispersed in dimethylformadie, DMF using high power sonication for 2 h and then hand-mixed with alumina nanopowder with particle size <50 nm; surface area 35–43 $m^2.g^{-1}$; melting point 2040 °C; and density 3.97 $g.cm^{-3}$) for 2 min. Since DMF was found to be a much more effective dispersant than ethanol for making stable, homogeneous CNT and composite dispersions. The liquid mixture was rotation ball milled for 8 h. It was then dried at 75 °C for 12 h on a heating plate in air, followed by in a vacuum oven at 100 °C for 3 days. The dried agglomerated mixture was ground and sieved and then placed again in the vacuum oven at 100 °C for another 4 days to thoroughly extract the solvent. Alumina and nanocomposite pellets were prepared by SPS in a HPD 25/1 furnace. A pressure of 100 MPa was applied concurrently with the heating (rate 300 °C/min) and released at the end of the sintering period, which was 3 min for all of the samples.

Cha et al., 2005, used MWNTs, fabricated by chemical vapor deposition for the preparation of the composites. The CNTs were acid-treated, oxidized at 190 °C for 4 h to remove catalysis particles and to generate functional groups on the surfaces of the nanotubes. The acid-treated CNTs were sonicated in distilled water for homogeneous dispersion. $Al(NO_3)_3 \cdot 9H_2O$ was added into a suspension with the CNTs and these were sonicated. This solution was vaporized and the powders that remained were oxidized in air atmosphere. The volume fractions of CNTs varied from 0 to 1.8 vol. %. During the calcination process, the chemical bonding between CNTs and the amorphous Al_2O_3 matrix was formed. SPS was used for densification. The composite powders were compacted in a graphite mold with a diameter of 8 mm and were heated by pulsed electric current at a vacuum pressure of 1 Pa. The heating rate was fixed for 100 °C/min. Alumina-CNT composite powders were synthesized in the catalytic decomposition of acetylene over alumina powders impregnated with iron catalysts by An et al., 2003. The alumina, MgO, and $Fe(NO_3)_3 \cdot 9H_2O$ powder were mixed by planetary ball milling to prepare the alumina powder impregnated with iron catalysts. The powder was mixed in ethanol by alumina balls. After ball milling, the ethanol solution was removed via rotary evaporation. The materials were then heated overnight followed by grinding into a fine powder. For the CNT synthesis, each alumina powder having four different iron catalyst contents were placed in an alumina boat mounted in a tube furnace.

Lim et al., 2005, used CNTs produced by the catalytic decomposition. The ceramic powders were mixed by planetary ball milling in ethanol then hot pressed for 1 h at 2123 K under 30

MPa. For some samples a modified tape casting method was employed for dispersing the CNTs. Tape casting slurry was prepared by using a planetary ball milling. Metylisobutyl-ketone, KD-1, polyvinyl-butyral, and di-butyl phathalate were used as solvent, dispersant, binder, and plasticizer, respectively. After the binder burnout, the sample was hot pressed at 2123 K under 30 MPa for 1 h.

2.5.3 Processing of Si_3N_4-based ceramic-CNT composites

Tatami et al., 2005, used highly pure and fine Si_3N_4 powder as a raw material together with Y_2O_3 AlN and TiO_2 as sintering aids. MWNTs were added as an electrical conductor. The amount of CNTs was changed from 0 to 12 wt. % of the total amount of the other raw powders. Three wt. % of dispersant was added to ethanol, whose ratio was 63.3 mg to 1 g of CNT to prepare a slurry using an ultrasonic vibrator. The slurry was mixed with the Si_3N_4–Y_2O_3–Al_2O_3–AlN–TiO_2 slurry with 2 wt. % of the dispersant and then prepared by ball milling. After mixing, the ethanol in the slurry was evaporated and the mixed powder was sieved to obtain granules. The granules were molded followed by cold isostatic pressing at 200 MPa. After dewaxing, the green bodies were gas pressure sintered (GPS) at 1800 °C for 2 h in 0.9 MPa N_2. For more densification, HIP was used at 1700 °C for 1 h under 100 MPa.

In the study of Balázsi (Balázsi et al., 2004) the compositions of the starting powder mixtures of the investigated materials were: 90 wt. % Si_3N_4, 4 wt. % Al_2O_3, and 6 wt. % Y_2O_3. In addition to batches of MWNTs, carbon black, graphite, and carbon fibers were added. The powder mixtures were milled in ethanol in a planetary type alumina ball mill for several hours. Each batch contained approximately 1 g alumina as contamination from balls and jars. After milling, the powder mixtures and CNTs were introduced in an ethanol bath and sonicated together. Samples were compacted by dry pressing at 220 MPa. The materials were sintered at 1700 °C in high purity nitrogen by a two-step sinter-HIP method using BN embedding powder.

3. Testing methods

3.1 Nanotubes and nanofibers characterization

The morphology and structure of nantubes and nanofibers were characterized by scanning and transmission electron microscopy. For examination of length, diameter and outer morphology usually field emission SEM was used. TEM and HR-TEM was employed to observe the crystal structure and graphene layers arrangement. The TEM specimens have been prepared by dispersing the CNFs in an acetone ultrasonic bath and dropping a suspension onto a carbon lace Cu grid. The TEM images and selected area electron diffraction patterns were recorded using an accelerating voltage of 200 kV.

3.2 Ceramic composites characterization

Microstructure studies of bulk composite materials were carried out also using high resolution SEM, TEM and HR-TEM. The SEM studies were done on samples that were prepared by diamond cutting, polishing and usually thermal etching in order to reveal the grain structure, topography and distribution of individual phases. Sometimes also fracture surfaces were observed as they can convey the information about 3D characteristics of the microstructure as well as about the cohesive and bonding characteristics of matrix grains and CNFs which determine the type of fracture and crack propagation.

X-ray diffraction (XRD) analysis was used for the determination of phase composition of final experimental composite materials.
Mechanical characterization was done in terms of measurement of basic mechanical properties of bulk composite materials such as hardness and fracture toughness by indentation method. Hardness is a measure of the material resistance against deformation by a hard body (indenter) driven into its surface. In ceramic materials it is measured most often by Vickers method where the indenter is a defined four-sided diamond pyramid. Vickers hardness (HV) is calculated according to the standard formula

$$HV = 1.8544\ F / d^2 \tag{1}$$

where F denotes the indentation load, d is the length of the indentation impact diagonal.
Fracture toughness (K_{IC}) expresses the resistance of the material against sharp crack propagation. In brittle materials it is possible to determine it from the length of the cracks emanating from the corners of the indents produced by sharp indenters (e.g. Vickers). This method has the advantage that it can be used to probe very small specimens of irregular shapes. This is often the case for new materials which are just being developed and the specimens required for other standardized and better defined methods are unavailable. The disadvantage of this type of tests is that the stress states creating the cracks are very complicated and difficult to describe analytically. There exist many phenomenological formulas relating fracture toughness to indentation crack lengths. For very brittle materials such as zirconia and alumina based ceramics, and also for very fine grained silicon nitride composites, the cracks were shown to have usually half-elliptical shape. In this case one of the most suitable is a formula proposed by (Anstis et al., 1981):

$$K_{IC} = \eta\ F\ (E/H)^{1/2} / c^{3/2} \tag{2}$$

where η is a geometric factor, E is the modulus of elasticity, H is the hardness expressed as pressure (i.e. in Pa), and c is the indentation radial crack half-length at the surface.
Wear behaviour of the experimental materials was studied in most cases in sliding against specific tribological partner or for self-mated pairs. In this sort of tests the surfaces were carefully prepared by polishing down to surface roughness below 1 μm. The tangential forces are measured during the test and friction coefficients are calculated. The worn surfaces are subsequently observed and the wear regimes, damage type and micromechanisms are identified. If possible the material losses due to wear are measured and then specific wear rates (W) can be calculated according to the standard (ISO 20808, 2004) as the volume loss (V) per distance (L) and applied load (F_p):

$$W = V / (L\ F_p) \tag{3}$$

The electrical conductivity measurements were carried out at ambient temperature using a two-point probe setup on an impedance analyzer at frequencies from 40 Hz to 40 kHz.

4. Properties

4.1 Microstructure

As mentioned before, carbon nanofibers can have many various crystallographic structures. A series of studies (Duszová et al., 2008a; Dusza et al., 2009; Puchý et al., 2010) was done on commercial grades of CNFs made by catalytic chemical vapour deposition (CCVD), which produces a good variety of high quality materials. The morphology of these types of carbon nanofibers is illustrated in the Figs. 3 and 4. The SEM images reveal that the CNFs are typically rope - like shaped with tubular structure and with the outer diameter of 50 - 250 nm and inner diameter from 20 nm to 230 nm. According to the SEM observation the length of the CNFs is up to several μm and the fibers are separated with the characteristic rope-shape. In several cases we found clusters of the fibers with crooked and entangled fibers, too. The majority of the fibers have a diameter from 200 to 300 nm. Beside the elongated carbon nanofibers (Fig. 3a) occasionally particle shaped carbon objects with the maximal size of several micron have been identified. Two basic types of nanofibers have been identified by SEM observation; conical hollow rope - type tubes, usually with a free ends and smooth surface and bamboo shaped nanofibers with waved surface, Fig. 3b. The third type of nanofibers with a straight line shape and rough surface was identified only randomly. Observation of the end of the rope - fibers revealed that the diameter of the hole in the fiber is different and the wall of the fibers consists from layers with different structure.

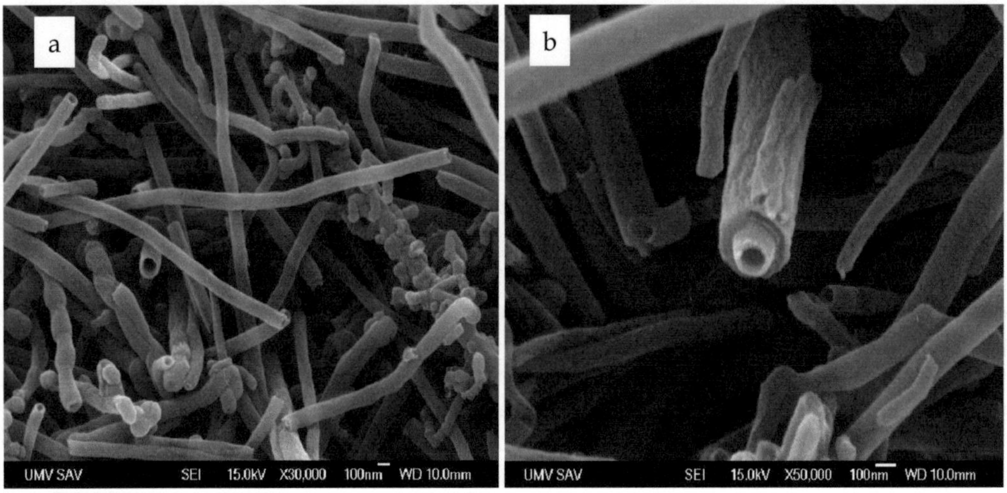

Fig. 3. Morphology of the carbon nanofibers, mix of rope-like and bamboo shaped nanofibers (a), example for smooth small and rough large diameter nanofibers (b), SEM, (Puchý et al., 2010).

The study of the surface morphology of the fibers revealed that the fibers with smaller diameters are smoother comparing to the fibers with larger diameters. Lee et al., 2007, investigated the surface morphology of hollow and solid as received, heat - treated and CVD surface treated fibers and found that the solid fibers exhibited rougher surface comparing to the hollow fibers in all forms. Measurement and statistical evaluation of the distribution of the diameters of nanofibers show that their diameter varied from 50 to 600 nm with the average diameter equal to 120 nm.

Results of TEM are in a good agreement of the results of SEM and it can be seen that in the case of straight pipe-shaped fibers the wall is smooth and uniform and consists from distinct sandwich of graphite layers. In this case the graphene layers are parallel to the axes of the fibers and form usually defect free material. The bamboo-shaped fibers are composed of

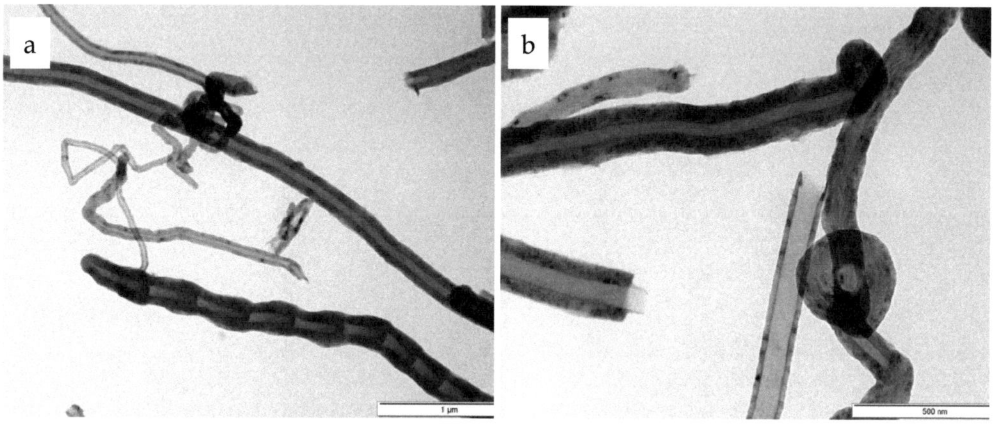

Fig. 4. Characteristic morphology of CNFs by TEM, rope like and bamboo - like CNFs (a), CNT with the root and broken CNT with longer inner layer (b), (Puchý et al., 2010).

multi - walled graphite structure and the bamboo sections have a size parallel with the axes of the fibers 2-3 times larger comparing to the outer diameter of the fibers. It seems that the carbon diffusion was not continuous at the all fibers during they growth, leading to a pulsed growth which translated into a periodic variation of the fibers diameter as it is illustrated in Fig. 4. Such bamboo-shaped fibers are composed of hollow segments of a size approximately 100 nm deliminated by curved stacking of carbon sheets. Longtin et al., 2007, prepared bamboo-type nanofibers by laser assisted catalytic vapor deposition applying two different catalyst preparation methods and the nanofibers analyzed using SEM, HREM and Raman spectroscopy. Using approximately 50 nm thick deposited catalyst layer within the pores of an alumina substrate results in an array of vertically aligned nanofibers with similar dimension as the analyzed nanofibers in the present work. The lattice fringe analysis of HREM indicated that pulsed growth at high rates leads to bamboo-type nanofibers having oriented graphitic domains similarly as in our case. Bamboo-shaped carbon-nitrogen nanotubes have been prepared by Lee et al., 2002, using pyrolysis of iron pentacarbonyl and acetylene mixture with ammonia and by pyrolisis of melamine with NaN_3-Fe-Ni. They showed that the nitrogen doping leads to the origin of the bamboo-like structure and results in the degraded crystallinity of graphite sheets. By TEM they found that in the bamboo-like CN nanofibers the wall thickness increases due to the generation of compartment layers, but the outer diameter remains constant. This is different comparing to the bamboo-like carbon nanofibers characterized in the present investigation when the outer diameter of the nanofibers increased during the generation of compartment layers due to the changed orientation of the graphite sheets.

The internal structure of the fibers is related to the orientation of the graphite sheets as illustrated in Figs 5, 6 (Puchý et al., 2010). The fiber illustrated in Fig. 5 has a wall thickness of 28 nm and hole diameter of 30 nm.

HREM shows that the graphene structure of the wall parallel to the axes in some cases is similar to the pipe-shaped fibers but sometimes these fibers consists of a wall from areas with differently oriented graphite layers. Many different orientations (Fig. 6) of such layers were identified, from parallel up to very high inclinations (Lee et al., 2007; Duszová et al., 2008a, etc.). Besides, some edge graphite planes can be formed in single or multi loops whereas some end planes are strongly undulated.

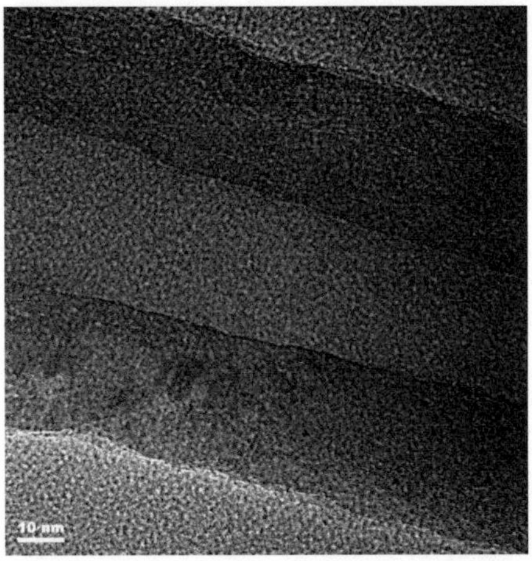

Fig. 5. HREM of a pipe - shaped conical hollow nanofiber with the wall thickness approximately 28 nm and hole diameter approx. 30 nm. The interlayer spacing of the graphite layers is approx. 0.35 nm.

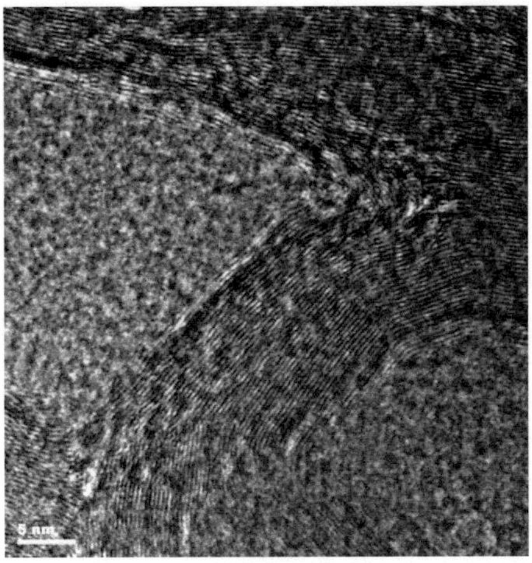

Fig. 6. HREM of a bamboo - shaped nanofiber with the wall thickness approximately 18 nm. The interlayer spacing of the graphite layers in this wall is approximately 0.33 nm (Puchý et al., 2010).

Hot pressed ZrO_2-CNFs composite were systematically studied in (Duszová et al., 2008a; Dusza et al., 2009) where the authors compared their microstructure to a monolithic zirconia prepared in the same way. The hot-pressed monolithic zirconia exhibits nearly full density but the density of the ZrO_2–CNF composite was lower due to the porosity in the composite connected mainly with the clusters of the carbon nanotubes present in its microstructure. XRD analysis revealed that the experimental materials consist of mainly tetragonal ZrO_2, however a small amount of monoclinic zirconia and yttrium oxide has been found, too.

In Fig. 7 the microstructure of the monolithic hot-pressed zirconia and the composite is illustrated at lower magnification. The monolithic sintered and hot-pressed zirconia consists of very small, submicron/nanometer-sized grains with randomly occurring defects in the form of very small-sized pores with dimensions of approximately 100–200 nm, mainly in the

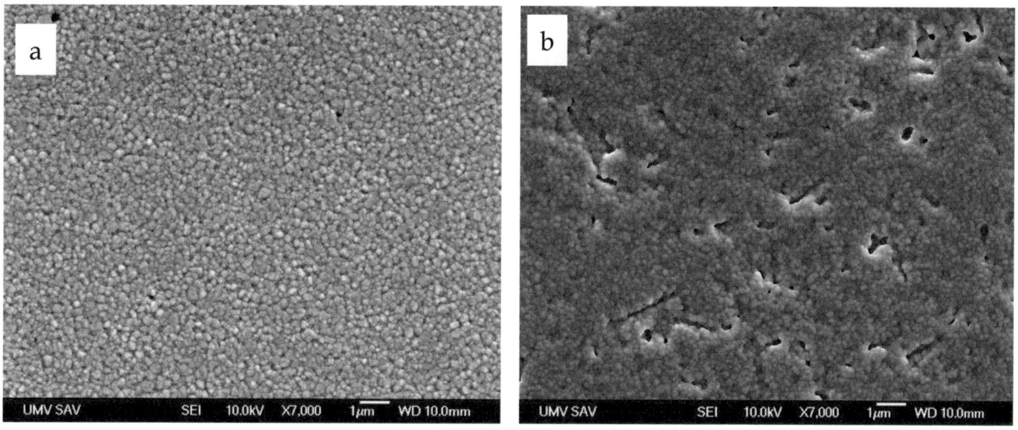

Fig. 7. SEM micrographs of the zirconia based experimental materials: microstructure of the HP monolithic zirconia (a) and HP ZrO_2 - CNFs composite at low magnification, thermally etched with the locations of the burned out CNFs during the thermal etching (b). (Dusza et al., 2009).

sintered material. The average grain size of the ZrO_2 in the sintered and hot-pressed zirconia is 190 nm and 162 nm, respectively. The microstructure of the composite consists of slightly smaller grained matrix with relatively well dispersed CNTs in the matrix, Fig. 7b (pores with their shape/orientation indicating the locations of the burned out CNTs during the thermal etching). The lower grain size of the zirconia together with the lower density of the composite compared to the monolithic material are evidence that the CNTs may hinder the sintering behavior of the zirconia-based composite.

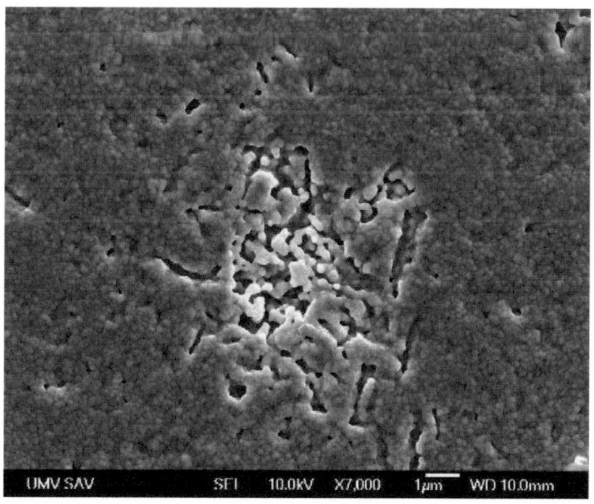

Fig. 8. Cluster of CNTs in the zirconia based composite, (Dusza et al., 2009).

Relatively high amount of clusters of the CNTs have been found on the polished surface and fracture surface of the samples indicating that the mixing procedure has to be improved. The size of the clusters varies from a few microns up to approximately 20 μm and porosity was always connected with these clusters (Fig. 8).
Balázsi with co-workers (Balázsi et al., 2006) developed CNT containing Si_3N_4 composites using hot isostatic pressing and compared them with the monolithic silicon nitride as well as

with composites containing carbon black and graphite prepared by identical procedure. They produced a series of composites with 1, 3, and 5 wt% of MWCNTs. All materials were carefully mixed, pressed at room temperature at 200 MPa and subsequently HIPed at 1700°C for various holding period (0 and 3 hrs) at 2 MPa or 20 MPa, respectively. They found that CNTs were preserved in the final structures. Problems with degradation of CNT, observed in other instances (Balázsi et al., 2003; Balázsi et al., 2004) were not observed in this newer work. The CNTs were located mainly in the inter-granular places and they were well attached to the silicon nitride grains as well as to each other. The proper separation and dispersion of carbon nanotubes proved to be a difficult task of composite preparation. The ball milling was followed by ultrasonic agitation of MWNT-powder mixtures for gaining a better homogeneity. By increasing the sonication time little advances have been achieved, but the tendency of nanotubes, the strong adherence and linking behavior to each other could not be totally suppressed. The CNTs in most cases remained in groups as nano- or micrometer sized islands in the matrix after sintering. By increasing the pressure from 2 to 20 MPa and holding time to 3 h a different microstructure could be developed. A grain growing process accompanied by a phase transformation was observed in 1 and 3 wt% CNT containing samples and large β-Si_3N_4 grains (several micrometers in length) formed. Generally, the CNT added samples had larger porosity than reference samples.

As regards the alumina based CMCs, this study focuses mostly on alumina-CNT composites prepared by SPS (Inam et al., 2010). For comparison also a monolithic Al_2O_3 ceramics was prepared by exactly same route.

The microstructure of the monolithic Al_2O_3 consists of relatively large, micron sized grains with randomly distributed submicrometric sized pores, located at the grain boundaries, Fig. 9a. Relatively large numbers of CNF clusters were observed on the polished and fracture surfaces of the Al_2O_3-3.5 % CNT composites, Fig. 9b. This material has much finer matrix with relatively well distributed carbon nanotubes with diameter of several nanometers and with very high aspect ratio. The smaller matrix grain size in the composite suggests that the carbon black and the CNTs hinder the grain growth during the sintering.

The CNTs in the Al_2O_3-CNT composite were located mainly in the intergranular places and were well attached to the alumina grains.

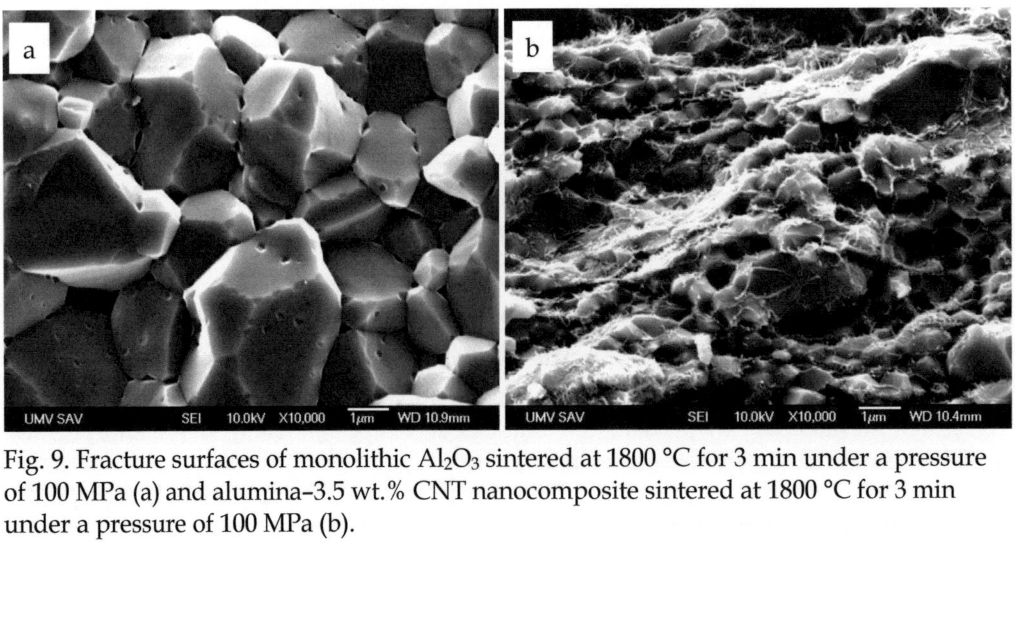

Fig. 9. Fracture surfaces of monolithic Al_2O_3 sintered at 1800 °C for 3 min under a pressure of 100 MPa (a) and alumina-3.5 wt.% CNT nanocomposite sintered at 1800 °C for 3 min under a pressure of 100 MPa (b).

4.2 Mechanical properties

As mentioned in the previous section, the CNT/Si_3N_4 prepared by Balázsi (Balázsi et al., 2006) had lower density (higher porosity) than the analogous reference sample. Consequently, the reinforcement processes (e.g. pullouts, crack bridging, crack deflection) were still not enough for a significant global strengthening. This resulted in decreasing values of modulus of elasticity and bending strength with increasing volume fraction of CNTs. By increasing the gas pressure (and holding time) the same level of densification can be achieved for composites with 1% CNT as for reference sample realised at low pressures (2 MPa). The modulus decreased from the value for silicon nitride (~250 GPa) linearly together with the apparent density regardless of the gas pressure used and for Si_3N_4-5%CNT with density around 2.2 g.cm^{-3} it reached values between 50 and 100 GPa.
The relationship between apparent density and three point bending strength of composites has similar trend to the modulus namely, the decrease of strength with increasing the CNT content. In this case however, the effect of pressure and holding time was more pronounced and the materials prepared using higher sintering pressure (and holding time) were stronger. In the case of samples with 1% CNT content, applying a higher sintering pressure a higher compaction level and an increase by 100%, from an average strength level of 300–600 MPa, was achieved. This sample was found to have the highest degree of densification connected to highest strength (even higher than for reference sample).

4.2.1 Hardness and fracture toughness

Even though the carbon nanotubes containing nanocomposites were proposed with the aim of improving mechanical performance, there is a trade-off between various properties. Typically, the introduction of carbon phases into a ceramics leads to lower hardness and its effect on the fracture toughness is ambiguous, even though there are some toughening mechanisms present in the crack, such as crack deflection, crack bridging and fibers pull-out.
Early examples represent hot pressed (2273 K) CNTs–SiC composites, investigated by Ma et al., 1998, which showed an increase of 10% in the strength and fracture toughness compared to SiC ceramics prepared with the same method. Alumina-CNT composites were also investigated in some depths by several authors. In contrast to reports of Peigney et al., 2000, and Flahaut et al., 2000, which did not observe an improvement of the mechanical properties of CNT–metal–alumina composites obtained by hot-pressing, Siegel et al., 2001, measured an increase in fracture toughness of 24% (from 3.4 up to 4.2 MPa.m$^{1/2}$) in MWCNT–alumina composites (10% in vol) prepared by hot-pressing. Zhan et al., 2003a, reported of dense alumina–SWNT composites possessing a toughness of 9.7 MPa.m$^{1/2}$, i.e. three times higher than that of pure alumina (3.3 MPa.m$^{1/2}$).
The reason given for the high performances of the as-prepared materials are the optimum dispersion of the CNTs and the process of spark plasma sintering performed at lower temperatures (1100 °C) with consequent lower damaging of the CNTs compared to traditional hot-pressing, (Borrell et al., 2010; Lupo et al., 2004).
Sun (Sun et al., 2005) studied the mechanical and fracture behavior of MWCNT/3Y-TZP composites containing 0.1–1.0 wt.% MWCNTs and SWCNTs prepared by SPS. They found that the addition of CNTs had a negative influence on the hardness of the composites and no influence (at 0.5 wt.%) or a negative influence on the fracture toughness. Comparing the influence of the addition of SWCNTs and MWCNTs on the fracture toughness, they found that the MWCNT are slightly more effective in the toughening. However, they used the

indentation technique for the toughness measurements, which is useful only for comparison purposes, cannot be considered a true material characterization. Similar results were found in (Ukai et al., 2006), where the failure properties of similar materials were studied. According to these works, the CNTs often agglomerate at the ZrO_2 boundaries, and the weak bonding between the CNTs and zirconia are the reason why the reinforcing effect of the CNTs is limited. The results of (Duszová et al., 2008a) found for zirconia with carbon nanofibers are in a very good agreement with these results (Fig. 10) and show that even the use of relatively coarse whisker-like CNFs is not effective in toughening the zirconia ceramic matrix at the current amounts used. On the other hand, it was noted that on the fracture surface/line different toughening mechanisms were found, mainly in the form of crack deflection at the CNFs. The reason for the relatively low indentation fracture toughness is probably the poor dispersion and therefore the limited toughening effect of the CNFs.
The indentation toughness of monolithic alumina is approximately 3.24 MPa.m$^{1/2}$, which is comparable to similar materials in the literature. The addition of 2 % carbon black increased the indentation toughness, but after increasing to 5 % the indentation toughness decreased to the similar level as of the monolithic material. The addition of 5 % CNTs increased the indentation toughness up to 4.14 MPa.m$^{1/2}$ which is the highest value obtained for the materials investigated.

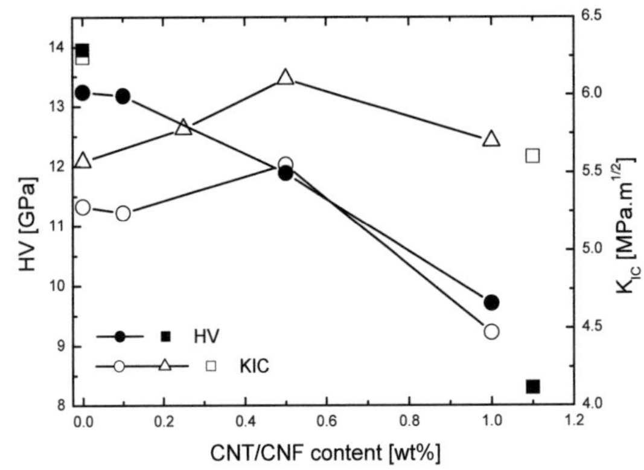

Fig. 10. Influence of the CNT/CNF addition on the Vickers hardness and indentation toughness of zirconia based composites. After Sun et al 2005 (circles), Ukai et al. 2006 (triangles), and Duszová et al 2008 (squares).

Fractography of the fracture surfaces of the composites revealed carbon based "bridges" between the alumina grains, which probably in the case of Al_2O_3 - CNT composite are strong enough to increase the resistance against the crack propagation. These results are in good agreement with the results of recent investigations of Ahmad (Ahmad et al., 2010). There, well-dispersed CNT-reinforced Al_2O_3 nanocomposites were prepared with reasonably high density using hot pressing. It was shown that 2 wt. % of CNT addition increased the hardness, flexural strength and fracture toughness of nanocomposites (from 3 to 4.3 MPa.m$^{1/2}$). Further CNT addition up to 5 wt. % slightly decreased the hardness, reduced the flexural strength but improved the toughness up to 4.5 MPa.m$^{1/2}$. The increase in toughness is believed to be associated with the strong interface connections between the CNT and the matrix, resulting in pullout resistance, bridged the crack gaps and hindered the crack propagation by exploiting CNTs elasticity, leading to improved fracture toughness.

4.2.2 Tribological properties

Fig. 11 shows examples of development of the friction coefficient over the test running distance (Hvizdoš et al., 2011a). In all cases was friction after short initial stage (in order of meters) rather stable and reproducible. For monolithic ZrO_2 the values of COF were slightly lower than similar results found in literature. The sliding wear behaviour of ZrO_2 using the ball on disk method and alumina as a friction partner was also investigated in (Suh et al., 2008). The applied loads ranged from 19.8 N to 98 N. There, the coefficient of friction varied between 0.6 and 0.7 depending on the sliding speed, and appeared to be independent on the applied load. Higher COF can be explained by the damage mechanisms - the alumina grains detached from the alumina surface and caused abrasion of zirconia. For lower applied load, however, the main wear mechanism in the zirconia is typically self-polishing by low intensity abrasion.

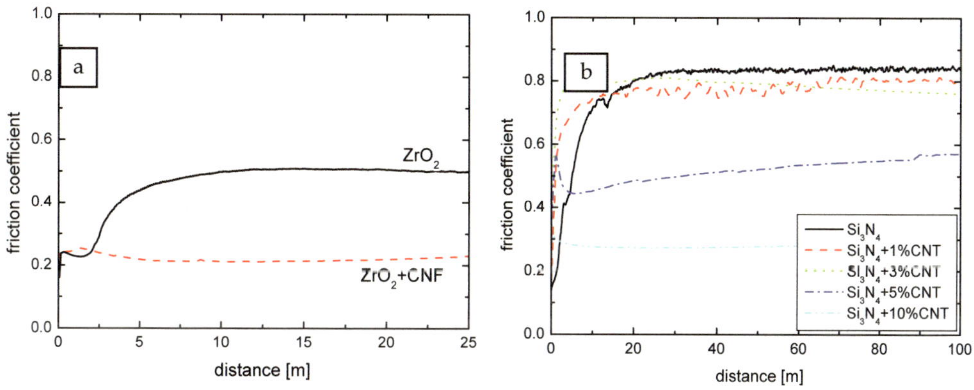

Fig. 11. Friction coefficient during the testing under similar conditions. (a) ZrO_2 based materials tested at 1 N load; (b) Si_3N_4 based materials at 1.5 N load, (Hvizdoš et al., 2011a).

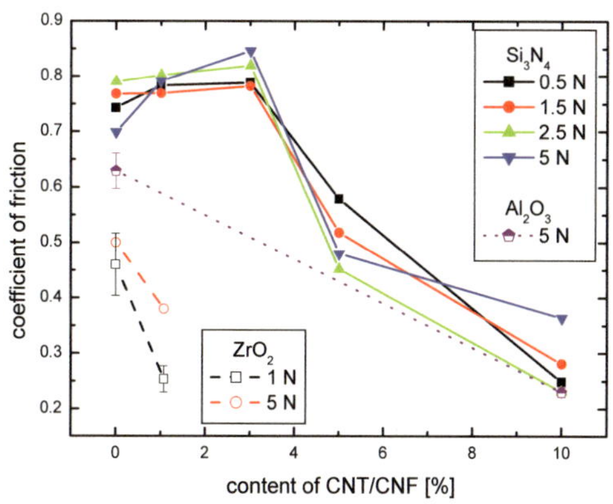

Fig. 12. Coefficient of friction as function of amount of carbon phases in the CMC materials (Hvizdoš et al., 2011a; Hvizdoš et al., 2011b).

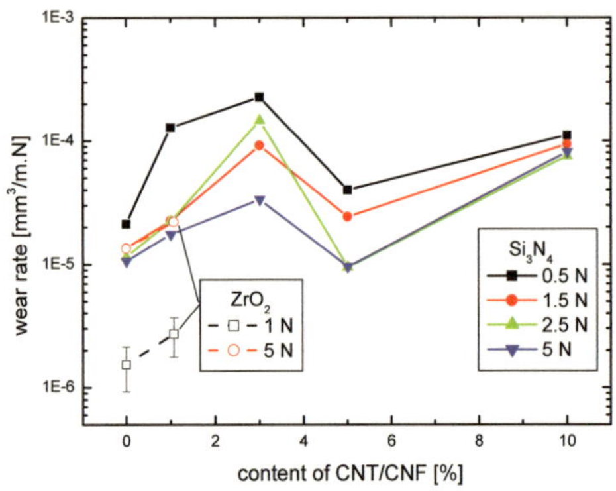

Fig. 13. Dependence of wear resistance of the experimental materials on their carbon content tested at various load levels.

Self-mated wear of monolithic silicon nitride was studied in various situations and can be summarized both in terms of the friction coefficient (Fig. 11b and 12, values 0.7-0.8) and wear rates (Fig. 13, values 1-2x10^{-5}mm^3/N.m) (Kašiarová et al., 2004; Hvizdoš et al., 2011a).

Fig. 12 summarizes the average values of friction for all materials. The Si_3N_4 based materials exhibited higher friction due to chemical similarity of both friction partners, particularly for low CNT contents. Here, the COF started to decrease only when the CNT content reached ~5 wt%. In Si_3N_4-10%CNT the COF values dropped to ~0.2 which was comparable to the values for ZrO_2-CNF. Similar values were found for self mated friction of alumina (Hvizdoš et al., 2011b). Here, only monolith and 10% CNT containing composite are given, but the tendency is very similar to that of silicon nitride materials and more than 60% reduction of friction coefficient was reported.

Fig. 13 shows the wear rates of various materials at different loads. Generally, the presence of CNT/CNF tends to decrease the wear resistance due to less than optimal microstructure. However, Hvizdoš et al., 2011a, found a sort of an optimum for fractions of around 5% CNT, where the wear resistance improved, probably thanks to significantly reduced friction. Then for Si_3N_4-10%CNT it only reached values similar to those for Si_3N_4-1%CNT.

Fig. 14 illustrates typical damage observed in the wear tracks. The wear testing of the monolithic materials of both types of materials left faint, quite smooth wear tracks with polished look, as it is shown in Fig 14a and 14c for sliding at 5N load. This is consistent with the low wear rates of materials with high quality microstructure.

Figs 14b and 14d illustrate the wear tracks of the composite materials and present the micromechanisms of the wear surface damage. Microscopic observations of the worn surface of the ZrO_2-CNF composite showed pulled out CNFs (Fig. 14b) and suggested the presence of a smeared layer of graphite and crushed CNFs, the so-called transferred film, which significantly lowers the friction coefficient, even though relatively low amount of carbon (1.07wt%) was incorporated into the microstructure. As Fig. 14d as well as the values of COF and wear rates suggest, the CNTs in silicon nitride based composites were much less effective in creating the transferred film. It may be due to their greater mechanical stability and better distribution throughout the microstructure. This could mean that their pull-out and also their destruction in the contact zone are more difficult.

In all cases by incorporating appropriate amount of carbon filaments the coefficient of friction between 0.2 and 0.3 was achieved. A similar lubricating effect by carbon phases was

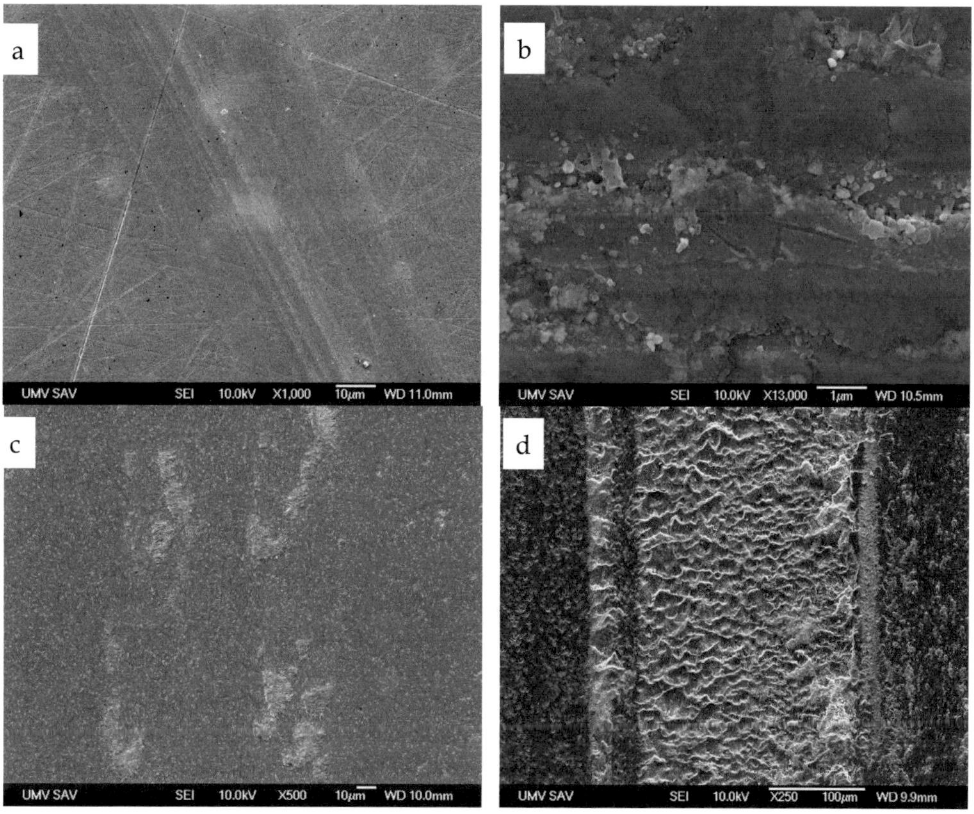

Fig. 14. Wear damage at normal load 5N: (a) smooth wear tracks in monolithic ZrO_2; (b) detail of the wear track in ZrO_2-CNF composite - pull-out of CNFs and smeared transferred film; (c) monolithic Si_3N_4 - smooth track with polished look; (d) Si_3N_4-3%CNT - surface abrasion.

reported for MWCNTs-containing Al_2O_3 composites by (An et al., 2003) and (Yamamoto et al., 2008a) and for SWCNT solids by (Yamamoto et al., 2008b). In (An et al., 2003) the authors reported a 40% decrease in friction coefficient in 12 wt.% MWCNT-Al_2O_3 composite down to the value of 0.3. Yamamoto with co-workers, for their composites reached the minimum of the friction coefficient (with the value of 0.3) at 4 mass% of MWCNT content. According to An et al. the rolling motion of CNTs at the interface between the specimen and the ball can probably lower the friction coefficient as well. This effect in the case of ZrO_2-CNF composite was not proven. Our results for silicon nitride based composites are in a very good agreement with the results reported for SWCNT solids sliding against Si_3N_4 where steady-state friction coefficients of 0.22-0.24 were found.

Alumina-CNT composites with CNT contents from 0 to 12 wt. % were fabricated by two different processing techniques by Lim et al. (Lim et al., 2005) in order to compare the effect of fabrication techniques on the tribological properties of CNT added alumina composites. Superior tribological properties were achieved by introduction of tape casting, followed by lamination and hot pressing. Wear loss decreased significantly and friction coefficients were maintained to be about the same by addition of the CNTs up to 12 wt. %. Agglomeration of CNTs, frequently observed with hot-pressed specimens, was significantly reduced and relatively uniform distribution of CNTs was obtained. The effective dispersion of CNTs contributed to the densification of composites and to superior mechanical and tribological properties of CNT added alumina composites.

4.3 Electrical properties

The conductivity of an individual CNT can be metallic or semiconducting, depending on the chirality of the CNT with axial component extremely high, 2×10^7 S/m (Ebbesen et al., 1996) comparable to that of silver, cooper or gold. Larger CNFs have properties similar to graphite. As it was shown in the section on microstructure, the CNTs and CNFs can be distributed along matrix grain boundaries and/or between matrix grains. When interconnected they can form continuous networks. Such networks in CNT/CNF nanocomposites can be electrically conductive thanks to various mechanisms that involve fluctuation-assisted tunnelling or variable range hopping between the individual nanofibers.

Inam et al., 2010, collected and summarized a number of studies of electrical properties of various CMCs prepared by hot-pressing and SPS. It seems that about 5 to 10 wt.% of carbon filament phases is necessary to ensure the percolation of electrical charges, and conductivities in orders of tens or hundreds S/m were found. Generally speaking, the carbon black, which was used for reference reasons, is less efficient, while larger CNFs or higher amounts of SWNT+MWNT form more conductive networks (588 S/m for HP alumina-5%CNF, and 853 S/m for magnesium aluminate-12.2%CNT, respectively).

Similar results were found recently for series of zirconia, alumina and silicon nitride based CMCs (Duszová et al., 2008b; Hvizdoš et al., 2011b). These results are summarized in Fig. 15. Here, for the materials with CNTs, the percolation threshold was between 3 and 5 wt.% for Si_3N_4 and between 2 and 5 wt% for alumina. For the latter composite the electrical conductivity increased from practically zero to the maximum value of 140 S/m for 10 wt.% of CNTs. Analogous materials prepared in the same way but with admixture of carbon black yielded electrical conductivity about an order of magnitude lower - e.g. for 5 wt% of additive it was 90.5±4.8 S/m for CNTs and 8.5±0.1 S/m for carbon black. This means that the size and shape of the mixed carbon black additives resulted in a limited graphite particle connection in the matrix, which was much less efficient in securing the percolation as it is illustrated schematically in Fig. 16.

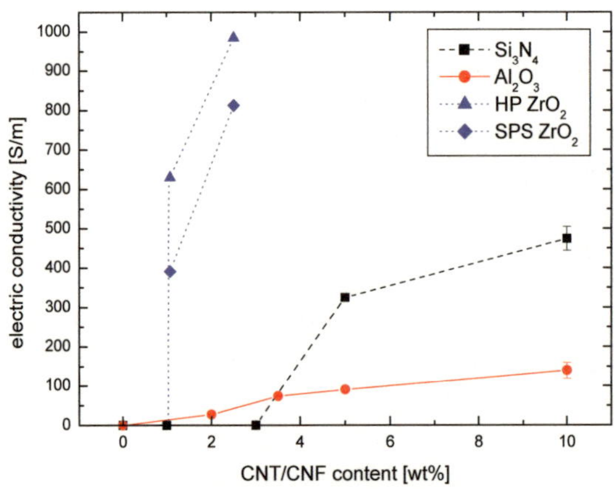

Fig. 15. Electrical conductivity - illustration of percolation theshold for different CMC-CNT /CNF composites (Duszová et al., 2008b; Hvizdoš et al., 2011b).

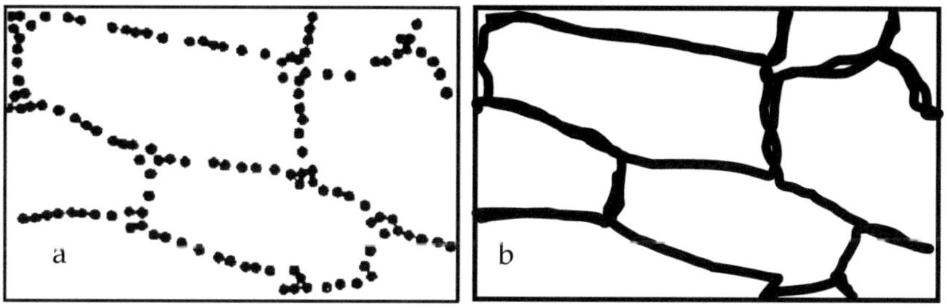

Fig. 16. Scheme of the microstructure of CMC with carbon black (a) and CNTs (b).

Carbon nanofibers were significantly more efficient, as shown in the case of ZrO_2-CNF composites (Fig. 15), where amounts as low as 1 to 3 wt.% of carbon nanofibers were sufficient to increase the electrical conductivity from ~10^{-12} S/m to 983± 200 S/m (hot pressed) or 813±273 S/m (SPSed), (Duszová et al., 2008b).

In summary it can be concluded that thanks to their fibrous nature and intergranular distribution, very little amount of carbon nanofibers (only about 1-5 wt%, depending on the basic material microstructure) is sufficient for the percolation effect to take place. In this way an increase in electrical conductivity by more than 13 orders of magnitude can be achieved.

5. Conclusion

The processing routes, microstructure characteristics, mechanical and functional properties of carbon nanofibers containing CMCs have been a subject of intensive study during last 10-15 years. The result show that introduction of CNTs/CNFs usually results in finer microstuctures. It also invariably leads to a drop in hardness. As for the fracture toughness, up to now the results are ambiguous. In most cases the fracture toughness decreases, particularly for higher fractions of carbon phases. Yet, the potential for improving in this area by toughening mechanisms such as fiber pull-out, crack bridging and deflection has been recognized. In this respect future work is necessary to optimize the interface between the matrix and the fibers, which will ensure good bonding of the two phases, yet allowing the toughening mechanisms to be active.

The tribological studies showed clear benefits of presence of carbon fibers. By a small amount of CNFs (as little as 1 wt%) the coefficient of friction of ZrO_2 can be lowered from 0.45 to 0.2 by formation of carbon film which acts as a solid lubricant. The CNTs are slightly less effective and higher amount is necessary to achieve similar results. The wear resistance generally decreases, which can be correlated to lower hardness and fracture toughness, but in several instances a sort of optimum was found (e.g. Si_3N_4-5wt%CNT) where the wear rate was similar to that of monolithic material but with much lower coefficient of friction.

The electrical conductivity can be increased very significantly and low amounts of CNFs or CNTs are sufficient to change an insulator to a functionalized conductive ceramics. Here again larger CNFs seem to be more effective and show clear potential for future development of carbon nanofibers containing ceramic matrix materials.

6. Acknowledgement

This work was done within the frame of the project „Centre of Excellence of Advanced Materials with Nano- and Submicron- Structure", which is supported by the Operational

Program "Research and Development" financed through European Regional Development Fund. The experiments were supported by the projects VEGA 2/0120/10 and COST Action MP0701. The authors would like to express their gratitude to C. Balázsi, J. Kuebler, G. Blugan, M. Reece, and others for the preparation of the experimental materials and to J. Morgiel and his co-workers for the TEM studies.

7. References

Ahmad, I.; Unwin, M.; Cao, H.; Chen, H.; Zhao, H.; Kennedy, A. & Zhu, Y.Q. (2010). Multi-walled carbon nanotubes reinforced Al_2O_3 nanocomposites: Mechanical properties and interfacial investigations, *Composites Science and Technology*, Vol. 70, No.4 (August 2010), pp. 1199–1206, ISSN 0266-3538

Al-Saleh, M.H. & Sundararaj, U. (2009). A review of vapor grown carbon nanofiber/polymer conductive composites, *Carbon*, Vol. 47, No.1 (January 2009), pp. 2–22, ISSN 0008-6223

An, J.W.; You, D.H. & Lim, D.S. (2003). Tribological properties of hot-pressed alumina-CNT composites, *Wear*, Vol. 255, No.1 (August 2003), pp.677-681, ISSN: 0043-1648

Anstis, G.R.; Chantikul, P.; Lawn, B. & Marshall, D. (1981). A critical evaluation of indentation techniques for measuring fracture toughness: I. Direct crack measurements, *Journal of the American Ceramic Society*, Vol. 64, No.9 (September 1981), pp. 533-538, ISSN 0002-7820

Balázsi, C.; Kónya, Z.; Wéber, F.; Biró, L.P. & Arató, P. (2003). Preparation and characterization of carbon nanotube reinforced silicon nitride composites. *Material Science and Engineering C*, Vol. 23, No. 6-8 (December 2003), pp. 1133–1137, ISSN 0928-4931

Balázsi, C.; Kónya, Z.; Kasztovszky, Z.; Wéber, F.; Vértesy, Z. & Biró, L.P. (2004). Examination of the carbon nanotube reinforced silicon nitride composites. In *Proc. of the 5th international conference on high temperature ceramic matrix composites (HTCMC-5),* New Frontiers and Horizons, Seattle, Washington, DC: American Ceramic Society; September 2004. pp. 107–112.

Balázsi, C.; Fényi, B.; Hegman, N.; Kovér, Z.; Wéber, F.; Vértesy, Z.; Kónya, Z.; Kiricsi, I.; Biró, L.P. & Arató, P. (2006). Development of CNT/ Si_3N_4 composites with improved mechanical and electrical properties, *Composites: Part B*, Vol. 37, No.6, pp. 418–424, ISSN 1359-8368

Belmonte, M.; González-Julián, J.; Miranzo, P. & Osendi, M.I. (2010). Spark plasma sintering: A powerful tool to develop new silicon nitride-based materials, *Journal of the European Ceramic Society*, Vol. 30, No. 14 (October 2010), pp. 2937–2946, ISSN 0955-2219

Borrell, A.; Fernandez, A.; Merino, C. & Torrecillas, R. (2010). High density carbon materials obtained at relatively low temperature by spark plasma sintering of carbon nanofibers. *International Journal of Materials Research,* Vol.101, No.1, pp. 112-116, ISSN 1862-5282

Callister, W.D. (2003). *Materials Science and Engineering: An Introduction*, 6th ed. Wiley, New York. ISBN 978-0471736967

Cha, S.I.; Kim, K.T.; Lee, K.H.; Mo, C.B. & Hong, S.H. (2005). Strengthening and toughening of carbon nanotube reinforced alumina nanocomposite fabricated by molecular level mixing process, *Scripta Materialia*, Vol. 53, No. 7 (October 2005), pp.793-797, ISSN 1359-6462

Datye, A.; Wu, K.-H.; Gomes, G.; Monroy, V.; Lin, H.-T.; Vleugels, J. & Vanmeensel, K. (2010). Synthesis, microstructure and mechanical properties of 3YTZP-MWNTs nanocomposite by direct in-situ growth of MWNTs on Zirconia particles. *Composite Science and Technology*, Vol.70, No.14, pp. 2086-2092, ISSN 0266-3538

Dusza, J. & Šajgalík, P. (2009). *Silicon nitride and alumina-based nanocomposites, Handbook of nanoceramics and their based nanodevices 2*, Stevenson Ranch, American Scientific Publishers, pp. 253-283, ISBN 1-58883-116-7

Dusza, J.; Blugan, G.; Morgiel, J.; Kuebler, J.; Inam, F.; Peijs, T.; Reece, M.J. & Puchý, V. (2009). Hot pressed and spark plasma sintered zirconia/carbon nanofiber composites, *Journal of the European Ceramic Society*, Vol. 29, No. 15 (December 2009), pp. 3177–3184, ISSN 0955-2219

Duszová, A.; Dusza, J.; Tomášek, K.; Morgiel, J.; Blugan, G. & Kuebler, J. (2008a). Zirconia/carbon nanofiber composite, *Scripta Materialia*, Vol. 58, No. 6 (March 2008), pp. 520-523, ISSN 1359-6462

Duszová, A.; Dusza, J.; Tomášek, K.; Blugan, G. & Kuebler, J. (2008b). Microstructure and properties of carbon nanotube/zirconia composite, *Journal of the European Ceramic Society*, Vol. 28, No. 5, pp. 1023-1027, ISSN 0955-2219

Ebbesen, T. W., Lezec, H. J., Hiura, H., Bennett, J. W., Ghaemi, H. F. & Thio, T., (1996). Electrical-conductivity of individual carbon nanotubes. *Nature*, Vol. 382, (July 1996), pp. 54–56, ISSN: 0028-0836

Endo, M.; Kim, Y.A.; Hayashi, T.; Fukai, Y.; Oshida, K.; Terrones, M.; Yanagisawa, T.; Higaki, S. & Dresselhaus, M.S. (2002). Structural characterization of cup-stacked-type nanofibers with an entirely hollow core, *Applied Physics Letters*, Vol. 80, No. 7 (February 2002), pp. 1267-1268, ISSN 0003-6951

Flahaut, E.; Peigney, A.; Laurent, Ch.; Marliére, C.; Chastel, F. & Rousset, A. (2000). Carbon nanotube-metal-oxide: microstructure, electrical conductivity mechanical properties, *Acta Materialia*, Vol. 48, No. 14 (September 2000), pp. 3803–3812, ISSN 1359-6454

Garmendia, N.; Santacruz, I.; Moreno, R. & Obieta, I. (2010). Zirconia-MWCNT nanocomposites for biomedical applications obtained by colloidal processing. *Journal of Materials Science-Materials in Medicine*, Vol.21, No.5, pp. 1445-1451, ISSN: 0957-4530

Garvie, R.C.; Hannink, R.H. & Pascoe, R.T. (1975). Ceramic steel?, *Nature*, Vol. 258, (December 1975), pp. 703–704, ISSN 0028-0836

Garvie, R.C.; Urbani, C.; Kennedy, D.R. & McNeuer, J.C. (1984). Biocompatibility of magnesia-partially stabilized zirconia (Mg-PSZ) ceramics, *Journal of Materials Science*, Vol. 19, No. 10, pp. 3224-3228, ISSN: 0022-2461

González-Julián, J.; Schneider, J.; Miranzo, P.; Osendi, M.I. & Belmonte, M. (2010). Tribologische Charakterisierung von Si_3N_4-CNT-Kompositkeramiken unter ungeschmierter und mediengeschmierter oszillierender Gleitbeanspruchung, in *Proc. „Tribologie-Fachtagung 2010"*, Deutsche Gesellschaft für Tribologie (GFT), Aachen, pp. 02/1 - 02/10

Hvizdoš, P.; Puchý, V.; Duszová, A. & Dusza J. (2010). Tribological behavior of carbon nanofiber-zirconia composite, *Scripta Materialia*, Vol. 63, No. 2 (July 2010), pp. 254–257, ISSN 1359-6462

Hvizdoš, P.; Duszová, A.; Puchý, V.; Tapasztó, O.; Kun, P.; Dusza, J. & Balázsi, C. (2011a). Wear behavior of ZrO_2-CNF and Si_3N_4-CNT nanocomposites, *Key Engineering Materials*, Vol. 465, pp. 495-498, ISBN-13 978-3-03785-006-0

Hvizdoš, P; Puchý, V; Duszová, A.; Dusza, J. (2011b). Tribological and electrical properties of ceramic matrix composites with carbon nanotubes, in *Engineering Ceramics 2011*, Smolenice, Slovakia, (May 8-12, 2011), in press

Iijima, S. (1991). Helical microtubules of graphitic carbon, *Nature*, Vol. 354, (November 1991), pp. 56-58, ISSN 0028-0836

Inam, F.; Yan, H.; Jayaseelan, D.D.; Peijs, T. & Reece, M.J. (2010). Electrically conductive alumina–carbon nanocomposites prepared by Spark Plasma Sintering, *Journal of the European Ceramic Society*, Vol. 30, pp. 153–157, ISSN: 0028-0836

ISO 20808:2004(E), "Fine ceramics - Determination of friction and wear characteristics of monolithic ceramics by ball-on-disc method", (2004).

Kašiarová, M.; Rudnayová, E.; Dusza, J.; Hnatko, M.; Šajgalík, P.; Merstallinger, A. & Kuzsella, L. (2004). Some tribological properties of a carbon-derived Si3N4/SiC Nanocomposite, *Journal of the European Ceramic Society*, Vol. 24, No.12, pp. 3431–3435, ISSN 0955-221

Krishnan, A.; Dujardin, E.; Treacy, M.M.J.; Hugdahl, J.; Lynum, S. & Ebbesen, T.W. (1997). Graphitic Cones and the Nucleation of Curved carbon Surfaces, *Nature*, Vol. 388, (July 1997), pp. 451–454, ISSN 0028-0836

Lee, S.; Kim, T.R.; Ogale, A.A. & Kim M.S. (2007). Surface and structure modification of carbon nanofibers. *Synthetic Metals*, Vol. 157, No. 16-17 (August 2007), pp. 644-650, ISSN 0379-6779

Lee, C.J.; Lyu, S.C; Kim, H.W.; Lee, J.H. & Cho, K.I. (2002). Synthesis of bamboo-shaped carbon–nitrogen nanotubes using C_2H_2–NH_3–$Fe(CO)_5$ system, *Chemical Physics Letters*, Vol. 359, No. 1-2 (June 2002), pp. 115-120, ISSN 0009-2614

Lim, D.S.; You, D.H.; Choi, H.J.; Lim, S.H. & Jang, H. (2005). Effect of CNT distribution on tribological behavior of aluminum-CNT composites. *Wear*, Vol. 259, No. 1-6 (July-August 2005), pp. 539-544, ISSN 0043-1648

Longtin, R.; Fauteux, C.; Goduguchinta, R. & Pegna, J. (2007). Synthesis of carbon nanofiber films and nanofiber composite coatings by laser-assisted catalytic chemical vapor deposition, *Thin Solid Films*, Vol. 515, No. 5 (January 2007), pp. 2958-2964, ISSN 0040-6090

Lupo, F.; Kamalakaran, R.; Scheu, C.; Grobert, N. & Ruhle, M. (2004). Microstructural investigations on zirconium oxide-carbon nanotube composites synthesized by hydrothermal crystallization, *Carbon*, vol. 42, no. 10, pp. 1995-1999, ISSN 0008-6223

Ma, R.Z.; Wu, J.; Wei, B.Q.; Liang, J. & Wu, D.H. (1998). Processing and properties of carbon nanotubes-nano-SiC ceramic. *Journal of Material Science*, Vol. 33, No. 21 (November 1998), pp. 5243–5463, ISSN 0022-2461

Mazaheri, M.; Mari, D.; Hesabi, Z.R.; Schaller, R. & Fantozzi, G. (2011). Multi-walled carbon nanotube/nanostructured zirconia composites: Outstanding mechanical properties in a wide range of temperature. Composites Science and Technology, in press, doi:10.1016/j.compscitech.2011.01.017

McEnaney, B. (2001). Carbon and graphites, mechanical properties of, *Encyclopedia of Materials: Science and Technology*, Amsterdam, New York, pp. 967–975, ISBN 978-0080431529

Melechko, A.V.; Merkulov, V.I.; McKnight, T.E.; Guillorn, M.A.; Klein, K.L.; Lowndes, D.H. & Simpson, M.L. (2005). Vertically aligned carbon nanofibers and related structures: Controlled synthesis and directed assembly, *Journal of Applied Physics*, Vol. 97, No.4 (February 2005), 041301, ISSN 0021-8979

Morrell, R. (1985). *Handbook of properties of technical and engineering ceramics*, Part 1., HMSO, London, ISBN 978-0114800529

Mukhopadhyay, A.; Chu, B.T.T.; Green, M.L.H. & Todd, R.I. (2010). Understanding the mechanical reinforcement of uniformly dispersed multiwalled carbon nanotubes in alumino-borosilicate glass ceramic. *Acta Materialia*, Vol.58, No.7 (April 2010), pp. 2685-2697, ISSN 1359-6454

Peigney, A.; Laurent, Ch.; Flahaut, E. & Rousset, A. (2000). Carbon nanotubes in novel ceramic matrix nanocomposites, *Ceramics International*, Vol. 26, No. 6 (July 2000), pp. 677–683, ISSN 0272-8842

Perera, D.S.; Tokita, M. & Moricca, S. (1998). Comparative study of fabrication of Si_3N_4/SiC composites by spark plasma sintering, *Journal of European Ceramic Society*, Vol. 18, No.4 (April 1998), pp. 401-404, ISSN: 0955-2219

Puchý, V.; Tatarko, P.; Dusza, J.; Morgiel, J.; Bastl, Z. & Mihaly, J. (2010). Characterization of carbon nanofibers by SEM, TEM, ESCA and Raman spectroscopy, *Kovove Materialy-Metallic Materials*, Vol. 48, No. 6, pp. 379-385, ISSN: 0023-432X

Ruoff, R.S.; Qian, D. & Liu, W.K. (2003). Mechanical properties of carbon nanotubes: theoretical predictions and experimental measurements. *Comptes Rendus Physique*. Vol. 4, No.9 (November 2003), pp. 993-1008, ISSN 1631-0705

Siegel, R.W.; Chang, S.K.; Ash, B.J.; Stone, J.; Ajayan, P.M.; Doremus R.W. & Schadler R.S. (2001). Mechanical behaviour of polymers and ceramics matrix nanocomposites, *Scripta Materialia*, Vol. 44, No.8 (May 2001), pp. 2061–2064, ISSN 1359-6462

Shi, S.L. & Liang, J. (2006). Effect of Multiwall Carbon Nanotubes on Electrical and Dielectric Properties of Yttria-Stabilized Zirconia Ceramic. *Journal of the American Ceramic Society*, Vol. 89, No. 11 (November 2006), pp. 3533–3535, ISSN 0002-7863

Shi, S.L. & Liang, J. (2007). Electronic transport properties of multiwall carbon nanotubes / yttria-stabilized zirconia composites, *Journal of Applied Physics*, Vol. 101, No.2, p. 023708, ISSN 0021-8979

Suh, M.S.; Chae, Y.H. & Kim, S.S. (2008). Friction and wear behavior of structural ceramics sliding against zirconia, *Wear*, Vol. 264, No. 9-10, (April 2008), pp. 800–806, ISSN 0043-1648

Sun, J.; Gao L.; Iwasa, M.; Nakayama, T. & Niihara, K. (2005). Failure investigation of carbon nanotube/3Y-TZP nanocomposites, *Ceramics International*, Vol 31, No. 8, pp. 1131-1134, ISSN 0272-8842

Tatami, J.; Katashima, T.; Komeya, K.; Meguro, T. & Wakihara, T. (2005). Electrically Conductive CNT-Dispersed Silicon Nitride Ceramics. *Journal of the American Ceramic Society*, Vol. 88, No. 10 (October 2005), pp. 2889–2893, ISSN 0002-7863

Ukai, T.; Sekino, T.; Hirvonen, A.; Tanaka, N.; Kusunose, T.; Nakayama, T. & Niihara, K. (2006) Preparation and electrical properties of carbon nanotubes dispersed zirconia nanocomposites, *Key Engineering Materials*, Vol.317-318, pp.661-664, ISSN 1662-9795

Yamamoto, G.; Omori, M.; Yokomizo, K.; Hashida, T. & Adachi, K. (2008a). Structural Characterization and Frictional Properties of Carbon Nanotube/Alumina Composites Prepared by Precursor Method, *Material Science and Engineering B*, Vol. 148, No. 1-3 (February 2008) pp. 265-269, ISSN 0921-5107

Yamamoto, G.; Hashida, T.; Adachi, K. & Takagi, T. (2008b). Tribological Properties of Single-Walled Carbon Nanotube Solids, *Journal of Nanoscience and Nanotechnology*, Vol. 8, No. 5 (May 2008), pp. 2665-2670, ISSN 1550-7033

Yu, M.F.; Lourie, O.; Dyer, M.J.; Moloni, K.; Kelly, T.F. & Ruoff, R.S. (2000). Strength and breaking mechanism of multiwalled carbon nanotubes under tensile load, *Science*, Vol. 287, No. 5453 (January 2000), pp. 637-640, ISSN 0036-8075

Zhan, G.-D.; Kuntz, J.D.; Wan, J & Mukherjee, A.K. (2003a). Single-wall carbon nanotubes as attractive toughening agents in alumina-based nanocomposites. *Nature Materials,* Vol.2, No. 1 (January 2003), pp. 38–42, ISSN 1476-1122

Zhan, G.-D.; Kuntz, J.D.; Garay, J.E. & Mukherjee, A.K. (2003b). Electrical properties of nanoceramics reinforced with ropes of single-walled carbon nanotubes. *Applied Physics Letters,* Vol. 83, No. 6 (August 2003), pp. 1228-1230, ISSN 0003-6951

13

Carbon Nanofibers: Evaluation of Life Cycle Environmental Impacts

Vikas Khanna and Nikki Campion
Department of Civil and Environmental Engineering, University of Pittsburgh, Benedum Hall, Pittsburgh, PA
United States of America

1. Introduction

The rapidly evolving field of nanotechnology has sparked a great level of interest and enthusiasm from the research and industrial community. Given the enormous promise that engineered nanomaterials and nanoproducts offer, it is essential to investigate the potential life cycle environmental impacts of emerging nanotechnologies for guiding their sustainable development. One rapidly emerging class of engineered nanoscale materials is carbon nanofibers (CNFs). While studies regarding the toxicological impacts of CNFs are needed, holistic understanding of the impact of CNFs and nanoproducts using systems analysis is central for their safe and sustainable development. Systems analysis with life cycle thinking offers a good tool for evaluating the broader impacts of emerging technologies at an early stage of research. This chapter will examine the needs, challenges, and promises of life cycle assessment (LCA) in the evaluation of CNFs and CNF-based products. A general introduction to LCA is explained in the first part of the chapter and the challenges and promises for the LCA of nanoproducts are identified. The use of LCA in guiding sustainable development of nanotechnology is illustrated in greater detail with a discussion on LCA of CNFs in section 2. Section 3 describes the life cycle energetic impacts of the use of CNF-based polymer composites in auto body panels as a potential replacement for traditional materials. A brief discussion of the toxicological impacts of CNFs is presented in section 4. Section 5 summarizes the major findings and provides directions for future research.

Nanotechnology is a fast emerging technology with large number of consumer goods containing engineered nanomaterials already in production [1, 2]. The potential impact of nanotechnology derives from the fact that unprecedented material properties are being discovered in nanoscale materials that enable a broad spectrum of novel applications. One such class of materials is carbonaceous nanomaterials. The last 25 years have seen the discovery of a variety of carbonaceous nanomaterials including fullerenes, nanotubes, nanohorns, and nanofibers [3]. Carbon Nanofibers (CNFs) belong to a new class of materials that have exceptional mechanical properties like high tensile strength (12000 MPa) that is approximately 10 times that of steel [4, 5]. Besides mechanical strength, CNFs possess novel electrical properties like high electrical conductivity. These properties of CNFs are being explored in a variety of ways by imparting functionalities in various intermediate and final value-added consumer products. Applications include the use of CNFs as reinforcements for high strength polymer nanocomposites, use in carbon-lithium batteries, and electrically conducting polymers [5]. High specific surface area of these fibers is an additional attribute that has been investigated for use of CNFs as catalyst support material especially for liquid phase reactions [4].

Although altered physicochemical properties make CNFs and CNF-based nanoproducts commercially attractive, they also raise concerns about the human and ecosystem impact of these materials [6-11]. While toxicological studies of CNFs are important, **holistic evaluation** of emerging CNF-based nanoproducts using **systems analysis** is critical for assessing their *environmental sustainability and impact on human health.* The need for combining a holistic systems view with reductionist research has been identified to reduce the chance of unpleasant surprises due to emerging technologies such as nanotechnology [12-14]. Despite the availability of methods for such holistic analysis, very few such studies have been carried out on nanoproducts [15]. Among systems analysis methods, life cycle assessment (LCA) has been popular and practical for understanding the environmental impact associated with a product or a process over its entire life cycle and offers great potential for understanding the life cycle environmental impacts of emerging nanomaterials and nanoproducts. The following sub-section provides a detailed discussion about the LCA methodology along with the challenges and benefits of LCA of emerging nanoproducts.

1.1 Life cycle assessment (LCA)

1.1.1 Methodology

LCA is an approach to quantify the environmental impact of a product or a process over its entire life cycle. The methodology is standardized via. ISO 14040 and 14044. The LCA methodology involves four steps [16, 17]. Figure 1 presents a schematic description of the different phases in a typical LCA. The four phases are briefly discussed below.

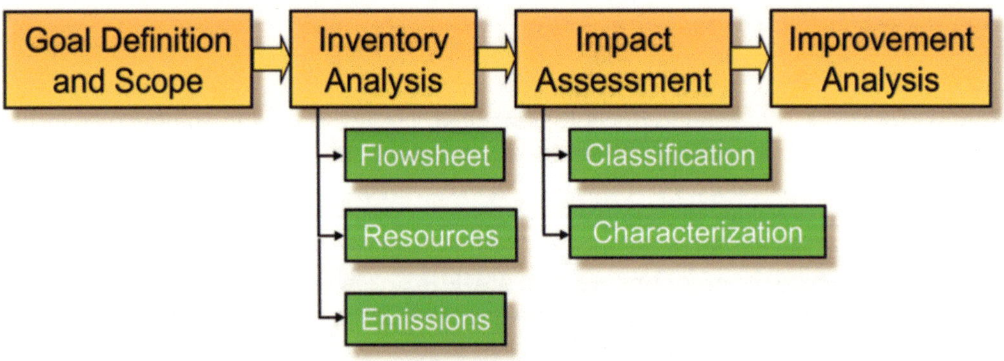

Fig. 1. Phases of a Typical LCA.

Goal and Scope Definition- The goal and scope definition involves defining the boundaries of the LCA study and specifies which processes are included in the study. In general, the scope of an LCA study consists of resource extraction, material and energy inputs synthesis, product manufacture, waste treatment, packaging, transportation, product use, and end-of-life phase of a product. Such an analysis is termed as "Cradle-to-grave". The purpose of the LCA study is clearly stated in this phase. This can be either comparative assessment of nanoproducts vs. alternatives, evaluation of a nanoproduct with different synthesis routes, and/or a single nanoproduct for a new application. This aids in defining the appropriate functional unit for the study resulting in a fair comparison. As an example, consider the life cycle environmental comparison of plastic vs. paper grocery bags. One paper bag may physically hold the same amount of groceries as two plastic bags. Thus the appropriate functional unit in this case should be the comparison of one paper bag vs. two plastic bags. Once the appropriate functional unit is defined, the process boundary should be specified clearly to define the scope of the LCA study. Proper care and vigilance are crucial during the process boundary selection to yield meaningful and interpretable results. In addition, doing multiple studies with different boundaries to determine the effect of the boundary limits can be helpful in certain situations. The life cycle practitioner should always state the purpose of the study and this usually helps in defining the boundary.

Inventory Analysis- This is the most resource-demanding phase of the LCA. It consists of identifying and collecting input and output data for each step in the life cycle of a product. There are several different ways of obtaining data for a typical LCA. Data sources typically include public or commercial LCA databases, published experimental data and LCA studies, propriety industrial data etc. National Renewable Energy Laboratory (NREL) Life Cycle Inventory database is an example of a public life cycle database. Several commercial LCA packages are also available. Some of the examples are SimaPro, Ecoinvent, GABI, and Umberto. Input data typically includes material and energy resources, labor, capital, and equipment. Output data consists of products, byproducts, and emissions of substances to air, water, and soil.

Impact Assessment- This phase involves classifying and characterizing emissions into various environmental impact categories to provide several indicators for analyzing the potential contributions of the resource extractions and wastes/emissions in an inventory to a number of potential impacts. The individual results for the different impact categories can be further normalized and weighed based on valuation techniques. Several Life Cycle Impact Assessment (LCIA) approaches exist and these have been described and critically assessed in detail in the literature [18]. The choice of which impact categories and hence impact assessment approach to use is a subjective one and at the discretion of the LCA practitioner.

The emissions from the complete life cycle of a product are available at the aggregate level. These are classified and characterized based on their impact into various impact categories often described as Midpoint Indicators. Common impact categories include Global Warming Potential, Acidification Potential, Eutrophication Potential, Human Toxicity Potential, Ozone Layer Depletion Potential and Photochemical Smog Formation Potential. Different chemicals may have a variety of different environmental impacts. For example, carbon dioxide results in global warming impact, Hydro Chloro-Fluoro Carbons have an adverse effect on ozone layer. Classification consists of combining different chemical impacts into a common metric. For example, 1 kg of methane has a global warming impact equivalent to 25 kg of carbon dioxide, similarly 1 kg of nitrous oxide has a global warming impact equivalent to 298 kg of carbon dioxide. Thus, the global warming potential characterization factors relative to carbon dioxide are 25 and 298 for methane and nitrous oxide respectively. These impacts are then expressed in terms of total equivalents of carbon dioxide to quantify the Global Warming Potential. Similar approach is followed for other impact categories. Characterization factors are available in the literature for a large number of chemicals and their impacts in various categories.

Interpretation- This is the phase in LCA where the results of the impact assessment are put together in a form that can be used directly to draw conclusions and make recommendations for improving the environmental performance of a product or the

process. Sensitivity and or uncertainty analysis is particularly useful before this phase to quantify the impact of important parameters on the LCA results.
Although a data intensive approach, LCA has found wide use in the industry and many companies have found it appealing to employ LCA techniques as a way of moving beyond environmental compliance to create win-win business opportunities by improving the quality of their products while minimizing the environmental impact of their industrial operations.

1.1.2 LCA: Challenges and benefits for emerging nanomaterial and nanoproduct

Despite the fact that the LCA methodology is standardized via ISO 14040 and 14044 standards and pertinent software and inventory databases are available, the methodology faces several challenges. Some of these include getting high quality life cycle inventory data, combining data and emissions in disparate units and at multiple spatial and temporal scales, dealing with high dimensionality data, and dealing with processes having a range of emissions [19, 20]. These challenges make it even more difficult to apply LCA to emerging technologies like nanotechnology. The need for the LCA of potential nanoproducts has been identified and discussed by researchers and various agencies [13, 21-24]. However, LCA of nanotechnology has progressed at a slow pace due to several formidable challenges. These are discussed below.

- Existing life cycle inventory databases are limited in scope and are useful for evaluating conventional products and processes.
- There is little publicly available information about the inputs and outputs of nanomanufacturing processes since most of the data is proprietary or available for laboratory scale operations.
- There is very little quantifiable data available on the human health and ecosystem impacts of engineered nanomaterials.
- Forecasting nanotechnology life cycle processes and activities is difficult since the technology is in its infancy and evolving rapidly.

LCA with its powerful toolbox can help address many critical concerns like material, energy, and environmental impact intensity of products, which complemented with the toxicological information of nanomaterials can help in sustainable development of nanotechnology. Some of the expected benefits of nanotechnology LCA are listed below:

- Identify phases in a nanoproduct's life cycle that have the maximum environmental impact and thus areas for improvement.
- Quantify how much of the energy savings and environmental impact during the use phase of nanoproducts are offset during their production phase.
- Identification of end-of-life scenarios and treatment options specific to nanoproducts.
- Evaluation of economic and environmental trade-offs of nanoproducts vs. conventional ones.

The data, results, and insights from LCA may be useful for identifying and allaying public concerns about emerging nanomaterials. It can also be used for risk assessment and as a screening tool for evaluating different technologies and product options.

1.1.3 Life cycle inventory for emerging nanoproducts- overview

The ISO methodology as outlined in ISO series 14040 and 14044 in general is applicable for evaluating potential nanoproducts and nanoprocesses. However, as mentioned in Section

2.2, the major hindrance is the severe lack of life cycle data for nanomanufacturing. This subsection describes the general approach that can be followed for compiling the LCA of potential nanoproducts in the light of the challenges mentioned in the preceding sections. Compiling life cycle inventory for nanomaterials and nanoproducts can be an excruciating and resource intensive task. This is because nanoprocesses are evolving rapidly and most of the industry data is proprietary. As a starting step, LCA practitioners can compile the life cycle inventory for nanoproducts based on laboratory experience and data available in the open literature. Wherever possible, plant specific data should be used. In the absence of any information, parallels can be drawn with similar technologies to get data on material and energy consumption. For example, the Chemical Vapor Deposition (CVD) process used for the synthesis of Carbon based nanomaterials, such as Single-Walled Carbon Nanotubes and CNFs is similar in nature to the CVD process used for chip manufacturing in the semiconductor industry. Missing data may be reconciled ensuring the satisfaction of conservation laws of mass and energy. Unlike traditional products and processes, additional information about the possible quantity and mode of release of nanoparticles is crucial for estimating the extent of human and ecological impact of nanoproducts. Various assumptions related to the boundary and selected processes should also be clearly stated. Since nanoproducts have only recently started penetrating the consumer market, information regarding the end-of-life of nanoproducts is missing. Sensitivity analyses based on different scenarios can provide useful insight and extreme bounds for the end-of-life impact of nanoproducts. The next section illustrates the use of life cycle approach in the environmental evaluation of one kind of engineered nanoparticles, CNFs.

2. Carbon nanofibers

2.1 CNFs: Process description

The avoidance of formation of carbon deposits has been of utmost importance in refinery applications like steam reforming of hydrocarbons, hydrocracking, hydrotreating etc. primarily because these carbon filaments leads to deactivation of the catalyst surface, blockage in reactor systems and reduction in heat transfer [4, 25]. However, in recent years, Carbon Nanofibers have received a great deal of attention both from the research and industrial community because of their novel properties that can be realized in a variety of ways in various applications. CNFs consist of monomolecular carbon fibers with diameters ranging from tens of nanometer to 200 nanometer. CNFs are characterized by high tensile strength (12000 MPa) and high Young's modulus (600 GPa) that is approximately 10 times that of steel [5]. Besides mechanical strength, CNFs possess novel electrical properties like high electrical conductivity. Potential applications of CNFs include their use for lighter and stronger polymer nanocomposites and in electronic components. Several academic and industrial research groups have directed their efforts towards synthesizing and optimizing the growth of these CNFs. The catalytic synthesis of CNFs consists of formation of these fibers on metallic catalysts. The process consists of reduction of the catalyst sample in a hydrogen-inert gas stream at a somewhat lower temperature followed by heating the catalyst up to the reaction temperature subsequent to which the reaction mixture consisting of hydrocarbon, hydrogen and inert gas is introduced into the system [26-28]. Several different metallic and bimetallic catalysts can be used. Most commonly used catalysts are iron, cobalt, nickel and copper both in bulk and supported form. Lower hydrocarbons like methane, ethylene, acetylene or benzene and CO are the common source of carbon material.

Despite great advances in synthesis methods and efforts at understanding the mechanism of nucleation and growth of CNFs, continuous production of CNFs has proven to be challenging and several issues need to be addressed. In recent years, a new method capable of synthesizing CNFs on a continuous scale called as the Vapor Grown Carbon Nanofibers (VGCNF) has emerged [29-32]. VGCNFs are produced by catalytic pyrolysis of hydrocarbons in the presence of a transition metal acting as the catalyst. Trace amounts of Sulfur is added to the feed to promote the formation of CNFs. Ferrocene dissolved in a suitable solvent or Iron Pentacarbonyl, $Fe(CO)_5$ are commonly used catalyst sources. The organometallic catalyst decomposes forming clusters of Fe particles that act as nuclei for the formation and further growth of CNFs. The fibers coming out of the reactor along with the off-gases are trapped or separated using a series of cyclone separators or trap mechanisms located downstream of the reactor. The entire reactor assembly is enclosed within an electric furnace to supply sufficient heat to maintain the pyrolysis temperature of 1100-1200^{0}C. The characteristics of the fibers such as their thickness and length and hence their properties can be controlled by carefully manipulating the reactor operating conditions.

2.2 Life cycle energy analysis

This section describes the results of the life cycle energy analysis of carbon nanofiber production. CNF synthesis from hydrocarbons is considered for LCA and quantifying the life cycle energy requirements. The system boundary employed for the LCA of CNFs is depicted in figure 2. The inventory data for this case study is obtained directly from the published literature and complemented with mass and energy balances. Figure 3 presents a direct comparison of the life cycle energy requirements for CNF synthesis with traditional materials such as aluminum, steel, and polypropylene.

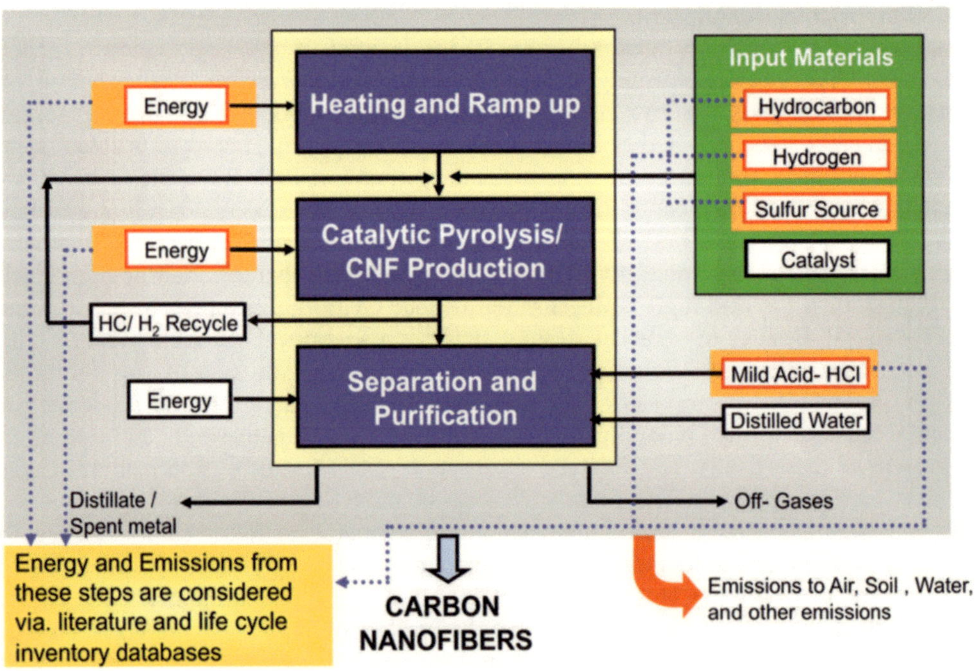

Fig. 2. LCA of Carbon Nanofibers: System Boundary.

Three different feed stocks i.e. methane, ethylene and benzene are considered to quantify the life cycle energy intensities of VGCNFs. Different scenarios for catalyst and carrier gas recycle rates are evaluated to address uncertainty and evaluate the bounds on the life cycle energy consumption and emissions. Figure 3 presents a comparison of the life cycle energy requirements for CNF synthesis with those of traditional materials namely Aluminum, Steel and Polypropylene. Figure 3 shows that the life cycle of CNFs is energy intensive compared to traditional materials with the life cycle energy requirements for CNF synthesis ranging from 2872 MJ/kg for benzene feedstock to around 10925 MJ/kg for methane. Process energy consumption constitutes a large fraction of the overall life cycle energy requirements ranging from 79 percent to 86 percent for benzene and methane feedstocks respectively. This is primarily the resources consumed, mainly coal, to produce the electrical energy required to maintain the high decomposition temperature (around 1,100 to 1,200°C). A further comparison with aluminum, steel and polypropylene reveals that the life cycle of CNFs is energy intensive with the life cycle energy requirements being 86, 30 and 119 MJ/kg for aluminum, steel and polypropylene respectively.

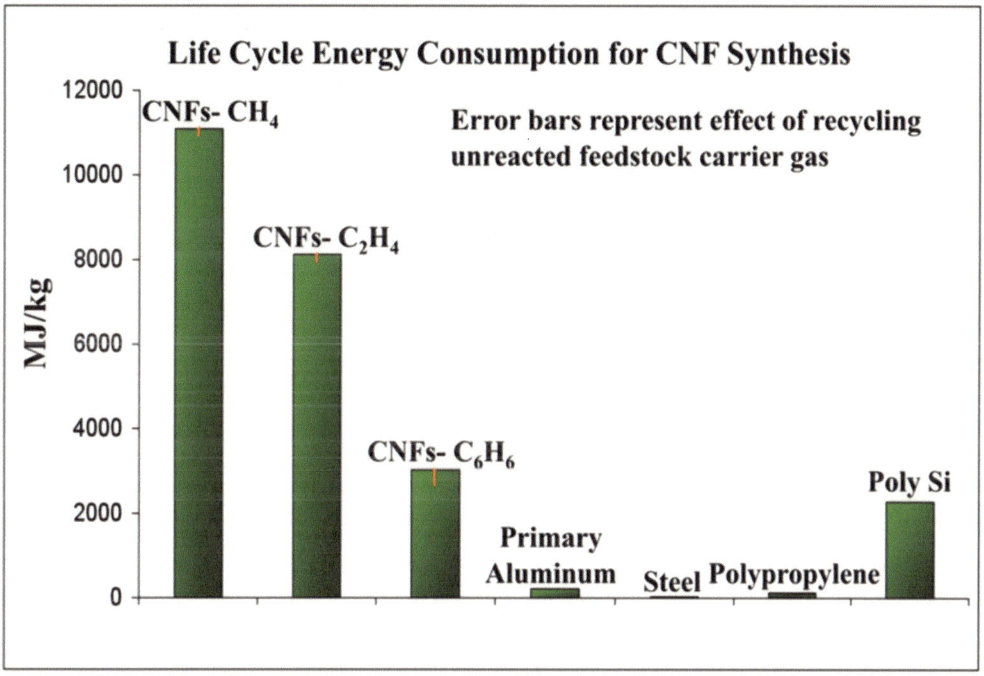

Fig. 3. Life Cycle Energy Analysis of Carbon Nanofibers.

The catalyst life cycle is not included due to the absence of life cycle inventory data for the catalyst thus resulting in lower bounds on the CNF life cycle energy requirements. Most industrial synthesis is carried out in the presence of a carrier gas that is generally H_2, NH_3 or some other inert gas. A sensitivity analysis is further carried out to study the influence of carrier gas recycle rates on the life cycle energy requirements. The results are displayed in Figure 3. Considering the methane feedstock, it is observed that even with complete recycling of the unreacted hydrocarbons and a 90% recycling of the hydrogen stream, the total energy consumption decreases from 10,925 MJ/kg to 10,778 MJ/kg. This corresponds to a 1.3% decrease in the overall life cycle energy requirements. In practice, such a high recycling rate is unlikely, and some of the unreacted materials may be burned for their fuel value. A similar trend is observed for ethylene feedstock, where complete recycling of the unreacted feed and 90% recycling of the hydrogen streams corresponds to a 2% reduction in the overall energy requirements. Thus, the process energy requirement still outweighs the energy savings due to material recycling. The energy required for separating unreacted hydrocarbons and hydrogen from the off-gases leaving the reactor is not accounted for here. Therefore, the numbers presented in figure 3 represent an upper bound on the energy savings from gas separation and system recycling.

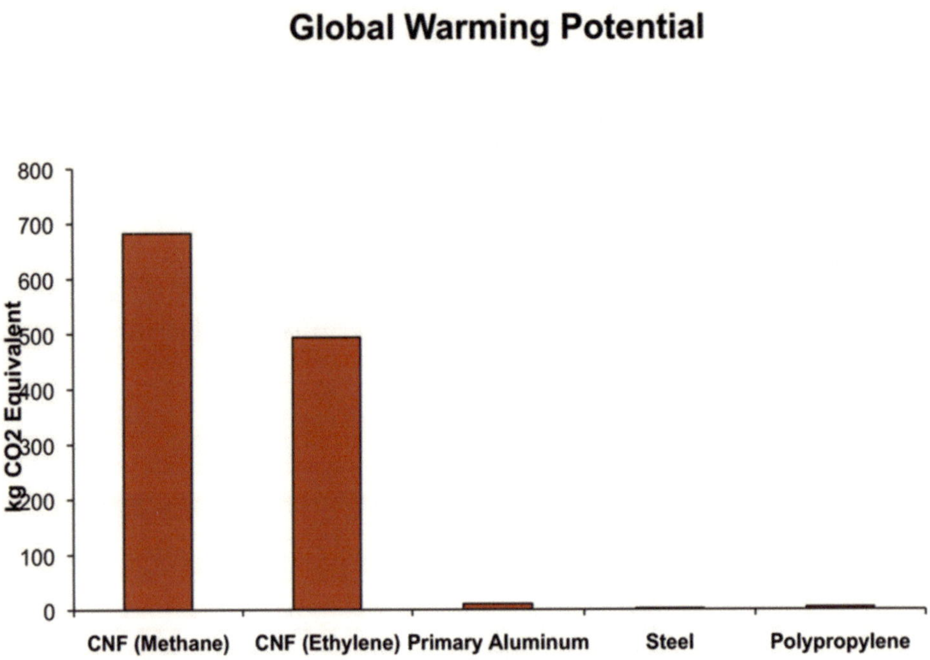

Fig. 4. Life Cycle Greenhouse Gas Emissions for CNF Synthesis.

On a per mass basis, the life cycle energy requirements for producing CNFs is 13 to 50 times what is required for producing primary aluminum. Several reasons can be attributed to this enormous difference. Firstly, processes required for manufacturing traditional materials are well-studied and have been optimized with respect to material and energy consumption over the last century or so. On the other hand, although researchers have been synthesizing CNFs on a laboratory scale, industrial synthesis of these nanoparticles remains a challenge. Secondly, industrial yields of these fibers are low even for continuous operations, i.e. in the range of 10-30 % by weight of the feedstock. This again entails research efforts for process improvement to increase the yield of these nanoparticles to make them competitive with alternatives for a given application. Unless, these materials are used in a widespread way especially in large volume applications like polymer nanocomposites, it will be challenging to exploit the economies of scale and hence make these materials cost competitive with other alternatives. Researchers are also exploring the possibility of other low temperature processes like the Plasma enhanced CVD for the synthesis of CNFs [33]. These processes are expected to be operating at room temperatures thus eliminating the need for process heating. However, converting these laboratory scale processes to full-fledged continuous industrial operation will be challenging. Needless to mention, LCA of these new processes should be an integral component of research efforts in order to ensure that savings in energy and material resources are not offset by increased consumption in other phases of the life cycle.

2.3 Environmental LCA of CNFs

Two base cases are evaluated for CNF synthesis, one with methane and the other with ethylene as the feedstock. Both cases are considered to have hydrogen as the carrier gas in accordance with the current industrial schemes. Figure 4 shows a higher Global Warming Potential (GWP) for both methane and ethylene based CNFs when compared with aluminum, steel, and polypropylene on an equal mass basis. A closer look at figure 4 reveals that the GWP potential of 1 kg methane based CNF is equivalent to about 65 kg of primary aluminum whereas 1 kg of ethylene based CNF has a GWP equivalent of about 47 kg of steel. Similar trends are observed for other environmental impact categories and are discussed elsewhere [34]. The impact numbers for CNFs represent only lower bounds whereas those for traditional materials are expected to be more precise since detailed and relatively more complete inventory data is available in the literature for traditional materials. Release and impact of CNFs on humans and ecosystem species during manufacturing are not accounted in the analysis. Although the results presented in figure 4 are obtained based on conservative set of assumptions, they still represent important step towards the need and development of life cycle inventory modules for nanomaterials and nanoproducts.

3. Carbon nanofiber polymer composites

The comparisons of CNFs in section 2 with traditional materials are presented on an equal mass basis to get insight into the life cycle energy intensities and environmental burden. However, products based on CNFs might be greener than alternatives for a given application. Quantity will be the deciding factor and hence specific nanoproducts and applications need to be evaluated. One of the potential applications is the use of CNFs as reinforcement for polymer composites resulting in high strength polymer nanocomposites.

Traditionally conventional fiber composites consisting of glass or micron sized carbon fibers have been most commonly used for composite applications. Although traditional fiber reinforced plastics have good in-plane mechanical properties that are governed by conventional fibers, they have poor properties in the transverse and thickness directions characterized mainly by the properties of the polymer matrix. Failure of the polymer matrix between the fibers can take place under impact or shear. CNFs can provide benefits here by directly reinforcing the polymer matrix between the long fibers thereby enhancing the strength in the transverse and thickness directions. The result is a high strength composite that combines the advantages of conventional long fibers and CNFs. Besides enhancing the mechanical strength, CNFs can also impart desirable level of electrical conductivity to the polymer composites. Electrically conductive polypropylene composites (electrical resistivity 10,000 ohm-cm) have been prepared with CNF loading levels of 3 volume percent in the polymer matrix [35]. Electrical resistivity values of as low as 100 ohm-cm have been reported for CNF reinforced epoxy composites at CNF loading content of 10 weight percent [36].

This section presents and discusses the life cycle energetic and Greenhouse Gas (GHG) impacts associated with the production and use of carbon nanofiber (CNF) PNCs. The use of PNCs in automotive body panels as a substitute for traditional materials is evaluated and compared with traditional materials. PNCs have enhanced mechanical properties, high strength-to-weight ratios, and are capable of offering specific functionalities such as desired level of electrical conductivity. These combinations of properties are making PNCs as one of the fastest growing plastic segment and an attractive alternative to conventional materials like steel and aluminum.

A typical life cycle of a PNC product is shown in figure 5. Several alternatives exist at each step in the complete life cycle. The selection of the polymeric resin depends on the application and hence the desired properties. Polypropylene (PP) and unsaturated polyester resin (UPR) are considered as the thermoplastic and thermosetting resins, respectively, as these have been widely studied in several nanocomposite experimental studies with respect to their mechanical and electrical properties. Specifically, both simple CNF and carbon nanofiber-glass fiber (CNF-GF) hybrid PNCs are evaluated and compared with steel for equal stiffness design. Life cycle inventory is developed based on published literature and best available engineering information.

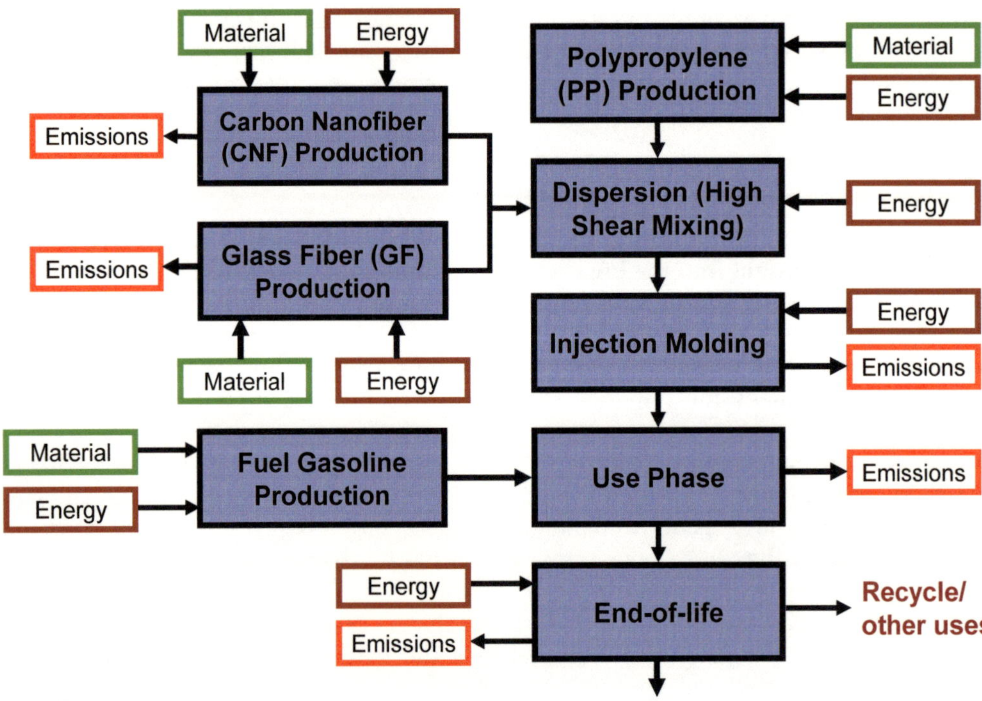

Fig. 5. Typical Life Cycle of a Polymer Nanocomposite.

The analysis is performed at two levels. In the first level of analysis, the functional unit is the cradle to gate life cycle comparison of a automotive body panel with CNF and CNF-GF reinforced polymer composites for equal stiffness design. The mass and the corresponding thickness of the PNC material are estimated using Ashby's approach for equal stiffness design [37, 38]. Use and end phase are not modeled for this analysis. In the second level of analysis, the life cycle energetic impact associated with the use of CNF and CNF-GF reinforced polymer composites vs steel in the body panels of a midsize car weighing 3300 lbs (1497 kilogram) are estimated with the functional unit being 150,000 vehicle miles (277,800 kilometers) traveled. End of life phase of the automobile is not included in this analysis.

This study is a cradle to gate study as it does not include the end of life issues specific to CNF reinforced polymer composite materials primarily because of the lack of quantifiable information about the recovery of CNF and GF from thermoplastic and thermoset polymer composite materials. Comparison of a PNC auto body panel for equal stiffness design evaluates the life cycle energetic impact first without including the use phase. The case study involving the use of CNF and CNF-GF reinforced PNCs vs. steel in the body panels of a midsize car goes a step further by including the vehicle use phase but still does not consider the end of life phase. Including the end of life considerations may not change the overall conclusions of this study.

The basis for comparison is the equal stiffness design of the components. This is justified on the basis that for structural and automotive applications, the component is assumed to have sufficient strength once it meets the stiffness criteria. Although this criteria may be true for automotive applications, but other considerations such as the impact properties may also influence the final material selection. Impact energy, an indicator of energy absorbing capacity is shown to decrease with the increasing content of CNFs in long fiber thermoplastic materials [35].

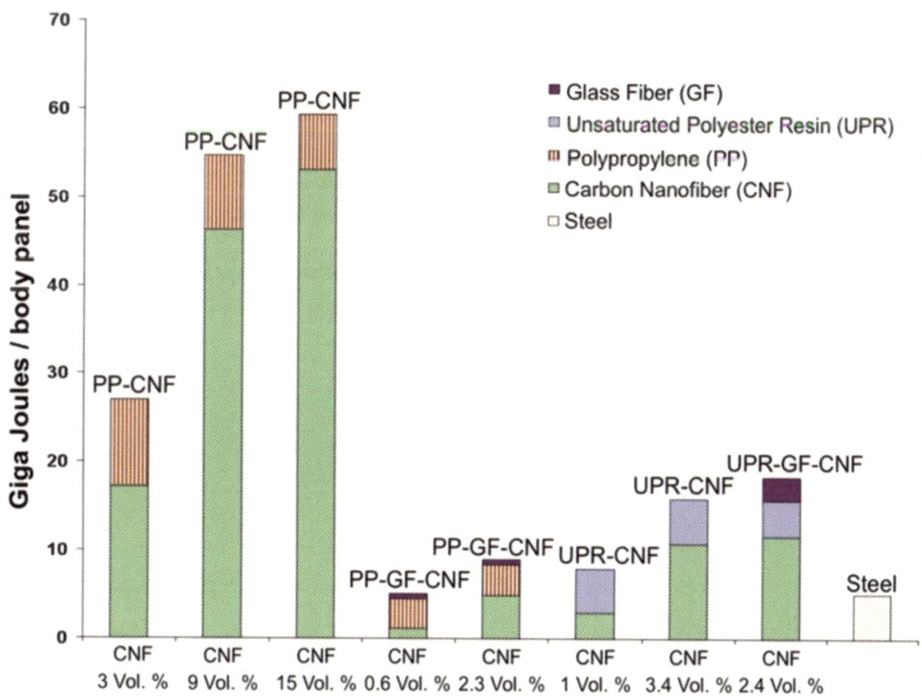

Fig. 6. Life cycle energy comparison of polymer nanocomposites with steel for equal stiffness design.

Primary weight savings for the automotive body panels are calculated directly using Ashby's approach. Secondary weight savings of 0.5 kg/kg of primary weight savings are considered resulting from downsizing of the chassis and auxiliaries. Based on the different values previously used in comparing aluminum-intensive vehicles with steel, a sensitivity analysis is performed by considering a range of 0.5-1 kg of secondary weight savings per kg of primary weight reduction [39-41]. Improvement in fuel economy relative to vehicle with steel body panels is calculated using the sedan equivalent estimation proposed and used previously [39, 42, 43].

$$FE_2 = FE_1\left(\frac{m_1}{m_2}\right)^{0.72}$$

where FE_i is the fuel economy of vehicle i and m_i is the mass of vehicle i.
The results of a cradle-to-gate (excludes use phase) life cycle comparison of PNCs vs. steel for equal stiffness of the components are presented in Figure 3. Figure 3 reveals that on a cradle-to-gate basis, CNF-based polymer composite auto body panels are 1.6-12 times more energy intensive when compared with steel. It is concluded that the product use phase might govern whether the high upstream manufacturing energy can be offset during the use phase to realize any life cycle energy savings.

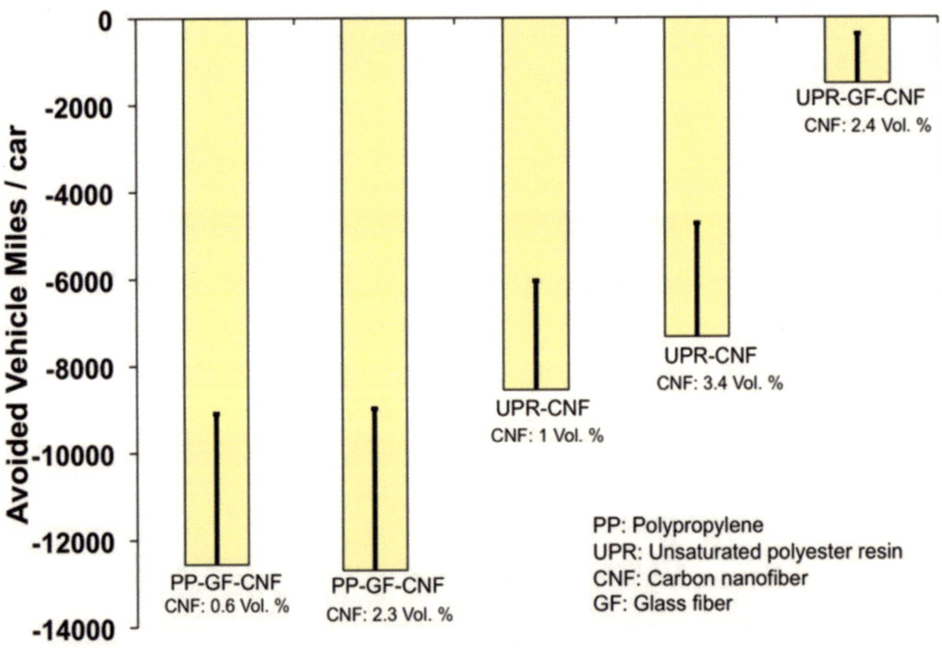

Fig. 7. Percentage savings in fuel gasoline consumption relative to steel only for use phase in auto body panels.

The use of CNF reinforced nanocomposites in body panels of light-duty vehicles is further evaluated. The life cycle of a midsize automobile with CNF reinforced nanocomposite body panels is evaluated and compared relative to conventional steel panels. The results are shown in figure 7. Overall for the range of CNF compositions and scenarios investigated, lifetime savings in fossil energy consumption are observed for PNC-based body panels relative to steel. Other factors such as cost, toxicity impact of CNF, and end-of-life issues specific to CNFs need to be considered to evaluate the final economic and environmental performance of CNF reinforced PNC materials.

The distribution of the life cycle energy consumption along different phases in the life cycle for CNF reinforced polymer composite automotive body panel is depicted in figure 8. Figure 8 does not include the use phase of vehicle. It is observed that majority of the life cycle energy impact results from the life cycle of carbon nanofiber production especially for polymer composites with higher loading ratios of CNFs. Among thermoset based PNCs, CNF constitute 38-68 percent of the cradle to gate life cycle energy consumption. The next highest impact results from the production of the polymeric resins followed by glass fibers.

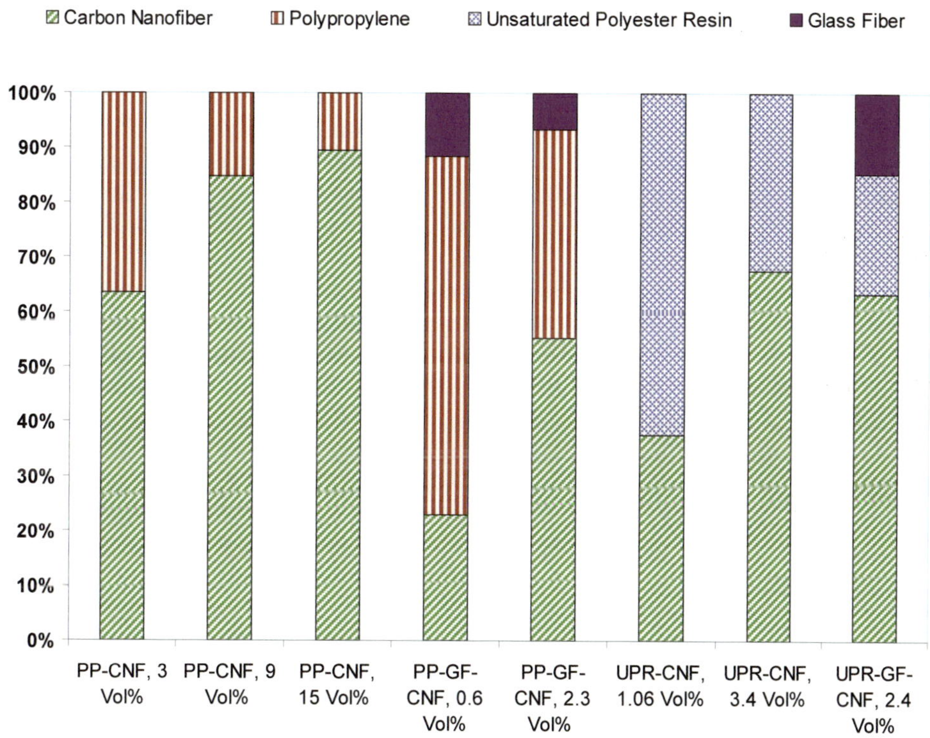

Fig. 8. Distribution of life cycle energy consumption for the manufacture of CNF based polymer composite automotive body panels. Vehicle use phase is not included.

4. Toxicological impact of CNFs

The previous sections highlighted the life cycle energetic impacts associated with CNFs and their use in automotive body panels. It also highlighted how the unique electrical, mechanical, and thermal properties [44], makes them extremely appealing to new technologies such as electronics, energy applications, and automotive applications [45]. However, the same properties that make CNFs attractive also raise concerns about their toxicity impacts. Because of the distinctive size and shape of CNFs, human toxicological reactions may vary as a result of the different modes of exposure, specifically inhalation, dermal, and consumption. The behaviors of nanomaterials are not the same as their bulk material behavior; therefore regulated material safety data sheets (MSDS) should not be regarded as a safety precaution when handling nanomaterials. This is especially important for the people who work in close proximity to these materials since they have a higher risk of inhalation, which could lead to significant respiratory effects [46].

There are very few toxicological studies on carbon nanotubes, and even less so for CNFs. There is still a need for better and more sophisticated studies that understand the health concerns of CNFs before they can be utilized for large volume applications. CNFs are known for their high aspect ratio, which represents a long and narrow shape [47, 48]. This unique shape and different aggregation scenarios, varied between bundle sizes and rope formations, can render different toxicological results, a problem for consistent toxicological understanding [48]. One research study compares carbon nanotubes to carbon fibers and carbon black particles, focusing on the ratio between nanoparticle size and surface area to the corresponding toxicity in cells. The toxicity results show that carbon black particles, a commercial example would be coal mine dust, and carbon fibers, which have a very similar resemblance to asbestos, were more toxic than the carbon nanotubes [47]. The study could only conclude that the toxicity and morphology reaction among the lung cells were not different depending on the carbon-based size, but that there was an obvious sensitivity reaction, where carbon black particles and carbon fibers had more cell response than carbon nanotubes [47].

Another study looked at both single-wall and multi-wall carbon nanotubes and their consequential dermal and pulmonary exposure routes [48]. In general, the aforementioned high aspect ratio is a large concern for the toxicity of nanomaterials, especially because as the particles and fibers get smaller, they have increased potential to cause harm [47, 48]. The study continues to divulge the issues of carbon particles and fibers, stating that fibers are much more pathogenic through inhalation and respiration modes of exposure than others [48]. Asbestos is the most widely researched fiber material and is used as a benchmark for toxicological studies on CNFs. Three characteristics were established in the study pertaining to CNFs and their toxicity potential in respiratory situations and have been supported by other prominent studies.

The size of fibers is important for determining how they will react and transport within the lung cells. Long fibers (>15μm) can have severe inflammatory reactions if they are narrow enough (<5μm) [48]. *Metal and chemical compositions* of individual fibers have their own toxicological reactivity within the cell [47-50]. Iron (Fe), cobalt (Co), and nickel (Ni) are the most widely used metal catalysts in the preparation of carbon nanotubes, which are built upon CNFs [50]. There is not conclusive evidence that states which metal catalyst induces the most adverse effects, but there is a difference between pure carbon nanotubes and carbon nanotubes with metal content [49, 50].

There is fear associated with the toxicology of CNFs because they have a strong resemblance to asbestos fibers and do not possess the quality to biodegrade [51]. Biopersistance is important evidence that CNFs can lead to adverse health effects, however it is difficult to quantify the behavior of these fibers as their size and chemical composition vary significantly [50]. Unfortunately, it is much easier to study the immediate effects of nanomaterials than the long-term or natural exposure effects.

Future research needs to take into account various nanomaterial point of views to gain a better, well rounded perspective of how nanomaterials react naturally. The medical world, including pharmaceuticals and medical devices comprised of nanomaterials, is one field that needs extra precaution [45, 52]. The life cycle of nanomaterials is also necessary for understanding how nanomaterial products and processes will influence the environment in end of life scenarios. The long-term effects on the environment are much harder to study and understand because of the little known knowledge of nanomaterial behavior and the difficulty of mimicking the natural environment in a laboratory. There is potential for environmental toxicity caused by nanomaterials, another reason why many are hesitant of the wide scale use of nanomaterials.

5. Summary

Engineered nanomaterials such as CNFs with their unique material properties offer the potential for reducing the material and energy intensity of engineered products and processes. While toxicological studies of engineered nanomaterials is essential, holistic understanding of nanomaterials and resulting nanoproducts via LCA is critical to evaluate the benefits of emerging nanoproducts before declaring them as superior to conventional products. LCA can help in providing metrics for environmental evaluation of potential nanoproducts vs. alternatives. LCA based on biophysical quantification of resource use can offer preliminary insight into the broader impact of emerging nanotechnologies. The LCA of CNFs discussed in this chapter highlighted the hugely energy intensive nature of the production processes of CNFs. However, it is difficult to reach any conclusions about the environmental impact of nanoproducts based on such engineered nanomaterials without taking into account their complete life cycle. A case study involving the use of CNF reinforced PNCs in the body panels of automobiles highlights that the use of PNCs with lower CNF loading ratios has the potential for net life cycle energy savings relative to steel owing to improved fuel economy benefits. Thus, evaluation of potential nanoproducts on functional unit basis is essential. Such comparisons can help identify if the increased upstream production energy requirements are offset by savings during the use phase of the product or not. Besides, for certain products such as automotive or aerospace components use phase might dominate the energy use across the complete life cycle and hence may offer net life cycle energy savings. In other cases, the manufacturing phase might play a more significant role and hence offset any potential energy or environmental benefits achieved in other life cycle phases. Besides, end-of-life challenges such as recovery and recycle specific to engineered nanomaterials might be encountered for certain nanoproducts. In the absence of detailed environmental fate and transport, scenario analysis might help in screening different alternatives. Besides, scenario analysis coupled with LCA can also provide novel insights about the long-term impact of nanomaterial use across the economy. For example, the quantity of individual nanomaterials in specific products might be small but overall can have huge material and energy impact if used in large volume applications. In such cases, the resource efficiency of the upstream nanomaterial manufacturing stage might alone govern the sustainability of potential nanotechnologies.

6. References

[1] Project on Emerging Nanotechnologies, W.W.C. Nanotechnology consumer products inventory [cited 2011 January 30]; Available from: http://www.nanotechproject.org/inventories/consumer.

[2] Schmid, K. and M. Riediker, Use of nanoparticles in Swiss industry: a targeted survey. *Environmental science & technology*, 2008. 42(7): p. 2253-2260.
[3] Gogotsi, Y., *Carbon nanomaterials*. 2006: CRC.
[4] De Jong, K.P. and J.W. Geus, Carbon nanofibers: catalytic synthesis and applications. *Catalysis Reviews*, 2000. 42(4): p. 481-510.
[5] Mordkovich, V., Carbon nanofibers: a new ultrahigh-strength material for chemical technology. *Theoretical Foundations of Chemical Engineering*, 2003. 37(5): p. 429-438.
[6] Dreher, K.L., Health and environmental impact of nanotechnology: toxicological assessment of manufactured nanoparticles. *Toxicol Sci*, 2004. 77(1): p. 3-5.
[7] Maynard, A.D., et al., Safe handling of nanotechnology. *Nature*, 2006. 444(7117): p. 267-9.
[8] Thomas, K. and P. Sayre, Research strategies for safety evaluation of nanomaterials, Part I: evaluating the human health implications of exposure to nanoscale materials. *Toxicol Sci*, 2005. 87(2): p. 316-21.
[9] Maynard, A., Nanotechnology: the next big thing, or much ado about nothing? *Annals of Occupational Hygiene*, 2006. 51(1): p. 1.
[10] Powers, K.W., et al., Research strategies for safety evaluation of nanomaterials. Part VI. Characterization of nanoscale particles for toxicological evaluation. *Toxicol Sci*, 2006. 90(2): p. 296-303.
[11] Thomas, T., et al., Research strategies for safety evaluation of nanomaterials, part VII: evaluating consumer exposure to nanoscale materials. *Toxicol Sci*, 2006. 91(1): p. 14-9.
[12] Initiative, T.N.N. Environmental, health, and safety research needs for engineered nanoscale materials. 2006 [cited 2011 January 31]; Available from: http://www.nano.gov/NNI_EHS_research_needs.pdf.
[13] Society, T.R. Nanoscience and nanotechnologies: opportunities and uncertainties. 2004 [cited 2011 January 31]; Available from: http://www.nanotec.org.uk/finalReport.htm.
[14] W. Kl¨opffer, M.A.C., P. Frankl, R. Heijungs, A. K¨ohler, and S.I. Olsen. Nanotechnology and Life Cycle Assessment. A systems approach to Nanotechnology and the environment Synthesis of Results Obtained at a Workshop Washington, DC 2–3 October 2006. 2007 [cited 2011 January 31].
[15] Theis, T. The Life Cycle of Nanomanufacturing Technologies. . in *Interagency Environmental Nanotechnology Grantees Workshop*. 2009. Tampa, FL.
[16] Curran, M.A., *Environmental life-cycle assessment*. 1996, New York: McGraw-Hill. 1 v. (various pagings).
[17] Graedel, T.E., *Streamlined life-cycle assessment*. 1998, Upper Saddle River, NJ: Prentice Hall. x, 310 p.
[18] Bare, J. and T. Gloria, Critical analysis of the mathematical relationships and comprehensiveness of life cycle impact assessment approaches. *Environ. Sci. Technol*, 2006. 40(4): p. 1104-1113.
[19] Rebitzer, G., et al., Life cycle assessment Part 1: Framework, goal and scope definition, inventory analysis, and applications. *Environment International*, 2004. 30(5): p. 701-720.
[20] United States. Environmental Protection Agency. Office of Air Quality Planning and Standards., *Life-cycle impact assessment : a conceptual framework, key issues, and summary of existing methods*. 1995, Research Triangle Park, NC: U.S. Environmental

Protection Agency, Office of Air Quality Planning and Standards. 1 v. (various pagings).
[21] Bauer, C., et al., Towards a framework for life cycle thinking in the assessment of nanotechnology. *Journal of Cleaner Production*, 2008. 16(8-9): p. 910-926.
[22] Russell, R. and R. Cresanti, Environmental Health and Safety Research Needs for Engineered Nanoscale Materials. 2006.
[23] Theis, T.L., et al., A life cycle framework for the investigation of environmentally benign nanoparticles and products. *physica status solidi (RRL)ñRapid Research Letters*.
[24] Dowling, A., et al., Nanoscience and nanotechnologies: opportunities and uncertainties. *London: The Royal Society & The Royal Academy of Engineering Report*, 2004.
[25] Rodriguez, N., A review of catalytically grown carbon nanofibers. *Journal of Materials Research;(United States)*, 1993. 8(12).
[26] Kim, M., N. Rodriguez, and R. Baker, The interaction of hydrocarbons with copper---nickel and nickel in the formation of carbon filaments. *Journal of Catalysis*, 1991. 131(1): p. 60-73.
[27] Lim, S., et al., Selective synthesis of thin carbon nanofibers: I. Over nickel-iron alloys supported on carbon black. *Carbon*, 2004. 42(8-9): p. 1765-1781.
[28] Takehira, K., et al., Catalytic growth of carbon fibers from methane and ethylene on carbon-supported Ni catalysts. *Applied Catalysis A: General*, 2005. 283(1-2): p. 137-145.
[29] Chambers, A., N. Rodriguez, and R. Baker, Influence of copper on the structural characteristics of carbon nanofibers produced from the cobalt-catalyzed decomposition of ethylene. *Journal of materials research*, 1996. 11(02): p. 430-438.
[30] Fan, Y.Y., et al., Tailoring the diameters of vapor-grown carbon nanofibers. *Carbon*, 2000. 38(6): p. 921-927.
[31] Tibbetts, G.G., et al., Role of sulfur in the production of carbon fibers in the vapor phase. *Carbon*, 1994. 32(4): p. 569-576.
[32] Tibbetts, G.G., D.W. Gorkiewicz, and R.L. Alig, A new reactor for growing carbon fibers from liquid-and vapor-phase hydrocarbons. *Carbon*, 1993. 31(5): p. 809-814.
[33] Minea, T., et al., Room temperature synthesis of carbon nanofibers containing nitrogen by plasma-enhanced chemical vapor deposition. *Applied physics letters*, 2004. 85: p. 1244.
[34] Khanna, V., B.R. Bakshi, and L.J. Lee, Carbon nanofiber production: Life cycle energy consumption and environmental impact. *Journal of Industrial Ecology*, 2008. 12(3): p. 394-410.
[35] van Hattum, F., et al., Conductive long fibre reinforced thermoplastics by using carbon nanofibres. *Plastics, Rubber and Composites, 35*, 2006. 6(7): p. 247-252.
[36] Choi, Y., et al., Mechanical and physical properties of epoxy composites reinforced by vapor grown carbon nanofibers. *Carbon*, 2005. 43(10): p. 2199-2208.
[37] Ashby, M., *Materials selection in mechanical design*. 2005: Butterworth-Heinemann.
[38] Ashby, M. and D. Jones, *Engineering materials 1. An introduction to their properties and applications*. 1980: Pergamon Press, Oxford, England.
[39] Lloyd, S.M. and L.B. Lave, Life cycle economic and environmental implications of using nanocomposites in automobiles. *Environ Sci Technol*, 2003. 37(15): p. 3458--3466.
[40] Stodolsky, F., et al., Life-cycle energy savings potential from aluminum-intensive vehicles. 1995.

[41] Das, S., The life-cycle impacts of aluminum body-in-white automotive material. *JOM Journal of the Minerals, Metals and Materials Society*, 2000. 52(8): p. 41-44.
[42] MacLean, H. and L. Lave, Environmental implications of alternative-fueled automobiles: Air quality and greenhouse gas tradeoffs. *Environ. Sci. Technol*, 2000. 34(2): p. 225-231.
[43] Lave, L., Conflicting objectives in regulating the automobile. *Science*, 1981. 212(4497): p. 893.
[44] Lam, C.-w., et al., A Review of Carbon Nanotube Toxicity and Assessment of Potential Occupational and Environmental Health Risks. *Critical Reviews in Toxicology*, 2006. 36(3): p. 189-217.
[45] Illuminato, I., G. Miller, and L. Matthes, Nanotechnology, Climate, and Energy: Overheated Promises and Hot Air? 2010, Friends of the Earth International US edition. p. 88.
[46] Colvin, V.L., The potential environmental impact of engineered nanomaterials. *Nat Biotech*, 2003. 21(10): p. 1166-1170.
[47] Magrez, A., et al., Cellular Toxicity of Carbon-Based Nanomaterials. *Nano Letters*, 2006. 6(6): p. 1121-1125.
[48] Smart, S.K., et al., The biocompatibility of carbon nanotubes. *Carbon*, 2006. 44(6): p. 1034-1047.
[49] Poland, C.A., et al., Carbon nanotubes introduced into the abdominal cavity of mice show asbestos-like pathogenicity in a pilot study. *Nat Nano*, 2008. 3(7): p. 423-428.
[50] Pulskamp, K., S. DiabatÈ, and H.F. Krug, Carbon nanotubes show no sign of acute toxicity but induce intracellular reactive oxygen species in dependence on contaminants. *Toxicology Letters*, 2007. 168(1): p. 58-74.
[51] Kostarelos, K., The long and short of carbon nanotube toxicity. *Nat Biotech*, 2008. 26(7): p. 774-776.
[52] Lacerda, L., et al., Carbon nanotubes as nanomedicines: From toxicology to pharmacology. *Advanced Drug Delivery Reviews*, 2006. 58(14): p. 1460-1470.

Part 3

Functional Applications

14

Functional Applications of Electrospun Nanofibers

Jian Fang[1], Xungai Wang[1,2] and Tong Lin[1]
[1]*Centre for Material and Fibre Innovation, Deakin University*
[2]*School of Textile Science and Engineering, Wuhan Textile University*
[1]*Australia*
[2]*China*

1. Introduction

With the rapid development of nanoscience and nanotechnology over the last two decades, great progress has been made not only in preparation and characterization of nanomaterials, but also in their functional applications. As an important one-dimensional nanomaterial, nanofibers have extremely high specific surface area because of their small diameters, and nanofiber membranes are highly porous with excellent pore interconnectivity. These unique characteristics plus the functionalities from the polymers themselves impart nanofibers with many desirable properties for advanced applications.
Several methods have been developed to fabricate nanofibers, such as template (Ikegame et al., 2003), self-assembly (Hong et al., 2003), phase separation (Ma and Zhang, 1999), melt-blowing (Ellison et al., 2007) and electrospinning (Doshi and Reneker, 1995, Lin et al., 2004, Lin et al., 2005a, Fang et al., 2007, Xue et al., 2009, Fang et al., 2010). Electrospinning has been regarded as the most promising approach to produce continuous nanofibers on a large scale and the fiber diameter can be adjusted from nanometers to micrometers (Li and Xia, 2004b).

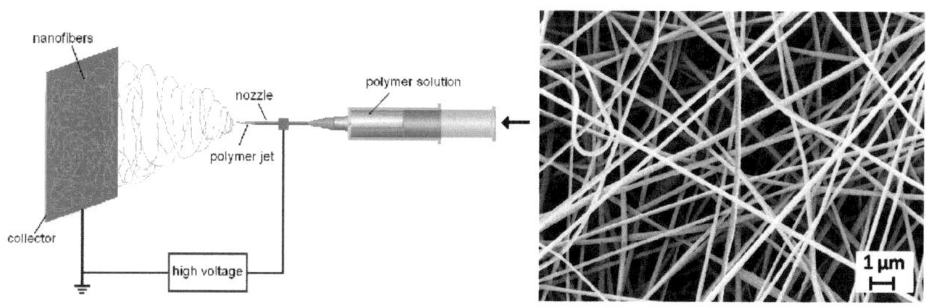

Fig. 1. Schematic for the single needle type electrospinning (Fang et al., 2008) and scanning electron microscopy (SEM) image of a nanofiber membrane.

Technically, electrospinning is a process that uses a strong electrical field to draw a polymer fluid into fine filaments. The basic electrospinning set-up, shown in Fig. 1, consists of a needle nozzle, a high voltage power supply, a container for spinning fluid and an electrode collector. When a viscous fluid is charged with a high voltage, the electrostatic force draws the fluid into a liquid jet. Because of the interaction between the jet and external electric field and charge repulsion inside the jet, the charged jet undergoes a bending or whipping instability to stretch it thinner. Solvent evaporation from the filaments results in solid fibers (Fig. 1). In most cases, the as-spun fibers deposit randomly on the electrode collector forming a non-woven nanofiber mat. Aligned nanofibers can also be produced using controlled fiber deposition techniques (Theron et al., 2001, Li et al., 2004). Until now, a large number of polymeric and inorganic materials have been electrospun into nanofibers. The nanofibers can have many different morphologies, such as porous-surface nanofibers (Madhugiri et al., 2004, McCann et al., 2005), core-sheath (Sun et al., 2003, Li and Xia, 2004a) and side-by-side structures (Lin et al., 2005b). Electrospun nanofibers have attracted considerable attention because of their unique properties, eases of fabrication and functionalization, and versatility in controlling the fiber diameter and morphology. The extremely fine electrospun nanofibers make them very useful in a wide range of advanced applications.

Although several reviews on nanofibers have been published (Li and Xia, 2004b, Greiner and wendorff, 2007, Thavasi et al., 2008, Cui et al., 2010), new applications are found continuously. In this chapter, we expand on an earlier review on nanofiber applications (Fang et al., 2008) and summarize the recent research advances in the application of electrospun nanofibers. The state-of-the-art applications in functional areas, such as biomedical, energy harvest and storage, and environmental protection are presented in detail. An outlook on the possible future directions is also provided.

2. Biomedical

2.1 Tissue engineering scaffolds

Tissue engineering is an emerging interdisciplinary and multidisciplinary research field which involves the use of living cells, manipulated through their extracellular environment or genetically to develop biological substitutes for implantation into the body and/or to foster remodeling of tissues in some active manners. A basic principle of the tissue engineering is illustrated in Fig. 2. The purpose of tissue engineering is to repair, replace, maintain, or enhance the function of a particular tissue or organ.

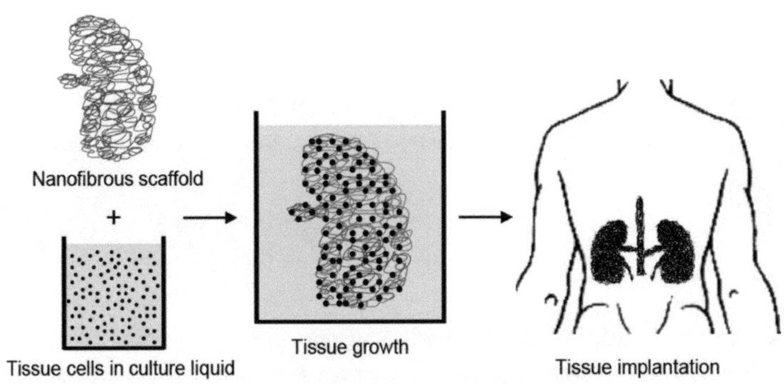

Fig. 2. Illustration of tissue engineering (Fang et al., 2008).

The core technologies intrinsic to this effort can be organized into three areas: cell technology, scaffold construct technology and technologies for *in vivo* integration. The scaffold construct technology focuses on designing, manufacturing and characterizing three-dimensional (3D) scaffolds for cell seeding and *in vitro* or *in vivo* culturing. For a functional scaffold, a few basic requirements have to be met (Ma, 2004). First, a scaffold should possess a high degree of porosity, with an appropriate pore size distribution. Second, a large surface area is needed. Third, biodegradability is often required, with the degradation rate matching the rate of neo-tissue formation. Fourth, the scaffold must possess the required structural integrity to prevent the pores of the scaffold from collapsing during neo-tissue formation, with the appropriate mechanical properties. Finally the scaffold should be non-toxic to cells and biocompatible, positively interacting with the cells to promote cell adhesion, proliferation, migration, and differentiated cell function. Among all biomedical materials under evaluation, electrospun nanofibrous scaffolds have exhibited great performance in cell attachment, proliferation and penetration, through both *in vitro* and *in vivo* trials.

2.1.1 Blood vessels

Blood vessels vary in sizes, mechanical and biochemical properties, cellar content and ultra-structural organization, depending on their location and specific function. It is required that the vascular grafts engineered should have desired characteristics. Blood vessel replacement, particularly a fine blood vessel (diameter<6 mm), has remained a great challenge.

A comparison of cell growth on an electrospun poly(L-lactide-co-ε-caprolactone) (P(LLA-CL)) nanofibers and a smooth P(LLA-CL) cast film revealed that the function of vascular endothelial cells (ECs) on the P(LLA-CL) cast film, rather than on the electrospun nanofiber mat, was enhanced (Xu et al., 2004). But because the electrospun nanofiber mats can give good support during the initial growth of vascular smooth muscle cells (Mo et al., 2004), smooth film combining with electrospun nanofiber mat could form a good 3D scaffold for blood vessel tissue engineering.

The effect of fiber diameter on endothelial cells culturing was examined using electrospun cellulose acetate (CA) fibers with three different diameter ranges, 0.01-0.2, 0.2-1 and 2-5 μm (Rubenstein et al., 2007). It was interesting to note that the endothelial cells showed a growth preference toward larger fibers. Similar results were also found on poly(ε-caprolactone) (PCL) nanofibers with five different diameters (Balguid et al., 2009), cell penetration increased with the increased fiber diameter and unobstructed cell delivery was observed only on the largest fibers (12.1 μm in diameter).

To mimic the morphological and mechanical characteristics of a native blood vessel scaffold, bilayered electrospun nanofiber architectures were fabricated. A scaffold consisting of a stiff and oriented poly(lactic acid) (PLA) outer nanofiber layer and a pliable and randomly oriented PCL inner nanofiber layer was electrospun. Such a hierarchical scaffold was reported to be able to support the attachment, spread and growth of mouse fibroblasts and human myofibroblasts (Vaz et al., 2005). A composite scaffold incorporating a highly porous poly(ester-urethane)urea (PEUU) inner layer and an external electrospun nanofiber layer was prepared for small diameter vascular grafts (Fig. 3a) (Soletti et al., 2010). The mechanical properties of the bilayered scaffold were comparable with native vessels, and the combination of the two layers enabled better cell integration and growth. Because of the porosity difference in the two layers, high cellular density was found in the inner layer (Fig. 3b).

Surface modification on electrospun nanofibers with natural proteins, such as collagen or gelatin, was found to be an effective way to promote ECs spreading and proliferation. Examples can be found for collagen-grafted PCL (Ma et al., 2005a) and gelatin-grafted polyethylene terephthalate (PET) nanofibers (Ma et al., 2005b). The modified-nanofiber mats

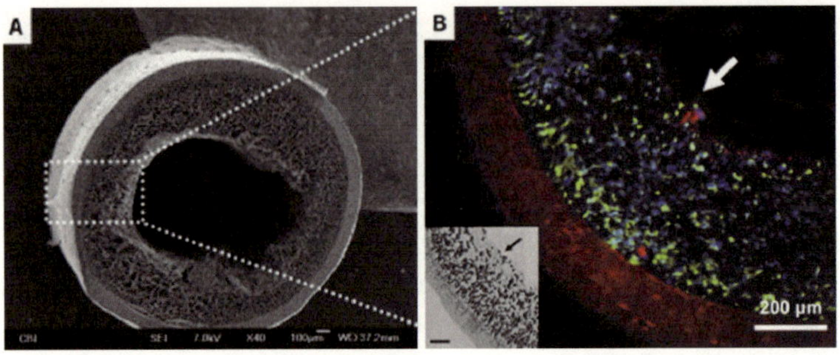

Fig. 3. The cross-sectional images of a double-layered small diameter PU vascular graft before (a) and after (b) cell culturing (Blue, nuclei; green, F-action; red, scaffold) (Soletti et al., 2010).

showed apparent enhancement in the spreading and proliferation of ECs compared with the non-grafted ones. Since a fibroblast growth factor (FGF-2) modulates cell growth, differentiation, migration and survival, the ability to bind FGF-2 to electrospun matrix was improved effectively by coating a bioactive recombinant fragment of perlecan on the surface of electrospun nanofibers (Casper et al., 2007).

To match the mechanical characteristic of natural blood vessel, which has a low-strain mechanical response to blood flow and prevents pulsatile energy from being dissipated as heat, elastin/polymer blend nanofibers were used (Sell et al., 2006). The preliminary cell culture studies showed that cells migrated through the full thickness of the elastin-containing grafts, but failed to migrate into the pure polymer nanofiber control scaffold. To mimic the ratio of collagen and elastin in native blood vessel, electrospun nanofibers of collagen/elastin/synthetic polymer (e.g. poly(lactic-co-glycolic acid) (PLGA), poly-L-lactic acid (PLLA), PCL) (45:15:40 w/w/w) blend were assessed (Lee et al., 2007). The as-spun nanofiber mat showed no cytotoxicity and was dimensionally stable, and its mechanical properties were similar to the native blood vessels.

In vivo assessments of cylindrical electrospun nanofiber constructs from collagen, gelatin, and synthetic biodegradable polymers such as poly(glycolic acid) (PGA), PLA and PGA/PLA copolymer were conducted (Telemeco et al., 2005). When implanted into the interstitial space of rat vastus lateralis muscle, collagen construct was rapidly and densely infiltrated by interstitial and endothelial cells, and functional blood vessels were evident within 7 days, while gelatin and synthetic nanofiber constructs were not infiltrated to any great extent and induced fibrosis.

2.1.2 Bones

Research on engineering bone tissues using electrospun nanofibers began with the work on muscles (Li et al., 2002). Cell growth on PLGA nanofibers indicated that nanofiber structure positively promoted cell-matrix and cell-cell interaction. For bone tissue engineering, nanofibers from PCL have been extensively studied (Yoshimoto et al., 2003, Ekaputra et al., 2009). It was observed that Mesenchymal stem cells (MSCs) penetrated into the PCL matrix accompanied with abundant extracellular matrix after 1 week of seeding, and mineralization and type I collagen occurred at 4 weeks. An *in vivo* experiment was also conducted by implanting a MSCs cultured PCL construct (4 weeks) in the omenta of rats (Shin et al., 2004). After 4 weeks, the constructs maintained the size and shape of the original scaffolds and had a bone-like appearance.

In addition to the pure PCL nanofibers, nanofibers from gelatin/PCL blend (Zhang et al., 2005a) and PCL composite with calcium carbonate nanoparticles and hydroxyapatite (HAp) nanoparticles (Fujihara et al., 2005) for bone scaffolds have been investigated. The addition of 50% gelatin to PCL improved both the fiber mechanical strength and surface wettability, therefore enhancing the cell attachment and growth on the scaffold (Duan et al., 2007). Also, the cells were observed to migrate up to 114 μm inside the scaffold within one week of culture. In contrast, PCL nanofibers containing these inorganic nanoparticles were found with higher osteoblast proliferation and differentiation.

A combination of PCL nanofibers and microfibers was used to develop multilayer scaffolds (Tuzlakoglu et al., 2005, Pham et al., 2006). This multilayer fiber mat contained a thin nanofiber upper layer and an inner microfiber fibrous structure. As the top nanofiber layer assisted cell attachment and spread and the inner microfiber layer provided large pores for cell migration, both cell attachment and migration were improved. Fiber diameter has shown obvious effect on cell culture. Gelatin nanofibers were electrospun with different fiber diameters of 110 and 600 nm for osteoblastic MG63 cell growth. The cells were found to have a similar attachment feature at the first 3 days and with the extended culture time they preferred the coarser fibers. In longer term experiments, the cells differentiated to a greater extent on the scaffolds made from the smaller diameter fibers (Sisson et al., 2010).

Bioactive glass (BG) has also been widely studied for bone regeneration. PCL/BG composite nanofibers considerably improved the differentiation of osteoblast cells as compared to pure PCL nanofibers (Kim et al., 2008a), and a short BG nanofiber/PCL matrix composite scaffold has shown greater mechanical stability and bioactivity than BG power/PCL composite (Jo et al., 2009).

Mechanical stimulation has been applied during bone regeneration, and the results showed that proper stimulation can enhance fracture healing due to the anabolic osteogenic effects (Rath et al., 2008). Mechanical stresses induced by embedded magnetic nanoparticles were applied onto PCL scaffolds to stimulate growth, proliferation and differentiation of preosteoblast MC3T3 cells (Kannarkat et al., 2010). *In vivo* studies (Ko et al., 2008, Schneider et al., 2009b) have also suggested the great potential of electrospun nanofiber scaffold in skeletal tissue engineering because of adequate mechanical properties and osteocondition.

2.1.3 Muscles

Collagen nanofibers were first used to assess the feasibility of culturing smooth muscle cell (Matthews et al., 2002). The cell growth on the collagen nanofibers was promoted and the cells were well integrated into the nanofiber network after 7 days of seeding. Smooth muscle cells also adhered and proliferated well on other polymer nanofiber mats blended with collagen (Stankus et al., 2004, Chen et al., 2010b), the incorporation of collagen into nanofibers was observed to improve fiber elasticity and tensile strength, and increase the cell adhesion.

The fiber surface wettability influences cell attachment. It was reported that when the PS nanofibers were treated with Argon plasma, the fiber wettability was increased significantly. As a result, cell attachment was increased by two folds (Baker et al., 2006). The alignment of nanofibers can induce cell orientation and promote skeletal muscle cell morphologenesis and aligned myotube formation. The cells cultured on the aligned

nanofibers exhibited an alignment factor of 0.74, compared with 0.19 on the randomly orientated scaffold (Baker et al., 2006).
Degradation of smooth muscle cell cultured PGA, PLGA and P(LLA-CL) nanofibers were investigated (Dong et al., 2010b). It has been found that cell culture accelerated the nanofiber scaffold degradation, especially for PGA. P(LLA-CL) nanofibers facilitate long-term (1-3 months) cell growth because of the slowest degradation rate.

2.1.4 Skins

As the largest body tissue, skin functions to prevent outside intrusion and regulate water retention and heat loss. Nanofiber mats from many different types of polymers have been evaluated for their cytocompatibility to fibroblast and/or keratinocytes. In most cases, the electrospun nanofiber mats exhibited good capability of supporting cell attachment and proliferation. Cells seeded on the nanofiber structure tended to maintain phenotypic shape and were guided to grow according to the nanofiber orientation.
Collagen-PCL core-sheath nanofibers were prepared by both one-step coaxial electrospinning and two-step spinning and solution coating process (Zhang et al., 2005b, Zhao et al., 2007). The presence of collagen surface shell on PCL nanofibers was found to support cell proliferation and encourage cell migration inside the scaffold. The collagen coating formed directly from the coaxial-electrospinning had higher cell proliferation efficiency than that from the solution coating method.
Nanofibers of polymer blends were also studied, and showed good cell attachment and proliferation. It was indicated that a small fraction of low molecular weight polyethylene glycol (PEG) in PLLA nanofibers increased the hydrophilicity. As a result, biological reactivity of fibroblast cells was improved (Bhattarai et al., 2006). The fiber wettability could also be improved by co-electrospinning of two different nanofibers (PLGA and chitosan/polyvinyl alcohol (PVA)) into the same fibrous matrix (Duan et al., 2006). The presence of hydrophilic nanofibers (chitosan/PVA) increased the absorption of nutrient fluid during cell culture and thus promoted fibroblast attachment, proliferation, migration and infiltration in the fiber matrix.
The effect of nanofiber alignment on cell adhesion and proliferation was studied (Zhong et al., 2006). When a rabbit conjunctive fibroblast was seeded on an aligned collagen nanofiber mat, lower cell adhesion but higher cell proliferation was observed on the aligned nanofibers, compared with that on the randomly orientated nanofiber mat. It has also been observed that the infiltration of epidermal skin cells was enhanced when aligned PLLA nanofibers were used (Kurpinski et al., 2010).
Cell culture on a 3D nanofiber mat containing large pores, was also studied. A 3D silk fibroin nanofiber matrix was prepared by directly depositing the newly-electrospun silk fibroin nanofibers in a methanol solution (Ki et al., 2007b). This 3D nanofiber matrix supported the cell proliferation inside the fiber matrix quite well. A 3D nanofiber mat was also prepared by adding a chemical blowing agent (BA) into nanofibers via electrospinning, and a post-electrospinning heat-treatment, leading to large pores within the nanofiber mat (Kim and Kim, 2007). These large pores could provide good access to the cells to penetrate inside the nanofiber mat. The relationship between inter-fiber distance and skin tissue regeneration has been investigated on electrospun gelatin scaffolds (Powell and Boyce, 2008). Cell infiltration was found only into the upper regions of the scaffold if the inter-fiber distance was smaller than 5.5 μm, and the distances between 5 to 10 μm have yielded most favorable skin substitute.

2.1.5 Neural tissues

Electrospun nanofibers represent effective guidance substrates for nervous tissue repair. *In vitro* cell culture study of neural stem cells (NSCs) on a PLLA nanofiber scaffold revealed that the nanofiber scaffold not only supported the NSC differentiation and neurite outgrowth, but also promoted NSC adhesion (Yang et al., 2004). Further work using aligned PLLA nanofiber or microfiber scaffolds showed that the directions of NSC elongation and its neurite outgrowth were parallel to the direction of fiber alignment (Yang et al., 2005). The aligned fibers in the scaffold can provide topographical guidance and enhance cell proliferation and neurite outgrowth (Mukhatyar et al., 2011), which has the potential for bridging long peripheral never gaps.

So far, the effects of fiber diameter on cell culturing are still unclear. PLLA nanofibers had higher neural stem cell differentiation than microfibers, but with similar cell orientation (Yang et al., 2005). Among three kinds of PES nanofibers with the diameters of 283, 749 and 1452 nm, cell proliferation and cell spreading increased with the decrease in fiber diameter (Christopherson et al., 2008). In another research on three kinds of PLLA electrospun fibers with different diameters, the neurite length was the shortest on the finest fibers (Wang et al., 2010).

Electrical stimulation has been incorporated during cell growth to improve cell function by preparing conductive nanofibrous scaffolds (Lee et al., 2009b). Polypyrrole (PPy) was grown on random and aligned PLGA nanofibers, and the electrically stimulated nanofibers resulted 40-50% longer neuritis and 40-90% more neurite formation compared to the one without any stimulation.

The feasibility of *in vivo* nerve regeneration using PLGA nanofibers was also investigated (Bini et al., 2004). After implantation of PLGA nanofiber guidance to the right sciatic nerve of rats, no inflammatory response was observed, and 5 out of 11 rats showed successful nerve regeneration one month after implantation. The *in vitro* test also confirmed that nerve stem cells adhered to, and differentiated on, the PLGA nanofiber mats (Bini et al., 2006).

2.1.6 Other tissue scaffolds

In addition to the above mentioned scaffolds, the studies of using nanofibers as scaffold to support other stem cells (Ito et al., 2005), and tissues, such as heart (Zong et al., 2005), cartilage (Matthews et al., 2003), ligament (Lee et al., 2005) and urinary tract (McManus et al., 2007) have been reported.

2.2 Wound healing

Wound healing is a native process of regenerating dermal and epidermal tissues. When an individual is wounded, a set of complex biochemical actions take place in a closely orchestrated cascade to repair the damage. These events can be classified into inflammatory, proliferative, and remodeling phases and epithelialization. Normally, body cannot heal a deep dermal injury. In full thickness burns or deep ulcers, there is no source of cells remaining for regeneration, except from the wound edges. As a result, complete re-epithelialization takes a long time and is complicated with scarring of the base (Marler et al., 1998).

Dressings for wound healing function to protect the wound, exude extra body fluids from the wound area, decontaminate the exogenous microorganism, improve the appearance and sometimes accelerate the healing process. For these functions, a wound dressing material should provide a physical barrier to a wound, but be permeable to moisture and oxygen. Electrospun nanofiber mat is a good wound dressing candidate because of its unique properties: the highly porous mat structure and well interconnected pores are particularly important for exuding fluid from the wound; the small pores and very high specific surface area not only inhibit the exogenous microorganism invasions, but also assist the control of fluid drainage; in addition, the electrospinning process provides a simple way to add drugs into the nanofibers for any possible medical treatment and antibacterial purposes.

A study on using electrospun polyurethane (PU) mat as wound dressing material revealed that the mat effectively exuded fluid from the wound, without fluid accumulation under the mat, and no wound desiccation occurred neither (Khil et al., 2003). The mat also showed a controlled water loss from evaporation, excellent oxygen permeability, and high fluid drainage ability, besides inhibiting the invasion of exogenous micro organism. Histological test indicated that the rate of epithelialization was increased and the dermis became well organized when the wounds were covered with the electrospun nanofiber mat. Silk nanofiber mats were electrospun with epidermal growth factor to promote wound healing process (Schneider et al., 2009a), and it has been found the functionalized mat increased the wound closure by 90%. An anti-cancer, anti-oxidant and anti-flammatory substance, curcumin, was loaded into PCL (Merrell et al., 2009) and CA (Suwantong et al., 2010) nanofibers for wound dressing. These dressings were superior in supporting human dermal fibroblast attachment and proliferation and exhibited high rate of wound closure.

An open wound healing test using an electrospun collagen nanofiber mat showed that the early-stage healing using collagen nanofiber mat was faster than that of using normal cotton gauze (Rho et al., 2006). In the first week, the wound surface in the cotton group was covered by fibrous tissue debris, below which dense infiltration of polymorphonuclear leukocytes and the proliferation of fibroblasts were formed. By comparison, the surface tissue debris in the collagen nanofiber group disappeared, and prominent proliferation of young capillaries and fibroblasts was found. Later stage healing processes were similar for both groups. PLGA/collagen nanofiber mats also showed an effective wound healing behavior with an active response to human fibroblasts in the early stage (Liu et al., 2010a).

In vivo wound healing of diabetic ulcers was investigated using electrospun block copolymer (PCL-PEG) and PCL. When the nanofibers were chemically modified with a recombinant human epidermal growth factor (rhEGF), the expression of keratinocyte-specific genes and EGF-receptor were enhanced (Choi et al., 2008). Systematic *in vivo* wound healing evaluations have been carried on PVA, PCL, polyacrylonitrile (PAN), poly(vinylidene fluoride-co-hexafluoropropene) (PVdF-HFP), PAN/PEU blend, and wool protein coated PVA and PCL nanofibers (Liu et al., 2008, Liu et al., 2010b). The results showed that wound healing performance was mainly influenced by the porosity, air permeability and surface wettability of the nanofiber mats. A mat with good hydrophilicity and high porosity facilitated the healing at the early stage. However, the fiber diameter and antibacterial activity had a limited effect on the healing efficiency.

Post-surgery tissue adhesion is a widely recognized problem for abdominal surgeries. It not only renders future operations more difficult but also causes other problems such as small bowel obstruction, female infertility, and chronic debilitating pain. An electrospun nanofiber mat containing antibiotic agents has been used as a barrier to prevent the post-surgery abdominal adhesions. It was found that the nanofiber mat eliminated post-surgery abdominal adhesion significantly, thus improving the healing process (Bolgen et al., 2007).

To decontaminate the bacteria invasion, biocides, such as silver (Hong, 2007, Chen and Chiang, 2010) and iodine complex (Ignatova et al., 2007) have been added to the electrospun nanofibers. It was reported that polyvinylpyrrolidone (PVP)-iodine complex (PVP-iodine) gradually released active iodine. Because of the broad-spectrum microbicidal activity of iodine, electrospun PVP-iodine nanofibers had external antibacterial, antimycotic and antiviral applications. Ag ions were incorporated into electrospun nanofibers via adding $AgNO_3$ into the polymer solution for electrospinning. To maintain a long term antibacterial activity and control the release of Ag ions, the Ag was embedded in the form of elementary state by a post-electrospinning treatment. Ag nanoparticles can also be directly incorporated into electrospun nanofibers via the electrospinning process. An Ag/PVA nanofiber mat exhibited excellent antimicrobial ability and good stability in moisture environment, as well as a quick and continuous release with good effectiveness (Hong, 2007).

Besides adding antibacterial additives, antimicrobial nanofibers can also be prepared by directly using antimicrobial polymers. For instance, polyurethanes containing quaternary ammonium groups were electrospun into nanofiber nonwovens, and the nanofibers showed very strong antimicrobial activities against Staphylococcus aureus and Escherichia coli (Kenawy et al., 2002).

2.3 Drug delivery and release control

Controlled release is an efficient process of delivering drugs in medical therapy. It can balance the delivery kinetics, immunize the toxicity and side effects, and improve patient convenience (Yih and Al-Fandi, 2006). In a controlled release system, the active substance is loaded into a carrier or device first, and then released at a predictable rate *in vivo* when administered by an injected or non-injected route.

As a potential drug delivery carrier, electrospun nanofibers have exhibited many advantages. The drug loading is very easy to implement via electrospinning process, and the high applied voltage used in electrospinning process had little influence on the drug activity. The high specific surface area and short diffusion passage length give the nanofiber drug system higher overall release rate than the bulk material (e.g. film). The release profile can be finely controlled by modulation of nanofiber morphology, porosity and composition.

Nanofibers for drug release systems mainly come from biodegradable polymers and hydrophilic polymers. Model drugs that have been studied include water soluble (Kenawy et al., 2002), poor-water soluble (Verreck et al., 2003a) and water insoluble drugs (Zeng et al., 2003). The release of macromolecules, such as DNA (Luu et al., 2003) and bioactive proteins (Zeng et al., 2005), from nanofibers was also investigated.

In most cases, water soluble drugs, including DNA and proteins, exhibited an early-stage burst (Zong et al., 2002). For some applications, preventing post-surgery induced adhesion for instance, and such an early burst release will be an ideal profile because most infections occur within the first few hours after surgery. A recent study also found that when a poorly water soluble drug was loaded into PVP nanofibers (Yu et al., 2009a), 84.9% of the drug can be released in the first 20 seconds when the drug-to-PVP ratio was kept as 1:4, which can be used for fast drug delivery systems. However, for a long-lasting release process, it would be essential to maintain the release at an even and stable pace, and any early burst release should be avoided. For a water insoluble drug, the drug release from hydrophobic nanofibers into buffer solution is difficult. However, when an enzyme capable of degrading nanofibers exists in the buffer solution, the drug can be released at a constant rate because of the degradation of nanofibers (Zeng et al., 2003). For example, when rifampin was encapsulated in PLA nanofibers, no drug release was detected from the nanofibers. However, when the buffer solution contained proteinase K, the drug release took place nearly in zero-order kinetics, and no early burst release happened. Similarly, initial burst release did not occur for poor-water soluble drugs, but the release from a non-biodegradable nanofiber could follow different kinetics (Verreck et al., 2003b). In another example, blending a hydrophilic but water-insoluble polymer (PEG-g-CHN) with PLGA could assist in the release of a poor-water soluble drug Iburprofen (Jiang et al., 2004). However, when a water soluble polymer was used, the poorly soluble drug was released accompanied with dissolving of the nanofibers, leading to a low burst release (Yang et al., 2007).

The early burst release can be reduced when the drug is encapsulated within the nanofiber matrix. When an amphiphilic block copolymer, PEG-b-PLA was added into Mefoxin/PLGA nanofibers, the cumulative amount of the released drug at earlier time points was reduced and the drug release was prolonged (Kim et al., 2004c). The reason for the reduced burst release was attributed to the encapsulation of some drug molecules within the hydrophilic block of the PEG-b-PLA. Amphiphilic block copolymer also assisted the dispersion and encapsulation of water-soluble drug into nanofibers when the polymer solution used an oleophilic solvent, such as chloroform, during electrospinning (Xu et al., 2005). In this case, a water-in-oil emulation can be electrospun into uniform nanofibers, and drug molecules are trapped by hydrophilic chains. The swelling of the hydrophilic chains during releasing assists the diffusion of drug from nanofibers to the buffer.

Coating nanofibers with a shell could be an effective way to control the release profile. When a thin layer of hydrophobic polymer, such as poly (p-xylylene) (PPX), was coated on PVA nanofibers loaded with bovine serum albumin (BSA)/luciferase, the early burst release of the enzyme was prevented (Zeng et al., 2005). Fluorination treatment (Im et al., 2010b) on PVA nanofibers introduced functional C-F groups and made the fiber surface hydrophobic, which dramatically decreased the initial drug burst and prolonged the total release time.

The polymer shell can also be directly applied via a coaxial co-electrospinning process, and the nanofibers produced are normally named "core-sheath" bicomponent nanofibers. In this case, even a pure drug can be entrapped into nanofiber as the core, and the release profile was less dependent on the solubility of drug released (Jiang et al., 2005, Huang et al., 2009). A research has compared the release behavior of two drug-loaded PLLA nanofibers prepared using blend and coaxial electrospinning (He et al., 2009). It was found that the blend fibers still showed an early burst release, while the threads made of core-sheath fibers provided a stable release of growth factor and other therapeutic drugs. In addition, the early burst release can also be lowered via encapsulating drugs into nanomaterial, followed by incorporating the drug-loaded nanomaterials into nanofibers. For example, halloysite nanotubes loaded with tetracycline hydrochloride were incorporated into PLGA nanofibers and showed greatly reduced initial burst release (Qi et al., 2010).

3. Environmental protection

Current environmental problems are caused by human activities in the last 150 years. They are having serious negative impacts on us and this is likely to continue for a very long time. Effective solutions are urgently needed to protect our environment. Due to their high specific surface area, electrospun nanofibers are expected to be used to collect pollutants via physical blocking or chemical adsorption.

3.1 Filtration

Filtration has been widely used in both households and industries for removing solid substances from air or liquid. In military, they are used in uniform garments and isolating bags, to decontaminate aerosol dusts, bacteria and even virus. Respirator is another example that requires efficient filtration ability. Similar function is also needed for some fabrics used in the medical area. For a fiber-based filter, the removal of particles is determined by different mechanisms. Large particles are blocked on the filter surface due to the sieve effect. Particles that are smaller than the surface-pores could still be collected by the fibers, through a mechanism such as interception, impaction, or static electrical attraction. Very fine particles could also be captured by the Brownian motion effect. The filtration efficiency is normally influenced by the filter physical structure (fiber fineness, matrix structure, thickness, pore size), fiber surface electronic properties, and its surface chemical characteristic (e.g. surface free energy). The particle collecting capability is also related to the size range of particles being collected. Besides the filtration efficiency, other properties such as pressure drop and flux resistance are also important factors to be evaluated for a filter media.

Electrospun nanofibers for filtration application have a long history. A company in US (Donaldson) has produced electrospun nanofiber-based filter products for industry, consumer and defense applications for more than 20 years, and its Ultra-web® nanofiber filter has been developed for nonwoven and filtration industry for a wide range of applications. Recently another company (AMSOIL) has also developed a nanofiber-based fuel filter for automobile applications. DuPont has electrospun fabric products for HVAC, automotive and liquid filtration, bedding protection and apparel applications.

Electrospun nanofiber mat provides dramatic increases in filtration efficiency at relatively small decreases in permeability. In comparison with conventional filter fibers at the same pressure drop, nanofibers with a diameter finer than half a micron have a much higher capability to collect the fine particles, because the slip flow around the nanofibers increases the diffusion, interception and inertial impaction efficiencies (Kosmider and Scott, 2002).

Both experimental measurements and theoretical calculations revealed that electrospun nanofiber mats were extremely efficient at trapping airborne particles (0.5~200 μm) (Gibson et al., 2001). A comparison study between a nylon-6 electrospun mat (thickness 100 μm, pore size 0.24 μm) and a commercial high efficiency particulate air (HEPA) filter (thickness 500 μm, pore size 1.7 μm) using 300 nm test particles indicated that the thin nanofiber mat had a slightly higher filtration efficiency (99.993%) than the HEPA filter (99.97%) (Barhate and Ramakrishna, 2007).

Besides solid particles, tiny liquid droplets within a liquid-liquid immiscible system could also be removed by a nanofiber mat (liquid-liquid coalescence filtration). PS nanofibers were electrospun from a recycled expanded-polystyrene (EPS), and mixed with micro glass fibers to form a filter media for removal of water droplets from a water-in-oil emulsion (Shin et al., 2005). The addition of small amount of PS nanofibers was reported to significantly improve the capture efficiency (from 68% to 88%), but the pressure drop of the filters was increased considerably.

Nanofibers were used as a supporting scaffold in ultrafiltration (UF) for oil/water emulsion separation. The reported UF mat has a three-layered composite structure consisting of a nonporous hydrophilic top layer, a crosslinked PVA electrospun nanofibrous mid-layer and a conventional nonwoven microfibrous substrate (Yoon et al., 2006, Tang et al., 2009). The electrospun nanofibrous substrate provided a well interconnected porous network with a

large specific surface area. The UF filter has a high flux rate and excellent organic solute rejection capability.
Long term filtration performance can be extended by using chemical (Qin and Wang, 2008) or physical (Homaeigohar et al., 2010) crosslinking treatments to stabilize the porous structure of nanofibrous mat. It was recently reported that filters made of multiple thin nanofiber layers had a much better filtration performance than single thick layer nanofiber mat in terms of the pressure drop during filtration (Zhang et al., 2010b).

3.2 Metal ion adsorption and recovery

The elevated level of metal ions has become a serious pollution in water resources, and gives a long term risk to human health and natural environment. These positively charged ions can be removed by many different methods, and among them adsorption through chemical and physical affinity is the most effective and economic one. In this case, electrospun nanofibers have great potential in collecting metal ions from a solution because of their high specific surface area, high porosity and controllable surface functionality.
Polymers with functional groups which have affinity to metal ions were directly electrospun into nanofibers for metal ion adsorption. Electrospun wool keratin/silk fibroin blend nanofibers were initially used to chelate absorb Cu(II) ions from water (Ki et al., 2007a). Wool protein has large number of hydrophilic amino acids with high affinity to metal ions, and the blend nanofibers showed a Cu(II) adsorption capability of 2.8 μg/mg with high recycling efficiency.
Two approaches have been used to improve the adsorption of metal ions on electrospun nanofibers: introducing functional materials to fiber surface using surface chemistry or coating techniques and increasing surface area to improve adsorption capability. For example, amidino diethylenediamine chelating groups was incorporated into PAN nanofiber using heterogeneous reaction with diethylenetriamine (Kampalanonwat and Supaphol, 2010), and the modified fibers had a pH dependent adsorption capacities of 150.6, 155.5, 116.5 and 60.6 mg/g for Cu(II), Ag(I), Fe(II) and Pb(II), respectively. The Cu(II) and Pb(II) adsorption capacities of amidoxime-modified PAN nanofibers were 52.70 and 263.45 mg/g (Saeed et al., 2008). Boehmite nanoparticles were incorporated into polycaprolactone and nylon-6 nanofibers for Cd(II) collection (Hota et al., 2008). In a latest research (Wu et al., 2010b), PVA/SiO_2 composite nanofibers were prepared and functionalized by mercapto groups. The specific surface area of the nanofibers was higher than 290 m^2/g and the nanofibers exhibited the highest Cu(II) adsorption capacity of 489.12 mg/g.

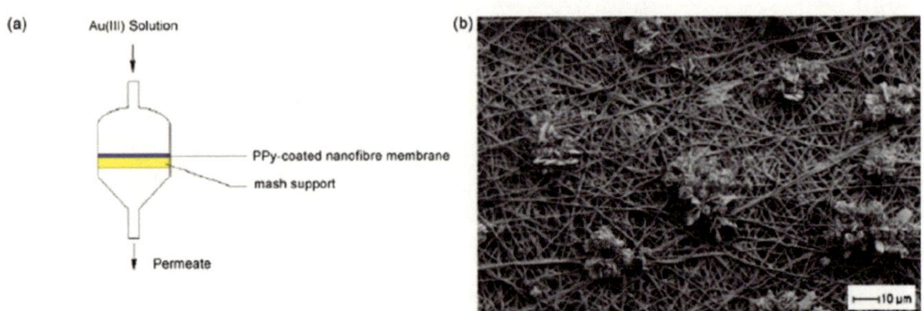

Fig. 4. (a) Apparatus for mat-based Au recovering, (b) SEM image of the PPy-coated nanofibers with aqueous $[Au(III)Cl_4]^-$ solution permeated (Wang et al., 2007).

Apart from the adsorption mechanism, directly converting metal ions into metal element on nanofibers was used for recovering precious metals. Wang et al (Wang et al., 2007) reported a simple and interesting method to grow polypyrrole on nanofiber surface using vapor-phase polymerization. A thin layer of PPy coating on the nanofibers had very little effect on the permeability of the nanofiber membrane, but showed a high efficiency to recover gold ions. The gold ions were converted into micron-sized gold particles (Fig. 4).

4. Catalyst and enzyme carriers

A carrier for catalyst in chemistry and biology is used to preserve high catalysis activity, increase the stability, and simplify the reaction process. An inert porous material with a large surface area and high permeability to reactants could be a promising candidate for efficient catalyst carriers. Using an electrospun nanofiber mat as catalyst carrier, the extremely large surface could provide a huge number of active sites, thus enhancing the catalytic capability. The well-interconnected small pores in the nanofiber mat warrant effective interactions between the reactant and catalyst, which is valuable for continuous-flow chemical reactions or biological processes. The catalyst can also be grafted onto the electrospun nanofiber surface via coating or surface modification.

4.1 Catalysts

Palladium (Pd)-loaded poly(acrylonitrile-acrylic acid) (PAN-AA) nanofibers were prepared by electrospinning a $PdCl_2$-containing PAN-AA solution into nanofibers and subsequently reducing the $PdCl_2$ embedded into Pd nanoparticles. The static catalytic activity of the Pd-loaded nanofibers for selective hydrogenation of dehydrolinalol was measured, to be about 4.5 times higher than that of Pd/Al_2O_3 catalyst (Demir et al., 2004). In a similar work, Pd-loaded PAN-AA nanofibers were confirmed to have high activity and good recycling property for hydrogenation of a-olefin at room temperature. The yield of the hexene to hexane catalyzed by the Pd/PAN-AA nanofibers was 4.7 times higher than that of Pd/γ-Al_2O_3 (Yu and Liu, 2007).

When a molecular catalyst was incorporated into nanofibers, the catalyst could leak out of the nanofibers during the catalysis reaction (Stasiak et al., 2007). Surface coating such as catalyst-loaded nanofiber with a thin layer of polymer considerably retained the catalyst in the nanofibers and the catalyst efficiency was improved at the same time. The catalysis performance was influenced by the type of shell polymer and coating thickness.

TiO_2 is a widely used catalytic material in commercial market. Electrospun PVA-platinum (Pt)/TiO_2 nanofibers were prepared for photocatalytic degradation of solid-state PVA (He and Gong, 2003). Compared with TiO_2 nanoparticles, electrospun TiO_2 nanofibers showed higher photocatalytic activities in terms of photocurrent generation by a factor of 3 because of the mesoporosity and nanoparticle alignment, which caused efficient charge separation through interparticle charge transfer along the nanofiber framework (Choi et al., 2010c). Fiber diameter was reported to play an important role in determining the photocatalytic activity of TiO_2 nanofibers, and the diameter of 200 nm was found to be optimal in photocatalytic performance (Li et al., 2010).

Doping has been used to further increase the photocatalytic activity of TiO_2 nanofibers. Materials such as Pt (Formo et al., 2009), gold (Pan and Dong, 2009), SnO_2 (Wang et al., 2009a) and vanadium (Zhang et al., 2010c) have been used to dope TiO_2 nanofibers for heterostructure using different approaches and enhanced photocatalytic behavior was achieved. SnO_2 with different morphologies and densities was formed on TiO_2 nanorods (Fig. 5a-d). These SnO_2/TiO_2 heterostructures were demonstrated to have high photocatalytic degradation efficiency and the photocatalytic performance was dependent on the SnO_2 morphology (Fig. 5e).

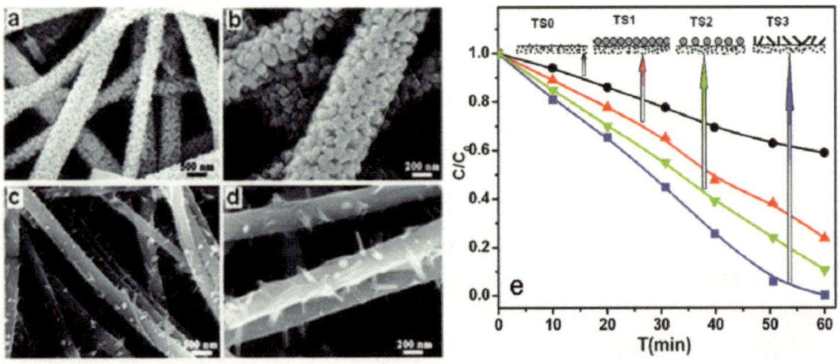

Fig. 5. SEM images of different SnO_2 morphologies on TiO_2 nanofibers (a-d) and the SnO_2 morphology dependent photocatalytic degradation behavior of Rhodamin B (e) (Wang et al., 2009a).

Apart from TiO_2 nanofibers, other inorganic nanofibers (such as ZnO (Lin et al., 2009) and $LaCoO_3$ (Dong et al., 2010a)) have also been investigated as catalysts. For example, ZnO based hierarchical nanofibers has been prepared by growing radically oriented ZnO nanowires on electrospun PLLA nanofibers to make flexible catalyst for continuous-flow photocatalytic water purification (Sugunan et al., 2010).

4.2 Enzymes

Chemical reactions using enzymes as catalysts have high selectivity and require mild reaction conditions. For easy separation from the reaction solution, enzymes are normally immobilized with a carrier. The immobilization efficiency mainly depends on the porous structure and enzyme-matrix interaction. To immobilize enzyme on electrospun nanofibers, many approaches have been used, including grafting enzyme on fiber surface, physical adsorption, and incorporating enzyme into nanofiber via electrospinning followed by crosslinking reaction.

To graft enzymes on nanofiber surface, the polymer used should possess reactive groups for chemical bonding (Wang and Hsieh, 2004, Wang et al., 2006b, Stoilova et al., 2010a). In some studies, polymer blends containing at least one reactive polymer were used (Jia et al., 2002, Kim et al., 2005a, Kim and Park, 2006). The immobilized enzymes normally showed a slightly reduced activity in aqueous environment compared with the un-immobilized native counterpart, but the activity in non-aqueous solution was much higher. For example, α-chymotrypsin was used as a model enzyme to bond chemically on the surface of electrospun PS nanofibers. The enzyme was measured to cover over 27.4% monolayer of the nanofiber surface, and the apparent hydrolytic activity of the enzyme-loaded was 65% of the native enzyme, while the activity in non-aqueous solution was over 3 orders of magnitude higher than that of its native enzyme under the same condition. In another study using PAN nanofibers to immobilize lipase, the tensile strength of the nanofiber mat was improved after lipase immobilization, and the immobilized lipase retained >90% of its initial reactivity after being stored in buffer at 30 °C for 20 days, whereas the free lipase lost 80% of its initial reactivity. Also the immobilized lipase still retained 70% of its specific activity after 10 repeated reaction cycles (Li et al., 2007). In addition, the immobilized enzyme also showed improved pH and thermal stabilities (Huang et al., 2007). Ethylenediamine was used to modify PAN nanofiber mat to introduce active and hydrophilic groups, followed by a chitosan coating for improvement of biocompatibility (Stoilova et al., 2010b).

Enzymes were incorporated into nanofibers via electrospinning, and subsequent crosslinking the enzymes incorporated effectively prevented their leaching. In the presence of PEO or PVA, casein and lipase were electrospun into ultra-thin fibers. After crosslinking with 4,4'-methylenebis(phenyl diisocyanate) (MDI), the fibers became insoluble, and the lipase en- capsulated exhibited 6 times higher hydrolysis activity towards olive oil than that of the films cast from the same solution (Xie and Hsieh, 2003). The crosslinked enzymes in nanofibers showed very high activity and stability. For example, the immobilized α-chymotrypsin in a shaken buffer solution maintained the same activity for more than two weeks (Herricks et al., 2005).

In addition to chemical bonding, the enzymes were also applied onto nanofibers simply via physical adsorption (Huang et al., 2006, Wang et al., 2006c). Polyacrylonitriles-2-methacryloyloxyethyl phosphoryl choline (PANCMPC) nanofiber was reported to have high biocompatibility with enzymes because of the formation of phospholipid micro-environment on the nanofiber surface. Lipase on the nanofibers showed a high immobilization rate, strong specific activity and good activity retention.

5. Sensors

Sensors have been widely used to detect chemicals for environment protection, industrial process control, medical diagnosis, safety, security and defense applications. A good sensor should have a small dimension, low fabrication cost and multiple functions, besides the high sensitivity, selectivity and reliability (Hall, 1998). High sensitivity and fast response require the sensor device having a large specific surface area and highly porous structure. Several approaches have been used to impart nanofibers with a sensing capability, such as using a polymeric sensing material to electrospin nanofibers, incorporating sensing molecules into nanofibers, or applying sensing material on nanofiber surface via coating/grafting technique.

Electrospun PAA nanofibers have been grafted with pyrene methanol (PM) as the sensing material to detect metal ions Fe^{3+} and Hg^{2+}, and an explosive 2,4-dinitrotuloene (DNT) in water (Wang et al., 2001, Wang et al., 2002a). Due to the quenching effect of these chemicals to the pyrene moieties, the fluorescent intensity of nanofibers had a linear response to the concentration of quenchers, and the nanofibers showed high sensitivities. Similarly, a PM-grafted poly(methyl methacrylate) PMMA nanofibers showed an order of magnitude higher sensitivity to target analyte DNT than its cast film counterpart (Wang et al., 2002b). Explosive 2,4,6-trinitrotoluene (TNT) vapor can be accurately detected using electrospun PAN nanofibers that were coated with a thin layer of conjugated polymer, poly(triphenylamine-alt-biphenylene vinylene) (TPA-PBPV) (Deng et al., 2009).

Fluorescence optical sensors were also prepared by a layer-by-layer electrostatic assembly technique to apply a conjugated polymer onto nanofiber surface for detection of methyl viologen and cytochrome c in aqueous solution (Wang et al., 2004) and porphyrin-doped silica nanofibers were used to trace TNT vapor (Tao et al., 2007). All those nanofiber sensors showed high sensitivity and rapid response. Besides fluorescent properties, conjugated

polymer embedded electrospun nanofibers were also reported to be able to sense volatile organic compounds (VOCs) based on optical absorption properties (Yoon et al., 2007).
A gas sensor using a specific absorption interaction between ammonia and poly(acrylic acid) (PAA) nanofibers was reported. The weight difference induced by the gas absorption was measured by a quartz crystal microbalance (QCM). This sensor was capable of detecting ppb level NH_3 in air, and the sensitivity was four times higher than that of the PAA cast film (Ding et al., 2005). The absorption of gas also leads to changes in FTIR absorption. PAN nanofibers containing metal oxide nanoparticles, such as iron oxide and zinc oxide, have been used to detect carbon dioxide (Luoh and Hahn, 2006). The addition of metal oxide nanoparticles enhanced the gas adsorption and thus improved the sensitivity. In a electrospun carbon nanofiber gas sensor (Im et al., 2010a), carbon black was blended into the fibers to increase the conductivity, and a porous structure was introduced to the fiber surface by etching under a basic condition. The sensory ability for NO and CO gases was reported to be improved dramatically, and the enhanced sensitivity was attributed to higher surface area and improved electrical conductivity due to the formation of carbon black network. It has been found that electrospun SnO_2 fibers with a smaller grain size showed much better sensitivity to CO and NO_2 gases. This was probably because the fibers containing smaller grains had higher resistance, which was the result of larger number of grain-grain interconnections (Park et al., 2011).
Electrical conductivity is an important property for sensor devices. Conducting nanofibers can be produced from semi-conducting oxides, conducting polymers and non-conductive polymers. Pure oxide nanofibers are normally produced by electrospinning a solution containing oxide sol-gel and polymer, followed by calcining treatment to remove the polymer. The detection of gas molecules using oxide nanofibers is based on the conductivity changes due to the doping effect of analyst gases to the oxide. A few oxide nanofibers have been assessed for detecting different gases, such as MoO_3 (Gouma, 2003) nanofibers for ammonia, WO_3 nanofibers for ammonia (Wang et al., 2006a) and NO_2 (Sawicka et al., 2005), TiO_2 nanofibers for NO_2 and H_2 (Kim et al., 2006). These sensors exhibited improved sensitivity, faster response and lower detection limit than that of sol-gel based films
Conducting polymer is another interesting sensor material. Electrospun polyaniline (PANi)/PS nanofibers containing glucose oxidase have been demonstrated to have a high sensitivity to glucose (Aussawasathien et al., 2005). PANi/PVP nanofibers also exhibited sensing ability to NO_2 (Bishop and Gouma, 2005). Organic/inorganic semiconductor Schottky nanodiode was fabricated by PANi nanofibers and inorganic n-doped semiconductor. The device has a rapid response and supersensitive to ammonia (Pinto et al., 2006). Poly(3,4-ethylenedioxythiophene)-poly(styrene sulfonate) (PEDOT:PSS) has also been blended into PVP nanofibers for chemical vapor detection (Choi et al., 2010a).
Besides the conducting and semi-conducting materials, insulating polymers were also used to fabricate electrical sensors. In this case, ions or conductive nano-fillers were added to improve the conductivity. When PEO nanofibers were doped with $LiClO_4$, the mat showed low conductivity and was sensitive to moisture, and the nanofiber mat was reported to have much higher sensitivity than its film-type counterpart (Aussawasathien et al., 2005). Carbon nanotubes/poly(vinylidene fluoride) (PVdF) composite nanofibers showed an increased straining sensing ability (as measured by voltage across the sensor), 35 times higher than that of the film counterpart (Laxminarayana and Jalili, 2005). In addition, electrospun nanofibers incorporated with carbon black showed sensitivity to VOCs (Kessick and Tepper, 2006). When the carbon black concentration was near the percolation threshold, the

composite fibers changed their resistance in volatile organic compounds. Using different polymer matrices, the sensor can be used to detect toluene, trichloroethylene, methanol, and dichloropentane vapors.

Humidity sensor is a very important device for environment tests. KCl doped ZnO (Qi et al., 2009b) and TiO_2 (Qi et al., 2009a) nanofibers have shown higher humidity sensitivity with faster response and recovery time compared to pure ZnO and TiO_2 nanofibers. $BaTiO_3$ nanofibers also exhibited excellent humidity sensing behavior because its complex impedance varied around three orders of magnitude in the whole humidity range (He et al., 2010).

6. Energy harvest and storage

Energy is essential for our modern civilization. The rapidly growing global energy demand has not only sped up the consumption of non-renewable fossil fuels, but also threatened regional stability. In addition to reduce the energy consumption using highly efficient technology, converting other energies into electrical power can considerably assist in alleviating the energy crisis. In this direction, nanotechnology is providing new solutions to solve the problems. It has been found that nanofibrous materials can have significantly higher energy conversion and storage efficiency than their bulk counterparts.

6.1 Solar cells

Solar cells utilize unlimited solar energy for power generation and have been considered as a major solution to current energy crisis. So far, single crystal and polycrystalline silicon based solar cells are dominating the commercial solar cell market. Dye-sensitized solar cells (DSSCs) and organic solar cells are still under development.

Conventional DSSCs have a dye-anchored mesoporous TiO_2 nanoparticle thin layer sandwiched between two conducting glass plates in the presence of an electrolyte. Since electrospun nanofibers are one-dimensional material with better electrical conductivity and higher specific surface area than nanoparticles, and large pore size in nanofibrous mat allows increased penetration of viscous polymer gel electrolyte, nanofibers have shown great application potential in DSSCs.

By using a thin layer of TiO_2 nanofibers as working electrode in DSSC, the photocurrent generation using polymer gel electrolytes was over 90% of liquid electrolyte (Song et al., 2004). DSSC devices (Priya et al., 2008) using polymer electrolytes based on electrospun PVdF-HFP nanofibers could achieve higher conversion efficiency and better long-term stability than conventional liquid electrolyte ones.

The largest problem of using inorganic nanofibrous layers in DSSC is the poor adhesion with conductive substrate. The nanofiber layers are easily separated after calcination due to different thermal expansion coefficients. Many approaches have been tested to address this issue. A TiO_2 nanoparticle layer was coated on conductive glass before nanofiber deposition to increase the adhesion between the glass and the TiO_2 layer (Onozuka et al., 2006). Hot pressing pre-treatment was applied on ZnO/PVA composite nanofiber mat before calcination (Kim et al., 2007b). A self-relaxation layer (Zhang et al., 2009) was also spontaneously formed during electrospinning to release the interfacial tensile stress generated in calcination.

To increase the short-circuit current, the electrospun TiO_2 electrode was treated with $TiCl_4$ aqueous solution to form an additional rutile TiO_2 layer on the fiber surface. Such a rutile

TiO_2 layer increased the fraction volume of active TiO_2 and inter-fiber connection, resulting in an increased photocurrent (Song et al., 2004, Lee et al., 2009a). Another method to improve charge generation and transport was through materials doping. Quantum dots (QDs) were decorated onto TiO_2 nanofibers to yield multiple carrier generation due to the quantum confinement effect (Sudhagar et al., 2009). The electron diffusion coefficient and mobility of the niobium-doped anatase TiO_2 nanofibers were an order of magnitude higher than those of the un-doped fibers (Archana et al., 2010).

Another advantage of electrospun nanofibers is that they can be used to form a transparent mat for counter electrode. This effort has become more and more urgent because of the rapidly growing price of tin and indium, which are typically used for making conductive glass. Copper nanofiber thin layers were prepared using electrospinning for transparent electrode in DSSC (Wu et al., 2010a). The resulted fibers had ultrahigh aspect ratios of up to 100,000 and fused crossing points with ultra-low junction resistance. The fibrous mat exhibited great flexibility and stretchability as well. Electrospun carbon nanofibers were also explored as counter electrode material and an energy conversion efficiency of 5.5% was achieved in the resultant solar cell (Joshi et al., 2010). The carbon nanofiber based DSSC had lower fill factor and overall performance, because of the higher total series resistance (15.5 $\Omega \cdot cm^2$) than that (4.8 $\Omega \cdot cm^2$) of Pt based traditional conductive substrate.

6.2 Fuel cells

Fuel cells are electrochemical devices capable of converting hydrogen or hydrogen-rich fuels into electrical current by a metal catalyst. There are many kinds of fuel cells, such as proton exchange mat (PEM) fuel cells, direct methanol fuel cells, alkaline fuel cells and solid oxide fuel cells (Sundmacher, 2010). PEM fuel cells are the most important one among them because of high power density and low operating temperature.

Pt nanoparticle catalyst is a main component in fuel cells. The price of Pt has driven up the cell cost and limited the commercialization. Electrospun materials have been prepared as alternative catalyst with high catalytic efficiency, good durability and affordable cost. Binary PtRh and PtRu nanowires were synthesized by electrospinning, and they had better catalytic performance than commercial nanoparticle catalyst because of the one-dimensional features (Kim et al., 2008c). Pt nanowires also showed higher catalytic activities in a polymer electrolyte membrane fuel cell (Kim et al., 2009).

Instead of direct use as catalyst, catalyst supporting material is another important application area for electrospun nanofibers. Pt clusters were electrodeposited on a carbon nanofiber mat for methanol oxidation, and the catalytic peak current of the composite catalyst reached 420 mA/mg compared with 185 mA/mg of a commercial Pt catalyst (Li et al., 2008b). Pt nanoparticles were immobilized on polyimide-based nanofibers using a hydrolysis process (Xuyen et al., 2009), and Pt nanoparticles were also loaded on the carbon nanotube containing polyamic acid nanofibers to achieve high catalytic current with long term stability (Nguyen et al., 2009).

Proton exchange mat is the essential element of PEM fuel cells and normally made of a Nafion film for proton conduction. Because pure Nafion is not suitable for electrospinning due to its low viscosity in solution, it is normally mixed with other polymers to make blend nanofibers. Blend Nafion/PEO nanofibers were embedded in an inert polymer matrix to make a proton conducting mat (Lee et al., 2009c), and a high proton conductivity of 0.06-0.08 S/cm at 15 °C in water and low water swelling of 12-23 wt% at 25 °C were achieved. Besides blend electrospinning, Nafion surface coating on polymeric nanofibers (Lin et al., 2010,

Molla and Compan, 2011a, Molla and Compan, 2011b) is another efficient way to obtain better fuel cell performance and reduce Nafion consumption.

6.3 Mechanical energy harvesters

Piezoelectric material can directly convert mechanical energies into electrical power and has shown great potential in powering low energy consumption devices. Piezoelectric power generators, also called nanogenerators in some literatures, are becoming an important source for renewable energy (Wang and Song, 2006, Wang, 2008). These nanogenerators are normally made of aligned inorganic nanowires and their preparation required preciously controlled conditions (Lu et al., 2009, Huang et al., 2010).

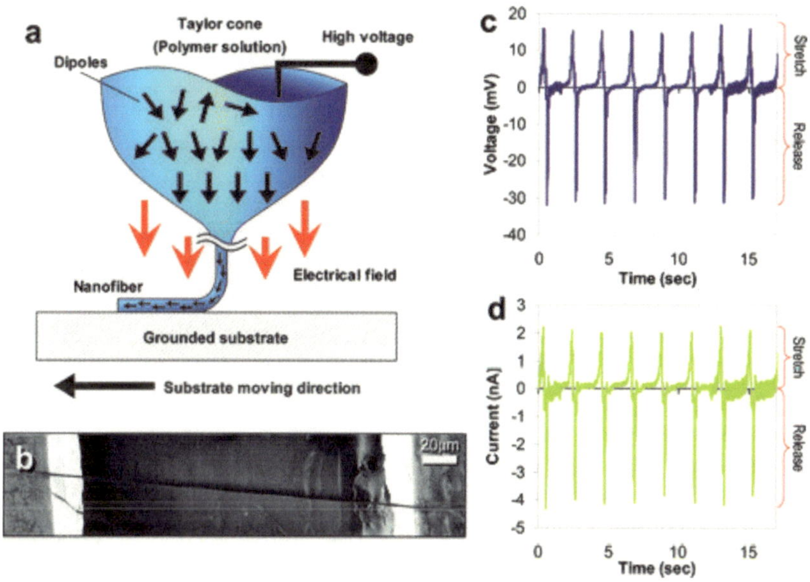

Fig. 6. (a) Scheme of near-field electrospinning process; (b) SEM image of a single PVdF nanofiber power generator; (c) and (d) voltage and current outputs from the generator under applied strain at 2 Hz (Chang et al., 2010).

Electrospinning can process different piezoelectric materials into nanofibrous structure and electrospun piezoelectric nanofibers have been demonstrated to have energy scavenging capability. PVdF nanofiber was directly deposited across two metal electrodes without extra poling treatment to make single fiber nanogenerator using a near field electrospinning technique (Fig. 6a and b) (Chang et al., 2010). Voltage (5-30 mV) and current (0.5-3 nA) outputs were generated when the fiber underwent repeated extensive deformation at 2 Hz (Fig. 6c and d). Rigid lead-zirconate-titanate (PZT) material has been made into electrospun nanofibers with highly improved flexibility for fabricating flexible power generator after a soft polymer packaging (Chen et al., 2010a). The output voltage and power under repeated stress application were 1.63 V and 0.03 μW, respectively. Aligned PVdF nanofibers were collected using a normal electrospinning process and the following poling treatment was used to improve piezo-responsiveness (Hansen et al., 2010). A hybrid nano system combining this fiber based power generator and a flexible enzymatic biofuel cell was demonstrated as a self-powered UV light sensor.

6.4 Lithium ion batteries

Lithium ion batteries are compact and rechargeable electrical energy storage devices. They have very high energy density. Like most other chemical batteries, porous structure is an essential requirement for the battery. A sponge-like electrode has high discharge current and capacity, and a porous separator between the electrodes can effectively stop the short circuit, but allows the exchange of ions freely. Solid electrolytes used in lithium ion battery are typically composed of a gel or porous host to retain the liquid electrolyte inside (Arora and Zhang, 2004). To have high ion conductivity, the host material, also called separator, should have high permeability to ions. A porous mat with well interconnected pores, suitable mechanical strength and high electrochemical stability could be a potential candidate.

Perfluoro polymers have been widely studied in polymer electrolyte because of their chemical and mechanical stability. An electrospun PVdF polymer electrolyte showed high uptake to electrolyte solution (320%—350%) and high ion conductivity (1.7×10^{-3} S/cm at 0 °C). The fibrous electrolyte also had high electrochemical stability of more than 5 V. The prototype cell (MCMB/PVdF based electrolyte/$LiCoO_2$) exhibited a very stable charge-discharge behavior (Kim et al., 2004b). When a thin layer of polyethylene (PE) was plasma polymerized onto PVdF nanofiber surface, a role of shutter by melting of the PE layer grafted was rendered to the nanofiber mat and improved the safety of battery (Choi et al., 2004). It was also found that the formation of interconnected web structure via heat treatment improved both the mechanical properties and dimensional stability of nanofiber mats (Choi et al., 2004, Gao et al., 2006).

Nanofiber mats from other types of polymers, such as PAN (Choi et al., 2005), were also studied as lithium battery separator. The electrospun PAN mat showed high ion conductivity and electrochemical stability. The prototype cell based on the electrospun PAN electrolyte separator with 1 M LiPF6-EC/DMC exhibited an initial discharge capacity of 145 mAh/g, and 94.1% of the initial discharge capacity after 150 cycles at a charge/discharge rate of 0.5 C/0.5 C. To improve ionic conductivity, charge-discharge capability and stability, SiO_2 nanoparticles (Kim et al., 2008b), PMMA (Ding et al., 2009) and PAN (Raghavan et al., 2010) were blended with PVdF-HFP nanofibers for making composite polymer electrolyte.

Besides being used as a separator, some electrospun nanofiber mats have been used as battery electrodes. For example, a carbonated electrospun nanofiber mat was used as anode in lithium ion battery, and the batteries showed a large reversible capacity of 450 mAh/g (Ahn et al., 2006). Some recent researches are focusing on improving battery performance by incorporating different inorganic nanomaterials into carbon nanofibers as electrodes. A anode material made of Fe_3O_4/C composite nanofibers had much better electrochemical performance with a high reversible capability of 1007 mAh/g at the 80th cycle and excellent rate capability (Wang et al., 2008). A Sn/C composite were encapsulated into hollow carbon nanofibers as an anode material for lithium batteries (Yu et al., 2009b) with a high reversible capacity of 737 mAh/g after 200 cycles at 0.5 C (480 mAh/g at 5 °C).

Also, $LiNi_{1/3}Mn_{1/3}Co_{1/3}O_2$ (Ding et al., 2008), vanadium oxide (Mai et al., 2010) and $LiFePO_4$/carbon composite nanofibers (Lee et al., 2010) have been investigated as cathode material for lithium ion batteries.

6.5 Supercapacitors

As an electrochemical device with a high power density and super high charging-discharging rate, supercapacitors (also known as double-layer capacitor) have been demonstrated with great potential in different emerging applications, such as power back-up for laptop or mobile phone and power source for hybrid electric vehicle. For a supercapacitor with high capacitive behavior, electrodes made of porous carbon materials are extremely important, and electrospinning technique has been used to prepare carbon nanofibrous mats with high specific surface area and controllable pore size for this purpose. Activated carbon nanofibers prepare from a PAN/DMF solution have shown a maximum specific capacitance of 173 F/g (Kim and Yang, 2003). The polybenzimidazole (PBI) based carbon nanofibers had specific surface areas ranging from 500 to 1220 m^2/g, and the fabricated double-layer capacitor exhibited specific capacitance between 35 ~ 202 F/g (Kim et al., 2004a).

The capacitance of electrospun nanofiber based supercapacitor can be enhanced by using composite carbon electrodes. $ZnCl_2$ (Kim et al., 2007a), carbon nanotubes (Guo et al., 2009) and nickel (Li et al., 2009), have been blended into procurer solutions for electrospinning and then carbonization. The $ZnCl_2$/C composite nanofibers exhibited a capacitance of 140 F/g with a specific surface area of 55 m^2/g when the carbon nanofibers were doped with 5 wt% $ZnCl_2$. The specific capacitance of an electrical double-layer capacitor with electrodes made of carbon nanotubes embedded carbon nanofiber reached as high as 310 F/g.

In a recent publication (Niu et al., 2011), bicomponent electrospinning was applied to prepare side-by-side nanofibers with PAN on one side and a thermoplastic polymer, PVP on the other side. The resultant carbon fibers after pyrolysis showed improved inter carbon fiber connections and crystallization. The capacitance of the electrochemical cell (Fig. 7a) made of these inter-bonded carbon nanofibers was much higher than that of the carbon nanofibers from polymer blends (Fig.7 b and c).

Apart from carbon, other inorganic nanofiber electrodes have also been examined for supercapacitor applications. For example, RuO_2 has been deposited on a Pt nanofiber mat to function as hybrid electrode and the supercapacitor exhibited a specific capacitance of 409.4 F/g with a capacity loss of only 21.4% from 10 to 1000mV/s (Choi et al., 2010b). Electrospun V_2O_5 nanofiber based supercapacitor had the highest specific capacitance of 190 F/g in an aqueous electrolyte and 250 F/g in an organic electrolyte when the nanofibers were annealed at 400 °C (Wee et al., 2010).

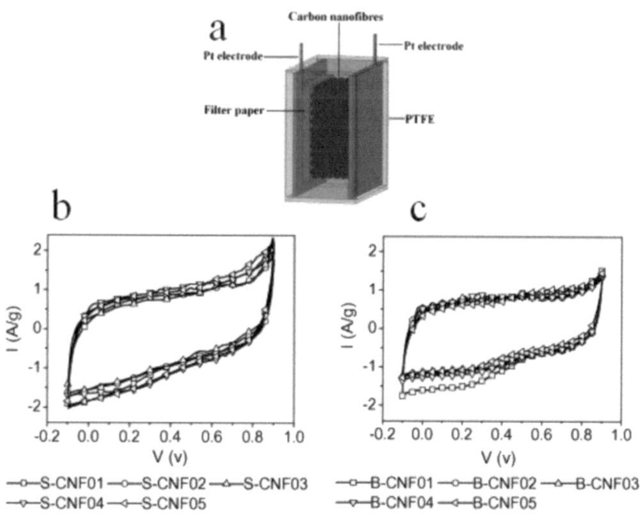

Fig. 7. (a) Illustration of electrochemical cell and the cyclic voltammetry curves of side-by-side (b) and blend (c) carbon nanofibers (Niu et al., 2011).

6.6 Hydrogen storage

Hydrogen has been widely known as an ideal alternative energy to solve energy crisis and global warming. Because high-pressure and cryogenic storage systems can't satisfy the criteria for on-board storage, the key issue for the current hydrogen energy research is hydrogen storage.

It has been reported that carbonaceous materials, such as active carbon, carbon nanotubes and graphite, are able to store hydrogen because of their large specific surface area, high pore volume and light weight. The hydrogen-storage capabilities of electrospun carbon fibers carbonated from different starting polymers (PAN (Kim et al., 2005b), PVdF (Hong et al., 2007), PANi (Srinivasan et al., 2010) and polycarbosilane (PCS) (Rose et al., 2010)) have also been assessed. Carbon nanofiber can be a better hydrogen storage material than other carbon materials because it has an optimized pore structure with controlled pore size (Im et al., 2009b).

To increase graphitization during carbonation, Fe (III) acetylacetonate was used as catalyst to prepare graphite nanofibers and the resulted nanofiber (surface areas of 60-253 m^2/g) had the H-storage capacity of 0.14-1.01 wt%. To increase the specific surface area of carbon nanofibers, different inorganic materials (normally metal (Im et al., 2009a) and metal oxides (Im et al., 2008a)) were mixed into polymer solutions for electrospinning. The resultant carbon nanofiber could have a specific surface area of 2900 m^2/g with an H-storage capacity of 3 wt%. Even though the hydrogen adsorption ability increases with the increasing the specific surface area, pore volume is also very important. It has been concluded that the most effect pore width is in the range from 0.6 to 0.7 nm (Im et al., 2008b), which is slightly larger than hydrogen molecule (0.4059 nm).

Exciting results have been obtained more recently. PANi fibers were prepared by electrospinning and showed a reversible hydrogen storage capacity of 3-10 wt% at different temperatures (Srinivasan et al., 2010). Highly porous carbide-derived carbon fibers with a specific surface area of 3116 m^2/g were prepared after pyrolysis and chlorination of electrospun polyacrbosilane fibers (Rose et al., 2010). The fibers have shown a very high hydrogen storage capability of 3.86 wt% at a low pressure (17 bar).

7. Other applications

Early studies on electrospun nanofibers also included reinforcement of polymers. As electrospun nanofiber mats have a large specific surface area and an irregular pore structure, mechanical interlocking among the nanofibers should occur. When a thin electrospun nylon-4, 6 nanofiber mat was added to epoxy, the composite showed transparency to visible light, and both the stiffness and strength were increased considerably compared with the pure epoxy film (Bergshoef and Vancso, 1999). Recently, polysulfone (PSF) (Li et al., 2008a) and polyetherketone cardo (PEK-C) (Zhang et al., 2010a) nanofibers were used to improve the toughness of carbon fiber/epoxy composite. When carbon nanofibers were dispersed into the PEK-C nanofiber phase, a synergistic effect to enhance the toughness and other mechanical properties was also observed (Zhang et al., 2011).

Electrospun polybenzimidazole (PBI) nanofibers have been used as fillers to reinforce epoxy and rubber (Kim and Eneker, 1999). An epoxy containing 15 wt% electrospun PBI nanofibers was found to have higher fracture toughness and modulus than the one containing 17 wt% PBI whiskers. Also the Young's modulus and tear strength of styrene-butadiene rubber

(SBR) containing the PBI nanofibers were higher than those of pure SBR. In addition, electrospun nylon PA 6 nanofibers were used to improve the mechanical properties of a BISGMA/TEGDMA dental restorative composite resins (Fong, 2004).

Electrospun nanofibers showed excellent capability to absorb sound. A leading nanofiber technology company, Elmarco, recently patented an electrospun nanofiber material that had unique sound absorption characteristics, with only about one-third of the weight of conventional sound absorption materials. It was able to absorb sounds across a wide range of frequencies, especially those below 1000 Hz.

Electrospun nanofibers have also shown application potential in field-effect transistor (FET). EFT behavior has been observed in camphor sulfonic-acid-doped electrospun PANi/PEO nanofibers (Pinto et al., 2003). Saturation channel currents were found at low-source-drain voltage with a hole mobility in the depletion regime of 1.4×10^{-4} cm^2/Vs. Electrospun nanofiber mat has also been demonstrated with the application ability for ultrafast identification of latent fingerprints (Yang et al., 2011).

8. Concluding remarks

The application areas of electrospun nanofibers have been expanding in recent years. Most studies in this area have been conducted on fibers produced on a very small scale, using a needle based electrospinning system. It is expected large-scale nanofiber production will be further improved by commercial technology providers (e.g. Elmarco and eSpin) and research efforts (Niu et al., 2009, Wang et al., 2009b) to provide large amount of high quality nanofibers and reduce the application cost.

So far, almost all of the applications are based on using nanofiber nonwoven mat. It is expected that more sophisticated structures made from nanofiber assemblies, i.e. nanofiber yarns (Ali et al., 2010) and fabrics, will be further developed for novel applications.

Further developments in melt electrospinning technology are needed to expand the range of polymers available for producing nanofibers, and to avoid the use of organic solvents for environmentally friendly nanofiber production. It is also worth noting that other non-electrospinning technologies, such as the melt-blowing and micro-fiber technologies used in the fiber/textile industry, have the potential of producing submicron fibers. Hybrid technologies are being developed also to mass produce nanofibers.

Based on the history and current achievements in electrospinning, electrospun nanofibers are expected to play more and more important role in many important application areas, such as renewable energy and environmental protection. Much work remains to realize the full application potential of nanofiber materials.

9. References

Ahn, Y.C., Park, S.K., G.T., K., Hwang, Y.J., Lee, C.G., Shin, H.S. & Lee, J.K. (2006). Development of high efficiency nanofilters made of nanofibers. *Current Applied Physice,* Vol.6, 1030-1035.

Ali, U., Zhou, Y., Wang, X. & Lin, T. (2010). Direct electrospinning of highly twisted, continuous nanofiber yarns. *The Journal of the Textile Institute,* accepted for publication (29 Dec 2010), DIO: 10.1080/00405000.2011.552254.

Archana, P.S., Jose, R., Jin, T., Vijila, C., Yusoff, M.M. & Ramakrishna, S. (2010). Structural and electrical properties of Nb-doped anatase TiO2 nanowires by electrospinning. *Journal of the American Ceramic Society,* Vol.93, No.12, pp 4096-4102.

Arora, P. & Zhang, Z. (2004). Battery separators. *Chemical Reviews,* Vol.104, No.10, pp 4419-4462.

Aussawasathien, D., Dong, J.H. & Dai, L. (2005). Electrospun polymer nanofiber sensors. *Synthetic Metals,* Vol.154, No.1-3, pp 37-40.

Baker, S.C., Atkin, N., Gunning, P.A., Granville, N., Wilson, K., Wilson, D. & Southgate, J. (2006). Characterisation of electrospun polystyrene scaffolds for three-dimensional in vitro biological studies. *Biomaterials,* Vol.27, No.16, pp 3136-3146.

Balguid, A., Mol, A., van, M.M.H., Bank, R.A., Bouten, C.V.C. & Baaijens, F.P.T. (2009). Tailoring fiber diameter in electrospun poly(.vepsiln.-Caprolactone) scaffolds for optimal cellular infiltration in cardiovascular tissue engineering. *Tissue Engineering, Part A,* Vol.15, No.2, pp 437-444.

Barhate, R.S. & Ramakrishna, S. (2007). Nanofibrous filtering media: Filtration problems and solutions from tiny materials. *Journal of Membrane Science,* Vol.296, No.1+2, pp 1-8.

Bergshoef, M.M. & Vancso, G.J. (1999). Transparent nanocomposites with ultrathin, electrospun Nylon-4,6 fiber reinforcement. *Advanced Materials,* Vol.11, No.16, pp 1362-1365.

Bhattarai, S.R., Bhattarai, N., Viswanathamurthi, P., Yi, H.K., Hwang, P.H. & Kim, H.Y. (2006). Hydrophilic nanofibrous structure of polylactide; fabrication and cell affinity. *Journal of Biomedical Materials Research, Part A,* Vol.78, No.2, pp 247-257.

Bini, T.B., Gao, S., Tan, T.C., Wang, S., Lim, A., Hai, L.B. & Ramakrishna, S. (2004). Electrospun poly(L-lactide-co-glycolide) biodegradable polymer nanofiber tubes for peripheral nerve regeneration. *Nanotechnology,* Vol.15, No.11, pp 1459-1464.

Bini, T.B., Gao, S., Wang, S. & Ramakrishna, S. (2006). Poly(L-lactide-co-glycolide) biodegradable microfibers and electrospun nanofibers for nerve tissue engineering: an in vitro study. *Journal of Materials Science,* Vol.41, No.19, pp 6453-6459.

Bishop, A. & Gouma, P. (2005). Leuco-emeraldine based polyaniline - poly-vinyl-pyrrolidone electrospun composites and bio-composites: A preliminary study of sensing behavior. *Reviews on Advanced Materials Science,* Vol.10, No.3, pp 209-214.

Bolgen, N., Vargel, I., Korkusuz, P., Menceloglu, Y.Z. & Piskin, E. (2007). In vivo performance of antibiotic embedded electrospun PCL membranes for prevention of abdominal adhesions. *Journal of Biomedical Materials Research, Part B: Applied Biomaterials,* Vol.81, No.2, pp 530-543.

Casper, C.L., Yang, W., Farach-Carson, M.C. & Rabolt, J.F. (2007). Coating electrospun collagen and gelatin fibers with perlecan domain I for increased growth factor binding. *Biomacromolecules,* Vol.8, No.4, pp 1116-1123.

Chang, C., Tran, V.H., Wang, J., Fuh, Y.-K. & Lin, L. (2010). Direct-write piezoelectric polymeric nanogenerator with high energy conversion efficiency. *Nano Letters,* Vol.10, No.2, pp 726-731.

Chen, J. & Chiang, Y. (2010). Bioactive electrospun silver nanoparticles-containing polyurethane nanofibers as wound dressings. *Journal of Nanoscience and Nanotechnology,* Vol.10, No.11, pp 7560-7564.

Chen, X., Xu, S., Yao, N. & Shi, Y. (2010a). 1.6 V nanogenerator for mechanical energy harvesting using PZT nanofibers. *Nano Letters,* Vol.10, No.6, pp 2133-2137.

Chen, Z.G., Wang, P.W., Wei, B., Mo, X.M. & Cui, F.Z. (2010b). Electrospun collagen-chitosan nanofiber: A biomimetic extracellular matrix for endothelial cell and smooth muscle cell. *Acta Biomaterialia,* Vol.6, No.2, pp 372-382.

Choi, J., Lee, J., Choi, J., Jung, D. & Shim, S.E. (2010a). Electrospun PEDOT:PSS/PVP nanofibers as the chemiresistor in chemical vapour sensing. *Synthetic Metals,* Vol.160, No.13-14, pp 1415-1421.

Choi, J.S., Leong, K.W. & Yoo, H.S. (2008). In vivo wound healing of diabetic ulcers using electrospun nanofibers immobilized with human epidermal growth factor (EGF). *Biomaterials,* Vol.29, No.5, pp 587-596.

Choi, S.-H., Hyun, T.-S., Lee, H., Jang, S.-Y., Oh, S.-G. & Kim, I.-D. (2010b). Facile synthesis of highly conductive platinum nanofiber mats as conducting core for high rate redox supercapacitor. *Electrochemical and Solid-State Letters,* Vol.13, No.6, pp A65-A68.

Choi, S.-S., Lee, Y.S., Joo, C.W., Lee, S.G., Park, J.K. & Han, K.-S. (2004). Electrospun PVDF nanofiber web as polymer electrolyte or separator. *Electrochimica Acta,* Vol.50, No.2-3, pp 339-343.

Choi, S.K., Kim, S., Lim, S.K. & Park, H. (2010c). Photocatalytic comparison of TiO_2 nanoparticles and electrospun TiO_2 nanofibers: Effects of mesoporosity and interparticle charge transfer. *Journal of Physical Chemistry C,* Vol.114, No.39, pp 16475-16480.

Choi, S.W., Kim, J.R., Jo, S.M., Lee, W.S. & Kim, Y.R. (2005). Electrochemical and spectroscopic properties of electrospun PAN-based fibrous polymer electrolytes. *Journal of the Electrochemical Society,* Vol.152, No.5, pp A989-A995.

Christopherson, G.T., Song, H. & Mao, H.-Q. (2008). The influence of fiber diameter of electrospun substrates on neural stem cell differentiation and proliferation. *Biomaterials,* Vol.30, No.4, pp 556-564.

Cui, W., Zhou, Y. & Chang, J. (2010). Electrospun nanofibrous materials for tissue engineering and drug delivery. *Science and Technology for Advanced Materials,* Vol.11, No.1, pp No pp. given.

Demir, M.M., Gulgun, M.A., Menceloglu, Y.Z., Erman, B., Abramchuk, S.S., Makhaeva, E.E., Khokhlov, A.R., Matveeva, V.G. & Sulman, M.G. (2004). Palladium nanoparticles by electrospinning from poly(acrylonitrile-co-acrylic acid)-$PdCl_2$ solutions. Relations between preparation conditions, particle size, and catalytic activity. *Macromolecules,* Vol.37, No.5, pp 1787-1792.

Deng, C., Gong, P., He, Q., Cheng, J., He, C., Shi, L., Zhu, D. & Lin, T. (2009). Highly fluorescent TPA-PBPV nanofibers with amplified sensory response to TNT. *Chemical Physics Letters,* Vol.483, No.4-6, pp 219-223.

Ding, B., Yamazaki, M. & Shiratori, S. (2005). Electrospun fibrous polyacrylic acid membrane-based gas sensors. *Sensors and Actuators, B: Chemical,* Vol.B106, No.1, pp 477-483.

Ding, Y., Zhang, P., Long, Z., Jiang, Y. & Gao, D. (2008). The morphology, structure and electrochemical properties of LiNi1/3Mn1/3Co1/3O2 prepared by electrospun method. *Journal of Alloys and Compounds,* Vol.462, No.1-2, pp 340-342.

Ding, Y., Zhang, P., Long, Z., Jiang, Y., Xu, F. & Di, W. (2009). The ionic conductivity and mechanical property of electrospun P(VdF-HFP)/PMMA membranes for lithium ion batteries. *Journal of Membrane Science,* Vol.329, No.1+2, pp 56-59.

Dong, B., Li, Z., Li, Z., Xu, X., Song, M., Zheng, W., Wang, C., Al-Deyab, S.S. & El-Newehy, M. (2010a). Highly efficient $LaCoO_3$ nanofibers catalysts for photocatalytic degradation of Rhodamine B. *Journal of the American Ceramic Society*, Vol.93, No.11, pp 3587-3590.

Dong, Y., Yong, T., Liao, S., Chan, C.K., Stevens, M.M. & Ramakrishna, S. (2010b). Distinctive degradation behaviors of electrospun polyglycolide, poly(L-lactide-co-glycolide), and poly(L-lactide-co-ε-caprolactone) nanofibers cultured with/without porcine smooth muscle cells. *Tissue Engineering, Part A*, Vol.16, No.1, pp 283-298.

Doshi, J. & Reneker, D.H. (1995). Electrospinning process and applications of electrospun fibers. *Journal of Electrostatics*, Vol.35, No.2&3, pp 151-60.

Duan, B., Wu, L., Li, X., Yuan, X., Li, X., Zhang, Y. & Yao, K. (2007). Degradation of electrospun PLGA-chitosan/PVA membranes and their cytocompatibility in vitro. *Journal of Biomaterials Science, Polymer Edition*, Vol.18, No.1, pp 95-115.

Duan, B., Yuan, X., Zhu, Y., Zhang, Y., Li, X., Zhang, Y. & Yao, K. (2006). A nanofibrous composite membrane of PLGA-chitosan/PVA prepared by electrospinning. *European Polymer Journal*, Vol.42, No.9, pp 2013-2022.

Ekaputra, A.K., Zhou, Y., Cool, S.M. & Hutmacher, D.W. (2009). Composite electrospun scaffolds for engineering tubular bone grafts. *Tissue Engingeering, Part A*, Vol.15, No.12, pp 3779-3788.

Ellison, C.J., Phatak, A., Giles, D.W., Macosko, C.W. & Bates, F.S. (2007). Melt blown nanofibers: Fiber diameter distributions and onset of fiber breakup. *Polymer*, Vol.48, No.20, pp 6180.

Fang, J., Lin, T., Tian, W., Sharma, A. & Wang, X. (2007). Toughened electrospun nanofibers from crosslinked elastomer-thermoplastic blends. *Journal of Applied Polymer Science*, Vol.105, No.4, pp 2321-2326.

Fang, J., Niu, H., Lin, T. & Wang, X. (2008). Applications of electrospun nanofibers. *Chinese Science Bulletin*, Vol.53, No.15, pp 2265-2286.

Fang, J., Wang, H., Niu, H., Lin, T. & Wang, X. (2010). Evolution of fiber morphology during electrospinning. *Journal of Applied Polymer Science*, Vol.118, No.5, pp 2553-2561.

Fong, H. (2004). Electrospun nylon 6 nanofiber reinforced BIS-GMA/TEGDMA dental restorative composite resins. *Polymer*, Vol.45, No.7, pp 2427-2432.

Formo, E., Yavuz, M.S., Lee, E.P., Lane, L. & Xia, Y. (2009). Functionalization of electrospun ceramic nanofibre membranes with noble-metal nanostructures for catalytic applications. *Journal of Materials Chemistry*, Vol.19, No.23, pp 3878-3882.

Fujihara, K., Kotaki, M. & Ramakrishna, S. (2005). Guided bone regeneration membrane made of polycaprolactone/calcium carbonate composite nano-fibers. *Biomaterials*, Vol.26, No.19, pp 4139-4147.

Gao, K., Hu, X., Dai, C. & Yi, T. (2006). Crystal structures of electrospun PVDF membranes and its separator application for rechargeable lithium metal cells. *Materials Science & Engineering, B: Solid-State Materials for Advanced Technology*, Vol.131, No.1-3, pp 100-105.

Gibson, P., Schreuder-Gibson, H. & Rivin, D. (2001). Transport properties of porous membranes based on electrospun nanofibers. *Colloids and Surfaces, A: Physicochemical and Engineering Aspects*, Vol.187-188, 469-481.

Gouma, P.I. (2003). Nanostructured polymorphic oxides for advanced chemosensors. *Reviews on Advanced Materials Science*, Vol.5, No.2, pp 147-154.

Greiner, A. & wendorff, J.H. (2007). Electrospinning: A fascinating method for the preparation of ultrathin fibers. *Angewandte Chemie, International Edition,* Vol.46, 5670-5703.

Guo, Q., Zhou, X., Li, X., Chen, S., Seema, A., Greiner, A. & Hou, H. (2009). Supercapacitors based on hybrid carbon nanofibers containing multiwalled carbon nanotubes. *Journal of Materials Chemistry,* Vol.19, No.18, pp 2810-2816.

Hall, E.A.H. (1998) Chemical sensors & biosensors for medical and biological applications.

Hansen, B.J., Liu, Y., Yang, R. & Wang, Z.L. (2010). Hybrid nanogenerator for concurrently harvesting biomechanical and biochemical energy. *ACS Nano,* Vol.4, No.7, pp 3647-3652.

He, C.-H. & Gong, J. (2003). The preparation of PVA-Pt/TiO2 composite nanofiber aggregate and the photocatalytic degradation of solid-phase polyvinyl alcohol. *Polymer Degradation and Stability,* Vol.81, No.1, pp 117-124.

He, C.-L., Huang, Z.-M. & Han, X.-J. (2009). Fabrication of drug-loaded electrospun aligned fibrous threads for suture applications. *Journal of Biomedical Materials Research, Part A,* Vol.89, No.1, pp 80-95.

He, Y., Zhang, T., Zheng, W., Wang, R., Liu, X., Xia, Y. & Zhao, J. (2010). Humidity sensing properties of $BaTiO_3$ nanofiber prepared via electrospinning. *Sensors and Actuators, B: Chemical,* Vol.146, No.1, pp 98-102.

Herricks, T.E., Kim, S.-H., Kim, J., Li, D., Kwak, J.H., Grate, J.W., Kim, S.H. & Xia, Y. (2005). Direct fabrication of enzyme-carrying polymer nanofibers by electrospinning. *Journal of Materials Chemistry,* Vol.15, No.31, pp 3241-3245.

Homaeigohar, S.S., Buhr, K. & Ebert, K. (2010). Polyethersulfone electrospun nanofibrous composite membrane for liquid filtration. *Journal of Membrane Science,* Vol.365, No.1-2, pp 68-77.

Hong, K.H. (2007). Preparation and properties of electrospun poly(vinyl alcohol)/silver fiber web as wound dressings. *Polymer Engineering and Science,* Vol.47, No.1, pp 43-49.

Hong, S., Kim, D., Jo, S., Kim, D., Chin, B. & Lee, D. (2007). Graphite nanofibers prepared from catalytic graphitization of electrospun poly(vinylidene fluoride) nanofibers and their hydrogen storage capacity. *Catalysis Today,* Vol.120, No.3-4, pp 413-419.

Hong, Y., Legge, R.L., Zhang, S. & Chen, P. (2003). Effect of amino acid sequence and pH on nanofiber formation of self-assembling peptides EAK16-II and EAK16-IV. *Biomacromolecules,* Vol.4, No.5, pp 1433-1442.

Hota, G., Kumar, B.R., Ng, W.J. & Ramakrishna, S. (2008). Fabrication and characterization of a boehmite nanoparticle impregnated electrospun fiber membrane for removal of metal ions. *Journal of Membrane Science,* Vol.43, No.1, pp 212-217.

Huang, C.T., Song, J.H., Tsai, C.M., Lee, W.F., Lien, D.H., Gao, Z.Y., Hao, Y., Chen, L.J. & Wang, Z.L. (2010). Single-InN-nanowire nanogenerator with upto 1 V output voltage. *Advanced Materials,* Vol.22, No.36, pp 4008-4013.

Huang, H., He, C., Wang, H. & Mo, X. (2009). Preparation of core-shell biodegradable microfibers for long-term drug delivery. *Journal of Biomedical Materials Research, Part A,* Vol.90, No.4, pp 1243-1251.

Huang, X., Ge, D. & Xu, Z. (2007). Preparation and characterization of stable chitosan nanofibrous membrane for lipase immobilization. *European Polymer Journal,* Vol.43, No.9, pp 3710-3718.

Huang, X., Xu, Z., Wan, L., Innocent, C. & Seta, P. (2006). Electro-spun nano fibers modified with phospholipid moieties for enzyme immobilization. *Macromolecular Rapid Communications,* Vol.27, No.16, pp 1341-1345.

Ignatova, M., Manolova, N. & Rashkov, I. (2007). Electrospinning of poly(vinyl pyrrolidone)-iodine complex and poly(ethylene oxide)/poly(vinyl pyrrolidone)-iodine complex - a prospective route to antimicrobial wound dressing materials. *European Polymer Journal,* Vol.43, No.5, pp 1609-1623.

Ikegame, M., Tajima, K. & Aida, T. (2003). Template synthesis of polypyrrole nanofibers insulated within one-dimensional silicate channels: Hexagonal versus lamellar for recombination of polarons into bipolarons. *Angewandte Chemie, International Edition,* Vol.42, No.19, pp 2154-2157.

Im, J.S., Kang, S., Lee, S. & Lee, Y. (2010a). Improved gas sensing of electrospun carbon fibers based on pore structure, conductivity and surface modification. *Carbon,* Vol.48, No.9, pp 2573-2581.

Im, J.S., Kwon, O., Kim, Y., Park, S. & Lee, Y. (2008a). The effect of embedded vanadium catalyst on activated electrospun CFs for hydrogen storage. *Microporous and Mesoporous Materials,* Vol.115, No.3, pp 514-521.

Im, J.S., Park, S.J., Kim, T. & Lee, Y.S. (2009a). Hydrogen storage evaluation based on investigations of the catalytic properties of metal/metal oxides in electrospun carbon fibers. *International Journal of Hydrogen Energy,* Vol.34, No.8, pp 3382-3388.

Im, J.S., Park, S.J., Kim, T.J., Kim, Y.H. & Lee, Y.S. (2008b). The study of controlling pore size on electrospun carbon nanofibers for hydrogen adsorption. *Journal of Colloid and Interface Science,* Vol.318, No.1, pp 42-49.

Im, J.S., Park, S.J. & Lee, Y.S. (2009b). Superior prospect of chemically activated electrospun carbon fibers for hydrogen storage. *Materials Research Bulletin,* Vol.44, No.9, pp 1871-1878.

Im, J.S., Yun, J., Lim, Y., Kim, H. & Lee, Y. (2010b). Fluorination of electrospun hydrogel fibers for a controlled release drug delivery system. *Acta Biomaterialia,* Vol.6, No.1, pp 102-109.

Ito, Y., Hasuda, H., Kamitakahara, M., Ohtsuki, C., Tanihara, M., Kang, I.-K. & Kwon, O.H. (2005). A composite of hydroxyapatite with electrospun biodegradable nanofibers as a tissue engineering material. *Journal of Bioscience and Bioengineering,* Vol.100, No.1, pp 43-49.

Jia, H., Zhu, G., Vugrinovich, B., Kataphinan, W., Reneker, D.H. & Wang, P. (2002). Enzyme-carrying polymeric nanofibers prepared via electrospinning for use as unique biocatalysts. *Biotechnology Progress,* Vol.18, No.5, pp 1027-1032.

Jiang, H., Fang, D., Hsiao, B., Chu, B. & Chen, W. (2004). Preparation and characterization of ibuprofen-loaded poly(lactide-co-glycolide)/poly(ethylene glycol)-g-chitosan electrospun membranes. *Journal of Biomaterials Science, Polymer Edition,* Vol.15, No.3, pp 279-296.

Jiang, H., Hu, Y., Li, Y., Zhao, P., Zhu, K. & Chen, W. (2005). A facile technique to prepare biodegradable coaxial electrospun nanofibers for controlled release of bioactive agents. *Journal of Controlled Release,* Vol.108, No.2-3, pp 237-243.

Jo, J., Lee, E., Shin, D., Kim, H., Kim, H., Koh, Y. & Jang, J. (2009). In vitro/in vivo biocompatibility and mechanical properties of bioactive glass nanofiber and poly(ε-caprolactone) composite materials. *Journal of Biomedical Materials Research, Part B,* Vol.91, No.1, pp 213-220.

Joshi, P., Zhang, L., Chen, Q., Galipeau, D., Fong, H. & Qiao, Q. (2010). Electrospun carbon nanofibers as low-cost counter electrode for dye-sensitized solar cells. *ACS Applied Materials & Interfaces,* Vol.2, No.12, pp 3572-3577.

Kampalanonwat, P. & Supaphol, P. (2010). Preparation and adsorption behavior of aminated electrospun polyacrylonitrile nanofiber mats for heavy metal ion removal. *ACS Applied Materials & Interfaces,* Vol.2, No.12, pp 3619-3627.

Kannarkat, J.T., Battogtokh, J., Philip, J., Wilson, O.C. & Mehl, P.M. (2010). Embedding of magnetic nanoparticles in polycaprolactone nanofiber scaffolds to facilitate bone healing and regeneration. *Journal of Applied Physics,* Vol.107, No.9, Pt. 2, pp 09B307/1-09B307/3.

Kenawy, E.-R., Bowlin, G.L., Mansfield, K., Layman, J., Simpson, D.G., Sanders, E.H. & Wnek, G.E. (2002). Release of tetracycline hydrochloride from electrospun poly(ethylene-co-vinylacetate), poly(lactic acid), and a blend. *Journal of Controlled Release,* Vol.81, No.1-2, pp 57-64.

Kessick, R. & Tepper, G. (2006). Electrospun polymer composite fiber arrays for the detection and identification of volatile organic compounds. *Sensors and Actuators, B: Chemical,* Vol.B117, No.1, pp 205-210.

Khil, M.S., Cha, D.I., Kim, H.Y., Kim, I.S. & Bhattarai, N. (2003). Electrospun nanofibrous polyurethane membrane as wound dressing. *Journal of Biomedical Materials Research, Part B: Applied Biomaterials,* Vol.67, No.2, pp 675-679.

Ki, C.S., Gang, E.H., Um, I.C. & Park, Y.H. (2007a). Nanofibrous membrane of wool keratose/silk fibroin blend for heavy metal ion adsorption. *Journal of Membrane Science,* Vol.302, No.1-2, pp 20-26.

Ki, C.S., Kim, J.W., Hyun, J.H., Lee, K.H., Hattori, M., Rah, D.K. & Park, Y.H. (2007b). Electrospun three-dimensional silk fibroin nanofibrous scaffold. *Journal of Applied Polymer Science,* Vol.106, No.6, pp 3922-3928.

Kim, B.C., Nair, S., Kim, J., Kwak, J.H., Grate, J.W., Kim, S.H. & Gu, M.B. (2005a). Preparation of biocatalytic nanofibers with high activity and stability via enzyme aggregate coating on polymer nanofibers. *Nanotechnology,* Vol.16, No.7, pp 382-388.

Kim, C., Ngoc, B.T.N., Yang, K.S., Kojima, M., Kim, Y.A., Kim, Y.J., Endo, M. & Yang, S.C. (2007a). Self-sustained thin webs consisting of porous carbon nanofibers for supercapacitors via the electrospinning of polyacrylonitrile solutions containing zinc chloride. *Advanced Materials,* Vol.19, No.17, pp 2341-2346.

Kim, C., Park, S.H., Lee, W.J. & Yang, K.S. (2004a). Characteristics of supercapacitor electrodes of PBI-based carbon nanofiber web prepared by electrospinning. *Electrochimica Acta,* Vol.50, No.2-3, pp 877-881.

Kim, C. & Yang, K.S. (2003). Electrochemical properties of carbon nanofiber web as an electrode for supercapacitor prepared by electrospinning. *Applied Physics Letters,* Vol.83, No.6, pp 1216-1218.

Kim, D.K., Park, S.H., Kim, B.C., Chin, B.D., Jo, S.M. & Kim, D.Y. (2005b). Electrospun polyacrylonitrile-based carbon nanofibers and their hydrogen storage. *Macromolecular Research,* Vol.13, No.6, pp 521-528.

Kim, G. & Kim, W. (2007). Highly porous 3D nanofiber scaffold using an electrospinning technique. *Journal of Biomedical Materials Research, Part B: Applied Biomaterials,* Vol.81, No.1, pp 104-110.

Kim, H.-W., Lee, H.-H. & Chun, G.-S. (2008a). Bioactivity and osteoblast responses of novel biomedical nanocomposites of bioactive glass nanofiber filled poly(lactic acid). *J. Biomed. Mater. Res., Part A,* Vol.85A, No.3, pp 651-663.

Kim, H.J., Kim, Y.S., Seo, M.H., Choi, S.M. & Kim, W.B. (2009). Pt and PtRh nanowire electrocatalysts for cyclohexane-fueled polymer electrolyte membrane fuel cell. *Electrochemistry Communications,* Vol.11, No.2, pp 446-449.

Kim, I.D., Hong, J.M., Lee, B.H., Kim, D.Y., Jeon, E.K., Choi, D.K. & Yang, D.J. (2007b). Dye-sensitized solar cells using network structure of electrospun ZnO nanofiber mats. *Applied Physics Letters,* Vol.91, No.16, pp 163109/1-163109/3.

Kim, I.D., Rothschild, A., Lee, B.H., Kim, D.Y., Jo, S.M. & Tuller, H.L. (2006). Ultrasensitive chemiresistors based on electrospun TiO2 nanofibers. *Nano Letters,* Vol.6, No.9, pp 2009-2013.

Kim, J.K., Cheruvally, G., Li, X., Ahn, J.H., Kim, K.W. & Ahn, H.J. (2008b). Preparation and electrochemical characterization of electrospun, microporous membrane-based composite polymer electrolytes for lithium batteries. *Journal Power Sources,* Vol.178, No.2, pp 815-820.

Kim, J.R., Choi, S.W., Jo, S.M., Lee, W.S. & Kim, B.C. (2004b). Electrospun PVdF-based fibrous polymer electrolytes for lithium ion polymer batteries. *Electrochimica Acta,* Vol.50, No.1, pp 69-75.

Kim, J.S. & Eneker, D.H. (1999). Mechanical properties of composites using ultrafine electrospun fibers. *Polymer Composites,* Vol.20, No.1, pp 124-131.

Kim, K., Luu, Y.K., Chang, C., Fang, D., Hsiao, B.S., Chu, B. & Hadjiargyrou, M. (2004c). Incorporation and controlled release of a hydrophilic antibiotic using poly(lactide-co-glycolide)-based electrospun nanofibrous scaffolds. *Journal of Controlled Release,* Vol.98, No.1, pp 47-56.

Kim, T.G. & Park, T.G. (2006). Surface functionalized electrospun biodegradable nanofibersfor immobilization of bioactive molecules. *Biotechnology Progress,* Vol.22, No.4, pp 1108-1113.

Kim, Y.S., Nam, S.H., Shim, H.S., Ahn, H.J., Anand, M. & Kim, W.B. (2008c). Electrospun bimetallic nanowires of PtRh and PtRu with compositional variation for methanol electrooxidation. *Electrochemistry Communications,* Vol.10, No.7, pp 1016-1019.

Ko, E.K., Jeong, S.I., Rim, N.G., Lee, Y.M., Shin, H. & Lee, B.-K. (2008). In vitro osteogenic differentiation of human Mesenchymal stem cells and In vivo bone formation in composite nanofiber meshes. *Tissue Engineering, Part A,* Vol.14, No.12, pp 2105-2119.

Kosmider, K. & Scott, J. (2002). Polymeric nanofibres exhibit an enhanced air filtration performance. *Filtration+Separation,* Vol.39, No.6, pp 20-22.

Kurpinski, K.T., Stephenson, J.T., Janairo, R.R.R., Lee, H. & Li, S. (2010). The effect of fiber alignment and heparin coating on cell infiltration into nanofibrous PLLA scaffolds. *Biomaterials,* Vol.31, No.13, pp 3536-3542.

Laxminarayana, K. & Jalili, N. (2005). Functional nanotube-based textiles: Pathway to next generation fabrics with enhanced sensing capabilities. *Textile Research Journal,* Vol.75, No.9, pp 670-680.

Lee, B.H., Song, M.Y., Jang, S.Y., Jo, S.M., Kwak, S.Y. & Kim, D.Y. (2009a). Charge transport characteristics of high efficiency dye-sensitized solar sells based on electrospun TiO2 nanorod photoelectrodes. *Journal of Physical Chemistry C,* Vol.113, No.51, pp 21453-21457.

Lee, C.H., Shin, H.J., Cho, I.H., Kang, Y.M., Kim, I.A., Park, K.D. & Shin, J.W. (2005). Nanofiber alignment and direction of mechanical strain affect the ECM production of human ACL fibroblast. *Biomaterials,* Vol.26, No.11, pp 1261-70.

Lee, J.Y., Bashur, C.A., Goldstein, A.S. & Schmidt, C.E. (2009b). Polypyrrole-coated electrospun PLGA nanofibers for neural tissue applications. *Biomaterials,* Vol.30, No.26, pp 4325-4335.

Lee, K.M., Choi, J., Wycisk, R., Pintauro, P.N. & Mather, P.T. (2009c). Nafion nanofiber membranes. *ECS Transctions,* Vol.25, No.1, pp 1451-1458.

Lee, S.H., Jung, M.J., Im, J.S., Sheem, K.Y. & Lee, Y.S. (2010). Preparation and characterization of electrospun $LiFePO_4$/carbon complex improving rate performance at high C-rate. *Research on Chemical Intermediates,* Vol.36, No.6-7, pp 591-602.

Lee, S.J., Yoo, J.J., Lim, G.J., Atala, A. & Stitzel, J. (2007). In vitro evaluation of electrospun nanofiber scaffolds for vascular graft application. *Journal of Biomedical Materials Research, Part A,* Vol.83, No.4, pp 999-1008.

Li, D., Wang, Y. & Xia, Y. (2004). Electrospinning nanofibers as uniaxially aligned arrays and layer-by-layer stacked films. *Advanced Materials,* Vol.16, No.4, pp 361-366.

Li, D. & Xia, Y. (2004a). Direct fabrication of composite and ceramic hollow nanofibers by electrospinning. *Nano Letters,* Vol.4, No.5, pp 933-938.

Li, D. & Xia, Y. (2004b). Electrospinning of nanofibers: Reinventing the wheel? *Advanced Materials,* Vol.16, No.14, pp 1151-1170.

Li, G., Li, P., Zhang, C., Yu, Y., Liu, H., Zhang, S., Jia, X., Yang, X., Xue, Z. & Ryu, S. (2008a). Inhomogeneous toughening of carbon fiber/epoxy composite using electrospun polysulfone nanofibrous membranes by in situ phase separation. *Composites Science and Technology,* Vol.68, No.3-4, pp 987-994.

Li, H., Zhang, W., Li, B. & Pan, W. (2010). Diameter-dependent photocatalytic activity of electrospun TiO2 nanofiber. *Journal of the American Ceramic Society,* Vol.93, No.9, pp 2503-2506.

Li, J., Liu, E.-h., Li, W., Meng, X.-y. & Tan, S.-t. (2009). Nickel/carbon nanofibers composite electrodes as supercapacitors prepared by electrospinning. *Journal of Alloys and Compounds,* Vol.478, No.1-2, pp 371-374.

Li, M., Han, G. & Yang, B. (2008b). Fabrication of the catalytic electrodes for methanol oxidation on electrospinning-derived carbon fibrous mats. *Electrochemistry Communications,* Vol.10, No.6, pp 880-883.

Li, S., Chen, J. & Wu, W. (2007). Electrospun polyacrylonitrile nanofibrous membranes for lipase immobilization. *Journal of Molecular Catalysis B: Enzymatic,* Vol.47, No.3-4, pp 117-124.

Li, W.J., Laurencin, C.T., Caterson, E.J., Tuan, R.S. & Ko, F.K. (2002). Electrospun nanofibrous structure: A novel scaffold for tissue engineering. *Journal of Biomedical Materials Research,* Vol.60, No.4, pp 613-621.

Lin, D., Wu, H., Zhang, R. & Pan, W. (2009). Enhanced photocatalysis of electrospun Ag-ZnO heterostructured nanofibers. *Chemistry of Materials,* Vol.21, No.15, pp 3479-3484.

Lin, H.L., Wang, S.H., Chiu, C.K., Yu, T.L., Chen, L.C., Huang, C.C., Cheng, T.H. & Lin, J.M. (2010). Preparation of Nafion/poly(vinyl alcohol) electro-spun fiber composite membranes for direct methanol fuel cells. *Journal of Membrane Science,* Vol.365, No.1-2, pp 114-122.

Lin, T., Wang, H., Wang, H. & Wang, X. (2004). The charge effect of cationic surfactants on the elimination of fibre beads in the electrospinning of polystyrene. *Nanotechnology,* Vol.15, No.9, pp 1375-1381.

Lin, T., Wang, H., Wang, H. & Wang, X. (2005a). The effect of polymer concentration and charge on the morphology of the electrospun polyacrylonitrile nanofibres. *Journal of Materials Science & Technology,* Vol.21, No.9, pp 9-12.

Lin, T., Wang, H. & Wang, X. (2005b). Self-crimping bicomponent nanofibers electrospun from polyacrylonitrile and elastomeric polyurethane. *Advanced Materials,* Vol.17, No.22, pp 2699-2703.

Liu, S.J., Kau, Y.C., Chou, C.Y., Chen, J.K., Wu, R.C. & Yeh, W.L. (2010a). Electrospun PLGA/collagen nanofibrous membrane as early-stage wound dressing. *Journal of Membrane Science,* Vol.355, No.1-2, pp 53-59.

Liu, X., Lin, T., Fang, J., Yao, G. & Wang, X. (2008). Electrospun nanofibre membranes as wound dressing materials. *Advances in Science and Technology,* Vol.57, No.Biomedical Applications of Smart Materials, Nanotechnology and Micro/Nano Engineering, pp 125-130.

Liu, X., Lin, T., Fang, J., Yao, G., Zhao, H., Dodson, M. & Wang, X. (2010b). In vivo wound healing and antibacterial performances of electrospun nanofibre membranes. *Journal of Biomedical Materials Research, Part A,* Vol.94, No.2, pp 499-508.

Lu, M.Y., Song, J.H., Lu, M.P., Lee, C.Y., Chen, L.J. & Wang, Z.L. (2009). ZnO-ZnS heterojunction and ZnS nanowire arrays for electricity generation. *ACS Nano,* Vol.3, No.2, pp 357-362.

Luoh, R. & Hahn, H.T. (2006). Electrospun nanocomposite fiber mats as gas sensors. *Composites Science and Technology,* Vol.66, No.14, pp 2436-2441.

Luu, Y.K., Kim, K., Hsiao, B.S., Chu, B. & Hadjiargyrou, M. (2003). Development of a nanostructured DNA delivery scaffold via electrospinning of PLGA and PLA-PEG block copolymers. *Journal of Controlled Release,* Vol.89, No.2, pp 341-353.

Ma, P.X. (2004) Tissue engineering. In encyclopedia of polymer science and technology, NJ, John Wiley & Sons.

Ma, P.X. & Zhang, R. (1999). Synthetic nano-scale fibrous extracellular matrix. *Journal of Biomedical Materials Research,* Vol.46, No.1, pp 60-72.

Ma, Z., He, W., Yong, T. & Ramakrishna, S. (2005a). Grafting of gelatin on electrospun poly(caprolactone) nanofibers to improve endothelial cell spreading and proliferation and to control cell orientation. *Tissue Engineering,* Vol.11, No.7/8, pp 1149-1158.

Ma, Z., Kotaki, M., Yong, T., He, W. & Ramakrishna, S. (2005b). Surface engineering of electrospun polyethylene terephthalate (PET) nanofibers towards development of a new material for blood vessel engineering. *Biomaterials,* Vol.26, No.15, pp 2527-2536.

Madhugiri, S., Sun, B., Smirniotis, P.G., Ferraris, J.P. & Balkus, K.J. (2004). Electrospun mesoporous titanium dioxide fibers. *Microporous and Mesoporous Materials,* Vol.69, No.1-2, pp 77-83.

Mai, L., Xu, L., Han, C., Xu, X., Luo, Y., Zhao, S. & Zhao, Y. (2010). Electrospun ultralong hierarchical vanadium oxide nanowires with high performance for lithium ion batteries. *Nano Letters,* Vol.10, No.11, pp 4750-4755.

Marler, J.J., Upton, J., Langer, R. & Vacanti, J.P. (1998). Transplantation of cells in matrixes for tissue regeneration. *Advanced Drug Delivery Reviews,* Vol.33, No.1,2, pp 165-182.

Matthews, J.A., Boland, E.D., Wnek, G.E., Simpson, D.G. & Bowlin, G.L. (2003). Electrospinning of collagen type II: a feasibility study. *Journal of Bioactive and Compatible Polymers,* Vol.18, No.2, pp 125-134.

Matthews, J.A., Wnek, G.E., Simpson, D.G. & Bowlin, G.L. (2002). Electrospinning of collagen nanofibers. *Biomacromolecules,* Vol.3, No.2, pp 232-238.

McCann, J.T., Li, D. & Xia, Y. (2005). Electrospinning of nanofibers with core-sheath, hollow, or porous structures. *Journal of Materials Chemistry,* Vol.15, No.7, pp 735-738.

McManus, M., Boland, E., Sell, S., Bowen, W., Koo, H., Simpson, D. & Bowlin, G. (2007). Electrospun nanofibre fibrinogen for urinary tract tissue reconstruction. *Biomedical Materials,* Vol.2, No.4, pp 257-262.

Merrell, J.G., McLaughlin, S.W., Tie, L., Laurencin, C.T., Chen, A.F. & Nair, L.S. (2009). Curcumin-loaded poly(ε-caprolactone) nanofibres: diabetic wound dressing with anti-oxidant and anti-inflammatory properties. *Clinical and Experimental Pharmacology and Physiology,* Vol.36, No.12, pp 1149-1156.

Mo, X.M., Xu, C.Y., Kotaki, M. & Ramakrishna, S. (2004). Electrospun P(LLA-CL) nanofiber: a biomimetic extracellular matrix for smooth muscle cell and endothelial cell proliferation. *Biomaterials,* Vol.25, No.10, pp 1883-1890.

Molla, S. & Compan, V. (2011a). Performance of composite Nafion/PVA membranes for direct methanol fuel cells. *Journal of Power Sources,* Vol.196, No.5, pp 2699-2708.

Molla, S. & Compan, V. (2011b). Polyvinyl alcohol nanofiber reinforced Nafion membranes for fuel cell applications. *Journal of Membrane Science,* Vol.372, No.1-2, pp 191-200.

Mukhatyar, V.J., Salmeron-Sanchez, M., Rudra, S., Mukhopadaya, S., Barker, T.H., Garcia, A.J. & Bellamkonda, R.V. (2011). Role of fibronectin in topographical guidance of neurite extension on electrospun fibers. *Biomaterials,* Vol.32, No.16, pp 3958-3968.

Nguyen, T.X., Kim, T.H., Geng, H.-Z., Lee, I.H., Kim, K.K. & Lee, Y.H. (2009). Three-dimensional architecture of carbon nanotube-anchored polymer nanofiber composite. *Journal of Materials Chemistry,* Vol.19, No.42, pp 7822-7825.

Niu, H., Lin, T. & Wang, X. (2009). Needleless electrospinning. I. A comparison of cylinder and disk nozzles. *Journal of Applied Polymer Science,* Vol.114, No.6, pp 3524-3530.

Niu, H., Zhang, J., Xie, Z., Wang, X. & Lin, T. (2011). Preparation, structure and supercapacitance of bonded carbon nanofiber electrode materials. *Carbon,* Vol.49, No.7, pp 2380-2388.

Onozuka, K., Ding, B., Tsuge, Y., Naka, T., Yamazaki, M., Sugi, S., Ohno, S., Yoshikawa, M. & Shiratori, S. (2006). Electrospinning processed nanofibrous TiO2 membranes for photovoltaic applications. *Nanotechnology,* Vol.17, No.4, pp 1026-1031.

Pan, C. & Dong, L. (2009). Fabrication of gold-doped titanium dioxide (TiO2:Au) nanofibers photocatalyst by vacuum ion sputter coating. *Journal of Macromolecular Science, Part B: Physics,* Vol.48, No.5, pp 919-926.

Park, J.Y., Asokan, K., Choi, S.W. & Kim, S.S. (2011). Growth kinetics of nanograins in SnO2 fibers and size dependent sensing properties. *Sensors and Actuators, B: Chemical,* Vol.B152, No.2, pp 254-260.

Pham, Q.P., Sharma, U. & Mikos, A.G. (2006). Electrospun poly(e-caprolactone) microfiber and multilayer nanofiber/microfiber scaffolds: Characterization of scaffolds and measurement of cellular infiltration. *Biomacromolecules,* Vol.7, No.10, pp 2796-2805.

Pinto, N.J., Gonzalez, R., Johnson, A.T., Jr. & MacDiarmid, A.G. (2006). Electrospun hybrid organic/inorganic semiconductor Schottky nanodiode. *Applied Physics Letters,* Vol.89, No.3, pp 033505/1-033505/3.

Pinto, N.J., Johnson, A.T., Jr., MacDiarmid, A.G., Mueller, C.H., Theofylaktos, N., Robinson, D.C. & Miranda, F.A. (2003). Electrospun polyaniline/polyethylene oxide nanofiber field-effect transistor. *Applied Physics Letters,* Vol.83, No.20, pp 4244-4246.

Powell, H.M. & Boyce, S.T. (2008). Fiber density of electrospun gelatin scaffolds regulates morphogenesis of dermal-epidermal skin substitutes. *Journal of Biomedical Materials Research, Part A,* Vol.84, No.4, pp 1078-1086.

Priya, A.R.S., Subramania, A., Jung, Y.-S. & Kim, K.-J. (2008). High-Performance Quasi-Solid-State Dye-Sensitized Solar Cell Based on an Electrospun PVdF-HFP Membrane Electrolyte. *Langmuir,* Vol.24, No.17, pp 9816-9819.

Qi, Q., Feng, Y., Zhang, T., Zheng, X. & Lu, G. (2009a). Influence of crystallographic structure on the humidity sensing properties of KCl-doped TiO2 nanofibers. *Sensors and Actuators, B: Chemical,* Vol.B139, No.2, pp 611-617.

Qi, Q., Zhang, T., Wang, S. & Zheng, X. (2009b). Humidity sensing properties of KCl-doped ZnO nanofibers with super-rapid response and recovery. *Sensors and Actuators, B: Chemical,* Vol.B137, No.2, pp 649-655.

Qi, R., Guo, R., Shen, M., Cao, X., Zhang, L., Xu, J., Yu, J. & Shi, X. (2010). Electrospun poly(lactic-co-glycolic acid)/halloysite nanotube composite nanofibers for drug encapsulation and sustained release. *Journal of Materials Chemistry,* Vol.20, No.47, pp 10622-10629.

Qin, X.H. & Wang, S.Y. (2008). Electrospun nanofibers from crosslinked poly(vinyl alcohol) and its filtration efficiency. *Journal of Applied Polymer Science,* Vol.109, No.2, pp 951-956.

Raghavan, P., Zhao, X., Shin, C., Baek, D., Choi, J., Manuel, J., Heo, M., Ahn, J. & Nah, C. (2010). Preparation and electrochemical characterization of polymer electrolytes based on electrospun poly(vinylidene fluoride-co-hexafluoropropylene)/polyacrylonitrile blend/composite membranes for lithium batteries. *Journal of Power Sources,* Vol.195, No.18, pp 6088-6094.

Rath, B., Nam, J., Knobloch, T.J., Lannutti, J.J. & Agarwal, S. (2008). Compressive forces induce osteogenic gene expression in calvarial osteoblasts. *Journal of Biomechanics,* Vol.41, No.5, pp 1095-103.

Rho, K.S., Jeong, L., Lee, G., Seo, B.M., Park, Y.J., Hong, S.D., Roh, S., Cho, J.J., Park, W.H. & Min, B.M. (2006). Electrospinning of collagen nanofibers: Effects on the behavior of normal human keratinocytes and early-stage wound healing. *Biomaterials,* Vol.27, No.8, pp 1452-1461.

Rose, M., Kockrick, E., Senkovska, I. & Kaskel, S. (2010). High surface area carbide-derived carbon fibers produced by electrospinning of polycarbosilane precursors. *Carbon,* Vol.48, No.2, pp 403-407.

Rubenstein, D., Han, D., Goldgraben, S., El-Gendi, H., Gouma, P.-I. & Frame, M.D. (2007). Bioassay chamber for angiogenesis with perfused explanted arteries and electrospun scaffolding. *Microcirculation*, Vol.14, No.7, pp 723-737.

Saeed, K., Haider, S., Oh, T.J. & Park, S.Y. (2008). Preparation of amidoxime-modified polyacrylonitrile (PAN-oxime) nanofibers and their applications to metal ions adsorption. *Journal of Membrane Science*, Vol.322, No.2, pp 400-405.

Sawicka, K.M., Prasad, A.K. & Gouma, P.I. (2005). Metal oxide nanowires for use in chemical sensing applications. *Sensor Letters*, Vol.3, No.1, pp 31-35.

Schneider, A., Wang, X.Y., Kaplan, D.L., Garlick, J.A. & Egles, C. (2009a). Biofunctionalized electrospun silk mats as a topical bioactive dressing for accelerated wound healing. *Acta Biomaterialia*, Vol.5, No.7, pp 2570-2578.

Schneider, O.D., Weber, F., Brunner, T.J., Loher, S., Ehrbar, M., Schmidlin, P.R. & Stark, W.J. (2009b). In vivo and in vitro evaluation of flexible, cottonwool-like nanocomposites as bone substitute material for complex defects. *Acta Biomaterialia*, Vol.5, No.5, pp 1775-1784.

Sell, S.A., McClure, M.J., Barnes, C.P., Knapp, D.C., Walpoth, B.H., Simpson, D.G. & Bowlin, G.L. (2006). Electrospun polydioxanone-elastin blends: potential for bioresorbable vascular grafts. *Biomedical Materials*, Vol.1, No.2, pp 72-80.

Shin, C., Chase, G.G. & Reneker, D.H. (2005). Recycled expanded polystyrene nanofibers applied in filter media. *Colloids and Surfaces, A: Physicochemical and Engineering Aspects*, Vol.262, No.1-3, pp 211-215.

Shin, M., Yoshimoto, H. & Vacanti, J.P. (2004). In vivo bone tissue engineering using Mesenchymal stem cells on a novel electrospun nanofibrous scaffold. *Tissue Engineering*, Vol.10, No.1/2, pp 33-41.

Sisson, K., Zhang, C., Farach-Carson, M.C., Chase, D.B. & Rabolt, J.F. (2010). Fiber diameters control osteoblastic cell migration and differentiation in electrospun gelatin. *Journal of Biomedical Materials Research, Part A*, Vol.94, No.4, pp 1312-1320.

Soletti, L., Hong, Y., Guan, J., Stankus, J.J., El-Kurdi, M.S., Wagner, W.R. & Vorp, D.A. (2010). A bilayered elastomeric scaffold for tissue engineering of small diameter vascular grafts. *Acta Biomaterialia*, Vol.6, No.1, pp 110-122.

Song, M.Y., Kim, D.K., Ihn, K.J., Jo, S.M. & Kim, D.Y. (2004). Electrospun TiO2 electrodes for dye-sensitized solar cells. *Nanotechnology*, Vol.15, No.12, pp 1861-1865.

Srinivasan, S.S., Ratnadurai, R., Niemann, M.U., Phani, A.R., Goswami, D.Y. & Stefanakos, E.K. (2010). Reversible hydrogen storage in electrospun polyaniline fibers. *International Journal of Hydrogen Energy*, Vol.35, No.1, pp 225-230.

Stankus, J.J., Guan, J. & Wagner, W.R. (2004). Fabrication of biodegradable elastomeric scaffolds with sub-micron morphologies. *Journal of Biomedical Materials Research, Part A*, Vol.70, No.4, pp 603-614.

Stasiak, M., Studer, A., Greiner, A. & Wendorff, J.H. (2007). Polymer fibers as carriers for homogeneous catalysts. *Chemistry--A European Journal*, Vol.13, No.21, pp 6150-6156.

Stoilova, O., Ignatova, M., Manolova, N., Godjevargova, T., Mita, D.G. & Rashkov, I. (2010a). Functionalized electrospun mats from styrene-maleic anhydride copolymers for immobilization of acetylcholinesterase. *European Polymer Journal*, Vol.46, No.10, pp 1966-1974.

Stoilova, O., Manolova, N., Gabrovska, K., Marinov, I., Godjevargova, T., Mita, D.G. & Rashkov, I. (2010b). Electrospun polyacrylonitrile nanofibrous membranes tailored for acetylcholinesterase immobilization. *Journal of Bioactive and Compatable Polymers,* Vol.25, No.1, pp 40-57.

Sudhagar, P., Jung, J.H., Park, S., Lee, Y.-G., Sathyamoorthy, R., Kang, Y.S. & Ahn, H. (2009). The performance of coupled (CdS:CdSe) quantum dot-sensitized TiO2 nanofibrous solar cells. *Electrochemistry Communications,* Vol.11, No.11, pp 2220-2224.

Sugunan, A., Guduru, V.K., Uheida, A., Toprak, M.S. & Muhammed, M. (2010). Radially oriented ZnO nanowires on flexible poly-L-lactide nanofibers for continuous-flow photocatalytic water purification. *Journal of the American Ceramic Society,* Vol.93, No.11, pp 3740-3744.

Sun, Z., Zussman, E., Yarin, A.L., Wendorff, J.H. & Greiner, A. (2003). Compound core-shell polymer nanofibers by co-electrospinning. *Advanced Materials,* Vol.15, No.22, pp 1929-1932.

Sundmacher, K. (2010). Fuel cell engineering: Toward the design of efficient electrochemical power plants. *Industrial & Engineering Chemistry Research,* Vol.49, No.21, pp 10159-10182.

Suwantong, O., Ruktanonchai, U. & Supaphol, P. (2010). In vitro biological evaluation of electrospun cellulose acetate fiber mats containing asiaticoside or curcumin. *Journal of Biomedical Materials Research, Part A,* Vol.94, No.4, pp 1216-1225.

Tang, Z., Qiu, C., McCutcheon, J.R., Yoon, K., Ma, H., Fang, D., Lee, E., Kopp, C., Hsiao, B.S. & Chu, B. (2009). Design and fabrication of electrospun polyethersulfone nanofibrous scaffold for high-flux nanofiltration membranes. *Journal of Polymer Science, Part B: Polymer Physics,* Vol.47, No.22, pp 2288-2300.

Tao, S., Li, G. & Yin, J. (2007). Fluorescent nanofibrous membranes for trace detection of TNT vapor. *Journal of Materials Chemistry,* Vol.17, No.26, pp 2730-2736.

Telemeco, T.A., Ayres, C., Bowlin, G.L., Wnek, G.E., Boland, E.D., Cohen, N., Baumgarten, C.M., Mathews, J. & Simpson, D.G. (2005). Regulation of cellular infiltration into tissue engineering scaffolds composed of submicron diameter fibrils produced by electrospinning. *Acta Biomaterialia,* Vol.1, No.4, pp 377-85.

Thavasi, V., Singh, G. & Ramakrishna, S. (2008). Electrospun nanofibers in energy and environmental applications. *Energy & Environmental Science,* Vol.1, No.2, pp 205-221.

Theron, A., Zussman, E. & Yarin, A.L. (2001). Electrostatic field-assisted alignment of electrospun nanofibres. *Nanotechnology,* Vol.12, No.3, pp 384.

Tuzlakoglu, K., Bolgen, N., Salgado, A.J., Gomes, M.E., Piskin, E. & Reis, R.L. (2005). Nano- and micro-fiber combined scaffolds: A new architecture for bone tissue engineering. *Journal of Materials Science: Materials in Medicine,* Vol.16, No.12, pp 1099-1104.

Vaz, C.M., Van Tuijl, S., Bouten, C.V.C. & Baaijens, F.P.T. (2005). Design of scaffolds for blood vessel tissue engineering using a multi-layering electrospinning technique. *Acta Biomaterialia,* Vol.1, No.5, pp 575-82.

Verreck, G., Chun, I., Peeters, J., Rosenblatt, J. & Brewster, M.E. (2003a). Preparation and characterization of nanofibers containing amorphous drug dispersions generated by electrostatic spinning. *Pharmaceutical Research,* Vol.20, No.5, pp 810-817.

Verreck, G., Chun, I., Rosenblatt, J., Peeters, J., Van Dijck, A., Mensch, J., Noppe, M. & Brewster, M.E. (2003b). Incorporation of drugs in an amorphous state into electrospun nanofibers composed of a water-insoluble, nonbiodegradable polymer. *Journal of Controlled Release,* Vol.92, No.3, pp 349-360.

Wang, C., Shao, C., Zhang, X. & Liu, Y. (2009a). SnO_2 nanostructures-TiO_2 nanofibers heterostructures: Controlled fabrication and high photocatalytic properties. *Inorganic Chemistry,* Vol.48, No.15, pp 7261-7268.

Wang, G., Ji, Y., Huang, X., Yang, X., Gouma, P.I. & Dudley, M. (2006a). Fabrication and characterization of polycrystalline WO_3 nanofibers and their application for ammonia sensing. *Journal of Physical Chemistry B,* Vol.110, No.47, pp 23777-23782.

Wang, H., Ding, J., Lee, B., Wang, X. & Lin, T. (2007). Polypyrrole-coated electrospun nanofibre membranes for recovery of Au(III) from aqueous solution. *Journal of Membrane Science,* Vol.303, No.1-2, pp 119-125.

Wang, H.B., Mullins, M.E., Cregg, J.M., McCarthy, C.W. & Gilbert, R.J. (2010). Varying the diameter of aligned electrospun fibers alters neurite outgrowth and Schwann cell migration. *Acta Biomaterialia,* Vol.6, No.8, pp 2970-2978.

Wang, L., Yu, Y., Chen, P.C., Zhang, D.W. & Chen, C.H. (2008). Electrospinning synthesis of C/Fe3O4 composite nanofibers and their application for high performance lithium-ion batteries. *J. Power Sources,* Vol.183, No.2, pp 717-723.

Wang, X., Drew, C., Lee, S.-H., Senecal, K.J., Kumar, J. & Samuelson, L.A. (2002a). Electrospun nanofibrous membrane for highly sensitive optical sensors. *Nano Letters,* Vol.2, No.11, pp 1273-1275.

Wang, X., Kim, Y., Drew, C., Ku, B., Kumar, J. & Samuelson, L.A. (2004). Electrostatic assembly of conjugated polymer thin layers on electrospun nanofibrous membranes for biosensors. *Nano Letters,* Vol.4, No.2, pp 331-334.

Wang, X., Lee, S., Drew, C., Senecal, K.J., Kumar, J. & Samuelson, L.A. (2001). Electrospun nanofibrous membranes for optical sensing. *Polymeric Materials Science and Engineering,* Vol.85, 617-618.

Wang, X., Lee, S., Ku, B., Samuelson, L.A. & Kumar, J. (2002b). Synthesis and electrospinning of a novel fluorescent polymer PMMA-PM for quenching-based optical sensing. *Journal of Macromolecular Science, Pure and Applied Chemistry,* Vol.A39, No.10, pp 1241-1249.

Wang, X., Niu, H. & Lin, T. (2009b). Needleless electrospinning of nanofibers with a conical wire coil. *Polymer Engineering and Science,* Vol.49, No.8, pp 1582-1586.

Wang, Y. & Hsieh, Y. (2004). Enzyme immobilization to ultra-fine cellulose fibers via amphiphilic polyethylene glycol spacers. *Journal of Polymer Science, Part A: Polymer Chemistry,* Vol.42, No.17, pp 4289-4299.

Wang, Z., Xu, Z., Wan, L., Wu, J., Innocent, C. & Seta, P. (2006b). Nanofibrous membranes containing carbon nanotubes: electrospun for redox enzyme immobilization. *Macromolecular Rapid Communications,* Vol.27, No.7, pp 516-521.

Wang, Z.G., Wang, J.Q. & Xu, Z.K. (2006c). Immobilization of lipase from Candida rugosa on electrospun polysulfone nanofibrous membranes by adsorption. *Journal of Molecular Catalysis B: Enzymatic,* Vol.42, No.1-2, pp 45-51.

Wang, Z.L. (2008). Self-powered nanotech. *Scientific American.*

Wang, Z.L. & Song, J. (2006). Piezoelectric nanogenerators based on zinc oxide nanowire arrays. *Science,* Vol.312, No.5771, pp 243-246.

Wee, G., Soh, H.Z., Cheah, Y.L., Mhaisalkar, S.G. & Srinivasan, M. (2010). Synthesis and electrochemical properties of electrospun V2O5 nanofibers as supercapacitor electrodes. *Journal of Materials Chemistry,* Vol.20, No.32, pp 6720-6725.

Wu, H., Hu, L., Rowell, M.W., Kong, D., Cha, J.J., McDonough, J.R., Zhu, J., Yang, Y., McGehee, M.D. & Cui, Y. (2010a). Electrospun metal nanofiber webs as high-performance transparent electrode. *Nano Letters,* Vol.10, No.10, pp 4242-4248.

Wu, S., Li, F., Wang, H., Fu, L., Zhang, B. & Li, G. (2010b). Effects of poly(vinyl alcohol) (PVA) content on preparation of novel thiol-functionalized mesoporous PVA/SiO2 composite nanofiber membranes and their application for adsorption of heavy metal ions from aqueous solution. *Polymer,* Vol.51, No.26, pp 6203-6211.

Xie, J. & Hsieh, Y. (2003). Ultra-high surface fibrous membranes from electrospinning of natural proteins: casein and lipase enzyme. *Journal of Materials Science,* Vol.38, No.10, pp 2125-2133.

Xu, C., Yang, F., Wang, S. & Ramakrishna, S. (2004). In vitro study of human vascular endothelial cell function on materials with various surface roughness. *Journal of Biomedical Materials Research, Part A,* Vol.71, No.1, pp 154-161.

Xu, X., Yang, L., Xu, X., Wang, X., Chen, X., Liang, Q., Zeng, J. & Jing, X. (2005). Ultrafine medicated fibers electrospun from W/O emulsions. *Journal of Controlled Release,* Vol.108, No.1, pp 33-42.

Xue, Y., Wang, H., Yu, D., Feng, L., Dai, L., Wang, X. & Lin, T. (2009). Superhydrophobic electrospun POSS-PMMA copolymer fibres with highly ordered nanofibrillar and surface structures. *Chemical Communications,* No.42, pp 6418-20.

Xuyen, N.T., Jeong, H.K., Kim, G., So, K.P., An, K.H. & Lee, Y.H. (2009). Hydrolysis-induced immobilization of $Pt(acac)_2$ on polyimide-based carbon nanofiber mat and formation of Pt nanoparticles. *Journal of Materials Chemistry,* Vol.19, No.9, pp 1283-1288.

Yang, D., Li, Y. & Nie, J. (2007). Preparation of gelatin/PVA nanofibers and their potential application in controlled release of drugs. *Carbohydrate Polymers,* Vol.69, No.3, pp 538-543.

Yang, F., Murugan, R., Wang, S. & Ramakrishna, S. (2005). Electrospinning of nano/micro scale poly(L-lactic acid) aligned fibers and their potential in neural tissue engineering. *Biomaterials,* Vol.26, No.15, pp 2603-2610.

Yang, F., Xu, C.Y., Kotaki, M., Wang, S. & Ramakrishna, S. (2004). Characterization of neural stem cells on electrospun poly(L-lactic acid) nanofibrous scaffold. *Journal of Biomaterials Science, Polymer Edition,* Vol.15, No.12, pp 1483-1497.

Yang, S., Wang, C. & Chen, S. (2011). A release-induced response for the rapid recognition of latent fingerprints and formation of inkjet-printed patterns. *Angewandte Chemie, International Edition,* Vol.50, No.16, pp 3706-3709.

Yih, T.C. & Al-Fandi, M. (2006). Engineered nanoparticles as precise drug delivery systems. *Journal of Cellular Biochemistry,* Vol.97, No.6, pp 1184-1190.

Yoon, J., Chae, S.K. & Kim, J.M. (2007). Colorimetric sensors for volatile organic compounds (VOCs) based on conjugated polymer-embedded electrospun fibers. *Journal of the American Chemical Society,* Vol.129, No.11, pp 3038-3039.

Yoon, K., Kim, K., Wang, X., Fang, D., Hsiao, B.S. & Chu, B. (2006). High flux ultrafiltration membranes based on electrospun nanofibrous PAN scaffolds and chitosan coating. *Polymer,* Vol.47, No.7, pp 2434-2441.

Yoshimoto, H., Shin, Y.M., Terai, H. & Vacanti, J.P. (2003). A biodegradable nanofiber scaffold by electrospinning and its potential for bone tissue engineering. *Biomaterials,* Vol.24, No.12, pp 2077-2082.

Yu, D.-G., Shen, X.-X., Branford-White, C., White, K., Zhu, L.-M. & Bligh, S.W.A. (2009a). Oral fast-dissolving drug delivery membranes prepared from electrospun polyvinylpyrrolidone ultrafine fibers. *Nanotechnology,* Vol.20, No.5, pp 055104/1-055104/9.

Yu, J. & Liu, T. (2007). Preparation of nano-fiber supported palladium catalysts and their use for the catalytic hydrogenation of olefins. *Acta polymerica sinica*, No.6, pp 514-518.

Yu, Y., Gu, L., Wang, C., Dhanabalan, A., van, A.P.A. & Maier, J. (2009b). Encapsulation of Sn@carbon nanoparticles in bamboo-like hollow carbon nanofibers as an anode material in lithium-based batteries. *Angewandte Chemie, International Edition,* Vol.48, No.35, pp 6485-6489, S6485/1-S6485/3.

Zeng, J., Aigner, A., Czubayko, F., Kissel, T., Wendorff, J.H. & Greiner, A. (2005). Poly(vinyl alcohol) nanofibers by eectrospinning as a protein delivery system and the retardation of enzyme release by additional polymer coatings. *Biomacromolecules,* Vol.6, No.3, pp 1484-1488.

Zeng, J., Xu, X., Chen, X., Liang, Q., Bian, X., Yang, L. & Jing, X. (2003). Biodegradable electrospun fibers for drug delivery. *Journal of Controlled Release,* Vol.92, No.3, pp 227-231.

Zhang, J., Lin, T. & Wang, X. (2010a). Electrospun nanofibre toughened carbon/epoxy composites: Effects of polyetherketone cardo (PEK-C) nanofibre diameter and interlayer thickness. *Composites Science and Technology,* Vol.70, No.11, pp 1660-1666.

Zhang, J., Wang, X. & Lin, T. (2011). Synergistic effects of polyetherketone cardo (PEK-C)/carbon nanofibre composite on eposy resins. *Composites Science and Technology,* accepted for publication (18 Mar 2011), DIO:10.1016/j.compscitech.2011.03.008

Zhang, Q., Welch, J., Park, H., Wu, C., Sigmund, W. & Marijnissen, J.C.M. (2010b). Improvement in nanofiber filtration by multiple thin layers of nanofiber mats. *Journal of Aerosol Science,* Vol.41, No.2, pp 230-236.

Zhang, W., Zhu, R., Liu, X., Liu, B. & Ramakrishna, S. (2009). Facile construction of nanofibrous ZnO photoelectrode for dye-sensitized solar cell applications. *Applied Physics Letters,* Vol.95, No.4, pp 043304/1-043304/3.

Zhang, Y., Ouyang, H., Lim, C.T., Ramakrishna, S. & Huang, Z.-M. (2005a). Electrospinning of gelatin fibers and gelatin/PCL composite fibrous scaffolds. *Journal of Biomedical Materials Research, Part B: Applied Biomaterials,* Vol.72, No.1, pp 156-165.

Zhang, Y.Z., Venugopal, J., Huang, Z.M., Lim, C.T. & Ramakrishna, S. (2005b). Characterization of the surface biocompatibility of the electrospun PCL-collagen nanofibers using fibroblasts. *Biomacromolecules,* Vol.6, No.5, pp 2583-2589.

Zhang, Z., Shao, C., Zhang, L., Li, X. & Liu, Y. (2010c). Electrospun nanofibers of V-doped TiO_2 with high photocatalytic activity. *Journal of Colloid and Interface Science,* Vol.351, No.1, pp 57-62.

Zhao, P., Jiang, H., Pan, H., Zhu, K. & Chen, W. (2007). Biodegradable fibrous scaffolds composed of gelatin coated poly(ε-caprolactone) prepared by coaxial electrospinning. *Journal of Biomedical Materials Research, Part A,* Vol.83, No.2, pp 372-382.

Zhong, S., Teo, W., Zhu, X., Beuerman, R.W., Ramakrishna, S. & Yung, L.Y.L. (2006). An aligned nanofibrous collagen scaffold by electrospinning and its effects on in vitro fibroblast culture. *Journal of Biomedical Materials Research, Part A,* Vol.79, No.3, pp 456-463.

Zong, X., Bien, H., Chung, C., Yin, L., Fang, D., Hsiao, B.S., Chu, B. & Entcheva, E. (2005). Electrospun fine-textured scaffolds for heart tissue constructs. *Biomaterials,* Vol.26, No.26, pp 5330-5338.

Zong, X., Kim, K., Fang, D., Ran, S., Hsiao, B.S. & Chu, B. (2002). Structure and process relationship of electrospun bioabsorbable nanofiber membranes. *Polymer,* Vol.43, No.16, pp 4403-4412.

15

The Potential of Biomimetic Electrospun-Nanofibrous Scaffolds for Bone Tissue Engineering

Ha Na Park[1], Jung Bok Lee[1], Ho-Jin Moon[1], Dae Hyeok Yang[1] and Il Keun Kwon[1,2]
[1]*Department of Maxillofacial Biomedical Engineering, School of Dentistry*
[2]*Institute of Oral Biology, School of Dentistry*
Kyung Hee University, Seoul,
Republic of Korea

1. Introduction

Bone is a dynamic organ that is constantly being remodeled to maintain a healthy skeleton. Bone preserves skeletal composition, structural integrity and regulates mineral homeostasis for a lifetime. Bone has three important functions: first, it provides support and sites of attachment for muscles, secondly, it protects vital organs such as bone marrow and brain, and it acts as a metabolic organ with major reserves of calcium and phosphate.(Feng & McDonald, 2011)[1] Bone tissue is a mineralized organic structure mostly composed of collagenous fibers and calcium phosphate in the form of hydroxyapatite (HAp), with embedded cell components like osteoblasts and osteoclasts. For bone tissue regeneration, there are many studies on bone tissue regeneration with unlimited application and without the risk of disease transmission using the tissue-engineered scaffolds. There are many natural (collagen, gelatin, chitosan, and hyaluronic acid), synthetic biodegradable [poly(α-hydroxyacids), poly(glycolic acid) (PGA), poly(lactic acid) (PLA), and poly(lactic acid-co-glycolic acid) (PLGA)] polymers, and bioceramics (bioglass, tri-calcium phosphate, and hydroxyapatite). To design scaffolds for bone regeneration, osteoconductive and osteoinductive properties, as well as non-cytotoxicity and osteogenic differentiation ambience, must be provided. Among various fabrication methods for tissue engineering scaffolds, electrospinning (ELSP) has been used to produce nonwoven, porous, nano to micro scale fibrous scaffolds. A range of novel materials and processing tools have been developed to mimic the native extracellular matrix (ECM) for potential applications as tissue engineering scaffolds and ultimately to restore degenerated functions of the tissue. These electrospun-nanofibers have attracted considerable interest in tissue regeneration, including bone reconstruction. In order to provide some kind of function to the nanofibers, drugs, peptides, or growth factors have been loaded into the sheets. Recently, hybrid/composite nanofibers using biodegradable polymers, bioactive inorganic materials that could supply suitable mechanical properties and bioactives for progenitor/stem cells have been produced.(figure 1) (Seong et al., 2010)[2]

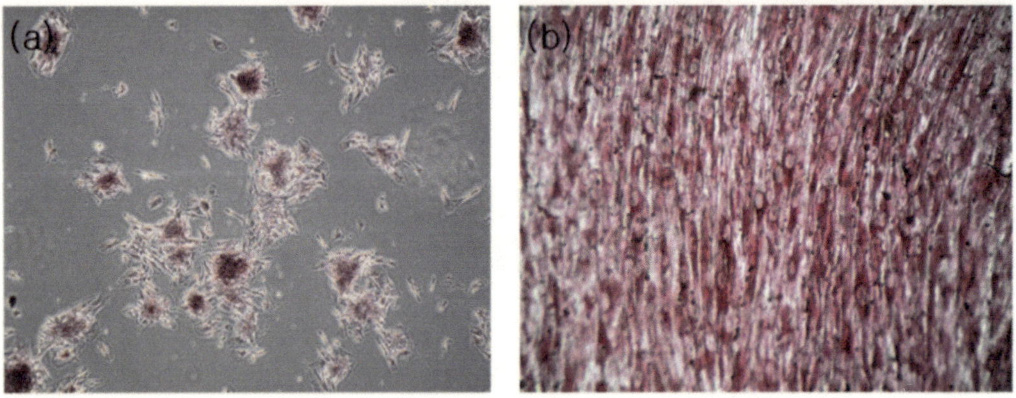

Fig. 1. Osteogenic differentiation of (a) dental pulp stem cells, and (b) adipose derived stem cells.

The nanofibers surface-functionalized with bone minerals (i.e. β-tricalcium phosphate, hydroxyapatite) or proteins/peptides are also a promising strategy for achieving therapeutic functions. Recent attempts to endow a 3D scaffolding technique to the electrospinning regime have shown some promise for engineering 3D tissue constructs. With the improvement in knowledge and techniques of bone-targeted nanofibrous matrices, bone tissue engineering is expected to be realized in the near future. In this chapter, we address the functionalization of electrospun-nanofibers with drugs, proteins/peptides, and growth factors, with particular focus on bone tissue regeneration. Some aspects of nanofiber design including architectural properties, surface functionalization and bio-functional materials selection are also addressed.

Electrospinning is an interesting and potentially profitable research area that provides outstanding opportunities for tissue regeneration, with the aim of repairing or replacing diseased or damaged tissue in the human body. Many studies have been done to better understand bone tissue trauma and diseases that result in disability and other problems. Therapeutic repair of skeletal tissues by electrospinning has raised great attention.

The repair of bone defects using artificial agents, which includes bone components and bone-associated cells, should be directed towards ensuring native bone recognizes and responds appropriately to form structural components.

2. Bone remodeling

Bone remodeling is the predominant metabolic process regulating bone structure and function. Bone remodeling involves the removal of old or damaged bone by osteoclasts (bone resorption) and the subsequent replacement of new bone formed by osteoblasts and is necessary to repair damaged bone and to maintain mineral homeostasis. The basic multicellular unit (BMU) which is known as the functional and anatomic site for bone remodeling carries this process out using various bone-lining cells, osteocytes, osteoclasts, and osteoblasts. (Parfitt, 1994; Seeman, 2009)[3] Across the BMU is a canopy of cells that creates a bone-remodeling compartment. The remodeling process involves four major distinct but overlapping phases. (a) The first phase of bone remodeling involves detection of an initiating remodeling signal. This signal can take direct mechanical strain for structural damage or hormone (estrogen or PTH) for homeostasis change. Then, in each specific site, activation of bone remodeling begins. (b) After activation of BMU, bone resorption is started by osteoclasts. Osteoblasts respond to signals generated by osteocytes or direct endocrine activation signals and recruit osteoclast precursors to the remodeling site. In this process, stromal-derived factors stimulate and this system allowed for production of two hematopoietic factors that are both necessary and sufficient for osteoclastogenesis. These are the TNF-related cytokine the receptor activator of nuclear factor κB (NF-κB) ligand (RANKL) and the monocyte/macrophage colony-stimulating factor (M-CSF) (Yasuda et al., 1998; Lacey et al., 1998)[4] and for the subsequent activation of RANK on the surface of hematopoietic precursor cells. (Nakagawa et al., 1998; Hsu et al., 1999)[5] Together, M-CSF and RANKL are required to induce expression of genes that typify the osteoclast lineage, including those encoding tartrate-resistant acid phosphatase (TRAP) leading to the development of mature osteoclasts. Then, reversal cells follow the osteoclasts, covering the newly exposed bone surface, and prepare it for deposition of replacement bone. Subsequently macrophages produce MMPs, the enzymes required for matrix degradation, and perform phagocytosis. (Newby et al., 2008)[6] (c) The remodeling process enters the next phase as osteoblast function begins to overtake bone resorption as the predominant event. Osteoblasts occupy the tail portion of the BMU by trenching the bone surface and deposit un-mineralized bone matrix known as osteoid and direct its formation and mineralization into mature lamellar bone. Osteoblast formation and function continue even after cessation of bone resorption to ensure a balance between bone removal and bone formation. (d) In the last phase, the osteoid becomes mineralized and this concludes the bone-remodeling cycle. Following mineralization mature osteoblasts go through apoptosis, return to a resting state with bone-lining cells or become embedded in the mineralized matrix, and differentiate into osteocytes. Therefore, the environment of bone surface is reestablished and maintained until the next bone remodeling process is initiated.

3. Nanofibrous scaffold using various polymers by electrospinning

Different kinds of materials such as synthetic materials, natural materials, and synthetic and natural blends, demonstrate a variety of mechanical properties, degradation rates, cell-material interactions, and biofunctions. In order to select the most proper characteristics for these polymers one must consider each material's suitability for bio-functionalization for use as bone tissue engineering scaffolds

3.1 ELSP using synthetic polymers

Synthetic materials have been commonly used in fabrication of fibrous sheets using electrospinning methods because of their good mechanical properties, ease of processing, and cost efficiency. Among the various synthetic materials, a family of poly(α-hydroxyl acid), polyester-based polymers, and their co-polymers, have been widely investigated due to their biodegradability, tissue integration effect, and easily controlled properties.(Yun et al., 2009; Seyedjafari et al., 2010; Haslauer et al., 2011; Lee et al., 2011; Burg et al., 2000)[7]
Yoshimoto et al. developed biodegradable PCL nanofiber scaffolds for bone tissue engineering. In this study the PCL scaffolds could regenerate specific defects and allowed sufficient cell penetration. Their study has shown that electrospun PCL scaffolds provide useful clinical application including an environment that supports mineralized tissue formation and may be a suitable candidate for the treatment of bone defects.(Yoshimoto et al., 2003)[8] Since Yoshimoto et al. investigation of PCL, several studies that use PCL for

bone tissue engineering have been done. (Erisken et al., 2008; Fujihara et al., 2005; Seol et al., 2010; Kolambkar et al., 2010)[9] Kolambkar et al. investigated the attachment, colonization, and osteogenic differentiation of two stem cell types, human mesenchymal stem cells (hMSCs) and human amniotic fluid stem (hAFS) cells, on electrospun PCL nanofibers. (Figure 2) They seeded cells on PCL nanofiber meshes and wrapped these around a three-dimensional (3D) collagen scaffold followed by 2 weeks of incubation *in vitro*. As a control, cells were seeded throughout the collagen scaffold alone. The collagen scaffold with cells seeded throughout had more cells on the exterior with numerous dead cells in the interior. When a cell-seeded mesh was wrapped around the scaffold, cells migrated on to the peripheral surface of the scaffold and displayed high viability. In addition cells colonized the top surface of the scaffold and migrated more than 500 μm into the scaffold from the mesh. This study provides support for the use of the PCL nanofiber mesh as a scaffold for cell culture *in vitro*, and a cell delivery agent for the repair of bone traumatic injury. (Kolambkar et al., 2010)[10]

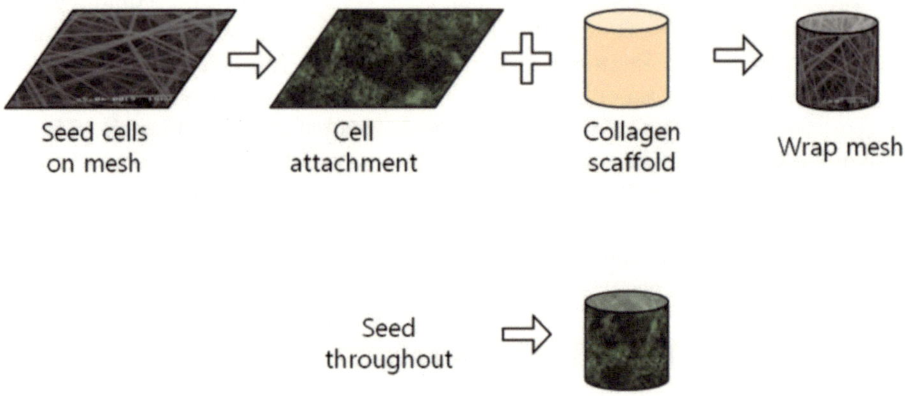

Fig. 2. Cell-seeded on nanofiber meshes to investigate the use of nanofiber meshes, cells were seeded on nanofiber meshes and wrapped around a three-dimensional (3D) collagen scaffold for 2 weeks *in vitro*.

Cho et al. studied the effect of hydrophilized PCL nanofiber mesh for guided bone regeneration (GBR). They fabricated a hybrid porous biodegradable membrane using PLGA membrane and electrospun PCL nanofiber mesh, PCL/PLGA hybrid membrane was compared with hydrophilized PLGA membrane, hydrophilized PCL nanofiber mesh and a commercialized GBR membrane for tensile and suture pullout strengths, model nutrient permeability, and bone regeneration behavior. The PCL/PLGA hybrid membrane showed the highest tensile and suture pullout strengths under both dry and wet conditions. In addition to these results, *in vivo* testing was performed and histology was done including H&E and MT staining, the fibrous connective tissues were filled into the bone defect and new bone formation was inhibited in the blank control group. However, in the membrane groups, the new bone was continuously grown, and the PCL/PLGA hybrid membrane showed compact and faster bone regeneration than other membrane groups, even though their permeability of nutrients was higher.(Figure 3,4) (Cho et al., 2009)[11]
Badami et al. studied the behaviors of MC3T3-E1 osteoblastic cells seeded on a variety of electrospun synthetic fibers including spreading, proliferation, and differentiation of

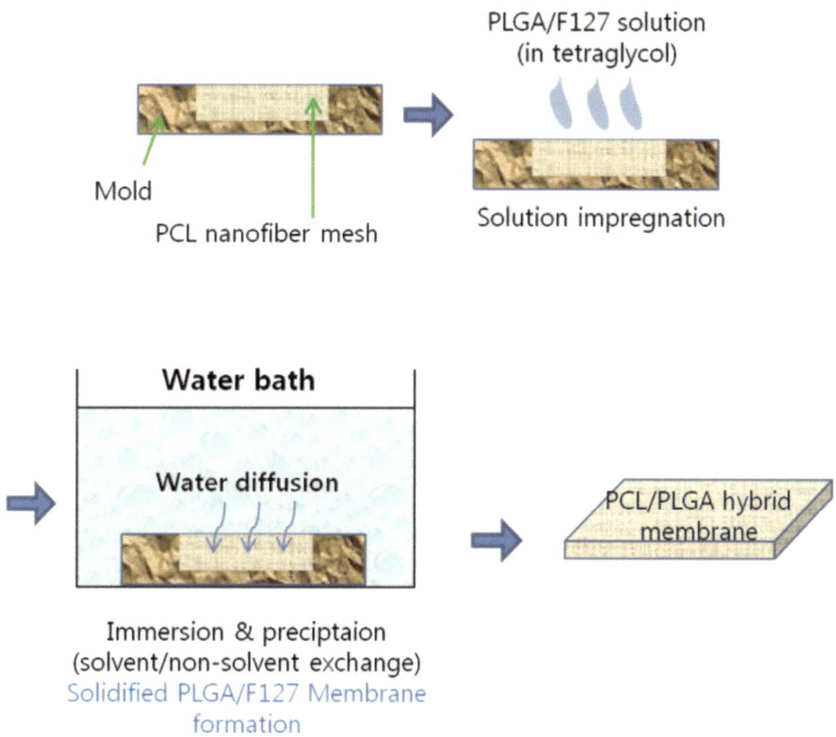

Fig. 3. Schematic of PCL/PLGA hybrid membrane fabrication process by an immersion precipitation method.

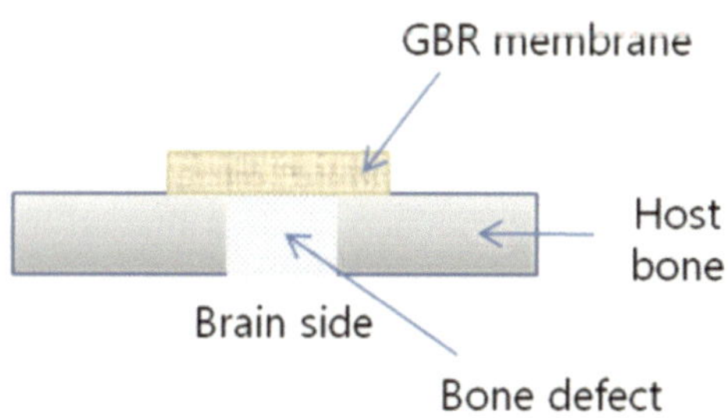

Fig. 4. Illustration of histological schematic.

MC3T3-E1. This behavior was found to depend on the fiber diameter. PLLA, PDLLA, PEG-PLLA, and PEG-PDLLA were electrospun and spin-coated on glass cover slips. The density of cultured cells on the fibrous sheet was equal to or greater than that on the smooth surface membranes in the presence of osteogenic factors. These researchers mentioned that surface morphology of the designed scaffold can promote spreading, orientation, and proliferation of osteoblast-like cells. (Badami et al., 2006)[12]
Aligned and randomly formed elastomeric poly(ester urethane)ureas (PEUURs) fibers were electrospun, and compared with spin-coated films by Bashur et al. They synthesized PEUUR by using PCL as the soft segment and a combination of urethane and urea as the

hard segment. Then, bone marrow stromal cells (BMSCs) were cultured on electrospun PEUUR meshes to determine the effect of fiber diameter and alignment on cell morphology, proliferation, and ligament gene expression. They demonstrated that these meshes support attachment and spreading of BMSCs. In addition, cell morphology was sensitive to fiber diameter and alignment, and adherent cells oriented parallel to fibers on aligned fiber PEUURs meshes. These results also showed that larger fibers suppressed expression of collagen α1, decorin, and tenomodulin and suggested that aligned electrospun submicron PEUURs fiber meshes may be preferable for ligament like tissue engineering. (Bashur et al., 2009)[13]

Synthetic materials have been widely used in electrospinning due to their easier to control pore size, fiber diameter, structural morphology, and better mechanical properties as compared to natural materials..

3.2 ELSP using natural materials

The most favorable advantage of natural polymers is their inherent biocompatibility established through the presence of specific cell recognition sites that are capable of binding cells. (Haslauer et al., 2011)[14] Natural polymers including collagen, silk, chitin, and their derivatives are the most popular materials for use as electrospun scaffolds. Collagen is one of the major extracellular matrix (ECM) morphologic components and bone tissue contains high levels of collagen type I. It is easily degraded and resorbed in the body and promotes cell adhesion. Silk is considered to be the most promising natural protein-type replacement for collagen in bone tissue engineering because of its biocompatibility, low degradation rate, and excellent mechanical strength. Chitosan possesses a number of useful characteristics such as structural similarity to glycosaminoglycan, improved bone formation, osteoconductivity, excellent biocompatibility, biodegradability, low immune response and good mechanical properties. However pure chitosan is difficult to electrospin due to its limited solubility, high viscosity, dielectric constant, and three-dimensional networks of strong hydrogen bonds. (Haslauer et al., 2011; Lee et al., 2011; O'Brien et al., 2004; Matthews et al., 2002; Beachley et al., 2010)[15] Silk fibroin (SF) has proven to be effective in biomedical applications because it has good oxygen and water vapor permeability, and biodegradability. Kim et al. fabricated SF fibrous membranes by the electrospinning method and evaluated their biocompatibility and its effect on bone regeneration using MC-3T3 cells. This studied found that the MC-3T3 cells were well attached and proliferated on the SF membrane. They also examined the activity of alkaline phosphatase (ALP) and calcification *in vitro* and *in vivo*. These results suggested that electrospun SF possessed good biocompatibility and lead to enhanced bone regeneration without any inflammatory reaction. (Kim et al., 2005)[16]

4. Functionalized fibrous sheets

In order to give biofunction to scaffolds, biodegradable polymers have been developed with various additives such as tricalcium phosphate (TCP), hydroxyapatite (HAp), calcium composite, carbon nanotubes (CNT), natural polymers, and proteins homogeneously incorporated into the scaffold.

4.1 Co-ELSP with natural polymer

Synthetic polymers such as polyester-based membranes are biodegradable, allow tissue integration and are easily controlled in terms of their physical and chemical properties.

However, they have poor cell response and bioactivity. In the case of natural polymers, they have good biocompatibility and biological activity. For these reasons, synthetic and natural polymer hybrid scaffolds have been investigated. (Haslauer et al., 2011; Lee et al., 2011; Bottino et al., 2011)[17]

For example, Haslauer et al. fabricated PCL fibers that were fully coated with collagen as formed by a core-sheath fiber. They examined the differentiation behavior of human adipose-derived stem cells (hASCs) on PCL/collagen sheath-core biocomponent fibrous scaffolds by proliferation, live/dead, and calcium quantity measurements. After two weeks of culture of hASCs on nanofibers, the results showed that viability, proliferation and osteogenesis of hASCs were promoted more thoroughly than on the PCL control nanofiber.(Figure 5) (Haslauer et al., 2011)[18].

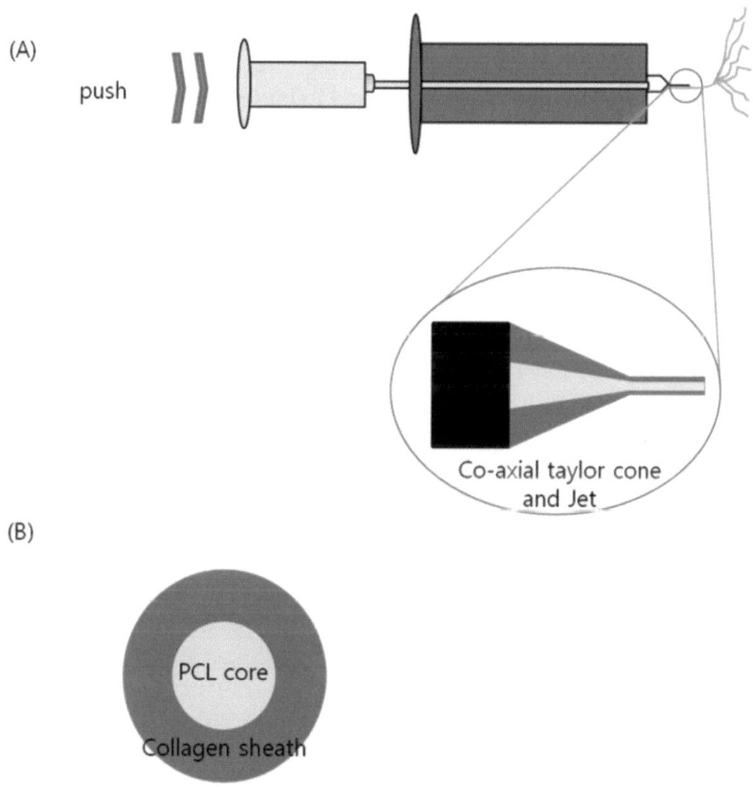

Fig. 5. Schematic illustration of core-sheath electrospinning system and its result fiber morphology.

4.2 ELSP with inorganic particles

In order to mimic natural ECM and bone component, or improve the mechanical properties of scaffolds, recent researchers have studied composites made with various particles including HAp, Ca^{2+}, and so on. Among various inorganic materials, HAp, Ca^{2+} composites and β-TCP were used in research for bone regeneration because those are the major components of bone. (Erisken et al., 2008; Fujihara et al., 2005; Seol et al., 2009; Ko et al., 2008; Seol et al., 2010)[19] HAp embedded in a collagen matrix has been extensively studied, since these are the cardinal elements of natural bone. The polymer nanofibers blended with HAp or β-TCP were able to bind to bone tissue easily because they have a similar chemical structure to the organic components of natural bone tissue. Many researchers have concluded that collagen and HAp composites are principal constituents for bone tissue engineering. (Seyedjafari et al., 2010; Nie et al., 2007; Mouthuy et al., 2010)[20] Many researchers used these particles either during electrospinning or incorporated them after electrospinning.

4.2.1 Composites made with components of natural bone

The combination of enhanced mechanical properties, biocompatibilities, and fibrous formabilities of the developed scaffolds is believed to have great potential for various biomedical engineering applications including bone tissue regeneration. Among various inorganic materials, HAp has been widely investigated due to its similarity to the mineral component of natural bone. This is considered an osteoconductive material allowing new bone regeneration. HAp has been widely applied in biomaterials as either a coating or filling material for bone tissue engineering. Electrospun fibers are hypothesized to play a role in sustaining mechanical properties, as well as allowing for biodegradability, and acting as an actual osteoconductive scaffold after being coated by hydroxyapatite.(Figure 6) (Nie et al., 2007; Mouthuy et al., 2010; Peng et al., 2011)[21]

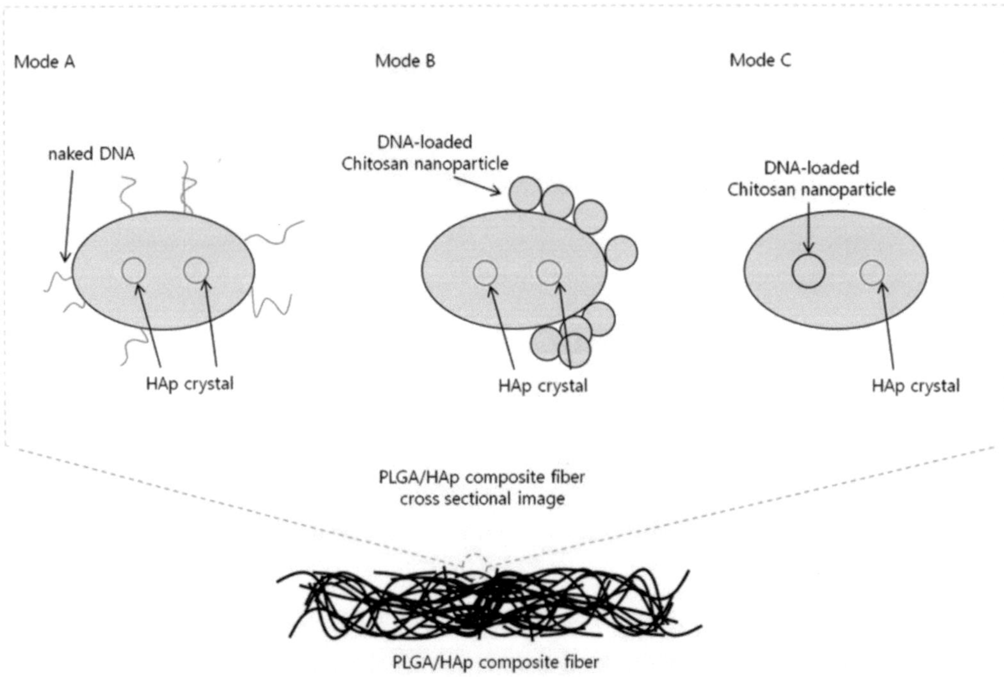

Fig. 6. Three DNA incorporation modes. (Mode A) PLGA/HAp composite fiber with naked DNA coated outside, (Mode B) PLGA/HAp composite fiber with DNA-loaded chitosan nanoparticle coated outside, and (Mode C) PLGA/HAp composite fiber with DNA-loaded chitosan nanoparticles encapsulated inside.

Bottino et al. designed and fabricated such a system via sequential multi-layering of scaffolds. They fabricated functionally graded membrane (FGM) consisting of a core layer (CL) and both functional surface layers (SLs). CL was consisted of three layers which were different ratios of PLCL, PLA, and gelatin. And SLs were layers of PLA, gelatin, and functional factors such as HAp or metronidazole benzoate (MET). MET was commonly used as an antibiotic in the treatment of periodontitis. This FGM has enhanced function including mechanical integrity, biodegradability and cell-membrane interactions. The multilayered electrospinning method could be interesting technique for generating the more predictable characteristics that can ultimately lead to promoted regeneration. (Figure 7) (Bottino et al., 2011)[22]

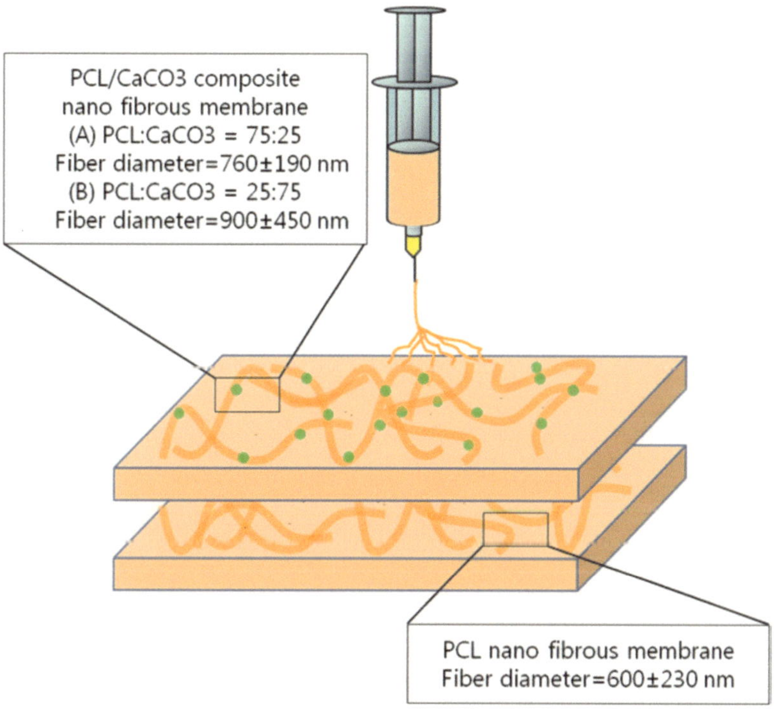

Fig. 7. Double-layered nano-fibrous GBR membranes to avoid rupture during cell seeding procedure. Two different composite membranes (i.e. (A) PCL:$CaCO_3$=75:25 and (B) PCL:$CaCO_3$=25:75) were prepared for osteoblast seeding.

PLLA and HAp composite scaffolds have been used widely for healing bone trauma and tissue regeneration. Seyedjafari et al. prepared PLLA nanofibrous scaffolds and deposited HAp nanoparticles (n-HAp) on the surface of these nanofibers. They investigated the capacity of n-HAp coated scaffolds for osteogenic induction using human cord blood derived unrestricted somatic stem cells (USSCs) for *in vitro* biocompatibility tests and subcutaneous implantation for *in vivo* animal tests. According to their results, these scaffolds had suitable mechanical properties for bone tissue engineering scaffolds and enhanced adhesion, proliferation, and osteogenic differentiation of USSCs. (Seyedjafari et al., 2010)[23]

PLLA fibrous scaffolds were also used by Whited et al. At first, they fabricated dual polymer composition scaffolds using PEO and PLLA. And then, PEO fibers were removed to increase pore size and porosity of the scaffolds effectively while coating the surface with a biomimetic apatite substrate. These scaffolds supported MC3T3-E1, pre-osteoblast proliferation, and osteogenic differentiation while accelerating cell infiltration into the scaffold and enhancing overall cell distribution. These results indicated that such a scaffold may facilitate enhanced osteoprogenitor maturation and colonization *in vitro*.(Whited et al., 2011)[24]

Other groups used PLLA for preparing nanofibers and studied the potential for use as bone tissue engineering scaffolds.

Rainer et al. studied about sternal bone healing using a HAp-functionalized electrospun scaffold. They developed PLLA/HAp nanofiber by electrospinning. The surface of the nanofiber showed that HAp particles were revealed onto the fiber and an intrinsic porosity with a uniform pore size was approximately 100nm. They tested the ability of this previously characterized scaffold to promote bone repair in a fractured animal model. The bone marrow of a rabbit was exposed, which is one of mesenchymal bone progenitors, in order to enable a direct interaction of the endogenous cells with a scaffold actively emanating a biological osteoinductive signaling. They then characterized the healing using axial reformatted images for treated and control groups at different time points and calculated bone density (BD) at each time point. They appropriately evaluated the major outcomes of their study by means of instrumental analysis and by virtue of the reported diagnostic accuracy of radiological imaging in sternal healing process which allowed for direct measurements of mineralization and fracture rim closure. They proposed the development of a cell-free system able to exploit and boost endogenous repair and stem cell resources, guiding them towards tissue restoration, because the use of autologous cells would present an additional limit in terms of clinical application. (Rainer et al., 2011)[25]

Some research groups have synthesized aligned nanofiber scaffolds within n-HAp particles using a metathesis method. Preliminary research was studied by Jose et al. They fabricated fibrous scaffolds based on PLGA and n-HAp and analyzed its mechanical properties. They dispersed the appropriate nanoparticles in the PLGA solution according to the result of their degradation tests. The fiber diameter was increased with the different amounts of n-HAp (1, 5, 10 and 20 wt.%). In this study, the increasing n-HAp concentration occurred along with an increase in the glass transition temperature. The presence of well-dispersed nano HAp particles reduced the chain mobility hence it could help to prevent shrinkage to some degree. It showed a sinusoidal trend with a slight decrease in modulus by a week due to the plasticizing effect of the medium which was followed by an increase because shrinkage, and a subsequent drop by 6th week due to degradation. (Jose et al., 2009)[26]

Yun et al. attempted to characterize the gene expression pattern during osteogenic differentiation of various stem cells on PLGA and HAp biocomposite nanofibrous scaffolds. PLGA and HAp biocomposite nanofibrous scaffolds were fabricated by electrospinning at a 5:1 blending ratio. They cultured primary adipose tissue-derived stem cells (hADSCs), bone marrow cells (MSCs), and dental pulp cells (PCs) from human in order to evaluate the biocompatibility of these scaffolds. Cytotoxicity of the scaffolds were tested using NIH-3T3 fibroblast cells. To regenerate new bone tissues, several biological properties are required for these scaffolds. Improved differentiation into osteogenic cells also improves the opportunity for bone and tooth regeneration and, continuous differentiations of stem cells into osteogenic cells were comparative investigated on this scaffold.(Yun et al., 2009)[27]

Peng et al. prepared HAp/PLLA random or aligned fibrous scaffolds and both types of HAp particles were oriented well along the long axes of fiber. The influence of the scaffold composition, HAp particle size (nano or micro) and fibrous morphology (aligned or randomized) were discussed. All of the HAp/PLLA scaffolds exhibited good biocompatibility and cell signaling property and better support for cell attachment, proliferation and differentiation than control, PLLA scaffolds. Fibrous morphology of the scaffold had a pronounced impact on the morphology of the cells in direct contact with the scaffold surface, but not on cell proliferation and differentiation. (Peng et al., 2011)[28]

Fujihara et al. and Seol et al. prepared PCL/Ca composite scaffolds.(Erisken et al., 2008; Fujihara et al., 2005; Ko et al., 2008)[29] Their results showed that these worked well as scaffolds and they discussed the impact of Ca ions, which are known to act as nucleation sites, as a promoter for the formation of low crystalline carbonate apatite crystals when released.(Figure 8).

Lee et al. investigated several PLGA/gelatin/HAp composite nanofibers and proved the applicable properties of these composites for bone tissue engineering. In this study, gelatin was shown to have good-interaction with cells when included with HAp as an electrospun fiber. They generated several nanofibrous PLGA/gelatin/HAp composites having various PLGA and gelatin blend ratios and they characterized the physical properties and morphologies of these nanofibrous composites. [Table 1] (Figure 9)

These composites showed different properties including cell adhesion and proliferation ability. (Figure 10) The results suggest that blending of gelatin and HAp with PLGA electrospun nanofibrous composites provides a good environment for bone regeneration. An optimal blend was found to occur at 3:7 of PLGA and gelatin. These composites with enhanced HAp exposure on the nanofiber surface, not only supported and provided sufficient adhesion for growth of MC3T3, osteoblast-like cells but also have properties desirable for use as a bone tissue scaffolds. (Lee et al., 2010)[30]

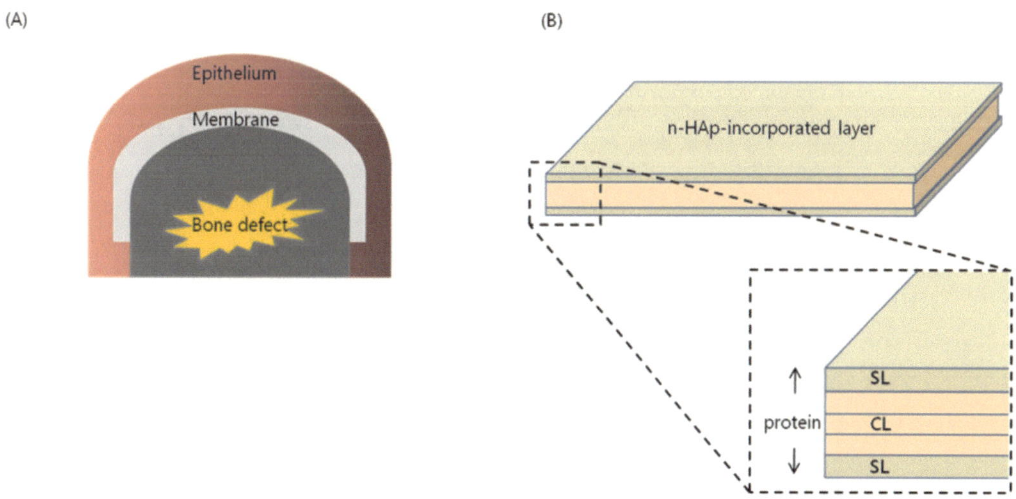

Fig. 8. Structural illustration of the functionally graded periodontal membrane(FGM). (A) Transversal illustration in guided bone regeneration. (B) Details of the core layer (CL) and the surfaces(SLs) morphology.

Mass ratio (PLGA : Gelatin : HAp)	Mean Diameter (μm)	Ultimate Tensile Stress (MPa)	Elongation at Break Point (%)	Young's Modulus (Mpa)
100 : 0 : 20	0.27 ± 0.08	2.6 ± 0.2	22.5 ± 4.0	86.7 ± 8.6
70 : 30 : 20	0.38 ± 0.14	3.3 ± 0.2	26.6 ± 9.3	79.8 ± 5.9
30 : 70 : 20	0.53 ± 0.17	4.0 ± 0.1	4.1 ± 0.7	294.8 ± 14.5
0 : 100 : 20	1.65 ± 0.13	1.8 ± 0.1	2.2 ± 0.8	142.1 ± 14.2

Table 1. Physical properties of PLGA/gelatin/HAp nanofibrous composites as various blending ratios.

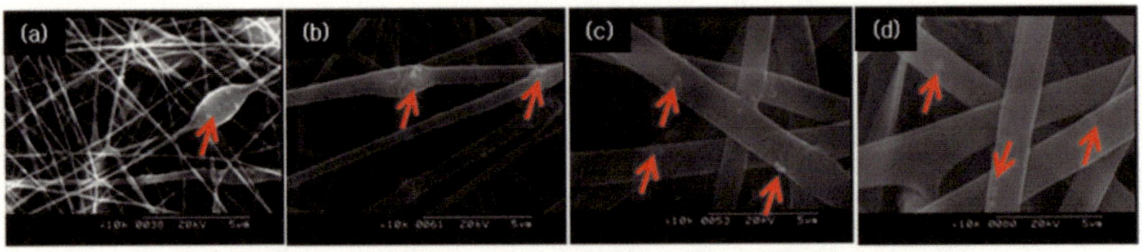

Fig. 9. Scanning electron microscopy and dispersed HAp composites. (a) PGH-1, (b) PGH-2, (c) PGH-3, (d) PGH-4.

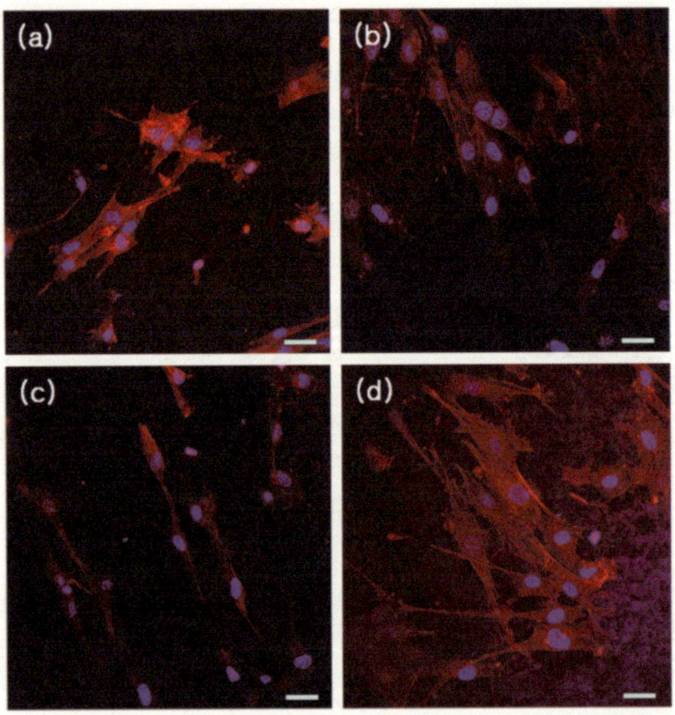

Fig. 10. Immunofluorescence of F-actin filaments in osteoblast-like cell(MC3T3) on nanofibrous PLGA/gelatin/HAp composites; (a) PGH-1, (b) PGH-2, (c) PGH-3, (d) PGH-4. Scale bars indicated 50 μm.

4.2.2 Non-components of natural bone

Kim et al. developed a nanocomposite made up of a bioactive glass in a nanofibrous form of PLA. (Figure 11)
Bioactive glasses were uniformly dispersed into the PLA nanofibers The nanocomposites induced rapid formation of a hydroxy carbonate apatite layer on the surface under simulated physiological medium. As the amount of bioactive nanofiber increased, the *in vitro* bioactivity of the nanocomposite was improved. The osteoblast response to the nanocomposites were assessed in terms of cell proliferation, differentiation, and mineralization. Osteoblasts attached and grew well on the nanocomposites and secreted collagenous proteins during the initialculturing periods. Herein the differentiation of cells and mineralized products of the composite nanofiber were observed to be significantly higher than that of pure PLA. (Kim et al., 2008)[31]
Multiwalled carbon nanotubes (MWCNTs) and demineralized bone powders (DBPs) have also been used as composites with scaffolds for bone tissue engineering. (Shao et al., 2011; Ko et al., 2008)[32]

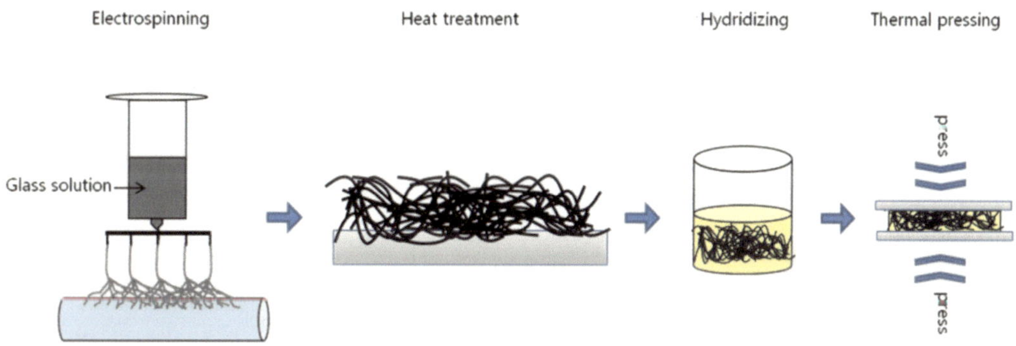

Fig. 11. Schematic illustration of the processing introduced to fabricate bioactive glass nanofiber and its nanocomposite result with PLA.

4.3 Co-ELSP with inorganic compounds

Generally, collagen accumulation, mineralization and osteoblast-specific gene expressions are also increased on electrospun scaffolds. Furthermore, some have fabricated functionalized electrospun scaffolds by using biomolecules, including calcium phosphate (Ca-P), HAp, and simulated body fluids (SBF). Rodriguez et al. generated a scaffold which mimicked bone structure and chemical composition using Ca-P and cellulose. As far as we know, most Ca composite have a similar chemical composition to the mineral constituents of bones. They obtained the cellulose fibrous scaffold by the electrospinning method, and the surface was modified with SBF and carboxymethyl cellulose (CMC) adsorption. They also investigated the bioactivity of electrospun regenerated cellulose scaffolds under physiological conditions and evaluated the enhancement of the biomineralization process due to CMC adsorption on these regenerated cellulose (RC) fibers. The biomimetic properties of Ca-P crystals on electrospun fibers were enhanced by the carboxyl groups added with CMC. This enhanced Ca-P mineralization on cellulose scaffolds can be accomplished by increasing the negative charge and by heating. Furthermore mineralization with SBF solution indicated that cellulose scaffolds can be a good candidate for use as a bioactive substrate for bone tissue engineering applications. (Rodriguez et al., 2011)[33] Rusu et al. produced HAp and chitosan hybrid fibrous composites using the electrospinning method. Pure chitosan is difficult to electrospin due to limited solubility, high viscosity, dielectric constant, and three-dimensional networks of strong hydrogen bonds. Owing to these poor electrospinnability and the adverse effects of the non-electrospinnable HAp nanoparticles and their aggregates, there have been very few attempts of using HA and chitosan for bone tissue engineering. (Rusu et al., 2005)[34]

4.4 Other treatment

4.4.1 Drug loading

Electrospun fibrous scaffolds have been proposed as drug release substrates by exploiting the direct dependence of the release rate of the drug deposited on the surface or imbedded in the fiber.

An osteogenic differentiation factor, dexamethasone (DEX), was incorporated into electrospun PCL nanofibers by Martins et al. The DEX incorporated into the polymeric carrier is in amorphous state and does not influence the typical nanofibers morphology. They studied *in vitro* drug released behavior and expression of osteogenic activity of the loaded growth/differentiation factor using human bone marrow mesenchymal stem cell. Overall, this data proposed that scaffold can be used as relevant drug carriers for bone tissue engineering strategies. (Martins et al., 2010)[35]

4.4.2 Hybrid scaffold fabricated by ELSP and other method

Erisken et al. demonstrate the fabrication and utilization of functionally graded non woven meshes of PCL incorporated with various concentrations of TCP nanoparticles using a hybrid twin-screw extrusion/electrospinning (TSEE) process. TSEE process allows various solid and liquid ingredients and their melting, dispersion, and pressurization together with electrospinning within the confines of a single process. They demonstrated *in vitro* that the ability to incorporate the β-TCP nanoparticles into PCL nanofibers in a controlled manner enabled better mimicking of the composition and structure of natural bone tissue. (Erisken et al., 2008)[36]

4.4.3 Conjugation with RGD

In 2010, poly(L-lactide) fiber meshes coated with a poly(lactide-co-ethylene oxide furmarate) (PLEOF) conjugated with Arg-Gly-Asp (RGD) peptide were reported by Xu et al. (Figure 12) Their scaffold is a fiber-reinforced hydrogel/HA laminate composite. Laminates of electrospun PLLA fiber mesh were synthesized. In this report, acrylamide-terminated RGD peptide (Ac-RGD) conjugated to the PLEOF/HA hydrogel phase could promote focal points of adhesion and lamination which increases the extent of mineralization of bone marrow stromal (BMS) cells and this resulted in significantly higher expression of osteopontin and osteocalcin as compared to the hydrogel or laminates without HAp or RGD. (Xu et al., 2010)[37]

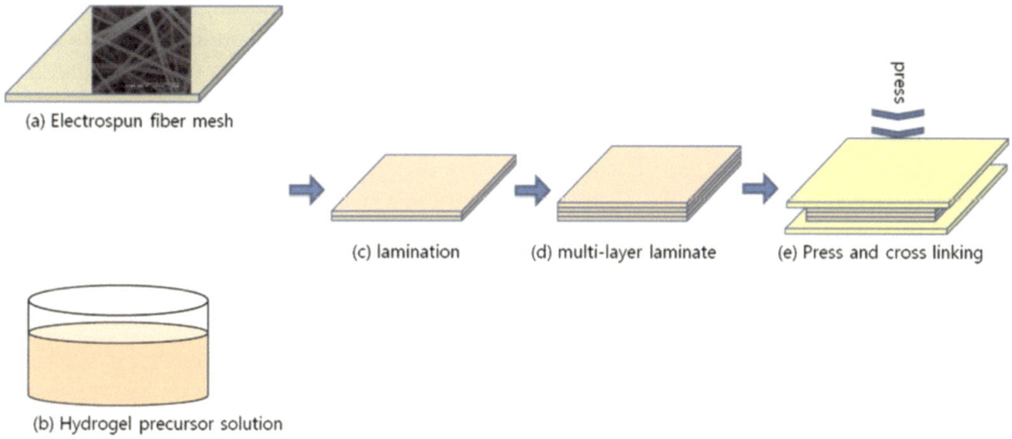

Fig. 12. Schematic diagram for the fabrication procedure of fiber-reinforced hydrogel/HA laminated composite. (a) The electrospun PLLA nanofiber mesh was placed on a Teflon plate, (b) the hydrogel precursor solution was brushed over the fibers, and (c) another fiber layer was placed on top of the precursor solution. This process was repeated to produce a multi-layer laminated composite (d). The precursor solution acted as a "glue" to hold together the fiber layers. Then, another Teflon plate was placed on top of the laminate (e), a pressure of 4.7 kPa was applied to the assembly to squeeze out the extra solution and it was allowed to crosslink (e). The hydrogel/HA precursor solution was prepared by mixing HA nanocrystals, BISAM crosslinker, Ac-GRGD cell adhesion peptide, APS initiator and TMEDA accelerator in PBS.

4.4.4 Fabrication of 3D fibrous mesh-form scaffolds using electrospun fibrous sheet

Mimicking the morphology of the natural extracellular matrix is considered a promising method for generating scaffolds for tissue engineering. Martins et al. developed hierarchical starch-based scaffolds which were fabricated by a combination of starch-polycaprolactone micro- and polycaprolactone nano-motifs. These scaffolds were produced by rapid prototyping (RP) and electrospinning method, respectively. The bulk of the composed/hierarchical scaffolds had enhanced cell retention and hierarchical scaffolds displayed significantly improved cell proliferation and osteoblastic activity, as assessed by alkaline phosphatase quantification. These observations support their hypothesis that the integration of nanofibers into 3D RP scaffolds substantially improves biological performance for use in bone tissue engineering. (Martins et al., 2009)[38]

5. Conclusion

The extracellular matrix (ECM) plays an important role in the function of bone growth. The major organic components of bone matrix are woven collagenous fibrils within hydrated polysaccharide chains such as hydroxyapatite (HAp), acting efficiently in response to external stress, and transmitting signals between cells and external stimulus via intracellular signaling cascades. Therefore, there has been a great deal of research in terms of type of materials, structural designs, and biofunctions which best mimic ECM using the electrospinning method.

Advances in nanotechnology have significantly impacted bone tissue engineering and bone regeneration. By electrospinning method, nano and micrometer-sized entities, structural and surface morphologies have been developed with unexpected properties, which can be used with various materials.

Development of more effective scaffold by means of electrospinning with the aim to aid in bone regeneration is in progress. Early scaffolds for bone tissue engineering were fabricated using electrospun polymers only. However the scaffolds provided not only mimic the ECM histoarchitecture but biological and mechanical support was needed at the same time during the healing process. This exerts a modulatory effect on cell colonization and orchestrates the bone regeneration response in a regulated manner. Inorganic particles implanted in the matrix exhibit high mechanical strength and good biocompatibility. Therefore in many studies Hap is widely used because it is similar to natural bone components and enhances the mechanical properties of scaffold. Finally nanofibers, contained drug or growth factor, can support cell growth and differentiation while at the same time maintaining physical structure.

The latest, in bone tissue engineering is the use of appropriate biomaterials including inorganic natural bone compounds, inorganic components to enhanced mechanical properties, and natural polymers. A number of studies have been done and are in progress in order to perfect the process of bone tissue regeneration.

6. Acknowledgment

This work was supported by the grant from the Ministry of Knowledge Economy of Korea (10035291), the Ministry of Education, Science and Technology of Korea (20110001689), and the Korea Science & Engineering Foundation (20100002088).

7. References

[1] Feng, X. & McDonald, M. (2011). Disorders of bone remodeling. Annu Rev Pathol, 6, 121-145.

[2] Seong, J. M., Kim, B. -C., Park, J. -H., Kwon, I. K., Mantalaris, A. & Hwang, Y. -S. (2010). Stem cells in bone tissue engineering. *Biomed Mater,* 5, 062001

[3] Parfitt, A. M. (1994). Osteonal and hemi-osteonal remodeling: the spatial and temporal framework for signal traffic in adult human bone. *J Cell Biochem,* 55, 273–286

Seeman, E. (2009). Bone modeling and remodeling. *Crit Rev Eukaryot Gene Expr,* 19, 219–233

[4] Yasuda, H., Shima, N., Nakagawa, N., Yamaguchi, K., Kinosaki, M., Mochizuki, S.-I., Tomoyasu, A., Yano, K., Goto, M., Murakami, A., Tsuda, E., Morinaga, T., Higashio, K., Udagawa, N., Takahashi, N. & Suda, T. (1998). Osteoclast differentiation factor is a ligand for osteoprotegerin/osteoclastogenesis inhibitory factor and is identical to TRANCE/RANKL. *Proc Natl Acad Sci,* 95, 3597–3602

Lacey, D. L., Timms, E., Tan, H. -L., Kelley, M. J., Dunstan, C. R., Burgess, T., Elliott, R., Colombero, A., Elliott, G., Scully, S., Hsu, H., Sullivan, J., Hawkins, N., Davy, E., Capparelli, C., Eli, A., Qian, Y. -X., Kaufman, S., Sarosi, I., Shalhoub, V., Senaldi, G., Guo, J., Delaney, J. & Boyle, W. J. (1998). Osteoprotegerin ligand is a cytokine that regulates osteoclast differentiation and activation. *Cell,* 93, 165–176

[5] Nakagawa, N., Kinosaki1, M., Yamaguchi, K., Shima, N., Yasuda, H., Yano, K., Morinaga, T. & Higashio, K. (1998). RANK is the essential signaling receptor for

osteoclast differentiation factor in osteoclastogenesis. *Biochem Biophys Res Commun*, 253, 395–400

Hsu, H., Lacey, D. L., Dunstan, C. R., Solovyev, I., Colombero, A., Timms, E., Tan, H. -L., Elliott, G., Kelley, M. J., Sarosi, I., Wang, L., Xia, X. -Z., Elliott, R., Chiu, L., Black, T., Scully, S., Capparelli, C., Morony, S., Shimamoto, G., Bass, M. B. & Boyle, W. J. (1999). Tumor necrosis factor receptor family member RANK mediates osteoclast differentiation and activation induced by osteoprotegerin ligand. *Proc Natl Acad Sci*, 96, 3540–3545

[6] Newby, A.C. (2008). Metalloproteinase expression in monocytes and macrophages and its relationship to atherosclerotic plaque instability. *Arterioscler Thromb Vasc Biol*, 28, 2108–2114

[7] Yun, Y. P., Kim, S. E., Lee, J. B., Heo, D. N., Bae, M. S., Shin, D. -R., Lim, S. -B., Choi, K. K., Park, S. J. & Kwon, I. K. (2009). Comparison of osteogenic differevtiation from adipose-derived stem cells, mesenchymal stem cells, and pulp cells on PLGA/Hydroxyapatite nanofiber. *Tissue Engineering and Regenerative Medicine*, 6, 336-345

Seyedjafari, E., Soleimani, M., Ghaemi, N. & Shabani, I. (2010). Nanohydroxyapatite-coated electrospun poly(L-lactide) nanofibers enhance osteogenic differentiation of stem cells and induce ectopic bone formation. *Biomacromolecules*, 11, 3118-3125

Haslauer, C. M., Moghe, A. K., Osborne, J. A., Gupta, B. S. & Loboa, E. G. (2011). Collagen-PCL sheath-core bicomponent electrospun scaffolds increase osteogenic differentiation and calcium accretion of human adipose-derived stem cells, *J Biomater Sci*, Vol 22, 13, 1695-1712

Lee, H., Yeo, M., Ahn, S. H., Kang, D. -O., Jang, C. H., Lee, H., Park, G. -M. & Kim, G. H. (2011). Designed hybrid scaffolds consisting of polycaprolactone microstrands and electrospun collagen-nanofibers for bone tissue regeneration, *J Biomed Mater Res Part B: Appl Biomater*, 97B, 263-270

Burg, K. J. L., Porter, S. & Kellam, J. F. (2000). Biomaterial developments for bone tissue engineering, *Biomater*, 21, 2347-2359

[8] Yoshimoto, H., Shin, Y. M., Terai, H. & Vacanti, J. P. (2003). A biodegradable nanofiber scaffold by electrospinning and its potential for bone tissue engineering, *Biomater*, 24(12), 2077-2082

[9] Erisken, C., Kalyon, D. M. & Wang, H. (2008). Funtionally graded electrospun polycaprolactone and β-tricalcium phosphate nanocomposites for tissue engineering applications, *Biomater*, 29, 4065-4073

Fujihara, K., Kotaki, M. & Ramakrishna, S. (2005). Guided bone regeneration membrane made of polycaprolactone/calcium carbonate composite nano-fibers, *Biomater*, 26, 4139-4147

Seol, Y. -J., Kim, K. -H., Kim, I. A. & Rhee, S. -H.(2010). Osteoconductive and degradable electrospun nonwoven poly(ε-caprolactone)/CaO-SiO2 gel composite fabric, *J Biomed Mater Res Part A*, 94A, 649-659

Kolambkar, Y. M., Peister, A., Ekaputra, A. K., Hutmacher, D. W. & Guldberg, R. E. (2010). Colonization and Osteogenic Differentiation of Different Stem Cell Sources on Electrospun Nanofiber Meshes, *Tissue Eng Part A*, 16(10), 3219-3230

[10] Kolambkar, Y. M., Peister, A., Ekaputra, A. K., Hutmacher, D. W. & Guldberg, R. E. (2010). Colonization and Osteogenic Differentiation of Different Stem Cell Sources on Electrospun Nanofiber Meshes, *Tissue Eng Part A*, 16(10), 3219-3230

[11] Cho, W. J., Kim, J. H., Oh, S. H., Nam, H. H., Kim, J. M. & Lee, J. H. (2009). Hydrophilized polycaprolactone nanofiber mesh-embedded poly(glycolic-co-lactic

acid) membrane for effective guided bone regeneration, *J Biomed Mater Res*, 91A, 400-407

[12] Badami, A. S., Kreke, M. R., Thompson, M. S., Riffle, J. S. & Goldstein, A. S. (2006). Effect of fiber diameter on spreading, proliferation, and differentiation of osteoblastic cells on electrospun poly(lactic acid) substrates, *Biomater*, 27, 596-606

[13] Bashur, C. A., Shaffer, R. D., Dahlgren, L. A., Guelcher, S. A. & Goldstein, A. S. (2009). Effect of fiber diameter and alignment of electrospun polyurethane meshes on mesenchymal progenitor cells, *Tissue Eng: Part A*, 15(9), 2435-2445

[14] Haslauer, C. M., Moghe, A. K., Osborne, J. A., Gupta, B. S. & Loboa, E. G. (2011). Collagen-PCL sheath-core bicomponent electrospun scaffolds increase osteogenic differentiation and calcium accretion of human adipose-derived stem cells, *J Biomater Sci*, Vol 22, 13, 1695-1712

[15] Haslauer, C. M., Moghe, A. K., Osborne, J. A., Gupta, B. S. & Loboa, E. G. (2011). Collagen-PCL sheath-core bicomponent electrospun scaffolds increase osteogenic differentiation and calcium accretion of human adipose-derived stem cells, *J Biomater Sci*, Vol 22, 13, 1695-1712

Lee, H., Yeo, M., Ahn, S. H., Kang, D. -O., Jang, C. H., Lee, H., Park, G. -M. & Kim, G. H. (2011). Designed hybrid scaffolds consisting of polycaprolactone microstrands and electrospun collagen-nanofibers for bone tissue regeneration, *J Biomed Mater Res Part B: Appl Biomater*, 97B, 263-270

O'Brien, F. J., Harley, B. A., Yannas, I. V. & Gibson, L. (2004). Influence of freezing rate on pore structure in freeze dried collagen GAG scaffolds, *Biomater*, 25, 1077-1086

Matthews, J. A., Wnek, G. E., Simpson, D. G. & Bowlin, G. L. (2002). Electrospinning of collagen nanofibers, *Biomacromol*, 3, 232-238

Beachley, V. & Wen, X. (2010). Polymer nanofibrous structures: Fabrication, biofunctionalization, and cell interactions, Prog Polym Sci, 35, 868-892

[16] Kim, K. -H., Jeong, L., Park, H. -N., Shin, S. -Y., Park, W. -H., Lee, S. -C., Kim, T. -I., Park, Y. -J., Seol, Y. -J., Lee, Y. -M., Ku, Y., Rhyu, I. -C., Han, S. -B. & Chung, C. -P. (2005). Biological efficacy of silk fibroin nanofiber membranes for guided bone regeneration, *J Biotech*, Vol. 120, 327-339

[17] Haslauer, C. M., Moghe, A. K., Osborne, J. A., Gupta, B. S. & Loboa, E. G. (2011). Collagen-PCL sheath-core bicomponent electrospun scaffolds increase osteogenic differentiation and calcium accretion of human adipose-derived stem cells, *J Biomater Sci*, Vol 22, 13, 1695-1712

Lee, H., Yeo, M., Ahn, S. H., Kang, D. -O., Jang, C. H., Lee, H., Park, G. -M. & Kim, G. H. (2011). Designed hybrid scaffolds consisting of polycaprolactone microstrands and electrospun collagen-nanofibers for bone tissue regeneration, *J Biomed Mater Res Part B: Appl Biomater*, 97B, 263-270

Bottino, M. C., Thomas, V. & Janowski, G. M.(2011). A novel spatially designed and functionally graded electrospun membrane for periodontal regeneration, *Acta Biomater*, Vol. 7, 216-224

[18] Haslauer, C. M., Moghe, A. K., Osborne, J. A., Gupta, B. S. & Loboa, E. G. (2011). Collagen-PCL sheath-core bicomponent electrospun scaffolds increase osteogenic differentiation and calcium accretion of human adipose-derived stem cells, *J Biomater Sci*, Vol 22, 13, 1695-1712

[19] Erisken, C., Kalyon, D. M. & Wang, H. (2008). Funtionally graded electrospun polycaprolactone and β-tricalcium phosphate nanocomposites for tissue engineering applications, *Biomater*, 29, 4065-4073

Fujihara, K., Kotaki, M. & Ramakrishna, S. (2005). Guided bone regeneration membrane made of polycaprolactone/calcium carbonate composite nano-fibers, *Biomater*, 26, 4139-4147

Seol, Y. -J., Kim, K. -H., Kang, Y. M., Kim, I. A. & Rhee, S. -H. (2009). Bioactivity, pre-osteoblastic cell responses, and osteoconductivity evaluations of the electrospun non-woven SiO2-CaO gel fabrics, *J Biomed Mater Res Part B: Appl Biomater*, 90B, 679-687

Seol, Y. -J., Kim, K. -H., Kim, I. A. & Rhee, S. -H.(2010). Osteoconductive and degradable electrospun nonwoven poly(ε-caprolactone)/CaO-SiO2 gel composite fabric, *J Biomed Mater Res Part A*, 94A, 649-659

Ko, E. K., Jeong, S. I., Rim, N. G., Lee, Y. M., Shin, H. & Lee, B. -K. (2008). *In vitro* osteogenic differentiation of human mesenchymal stem cells and *in vivo* bone formation in composite nanofiber meshes, *Tissue Eng Part A*, Vol. 14, No. 12, 2105-2119

[20] Seyedjafari, E., Soleimani, M., Ghaemi, N. & Shabani, I. (2010). Nanohydroxyapatite-coated electrospun poly(L-lactide) nanofibers enhance osteogenic differentiation of stem cells and induce ectopic bone formation. *Biomacromolecules*, 11, 3118-3125

Nie, H. & Wang, C. –H. (2007). Fabrication and characterization of PLGA/HAp composite scaffolds for delivery of BMP-2 plasmid DNA, *J Control Release*, Vol. 120, 111-121

Mouthuy, P. -A., Ye, H., Triffitt, J., Oommen, G. and Cui, Z. (2010). Physico-chemical characterization of functional electrospun scaffolds for bone and cartilage tissue engineering, *Proc IMechE*, Vol. 224, No. 12, 1401-1414

[21] Nie, H. & Wang, C. –H. (2007). Fabrication and characterization of PLGA/HAp composite scaffolds for delivery of BMP-2 plasmid DNA, *J Control Release*, Vol. 120, 111-121

Mouthuy, P. -A., Ye, H., Triffitt, J., Oommen, G. and Cui, Z. (2010). Physico-chemical characterization of functional electrospun scaffolds for bone and cartilage tissue engineering, *Proc IMechE*, Vol. 224, No. 12, 1401-1414

Peng, F., Yu, X. & Wei, M. (2011). *In vitro* cell performance on hydroxyapatite particles/poly(L-lactic acid) nanofibrous scaffolds with an excellent particle along nanofiber orientation, *Acta Biomater*, Vol. 7, No. 6, 2585-2592

[22] Bottino, M. C., Thomas, V. & Janowski, G. M.(2011). A novel spatially designed and functionally graded electrospun membrane for periodontal regeneration, *Acta Biomater*, Vol. 7, 216-224

[23] Seyedjafari, E., Soleimani, M., Ghaemi, N. & Shabani, I. (2010). Nanohydroxyapatite-coated electrospun poly(L-lactide) nanofibers enhance osteogenic differentiation of stem cells and induce ectopic bone formation. *Biomacromolecules*, 11, 3118-3125

[24] Whited, B. M., Whitney, J. R., Hofmann, M. C., Xu, Y. & Rylander, M. N. (2011). Pre-osteoblast infiltration and differentiation in highly porous apatite-coated PLLA electrospun scaffolds, *Biomater*, Vol. 32, No. 9, 2294-2304

[25] Rainer, A., Spadaccio, C., Sedati, P., Marco, F. D., Carotti, S., Lusini, M., Vadala, G., Martino, A. D., Morini, S., Chello, M., Covino, E., Denaro, V. & Trombetta, M. (2011). Electrospun hydroxyapatite-functionalized PLLA scaffold: Potenial applications in sternal bone healing, *Ann Biomed Eng*, Vol. 39, No. 7, 1882-1890

[26] Jose, M. V., Thomas, V., Johnson, K. T., Dean, D. R. & Nyairo, E. (2009). Aligned PLGA/HA nanofibrous nanocomposite scaffolds for bone tissue engineering, *Acta Biomater*, 5, 305-315

[27] Yun, Y. P., Kim, S. E., Lee, J. B., Heo, D. N., Bae, M. S., Shin, D. -R., Lim, S. -B., Choi, K. K., Park, S. J. & Kwon, I. K. (2009). Comparison of osteogenic differevtiation from adipose-derived stem cells, mesenchymal stem cells, and pulp cells on PLGA/Hydroxyapatite nanofiber. *Tissue Engineering and Regenerative Medicine*, 6, 336-345

[28] Peng, F., Yu, X. & Wei, M. (2011). *In vitro* cell performance on hydroxyapatite particles/poly(L-lactic acid) nanofibrous scaffolds with an excellent particle along nanofiber orientation, *Acta Biomater*, Vol. 7, No. 6, 2585-2592

[29] Erisken, C., Kalyon, D. M. & Wang, H. (2008). Funtionally graded electrospun polycaprolactone and β-tricalcium phosphate nanocomposites for tissue engineering applications, *Biomater*, 29, 4065-4073

Fujihara, K., Kotaki, M. & Ramakrishna, S. (2005). Guided bone regeneration membrane made of polycaprolactone/calcium carbonate composite nano-fibers, *Biomater*, 26, 4139-4147

Ko, E. K., Jeong, S. I., Rim, N. G., Lee, Y. M., Shin, H. & Lee, B. -K. (2008). *In vitro* osteogenic differentiation of human mesenchymal stem cells and *in vivo* bone formation in composite nanofiber meshes, *Tissue Eng Part A*, Vol. 14, No. 12, 2105-2119

[30] Lee, J. B., Kim, S. E., Heo, D. N. & Kwon, I. K. (2010). *In vitro* characterization of nanofibrous PLGA/gelatin/Hydroxyapatite composite for bone tissue engineering, *Macromol Res*, Vol. 18, No. 12, 1195-1202

[31] Kim, H.-W., Lee, H. -H. & Chun, G. -S. (2008). Bioactivity and osteoblast responses of novel biomedical nanocomposites of bioactive glass nanofiber filled poly(lactic acid), *J Biomed Mater Res*, 85A, 651-663

[32] Shao, S., Zhou, S., Li, L., Li, J., Luo, C., Wang, J., Li, X. and Weng, J. (2011). Osteoblast function on electrically conductive electrospun PLA/MWCNTs nanofiber, *Biomater*, 32, 2821-2833

Ko, E. K., Jeong, S. I., Rim, N. G., Lee, Y. M., Shin, H. & Lee, B. -K. (2008). *In vitro* osteogenic differentiation of human mesenchymal stem cells and *in vivo* bone formation in composite nanofiber meshes, *Tissue Eng Part A*, Vol. 14, No. 12, 2105-2119

[33] Rodriguez, K., Renneckar, S. and Gatenholm, Paul. (2011). Biomimetic calcium phosphate crystal mineralization on electrospun cellulose-based scaffolds, *ACS Appl Mater Interfaces*, 3, 681-689

[34] Rusu, V. M., Ng, C. -H., Wilke, M., Tiersch, B., Fratzl, P. & Peter, M. G. (2005). Size-controlled hydroxyapatite nanoparticles as self-organized organic-inorganic composite materials, *Biomater*, 26, 5414-5426

[35] Martins, A., Duarte, A. R. C., Faria, S., Marques, A. P., Reis, R. L. & Neves, N. M. (2010). Osteogenic inductions of hBMSCs by electrospun scaffolds with dexamethasone release functionality, *Biomater*, 31, 5875-5885

[36] Erisken, C., Kalyon, D. M. & Wang, H. (2008). Funtionally graded electrospun polycaprolactone and β-tricalcium phosphate nanocomposites for tissue engineering applications, *Biomater*, 29, 4065-4073

[37] Xu, W., Ma, J. & Jabbari, E. (2010). Material properties and osteogenic differentiation of marrow stromal cells in fiber-reinforced laminated hydrogel nanocomposites, *Acta Biomater*, 6, 1992-2002

[38] Martins, A., Chung, S., Pedro, A. J., Sousa, R. A., Marques, A. P., Reis, R. L. & Neves, N. M. (2009). Hierarchical starch-based fibrous scaffold for bone tissue engineering applications, *J Tissue Eng Regen Med*, 3, 37-42

16

Electrospun Nanofibers in Tissue Engineering

Mitchell R. Ladd, Tanner K. Hill, James J. Yoo and Sang Jin Lee
Wake Forest School of Medicine,
United States of America

1. Introduction

The field of tissue engineering and regenerative medicine is a fast growing scientific field. Many diseases and injuries that result in the loss of organ or tissue functions currently lack treatments which restore those functions and the patient to a desirable quality of life. Many current treatments could greatly benefit from incorporation of bioengineered organs and tissues, and investigation into this field shows great promise for modern medicine. Tissue engineering hypothesizes that by incorporating appropriate cells in the context of a three-dimensional scaffold and then implanting the cell-scaffold construct into an injury or defect, that the cells and scaffold will provide both active and passive healing properties. Various scaffolding systems exist and are generally made from natural and/or synthetic polymers. Scaffolds can be fabricated by many methods including solvent casting and particulate leaching, melt molding, rapid prototyping, phase separation, and many others (Yang et al., 2001). Of particular interest to the subject of nanofibers is electrospinning, which has seen widespread use in the field of tissue engineering due to the ease of use, scalability, adaptability, and capacity to form fibers on both the micro and nano scale (Sill & von Recum, 2008). In constructing scaffolds for tissue engineering it is ideal to provide cells with an environment which closely resembles their native extracellular matrix (ECM). The spinning of nanofibers allows for a connected and porous scaffold which can mimic the ECM of many tissue types structurally, chemically, and mechanically.

The ECM is a largely proteinaceous cellular environment that varies greatly between tissue types. In most tissue types the ECM is composed of a highly interconnected network of proteins such as collagen and elastin, and proteoglycans such as perlecan (Lodish et al., 2008). These molecules link together to form a functional environment in which cells live, move, receive and transmit signals, and which provides structural support to the cells, tissue, and organ as a whole. The composition of molecules in the ECM has a strong influence on the structural properties of the tissue which it helps compose. For example, in tissues which must stretch and bend, such as the muscle of the heart, elastin is a necessary ECM component which lends elasticity to the tissue when found in sufficient quantities (Lodish et al., 2008). Similar examples may be found in all tissue types, exemplifying the common theme of structure supplying function found throughout biology. The ECM is approximately nano scale and fibrous, though the fiber orientation depends on the tissue type. For example, dermis has randomly oriented fibers so as to provide structural support when stretched in different directions, while ligaments have a highly directional ECM so as to provide support in the direction of stress.

Cells respond to and adjust their environment in a variety of ways. Important to tissue engineering is the fact that the composition of the ECM can inform the cell as to what tissue type it is in, and hence can influence development and differentiation (Behonick & Werb, 2003). This fact is highly important for tissue engineering because rarely, if ever, are scaffolds seeded with terminally differentiated cells. Usually, a stem cell with the capacity to differentiate into a desired cell type is used, and thus control of the pathway of differentiation is important to the tissue engineer. Cells can be differentiated by the addition of soluble bioactive factors such as growth factors, but the effect of the physical environment on cellular differentiation also has a strong influence, and thus choosing scaffold properties which mimic the native ECM is desirable in order to trigger cellular differentiation toward a target cell type. During development of the chondrocyte, for example, fibronectin is deposited and then binds to cell surface receptors, triggering condensation of mesenchymal precursor cells into cell masses (Behonick & Werb, 2003). This deposited ECM will eventually come to form the boundaries between cartilage and other tissue types, just as ECM deposited in other tissues and organs establish their own boundaries. Some ECM proteins can further influence cellular development by binding growth factors and morphogens that are secreted by surrounding cells or that diffuse from a distance, providing chemical as well as physical influences on surrounding cells (Behonick & Werb, 2003).

2. The process of electrospinning

While many other scaffold fabrication methods are used in tissue engineering applications, few yet provide scaffolds with the critical similarities to the ECM that electrospinning of nanofibers can provide. Many fabrication processes allow for the inclusion of naturally derived materials such as collagen into the scaffold, but some, such as melt molding, require high temperatures which may denature proteins. Others, such as solvent casting and particulate leaching, provide scaffolds with very high porosity but potentially low interconnectivity, and may not allow for easy shaping (Yang et al., 2001). Production of nanofibers by electrospinning however provides excellent interconnectivity and control of porosity through adjustment of fiber size (Pham et al., 2006). This interconnectivity can allow for integration of cells into the scaffold if the pore size is large enough (Ju et al., 2010) and also allows for dissolution of soluble factors and nutrients through the scaffold. These properties closely mimic natural properties of the ECM and the parameters of the electrospinning process can be controlled so as to adjust to features of specific tissue types. For example, arteries are composed of a layer called the intima; an endothelial cell (EC) layer lining the lumen which sits on a basal lamina. Surrounding this layer is the tunica media which is a layer of connective tissue and smooth muscle cells (SMC). Ju et al. showed that it was possible to create a bilayered vascular scaffold which supported EC growth on an electrospun intimal layer and SMC growth on an electrospun media layer. These properties make electrospinning an incredibly versatile scaffold fabrication process, which is why it has been used for tissue engineering applications ranging from bone and blood vessels to tendons and nerve conduits.

Electrospinning in tissue engineering has come to prominence in the previous two decades. Used in many disciplines, electrospinning is a relatively simple and inexpensive scaffold fabrication process. It relies on a high voltage direct or alternating current source to charge a polymer solution or melt contained in a syringe. In tissue engineering, a solution is typically used in order to avoid high temperatures and additional equipment. Opposite the syringe is a grounded fiber collector to which the charged polymer is attracted as the polymer solution is pumped out at a controlled rate. At a critical voltage, the buildup of like charges in the solution near the tip of the syringe leads to a self-repulsive force greater than the surface tension of the liquid which causes the solution to erupt into the four distinct regions observed by Reneker and Chun (1996). These four regions form a polymer current, which terminates on the grounded collector, releasing its charge and forming solid polymer fibers. The first region is the base where the polymer jet emerges from solution and tapers away to form the Taylor cone, which points in the direction of the grounded collector (Greiner & Wendorff, 2007). Following this region is the jet which traverses much of the distance between the syringe and the collector and in which the polymer strand stretches and the solvent largely evaporates, causing the fiber diameter to decrease and solidify. The third region is often called the splay and was originally thought to be composed of many strands which bifurcate off from the main strand of the jet. However, the splay was actually shown to consist of a single rapidly whipping strand of polymer which forms an opening spiral pattern as it approaches the collector. The fourth region is the grounded collector where the polymer releases its charge and any remaining solvent evaporates, leaving only thin polymer fibers.

2.1 Electrospinning parameters

A primary benefit offered by electrospinning is the wide range of parameters which can be controlled (Table 1). These parameters can be further subdivided into three categories: processing parameters, solution parameters and ambient parameters. Processing parameters include applied voltage, distance from the syringe to the collector, flow rate, needle gauge and shape, collector shape, movement, and construction. It has been shown that increasing voltage tends to cause a recession of the Taylor cone and an increase in beading of the fibers in a poly(ethylene oxide) (PEO) and water solution (Deitzel et al. 2001). Others have seen bead formation only at low voltages, with a decrease in bead formation and increase in fiber diameter with increasing voltage in some polymer solutions and still other effects in different polymer solutions, leading to the conclusion that optimal conditions vary between different polymer/solvent systems (Sill & von Recum, 2008). The effect of flow rate on fiber morphology has not been studied intensively, but Megelski et al. (2002) have found that increasing flow rate tends to increase pore size, but may lead to beading effects due to the inability of the solvent to evaporate before it reaches the collector. The distance from the tip to the collector also influences fiber morphology. Decreasing the jet length has been found to increase fiber diameter as well as increase bead formation (Doshi & Reneker, 1995; Megelski et al., 2002). An extremely important aspect of electrospinning in tissue engineering is the geometry of the scaffold. This can be manipulated by the shape of the collector as well as rotational and translational movement. Flat electrospun mats can be formed by spinning onto a flat collector, while tubular shaped constructs can be formed by spinning onto a cylindrically shaped collector undergoing rotational, and optionally translational, movement. Fibers can also be spun such that they are aligned which can be useful depending on the application (Theron et al., 2001; Xu et al., 2004).

Solution parameters include polymer concentration, polymer molecular weight, solution conductivity, solution volatility, and solution surface tension. Polymer concentration has a great influence on the electrospinning process, affecting both fiber morphology and other parameters such as surface tension (Sill & von Recum, 2008). A sufficient amount of polymer

must be dissolved in the solvent in order for fiber formation to occur as the polymer strands must have sufficient chain overlap in order to physically link and form fibers. In general, increasing polymer concentration typically increases fiber diameter (Gupta et al., 2005). Surface tension is affected by polymer concentration, solvent, and addition of other chemical entities. It has been shown to have varying effects on fiber diameter (Pham et al., 2006) and may affect the critical voltage at which the spinning process begins.
Ambient parameters have not been studied extensively, but it has been seen that temperature may influence fiber diameter by changing the viscosity of the polymer jet (Pham et al., 2006). Humidity may also affect fiber morphology, especially in polymers which are dissolved in aqueous solution, as excessive humidity may prevent solvent evaporation. While, these parameters may not be easily manipulated in many common lab settings, they must be controlled as much as possible in order to produce consistent results.

Fabrication parameters	Effect on fiber morphology
Solution concentration/viscosity	↓ - bead formation, ↑ - larger fibers
Conductivity/solution charge density	↑ - uniform bead-free fibers, smaller fibers
Surface tension	No conclusive link
Polymer molecular weight/viscosity	↑ - reduced the number of beads
Dipole moment and dielectric constant	Successful spinning
Flow rate	↓ - smaller fibers, ↑ - larger fibers
Field strength/voltage	↓ - larger fibers, ↑ - smaller fibers
Distance between needle tip and collector	↓ - smaller fibers, ↑ - larger fibers
Needle tip design and diameter	↑ needle diameter - larger fibers
Collector geometry	Controlled fiber orientation, construct shape
Ambient parameters (temperature and humidity)	↑ temperature - ↓ viscosity, ↑ humidity - appearance of circular pores on the fibers

Table 1. Fabrication parameters for controlling the configuration of electrospun fibers.

3. Uses of electrospun nanofibers in tissue engineering

Electrospun fibers have been used in practically all areas of tissue engineering. Tissue engineering itself is a diverse field which aims to restore or replace the functions of tissues and organs damaged by injury and disease. In this section we will focus our topic by selecting tissue types which have been studied sufficiently using electrospun nanofibers.

3.1 Osseous tissue

The engineering of bone tissue presents unique problems due to the mechanical properties of hard tissue and the incorporation of minerals and inorganic molecules. With the exception of the ossicles, bones provide either structural or protective support, functioning in motion, maintenance of shape, formation of hinges and shields to protect vital organs. Bones also function as a reservoir for marrow, containing and protecting the hematopoietic stem cells (HSC) which produce all of our red and white blood cells. The focus of this section will be on the engineering of osseous tissue, which is the hard, calcified material stereotypically thought of as "bone". Osseous tissue is divided into two subtypes, compact and spongy, which are synonymous with cortical and cancellous bone. Compact bone is more dense than spongy, and tends to be stronger, more dense, and less vascular. Spongy bone contains marrow and is found inside compact osseous tissue and at the ends of long bones. Electrospun nanofiber meshes do not, on their own, possess mechanical properties similar to that of osseous tissue, but they have nonetheless been studied extensively in bone tissue engineering for their ability to support cell growth and differentiation into osteogenic phenotypes. A key cellular phenotype is the osteoblast, which is the cell type that lays down the ECM of osseous tissue and is thus the target cell type for recruitment and differentiation in bone reconstruction. Also found in osseous tissue are osteocytes and osteoclasts, which function in tissue maintenance and resorption respectively (Plotkin et al., 1999). All aforementioned cell types are found in mature osseous tissue, but tissue engineers focus on using osteoblasts because of their production of new ECM is essential for restructuring an electrospun mesh into a tissue capable of functioning as bone.

Several broad categories of nanofibers have been used in osseous tissue engineering: synthetic polymers such as poly(ε-caprolactone) (PCL), poly(L-lactic acid) (PLLA) and their copolymers, natural polymers such as collagen, composites of natural and synthetic polymers, inorganic materials such as bioactive glass and hydroxyapatite (HA), and composites of inorganic materials with synthetic and natural polymers (Jang et al., 2009). Results have shown that fibers incorporating materials naturally found in the ECM tend to be more effective at stimulating desired growth of cells than purely synthetic polymers, which is likely due the increased cell binding moieties found on natural materials. Nevertheless, studies using purely synthetic polymers have yielded useful information regarding the effect of surface topology on osteogenic differentiation. Badami et al. (2006) have shown that osteoprogenitor cell populations cultured on PLA, poly(ethylene glycol) (PEG) and their copolymers, with mean fiber diameters from 140 nm to 2.1 μm, had cell densities equal to or greater than cell populations cultured on glass plates in the presence of osteogenic culture medium. However, cell densities were actually observed to be lower when osteogenic medium was removed, illustrating the importance of proper chemical signaling, via ECM and soluble factors, in the growth and differentiation of all cells. An example of this is a study conducted using human mesenchymal stem cells (MSC), which are capable of differentiation into osseous tissue lineages, grown on electrospun silk fibroin and PEO, mixed with bone morphogenic protein-2 (BMP-2), HA nanoparticles or both. A synergistic effect was seen in calcium production per unit DNA (μg Ca/ng DNA) when both BMP-2 and HA nanoparticles where present in the scaffolds (Li et al., 2006). From this it can be concluded that providing cells with both chemical cues in the form of growth factors and a like-native substrate in the form of a nanofiber matrix incorporating HA is superior to either factor alone. This reaffirms the idea that in order to trigger stem cell differentiation toward a desired lineage, the native environment of that lineage must be reconstructed as closely as possible. Many attempts have been made at this reconstruction, including some interesting combinations of post-electrospinning mineralization. One successful attempt utilized sequential addition of acetic acid, $CaCl_2$, and H_3PO_4 onto electrospun collagen and electrospun poly(lactide-*co*-glycolide) (PLGA) nanofibers to create mineralized scaffolds for use in osseous tissue engineering (Liao et al., 2008). Their process resulted in noticeably mineralized nanofibers, effectively changing the surface topography of the matrix. Collagen nanofibers tended to have a more uniform distribution of mineralization than PLGA, and the mineralization tended to form where fibers crossed. Examination of the mineral phase formed during this process showed that the scaffolds were largely coated with HA and brushite, making this process an attractive method for post-electrospinning fiber modification in bone tissue engineering.

Another study illustrating the idea of post-electrospinning mineralization utilized a combination of electrospinning poly(lactide-*co*-ε-caprolactone) (PLCL)/gelatin and electrospraying HA nanoparticles (Gupta et al., 2009). Polymer solutions were electrospun onto glass coverslips taped to a rotating mandrel while a HA solution was electrosprayed concurrently onto the fibers. In comparison to fibers created using PLCL/gelatin/HA blends, the morphology of concurrently electrospun and electrosprayed fibers showed significantly more mineralization and topological heterogeneity. Additionally, when human fetal osteoblast cells were cultured on the PLCL/gelatin/HA electrosprayed scaffolds they showed higher alkaline phosphatase activity, an osteoblast marker, than PLCL/gelatin/HA blended scaffolds. These results not only reiterate the need for creating ECM mimicking scaffolds, but also illustrate that the development of new scaffold fabrication methods are still highly important in the field of tissue engineering.

Of particular interest in the area of adapting electrospinning for new techniques is the growing field of gene therapy. The ability to force expression of target genes in cells has recently garnered much attention, and a prime example in bone tissue engineering is the incorporation of a BMP-2 plasmid into PLGA/HA electrospun scaffolds (Nie & Wang, 2007). This example is especially interesting because it uses three different methods of DNA incorporation, giving rise to three different release profiles. Transfection by plasmid vectors is advantageous because, unlike viral vectors, there is little risk of incorporation of the DNA into the genome. Plasmid DNA was incorporated into the scaffolds by binding naked DNA onto the PLGA/HA nanofiber surface, binding DNA/chitosan nanoparticles to fiber surface, and incorporating DNA/chitosan nanoparticles into the fibers themselves. The addition of nanoparticles into the polymer solution significantly changed fiber morphology from nano scale to micro scale. Of significant interest, however, are the release profiles of DNA from the fibrous scaffolds. A fast, burst release occurred when naked DNA was added to the fiber surface, with 50% release after only 1-2 days. A slower controlled release observed when DNA/chitosan nanoparticles were added to pre-electrospun fiber surfaces, with 50% release after about 3-9 days, depending on the HA content. Incorporating DNA/chitosan nanoparticles into the fibers themselves further delayed release, with 50% release occurring after approximately 18-31 days. The differences in release profiles can be explained by examining the physical condition of the DNA. Naked DNA is capable of simple diffusion out of the scaffolds because it is small compared to the fibers and pore size, and can change shape. DNA encapsulated in chitosan nanoparticles however cannot diffuse freely due to their ~100 nm diameter size, and so these particles take longer to diffuse out. DNA/chitosan nanoparticles electrospun inside PLGA/HA fibers require fiber degradation to occur for release and thus their release is the slowest. Human MSC were found to grow on the scaffolds, though not as well as on tissue culture plates. *In vitro* expression of BMP-2 was used to measure transfection efficiency, and it was found that DNA/chitosan nanoparticles bound to the outside of nanofibers had the highest transfection rates, but also the highest toxicity. However, significant transfection was seen in the groups where nanoparticles were incorporated into the fibers, and this group also had the highest viability. These results are significant as they show the ability of nanofibers to act as more than passive substitutes for the ECM; they can also become active manipulators of cellular expression and differentiation.

3.2 Cartilage

Cartilage damage can result from chronic wear, injury and disease and plays a strong role in osteoarthritis. Of primary concern is articular cartilage that is found in the pivoting areas of joints. Articular cartilage is largely avascular and is comprised of mostly collagen, elastin, and proteoglycans, with a relatively low cell density. As a result, lesions in articular cartilage do not heal well, if at all, frequently lead to joint pain, and may ultimately lead to irreversible damage in larger areas of cartilage and the underlying bone (Hunziker, 2002). Many tissue engineering approaches to cartilage regeneration have been examined, including the use of hydrogels, collagen sponges, gelatin microspheres and other scaffolding systems (Stoop, 2008). Different types of electrospun nanofibers have been examined as well, ranging from pure synthetics to composites and natural polymers.

Attempts to use a purely synthetic approach to engineer cartilage have employed PLGA nanofibers and primary porcine chondrocytes to test for cell viability, proliferation, and maintenance of a chondrocyte-like phenotype (Shin et al., 2006). The maintenance of chondrocyte phenotype was given particular attention, as chondrocytes tend to de-differentiate during culture. Cells were grown on both electrospun PLGA nanofiber mats and PLGA membranes were used as a control. By measuring the DNA and glycosaminoglycan (GAG, a common ECM component) content it was seen that the nanofiber scaffolds supported a significantly higher cell number and greater ECM synthesis than membranes. Intermittent hydrostatic pressure was used to condition the cellularized scaffolds and was seen to greatly improve both cell proliferation and GAG synthesis over unconditioned electrospun scaffolds. These results show again the great effect that mechanical stimulation and scaffold topography can have on cell phenotype and proliferation.

While chondrocytes are commonly used for cartilage tissue engineering, they remain difficult to harvest due to the low density found in native cartilage. MSC are capable of differentiating into many cell lineages, and have been found to change into chondrogenic lineages when grown in high densities in cell pellets (Li et al., 2005). However, cell pellets do not possess good mechanical properties for implantation, and so the authors sought to use PCL nanofiber scaffolds to support MSC proliferation and differentiation into a chondrocyte lineage. Cells were grown on PCL nanofiber mats or as cell pellets formed by centrifuging, and both methods used either chondrogenic media or media containing TGF-β1 to induce differentiation to chondrocyte lineages. By examining the expressed RNA and rate of GAG synthesis the authors found that the 3D nature of the nanofiber scaffolds caused a significantly higher degree of differentiation toward the desired phenotype, showing that the alternate cell source, i.e. bone marrow derived MSC, can be successfully differentiated into chondrocyte-like cells for potential *in vivo* use. An interesting incorporation of nanofibers into a tissue engineered construct is the use of nanofibers not as a scaffold unto itself, but as a topological modification to a microfiber scaffold. Thorvaldsson et al. (2008) showed that it was possible to coat single PLLA microfibers with PCL nanofibers, giving additional ECM-like physical characteristics to the scaffold to aid the growth of chondrocytes. The results of this study show qualitatively that the addition of a nanofiber topography onto a microfiber scaffold may increase chondrocyte adhesion and provide a structure closer to the native ECM than either nanofibers or microfibers alone can provide.

3.3 Blood vessels

Tissue engineering of blood vessels is primarily focused on development of small diameter vascular grafts. Small diameter arteries, arteries of less than 5 mm in diameter, include the coronary arteries and many peripheral arteries found in the arms and legs. Both coronary artery disease and peripheral vascular disease are caused by atherosclerosis and can lead to serious complications including ischemic heart failure in the case of coronary artery disease and ischemic ulcers and tissue death leading to amputation in the case of peripheral vascular disease. Currently, autologous veins are the first choice for grafting when arterial bypass is necessary. However, this procedure is not optimal because it necessitates two surgeries, one to harvest the donor vein and one for implantation, and can lead to donor site morbidity. Additionally, an autologous graft may not be available due to prior surgery, disease, or trauma. Synthetic materials such as poly(ethylene terphthalate) (Dacron®) and expanded poly(tetrafluoroethylene) (ePTFE) have proven highly successful for large diameter grafts, but perform poorly in small diameter grafts due to thrombosis and anastomotic hyperplasia (Chan-Park et al., 2009). A non-autologous graft would thus be of great socioeconomic value and has been the subject of much investigation. Electrospinning lends itself well to the construction of vascular scaffolds due to the ease of shaping tubular constructs using rotation and translational motion.

One of the advantages of tissue engineering blood vessels is the relative simplicity of vascular structures. While blood vessels do possess the extremely high level of complexity found in all cells, not to mention the additional level of complexity created by a layered tissue structure, intercellular communication, and specific types of ECM, they are nonetheless significantly less complex structures than organs such as the kidney and brain. Two primary characteristics which tissue engineered vascular grafts must possess are proper mechanical properties to withstand physiological conditions and a confluent endothelialized lumen to resist thrombosis. Efforts to determine potential scaffold materials have been examined using electrospinning of synthetic/natural composites (Lee et al., 2007). Scaffolds were electrospun using 15% elastin, 45% collagen and 40% synthetic polymer, either PLGA, PLLA, PCL or PLCL, dissolved in 1,1,1,3,3,3-hexa-iso-fluoropropanol (HFIP) and crosslinked after electrospinning. It was found that mechanical properties differed between scaffolds containing different synthetic polymers. A time course study of scaffold behavior in culture medium conditions showed that scaffold deformation occurred in all scaffold types, but significant decreases in patency was observed in PLGA/collagen/elastin and PLCL/collagen/elastin scaffolds after 28 days, while PLLA/collagen/elastin and PCL/collagen/elastin maintained their patency. The scaffolds were also shown to support cell adhesion and growth, indicating that they could potentially be fully cellularized. These data show that it is possible to tailor the mechanical properties of vascular scaffolds by selecting the appropriate quantities and types of synthetic and natural polymers. It also highlights the potential decrease in patency of the scaffold when exposed to physiological conditions, which could lead to graft failure. Further studies were conducted to show that a PCL/collagen graft has native-like mechanical properties of compliance, burst pressure, and suture retention strength (Lee et al., 2008) (Figure 1). These properties are necessary for a vessel to have proper flow conductance, withstand physiological pressures, and be surgically implanted without risk of failure. Scaffolds were constructed by electrospinning 1:1 mass ratio of PCL/collagen to form scaffolds with an internal diameter of 4.75 mm. Burst pressure was measured by increasing fluid pressure inside the scaffold until failure occurred and was found to be approximately 4915 mmHg, significantly higher than any blood pressure experienced *in vivo*. The scaffolds had better suture retention strength than the accepted standard, and compliance comparable to native blood vessels. When the scaffolds were subjected to flow conditions they maintained structural integrity well, losing 14% of their ultimate tensile strength after 4 weeks compared to 3% in static conditions. Additional studies were performed using electrospun PCL/collagen nanofibers to test for growth of EC and SMC growth and proliferation. These results clearly show that it is possible to create electrospun nanofiber scaffolds with mechanical properties required of implantable grafts which are capable of supporting EC and SMC adhesion and proliferation (Lee et al., 2008).

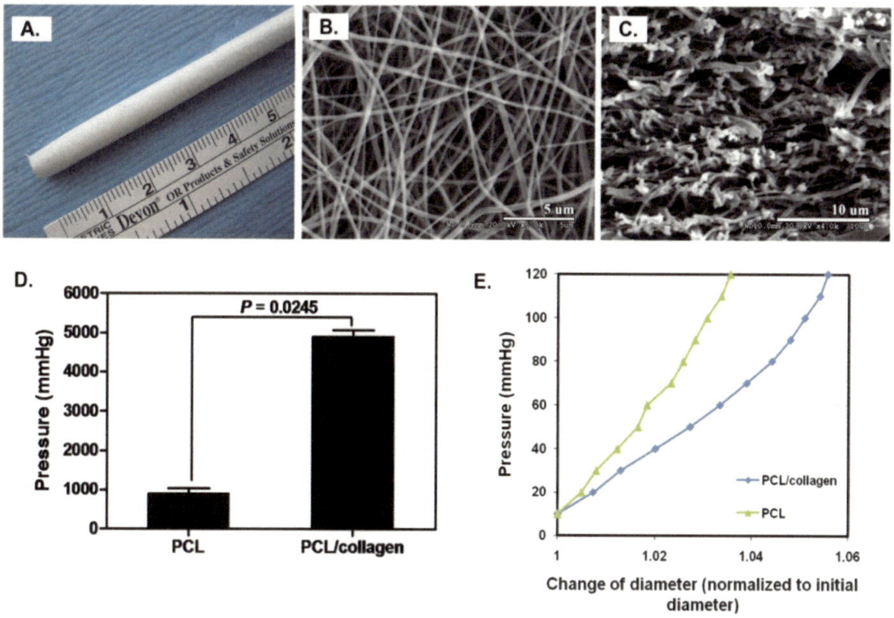

Fig. 1. (A) The gross appearance and SEM images of electrospun PCL/collagen composite scaffolds: (B) surface (×6K) and (C) cross-sectional (×4K) morphologies. (D) Burst pressure strength of electrospun PCL/collagen scaffolds compared to electrospun PCL scaffolds (*$P<0.05$). (E) Compliance curves for electrospun PCL/collagen scaffolds compared to electrospun PCL scaffolds. Reproduced with permission from all authors, *Biomaterials*, 2008, **29**, 19, 2891. © Elsevier 2008 (Lee et al., 2008).

In a second study, these PCL/collagen scaffolds were tested in an animal model to determine ability of EC and SMC seeded grafts to function *in vivo* (Tillman et al., 2009). Scaffolds were seeded with sheep primary EC and SMC and bioconditioned for nine days to acclimate the cells to *in vivo* flow conditions. A subset of scaffolds was then tested for platelet adhesion by an *ex vivo* system in sheep to flow blood through the scaffolds. Scaffolds without an endothelialized lumen were seen to have significantly more platelet adhesion than endothelialized scaffolds, suggesting that thrombus formation would be significantly less likely to occur when scaffolds are seeded with EC and preconditioned for flow. A second subset of scaffolds was used in a rabbit aorto-iliac bypass surgery and left *in vivo* for 4 weeks. In addition, the grafts showed little to no loss of patency over the course of implantation in rabbits as measured by ultrasonography. Tensile strength of the grafts was reduced after a 1-month explant, though remained comparable to native arteries. This study is an exciting example of the potential of nanofibers in tissue engineering. While much longer term animal and ultimately human studies will be needed, these results show that it is possible to create nano-structured vascular grafts capable of being cellularized and implanted into living animals to replace the function of native vessels (Tillman et al., 2009).

Further studies have examined the potential to create vascular grafts which combine both nano and microfibers to allow for better integration of SMC and development of multiple SMC layers (Ju et al., 2010). An internal layer of PCL/collagen nanofibers was electrospun to give fibers approximately 270 nm in diameter for the purpose of supporting a luminal EC monolayer. Surrounding this layer, PCL/collagen fibers with diameters of approximately 1 µm, 2.4 µm and 4.45 µm in different groups were added in order to increase the pore size and thus the capacity of the outer layer to support SMC ingrowth. Scaffolds were found to support EC on the luminal nanofiber layer and SMC growth on the outer microfiber layer. It was also found that SMC integration was significantly increased when 2.4 µm and 4.45 µm fibers were used, as these fiber sizes provide pore sizes large enough for cells to grow into, a feature not generally observed with pure nanofibers. These data illustrate that combinations of micro and nano scale structures can provide scaffold properties not observed when only one or the other is used alone (Ju et al. 2010).

3.4 Nerve

Tissues of the nervous system are often damaged by traumatic injury. Damaged nerves are slow to regenerate, and damage to the central nervous system or large nerves in the peripheral nervous system often never heal. This is evidenced by the inability of para- and quadriplegic sufferers to regain function of their extremities after spinal cord injury and permanent loss of function or sensation of an extremity due to a peripheral nerve injury. Of the two main divisions of the nervous system, tissue engineers have primarily focused on the regeneration of the peripheral nervous system. As such, successful approaches to nerve regeneration have typically incorporated a tubular scaffold which provides both an artificial ECM and directionality through which the peripheral nerve can grow. This regrowth has not been seen to exceed a few centimeters, but remains a great success of tissue engineering to date.

In order to determine appropriate substrates for nerve tissue regrowth it is necessary to test the growth of neuronal or other cells that have the capacity to differentiate into neurons. One study examined the behavior of C17.2 mouse nerve stem cells on electrospun PCL/gelatin nanofiber scaffolds with both random and aligned fiber orientations (Ghasemi-Mobarakeh et al., 2008). It was found that not only did PCL/gelatin fibers support cell growth better than PCL only fibers, but that when fibers were aligned the stem cells had neurite outgrowth along the direction of the fibers, suggesting that fiber alignment may be beneficial in directing new growth of nerve cells. However, sources of nerve stem cells may not be available in human adults so alternative sources will likely be necessary. MSC have been examined for their potential to differentiate into neural lineages on PLCL/collagen nanofiber scaffolds (Prabhakaran et al., 2009). PLCL/collagen scaffolds alone are not sufficient to induce differentiation into a neural lineage. It was necessary to induce this differentiation over the course of 35-37 days using a cocktail of various growth factors while the cells grew on the scaffolds. Cells grown on the PLCL/collagen scaffolds and treated with the growth factor cocktail were qualitatively found to have neuronal-like phenotypes to a greater extent than control groups, and also to have higher growth rates. This suggests that PLCL/collagen nanofibers have a beneficial effect on culturing a population of neuronal-like cells from MSC, thus making them a potentially useful source of cells for neural tissue engineering.

Nanofibers have also been used to form tubular conduits for regrowth of damaged nerves *in vivo*. One study examined the ability of a tubular construct made of both micro- and nanofibers to promote growth of rat sciatic nerve after a 10 mm section of nerve was removed (Panseri et al., 2008). PCL microfibers and PLGA-*co*-PCL nanofibers were electrospun separately into flat sheets. The microfibers were then deposited around a copper wire to create the lumen of the conduit, and nanofibers were deposited around the microfibers to create a tight mesh barrier. The experimental group had sections of sciatic nerve removed and the conduit sutured in place between the two dissected ends. The control group had their nerves either transected or had equal sized sections removed without the addition of the conduits. The injuries were left unrepaired in the control group. Re-connection of nerves in the control group did not occur in any subjects, as expected. Approximately 40% of conduits collapsed as a result of displacement and body movement. However, in all rats that did not suffer conduit collapse tissue regrowth and neurite outgrowth was observed inside the conduit. Markers of neuronal cells were observed through the length of the conduits as well as myelinated axons. Autophagy occurred in a set of animals tested, but those which did not experience it experienced increasing sensitivity in the experimental limb over the four month trial, showing a steadying increase in nerve function. The majority of the treated animals were also seen to experience motor function in the treated limb after four months. These results highlight the capacity of nanofibers to act as a simple barrier to assist in significant regeneration of peripheral nervous tissue and function.

3.5 Skin

The development of engineered skin is another goal of tissue engineering. There are numerous skin injuries, especially burns, which create large surface area wounds that require skin coverage. Currently, the primary mode of treatment is autologous skin grafting, but other regenerative medicine technologies are beginning to be employed. This is because autologous treatment frequently requires multiple graft procedures for large wounds and always results in donor site morbidity. Moreover, in extreme cases, donor sites are not always available for grafting. Tissue engineering would like to develop an off-the-shelf engineered skin or wound dressing that would prevent the need to use donor sites for grafting and that would preclude the need for multiple surgeries. An in depth discussion of normal skin anatomy and physiology is outside the scope of this chapter, but for more details the reader is referred to Moore & Dalley's Clinically Oriented Anatomy among other sources (Wong & Chang, 2009; Moore & Dalley, 1999; Ross et al., 2003). Skin is composed of three layers: epidermis, dermis, and hypodermis. The epidermis is avascular and composed of squamous epithelial cells call keratinocytes. There is a basal layer of keratinocytes that are stem cells which divide asymmetrically. After a cell division, one cell will remain in the basal layer while the other will be pushed upward. As the keratinocytes progress from the basal layer to the apical layer of the squamous epithelium, they produce an increasing amount of keratin. Once the cells reach the surface, they are filled with keratin and are eventually sloughed off. The epidermis provides the barrier function of the skin. Below the epidermis is the dermis. The dermis is loose connective tissue that consists of large amounts of ECM, namely type I and III collagen and elastin, although there are other components. The primary cell type of this layer is the fibroblast which maintains the dermis. Other structures in the dermis include blood vessels, lymphatics, glands, cutaneous nerves, and hair follicles. The hypodermis is primarily loose connective tissue with lots of adipose tissue. This layer of skin serves primarily as an energy storage site and as insulation (Moore & Dalley, 1999; Ross et al., 2003; Wong & Chang, 2009).

Electrospinning has been used as an approach for developing scaffolds for engineered skin and wound dressings, although less than other scaffold strategies (Priya et al., 2008). Nevertheless, several investigators have used electrospinning for skin tissue engineering. Of these, most of the reports have been *in vitro* evaluations (Kumbar et al., 2008; Powell & Boyce, 2009; Sun et al., 2005; Yang et al., 2009; Zhou et al., 2008). Sun et al. (2005) used electrospun polystyrene (PS) as a model system for investigating what effect the spatial cues of an electrospun system would be on engineered skin formation. They also evaluated the effect of culturing cells in an air-liquid interface (a common culture technique for skin culture) compared to submerged scaffolds. The electrospun PS scaffolds had fiber diameters of approximately 10 μm. They found that when keratinocytes, fibroblasts, and EC were co-cultured, they could self organize without any extra cues other than the three-dimensional electrospun scaffold and the air-liquid interface.

One study used electrospun PLGA as wound dressings for skin wounds (Kumbar et al., 2008). The study investigated the effect of fiber size on fibroblasts. The authors investigated scaffolds with fiber diameters ranging from 150 nm to 6000 nm to evaluate scaffold-seeded fibroblast behavior in terms of proliferation, morphology, and gene expression. They found that fibroblasts proliferated best and had spread morphology in the fiber range of 350 nm to 1100 nm. Moreover, scaffolds in this range had a higher expression of type III collagen. All the scaffolds showed fibroblast expression of type I collagen and elastin, but scaffolds in the 350-1100 nm range did not show a great difference from other fiber sizes tested, i.e. type I collagen and elastin did not show a fiber-dependent trend in expression. Finally, all scaffolds exhibited tensile moduli within the range of normal human skin (Kumbar et al., 2008).

Another study investigated the electrospinning of carboxyethyl chitosan/poly(vinyl alcohol) (PVA) scaffolds for future use as skin wound dressings (Zhou et al., 2008). Chitosan has good biological properties such as biodegradability, biocompatibility, is antibacterial, and promotes wound healing. Thus, the authors wanted to find a way to electrospun N-carboxyethyl chitosan in an aqueous solution (typically materials dissolved in aqueous medium are more difficult to electrospin due to the low volatility of water). Upon initial attempts, they were not able to electrospin chitosan, however upon blending with various ratios of PVA, they were able to form fibers. These fibers had average diameters ranging from 131 nm to 456 nm. After fabrication, the scaffolds were crosslinked with glutaraldehyde vapor so that they would not dissolve upon wetting. The results showed that the scaffolds containing chitosan could be formed by electrospinning in an aqueous solution, that the scaffolds could be crosslinked to prevent re-dissolving, and that the scaffolds were cytocompatible when tested with mouse fibroblasts. Thus, the study concluded that scaffold might be useful for skin regeneration (Powell & Boyce, 2009).

Powell & Boyce (2009) used electrospun collagen and PCL blends as scaffolds for forming engineered skin. They added PCL to electrospun collagen in order to improve its handling properties, i.e. to minimize size reduction upon wetting and increase mechanical strength. Upon comparing different mass percentages of PCL blended with collagen to PCL alone and collagen alone they found that mechanical properties were not significantly improved over collagen alone until there was at least 10% PCL in the scaffold. However, the scaffold containing 10% PCL had the lowest elongation. These scaffolds were co-cultured with fibroblasts and keratinocytes using an air-liquid interface. Fibroblasts were seeded 24 hr before keratinocytes. When evaluating the mechanical properties of seeded scaffolds at later time points, few differences between groups were found. The main finding was that the 30% PCL scaffold had the weakest mechanical properties. As the PCL content increased, the scaffolds were less able to accommodate the seeded cells showing a lower density of fibroblasts and un-stratified dermal and epidermal layers without a basal cell layer. In contrast, scaffolds that contained 10% PCL or less had well stratified dermal and epidermal layers with the presence of a basal cell layer. The investigators postulated that at the higher PCL concentrations, the fibroblasts were unable to form a confluent layer to accommodate the keratinocytes which thus resulted in an unorganized system. This unorganized system accounted for the lower mechanical properties observed. Interestingly, the 100% PCL scaffold had similar mechanical properties to the lower PCL content scaffolds even though it did not form well-organized engineered skin. The authors attributed this to the PCL itself resisting most of the load. Surface electrical capacitance, which is a measure of the hydration of the epidermal layer, was also measured. As the keratinocyte layer started to keratinize, it started to dry and act as a barrier which is a natural function of skin. The authors found that scaffolds with a lower percent of PCL had better capacitance values with the pure collagen scaffold and 1% PCL scaffolds approaching capacitance values similar to normal skin. Finally, the authors showed that adding PCL to electrospun collagen reduces the amount of scaffold contraction after wetting, although except for the 30% PCL and 100% PCL scaffolds, this reduction was relatively modest (Powell & Boyce, 2009).

Yang et al. (2009) fabricated electrospun scaffolds consisting of PLGA and type I collagen with high porosities and large pore sizes. The fiber diameter of the scaffolds was approximately 1 μm regardless of collagen concentration (collagen concentrations used were 0%, 5%, 10%, and 30%). To improve pore size by controlling fiber deposition density, a mandrel that consisted of metal rods parallel to one another that inserted into a ring at either side was used. Thus, the mandrel had the basic shape of a cylindrical mandrel, but instead of having a full surface it only had spokes of metal. This allowed fibers to deposit between the metal rods at a lower density than they would on a solid surface thus increasing the porosity and pore size of the scaffolds. This mandrel was dubbed the metal frame cylinder. By increasing the percentage of collagen in the scaffolds, the pore size and porosity decreased. These scaffolds had pore sizes in the range of 90-130 μm. The scaffolds with 30% collagen had the highest human dermal fibroblast cell attachment and viability. The seeded fibroblasts also produced more type I collagen on scaffolds containing higher concentrations of collagen. Finally, the addition of collagen decreased the mechanical properties of the electrospun scaffolds. Nevertheless, the authors concluded that adding collagen to the electrospun scaffolds improved cell attachment, proliferation, type I collagen production, and that the high porosity in the scaffolds allowed for fibroblasts to migrate into the scaffold making them good candidates for tissue engineering in general (Yang et al., 2009).

Even though there are several studies investigating electrospun scaffolds for skin tissue engineering, to our knowledge there have yet to be any publications that investigate the use of these scaffolds *in vivo*. In contrast, numerous tissue engineering strategies that employ non-electrospun scaffolds have been studied *in vivo* (Priya et al., 2008). Regardless, electrospinning has become a simple and economical method to generate scaffolds that mimic the ECM of tissues. Thus, electrospun scaffolds have even found use in skin tissue engineering. While the work presented here is promising, in order for it to be clinically translatable, animal studies will be necessary. The type of study will depend on the intended application of the engineered skin, but most likely the initial studies would investigate the ability of electrospun scaffolds to facilitate skin regeneration in a wound excision model. If these attempts are successful, then investigations should focus on using engineered skin constructs to treat burn wounds which are much more challenging than excisional wounds due to the large degree of necrosis and inflammation in a burn wound. As with any technology, animal studies should start in smaller animals such as rodents and progress to more clinically relevant models such as pigs. Perhaps future studies will investigate the use of electrospun scaffolds for engineered skin constructs *in vivo*.

3.6 Ligaments and tendons

Tendon and ligament injuries are common. Often, clinical repair of damaged tendons or ligaments require autografting from a donor site which, as with any autograft, results in donor site morbidity. Thus, there is much interest in developing engineered tendon and ligament for the repair of these injuries such that donor site morbidity can be avoided. The structure of tendons and ligaments, while fairly complex, is primarily composed of sparse specialized fibroblastic cells in an abundant type I collagen matrix. Tendon and ligament tissue engineering have been attempted for some time. It was initially thought that these tissues would be simple to engineer due to their low cell content and vascularity (which is also why tendons and ligaments frequently do not heal well (Strauss et al., 2007, as cited in Sahoo et al., 2010b), however, this has not proven to be true. While there have been a plethora of studies that try to engineer both tendon and ligament, few have used electrospinning as the fabrication method for their scaffold. As will be shown, this is primarily due to the lack of strength of most electrospun scaffolds.

Lee et al. (2005) used electrospun polyurethane (PU) to study the effect of fiber alignment and cyclic loading on the ECM generation of human ligament fibroblasts. Electrospinning was used in order to mimic the "fine fibrous structure of connective tissues." Electrospun mats were created where the PU fibers were either aligned or randomly oriented. The aligned fibers had an average diameter of 657 nm. The scaffolds were then seeded with human ligament fibroblasts. In addition to seeding, some scaffolds were exposed to cyclic strain. For aligned fibers, strain was applied in both the direction parallel to alignment and perpendicular to alignment. Not surprisingly, the aligned scaffolds had higher stiffness and strength. On the aligned structures, the seeded cells were oriented in the direction of alignment and produced more collagen than on the randomly oriented scaffolds. Cyclic stretching increased collagen content in both the scaffolds that were stretched parallel to the fiber orientation and those stretched perpendicular to the fiber orientation, although stretching in the parallel direction had the highest collagen. Stretching of randomly oriented scaffolds did not increase collagen content over either the aligned or randomly oriented statically seeded control groups. Thus, the authors concluded that aligned nanofiber scaffolds are a biomimetic scaffold that can recapitulate an environment similar to the native ligament (Lee et al., 2005). As has been illustrated previously in Section 3, aligned fibers, which are easy to form with electrospinning, have proven to be beneficial topographical cues for various tissue types.

Sahoo et al. have published a series of studies investigating the development of a composite scaffold for ligament tissue engineering consisting of a knitted component covered with electrospun fibers (Sahoo et al., 2006, 2007, 2010a, 2010b). Sahoo et al. (2006) combined electrospinning with knitted scaffolds in order to take advantage of a knitted scaffold's strength and the high surface-to-volume ratio and ECM scale of electrospun fibers. They electrospun PLGA nanofibers onto knitted PLGA scaffolds. The electrospun fibers had diameters ranging from 300-900 nm. Bone marrow-derived MSC were chosen as their cell source because ligament fibroblasts are scarce and have low proliferation rates whereas MSC are easy to obtain, have good proliferative properties, can differentiate into various mesenchymal lineages, and produce ECM with type I collagen. When these scaffolds were seeded with a cell suspension of MSC and compared to knitted scaffolds without electrospun fibers that had cells seeded in a fibrin gel matrix, the scaffolds with electrospun fibers had better proliferation and higher expression of type I collagen, decorin, and biglycan (Sahoo et al., 2006). Sahoo et al. (2007) compared knit PLGA and PLLA scaffolds coated with PCL, electrospun PLGA fibers, or collagen to woven fibers. They found that while woven fibers and knitted PLGA coated with PCL had better strength than the other groups, they had poor cell attachment and proliferation. Knitted PLGA with electrospun PLGA fibers had the best cell proliferation. Knitted PLLA scaffolds did not fair as well in terms of cell proliferation and attachment. Thus, the best scaffold in terms of biological properties was a knitted PLGA scaffold with electrospun PLGA fibers. However, these scaffolds degraded rapidly *in vitro* and lost their properties within 3-4 weeks. Thus, the authors proposed that a different material suck as silk would be better to use as the knitted scaffold material (Sahoo et al., 2007). Sahoo et al. (2010a) published one study describing the incorporation of basic fibroblast growth factor (bFGF) into electrospun PLGA fibers which was released over the course of 1 week. bFGF was used because it is known to stimulate MSC to proliferate, self-renew, and differentiate into fibroblastic cells (Sahoo et al., 2010a). Bovine serum albumin, used as a carrier protein, and bFGF were blended with PLGA and dissolved in HFIP. Fibers were electrospun from this solution. The authors showed that the fibers contained bFGF and that bFGF was indeed released from the fibers over the course of a week. Released bFGF was bioactive, promoted cell proliferation and differentiation to a fibroblastic state, increased expression of tendon/ligament specific proteins, and increased the deposition of collagen and tenascin-C (Sahoo et al., 2010a). As a continuation of this work, Sahoo et al. incorporated the bFGF-releasing PLGA nanofibers with a knitted silk scaffold (Sahoo et al., 2010b). The knitted silk scaffold provided better mechanical strength with *in vitro* degradation than previous scaffolds fabricated by this group. Moreover, after 3 weeks in culture with MSC, the bFGF-releasing group had higher strength than unseeded or bFGF-negative scaffolds. The bFGF-releasing group also had a higher stiffness, although it was not statistically significant. Finally, the authors showed that these novel scaffolds promoted and facilitated MSC attachment, proliferation and tenogenic differentiation. Expression of ECM proteins type I collagen, type III collagen, biglycan and fibronectin was increased on the bFGF-releasing group compared to other groups after 14 days. Interestingly, at 7 days, the expression of the collagen proteins was lower on the bFGF-releasing scaffolds than on the bFGF-negative scaffolds, but was higher at 14 days. The authors hypothesized that this was due to the fact that the bFGF initially promoted the proliferation of undifferentiated MSC, and then later their tenogenic differentiation (Sahoo et al., 2010b). Thus, Sahoo et al. in their series of studies have developed interesting scaffolds that incorporate electrospun fibers to serve as a biomimetic topographical cue and as a depot for growth factor release (Sahoo et al. 2006,

2007, 2010a, 2010b). This set of studies demonstrates that electrospun fibers can be used as an adjunct to other scaffolds to improve ECM mimicry and as a depot for drug or growth factor release.

Vaquette et al. (2010) used a similar approach to Sahoo et al. In this study, aligned PLCL fibers were electrospun (although in this case they were microfibers and not nanofibers) onto knit silk scaffolds. They showed that these scaffolds had stress-strain behavior similar to native ligament, that rat MSC, when seeded on the scaffolds, proliferated on the scaffolds and aligned in the direction of the fibers. They also found that MSC seeded on these scaffolds produced type I and III collagens in higher amounts than MSC cultured on tissue culture plates. The authors concluded that they created scaffolds with similar properties to ligaments that would be suitable for use in a bioreactor and neo-ligament formation in the future (Vaquette et al., 2010).

Another study used electrospun nanofibers as a scaffold for rotator cuff repair (Moffat et al., 2009). In this study, Moffat et al. compared the culture of human rotator cuff tendon fibroblast culture on aligned versus unaligned electrospun PLGA scaffolds. The authors found that aligned scaffolds were stronger and stiffer than unaligned scaffolds. Additionally, fibroblasts cultured on the aligned scaffolds attached along the axis of alignment whereas on unaligned scaffolds they were randomly oriented. Fibroblasts on aligned scaffolds had higher expression of α2 integrin suggesting that aligned matrices better mimic native collagen than unaligned matrices (this is because α2 integrin complexes with β1 integrin in order to bind to collagen). Moreover, type I and type III collagens were produced by fibroblasts on both aligned and unaligned scaffolds, but the collagens deposited on the aligned scaffold were done so in the direction of alignment. Finally, the aligned scaffolds maintained mechanical properties better than unaligned scaffolds when they were acellular or seeded. More importantly, despite the decrease in mechanical properties due to hydrolytic degradation of PLGA, the scaffolds maintained physiologically relevant mechanical properties over the 14 day-culture period compared to those known in the literature for rotator cuff tendons. Thus, the authors suggest that aligned nanofiber scaffolds have potential for tendon regeneration, especially rotator cuff tendons (Moffat et al., 2009). This is another example of where electrospun scaffold mimic the ECM of native tissue.

In summary for tendon and ligament tissue engineering, electrospinning has primarily been used as an adjunct to other fabrication techniques. This is because electrospun scaffolds are typically mats and relatively thin and so do not exhibit the strength needed to serve as scaffolds for these tissue types. It certainly may be possible to use electrospun mats to start the development of tissue engineered tendons or ligaments, for instance by utilizing cyclic stretching in a bioreactor, but in general, they seem unlikely to serve as a material-only replacement due to their lack of strength. For this reason, nearly all of the studies described above combined electrospinning with another fabrication technique (in these cases knitted scaffolds) to take advantage of the nanofiber architecture of spun fibers and the strength and porosity of knitted scaffolds (Ouyang et al., 2003 as cited in Sahoo et al. 2006; Sahoo et al., 2006). The exception is the study by Moffat et al. which claimed that physiologically relevant mechanical properties for rotator cuff tendons were maintained using electrospun PLGA scaffolds alone (Moffat et al., 2009). Perhaps using aligned fibers will improve the ability of electrospun scaffolds to be used alone instead of in a composite. Moreover, rotator cuff tendons are broad and flat which makes an electrospun membrane alone more suitable for this application than for other tendon applications where tendons are more rounded.

3.7 Ligament/tendon-bone interfaces

More recently, researchers have started to investigate ways to form composite tissues which will be important in linking tissues that interact with one another. An example of this is the ligament/tendon to bone interface. These interfaces are particularly relevant when considering the clinical repair of torn or ruptured tendons or ligaments such as the rotator cuff and anterior cruciate ligament (ACL). Both of these injuries are common in the United States with 75,000 rotator cuff repairs (Vitale et al. as cited in Moffat et al., 2009) and 100,000 ACL reconstruction surgeries performed annually (American Academy of Orthopaedic Surgeons, as cited in Spalazzi et al., 2008).

Two groups have used electrospun scaffolds in attempt to engineer this tissue interface (Spalazzi et al., 2008; Li et al., 2009). Spalazzi et al. (2008) used electrospun PLGA to create a mechanoactive scaffold by taking advantage of the contraction that occurs in PLGA upon wetting. They hypothesized that applying a compressive load to a tendon would promote its metaplasia into a fibrocartilage-like tissue. This would be useful because tendon autografts are frequently used to repair anterior cruciate ligaments, however, when tendons are used alone instead of bone-patellar tendon-bone (BPTB) grafts, they do not integrate with the bone. This is because unlike BPTB grafts, tendon autografts do not contain an attached segment of bone. Thus, to aid integration with bone, the authors wanted to find a way to cause tendons to remodel towards a fibrocartilage phenotype. To accomplish this, aligned nanofibers were fabricated that were then either wrapped around patellar tendon alone, or first around a cylindrical scaffold composed of sintered PLGA microspheres that was then wrapped around the tendon. After the scaffolds were wetted, the electrospun fibers naturally contracted which resulted in compressive loading of the tendon, or in case of the composite cylinder-electrospun scaffold, the compression of the cylindrical scaffold and subsequent compression of the patellar tendon. DNA content, histology, GAG content, and expression of fibrocartilage specific markers type II collagen, aggrecan, and TGF-β3 were evaluated. In the case of the electrospun scaffold wrapped around the tendon alone, remodeling occurred by 5 days as evidenced by dense matrix with high cellularity, but the tendon had recovered its normal crimp pattern by 14 days. When the cylindrical-electrospun composite scaffold (or mesh+collar scaffold) was used, the increase in density and cellularity remained at 14 days. In the mesh+collar scaffold compressed tendons collagen fiber size decreased and remained parallel to the long axis of the tendon whereas non-loaded tendon had a loss of collagen organization by 14 days. GAG content was higher in the mesh+scaffold loaded tendons than uncompressed tendon by day 1 although the unloaded tendon had higher DNA content. Finally, after 24 hours, the mesh+scaffold compressed tendons had increased type II collagen, aggrecan, and TGF-β3 expression compared to day 0 and compared to unloaded controls. Thus, the authors proposed that these scaffolds could be used clinically by sliding tendon autografts through the mesh+scaffold "collars" before using them to repair anterior cruciate ligaments. After placement, the mesh+scaffold would contract, transferring a compressive load to the tendon causing tendon metaplasia to fibrocartilage. Moreover, because the scaffolds are made of PLGA, they would degrade leaving behind re-organized tissue. Finally, the authors also proposed that the scaffolds could be further modified to provide growth factors to aid in the metaplasia. All of this is hypothesized to ultimately result in tendon integration with bone and thus improve outcomes of ACL repair with tendon autografts that lack a bone component (Spalazzi et al., 2008).

Li et al. (2009) used electrospun scaffolds in which a gradient of calcium phosphate was formed to mimic tendon-to-bone interfaces. Electrospun PLGA or PCL scaffolds were plasma treated or coated with gelatin respectively and a gradient of calcium phosphate mineral was applied along the length of the scaffold using ten times simulated body fluid. A mineral gradient was formed in these scaffolds as evidenced with SEM and decreasing mineral content was observed when moving from one end of the scaffold to the other. Moreover, the gradient had functional consequences in that it created a mechanical property gradient across the scaffold where the mineralized side had a higher modulus than the non-mineralized side and there was a gradual decrease in modulus from the mineralized end to the non-mineralized end. Finally, MC3T3 cells, a preosteoblastic cell line, were seeded on the scaffolds and showed preferential attachment and spreading on the more mineralized regions of the scaffolds. The cell density of these cells decreased with decreasing mineral content. Thus the authors proposed that these scaffolds may be useful for engineering tendon-bone interfaces (Li et al., 2009).

As tissue engineering moves forward, interfaces between different tissue types will become increasingly important. The first study described above uses electrospinning as a potential an adjuvant to current clinical therapies to improve bone-tendon integration and thus clinical outcomes (Spalazzi et al., 2008). The second study demonstrates the electrospinning can be used as a technique to facilitate the engineering of these transitions. Here, the authors create a mineral gradient along the length of an electrospun scaffold in order to facilitate the formation of the bone-to-tendon transition and promote rotator cuff tendon repair in future studies (Li et al., 2009). Thus, electrospun nanofibers will likely continue to find novel uses in the field of composite tissue engineering.

3.8 Skeletal muscle and muscle-tendon junctions

When skeletal muscle tissue loses function due to a congenital condition or an acquired condition such as a traumatic injury or tumor excision, there is no effective clinical treatment for replacing that function (Moon et al., 2008). Tissue engineering once again offers the opportunity to one day develop clinical treatments to repair damaged skeletal muscle and restore function. Our laboratory has used electrospinning to develop scaffolds for skeletal muscle tissue engineering (Choi et al., 2008). Electrospun scaffolds were made of a blend of PCL and collagen to investigate the effect that aligned fibers would have on skeletal myotube formation. Fibers were aligned by increasing the rotation speed of a plate mandrel. The aligned-fiber scaffolds enhanced skeletal muscle cell alignment and myofiber length compared to randomly oriented electrospun scaffolds. This suggested, as has already been pointed out in Section 3, that aligned-fiber scaffolds may be advantageous for organizing tissue engineered tissues, in this case skeletal muscle. The scaffolds from this study may one day be useful for forming functional tissue that could be implanted into a patient (Choi et al., 2008).

More recently, there has been increasing interest in composite tissue engineering, which as described in the previous section will become increasingly important as the field of tissue engineering advances. Specifically, we have begun to investigate scaffolds that would be appropriate for muscle-tendon junction tissue engineering (Ladd et al., 2010). Muscle and tendon tissue have different design requirements for scaffolds, but we wanted to create a single scaffold that could accommodate both tissue types. To do this, we decided to design scaffolds that mimic the trends in mechanical properties observed in native MTJ. Our working hypothesis is that a scaffold that mimics these trends would promote the formation of MTJ tissue. We employed a co-electrospinning fabrication technique to begin developing such scaffolds. We showed that by using PCL and PLLA blended with type I collagen, that we could develop scaffolds with regional differences in mechanical properties and that these differences mimicked the trends observed in native MTJ (Ladd et al., 2010). Figure 2 showing the result of co-electrospinning to create a scaffold with regional variations in mechanical properties that mimic the trends observed in native muscle-tendon junction. While our working hypothesis remains to be tested, this is another example of where electrospinning is being used to create novel scaffolds for tissue engineering.

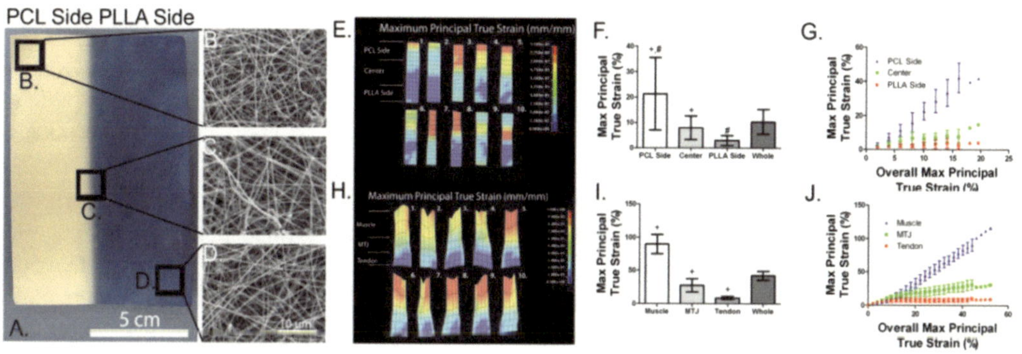

Fig. 2. (A) Image showing the three regions of the scaffold: PCL side, Center, and PLLA side with methylene blue dye added. (B-D) SEM images from the different regions of scaffold showing fiber morphology and diameter (all images ×4k). (B) PCL side (550±98 nm), (C) center (504±93 nm), (D) PLLA side (452±40 nm). (E) Color maps of the true strain profiles of the scaffolds when undergoing uniaxial tensile testing. Maximum principal strain is displayed and the scale is from 0 to 0.25 mm/mm strain. (E-G) The quantification (E) by showing the differences in max strain from each region. Maximum principal true strain for the regions of the scaffold at each overall strain interval for the whole scaffold. (H-J) Analagous data for native MTJ tissue tested in the same manner. Note that (H) has a scale from 0 to 1 mm/mm strain. From these figures, it is clear that the scaffolds mimic the trends in mechanical properties observed in native tissue. Matching symbols indicate statistical difference from one another. Significance was defined as *P<0.05. Reproduced with permission from all authors, *Biomaterials*, 2010, **32**, 6, 1549. © Elsevier 2010 (Ladd et al., 2010).

In summary, electrospinning provides a unique fabrication technique that could prove to be useful for both skeletal muscle tissue engineering and composite tissue engineering. The ability to form and control the orientation of nanofibers allows scaffolds to incorporate topographical cues to help align skeletal myofibers. The ability to incorporate multiple material types into a single scaffold allows one to create scaffolds with regional variations in mechanical properties that may be useful for MTJ tissue engineering and also may prove useful for other types of composite tissue engineering. Undoubtedly, more work is needed to create clinically useful applications despite its successful use in the laboratory setting. Nevertheless, electrospinning remains as a useful tool in the tissue engineer's armamentarium.

4. Current challenges with using electrospinning in tissue engineering

As with any fabrication method, there are advantages and disadvantages to using electrospinning as a means of scaffold fabrication for tissue engineering. Many of the advantages have already been highlighted in previous sections such as the mimicking of native ECM and the ability to adapt the technique for versatile applications. Here, we discuss several limitations of the electrospinning and strategies to work towards overcoming them.

An inherent limitation to using electrospun nanofibrous scaffolds in tissue engineering applications is the small pore size. Nanofiber scaffolds typically result in pore sizes on the sub-micron to few micron scale. Because all mammalian cells range in size from several micrometers to several hundred micrometers in size, these cells typically cannot penetrate the scaffold. This poses a problem for tissue engineers because we often aim to form fully cellularized three dimensional tissue constructs. There have been attempts to improve the pore size of electrospun nanofiber scaffolds. The two primary methods are to co-electrospin a set of sacrificial fibers or adding porogens (Baker et al., 2008; Nam et al., 2007). Sacrificial fibers are fibers that are co-electrospun with the main scaffold component, but that are easily dissolved after fabrication leaving behind larger pores. For instance, Baker et al. (2008) used PEO, a water soluble polymer, as the sacrificial component of electrospun scaffolds. After dissolving the PEO, the pore size of the electrospun scaffolds increased. Porogens, primarily salt, have also been added to electrospun scaffolds as they are being electrospun (Nam et al., 2007). Typically the porogens are dropped onto a mandrel or collecting plate while the fibers are being deposited. Then, after fabrication, the porogens can be removed by dissolution. These approaches have improved pore size and allowed for improved cellular infiltration. However, these approaches may compromise the mechanical properties and dimensional stability of the scaffold when the fibers dissolve and collapse. Microintegration is another approach developed by Stankus et al. (2006) for producing a highly cellularized construct. However, this process may limit homogeneity, sterility, and cell survival, due to the long-term processing times required.

Another inherent limitation of electrospinning is scaffold thickness, especially when spinning onto a cylindrical mandrel. Electrospun scaffolds tend to be thin, typically less than one millimeter in many cases (Inoguchi et al., 2006; Lee et al., 2008). Scaffolds with thicknesses greater than one millimeter are possible to fabricate, but are less commonly seen in the literature (Pham et al., 2006b). Moreover, in our experience, achieving millimeter or greater thicknesses requires the use of a large volume of polymer solution. The inability to obtain thicker scaffolds may be dictated by the strength of the electric field. When spinning is initiated, there is an electrostatic attraction from the charged needle tip to the grounded mandrel. As spinning continues and fibers are deposited on the mandrel in layers. We presume that the layers act as an insulator, thus preventing sufficient current flow from the newly deposited fibers to the grounded mandrel and resulting in a buildup of charge on the outer fibers. This charge buildup would decrease the electrostatic force the charged polymer feels from the grounded mandrel, resulting in the fibers depositing on the mandrel with less efficiency, an increased attraction of the fibers for other objects, and a decrease in mat density due to loose fiber packing. In other words, the fiber deposition efficiency decreases as fibers are deposited on the mandrel and thus in order to obtain thicker scaffolds, one must use a large volume of polymer solution. McCullen et al. (2010) attempted to increase scaffold thickness using multi-layering electrospun scaffolds with a collagen gel for bone tissue engineering. After fabricating multi-layered electrospun scaffolds, micro-scaled pores were created by a laser ablation technique. Thus, they were able to maintain the unique nanofibrous structure that supports cellular interactions and to fabricate micro-scaled features within the scaffold.

An additional limitation of electrospinning is the inability to precisely control where the fibers are deposited (Teo & Ramakrishna, 2005). As fibers whip towards a collecting plate or mandrel, they deposit randomly unless there is some other force to cause them to orient differently. To try to overcome this limitation, researchers have used large cylindrical mandrels with high rotation speeds to cause fiber alignment (Choi et al., 2008; Moffat et al., 2009) or used parallel electrodes to apply an electric field that would force the fiber to align across the plates or in other orientations (Li et al., 2003; Teo & S Ramakrishna, 2005; Li et al., 2005; Yang et al., 2009). Still, electrospinning would be an even more powerful technique if one could precisely control where the polymer jet would land such that computer-assisted design of electrospun scaffolds would be possible. This would allow researchers to easily create nanofibers and create a much broader range of scaffold geometries and thicknesses. Unfortunately, no one has discovered a way to do this to date. The closest anyone has come to approaching this ability was in a study by Kameoka et al. (2003) where they created a scanning tip electrospinning source that used an arrow-shaped tip, polymer droplet, and high voltage to create a Taylor cone and spin fibers. They were able to control deposition to a region of about 5 mm in size. While some researchers have found creative ways to make electrospun scaffolds of differing geometries (besides tubes and sheets), the inability to precisely control the deposition location of the polymer jet limits the variety of possible geometries.

Most researchers tend to have electrospinning systems which still require a manual setup, i.e. manual positioning of the needle tip from the collector. Moreover, often humidity and temperature are not well-controlled and these ambient parameters can have an effect on the fibers. Because these parameters are often not precisely controlled, there is an increase in the variation of properties and fiber morphologies from scaffold to scaffold. In order to improve the repeatability and homogeneity between spinning batches, modern technology should be incorporated into electrospinning systems to more precisely control these parameters. For instance, without much difficulty, a computer-controlled spinning apparatus that is self-contained to control temperature and humidity and that precisely controls air gap distances, mandrel rotation speed, voltage, and spinneret flow rate could be developed. Such a system would allow for more consistent scaffold fabrication and serve as an improved system to continue to investigate how these spinning parameters affect final fiber morphology.

5. The future of electrospinning in tissue engineering and regenerative medicine

Electrospinning has evolved as a powerful tool in various tissue engineering applications. This fabrication technology provides the ability to control biomaterial composition, fiber diameter, fiber alignment, geometry, and drug/protein incorporation into a scaffold. Nano-scaled fibers generated by electrospinning are able to improve the cellular interactions of a wide variety of cell types; moreover, the cells are able to maintain their phenotypic and functional characteristics on nano-scaled fibrous scaffolds. Additionally, a growing body of evidence demonstrates that nano-to micro-scaled topography plays an important role in controlling the adhesion, proliferation, differentiation, and survival of cells in culture.

Therefore, electrospun nano-scaled fibrous scaffolds can serve as a tool for studying the topographical aspects of cellular interactions that would lead to improved tissue formation. Furthermore, these electrospun scaffolds can be functionalized by adding biochemical and mechanical cues to enhance cellular interactions for tissue engineering applications. However, specific interactions between cells and electrospun scaffolds functionalized through surface modifications and bioactive factor incorporation are still poorly understood. Understanding of the specific cues that enhance cell adhesion, proliferation, and guidance of cells seeded on a scaffold, as well as cues that could affect host cell infiltration, differentiation, and vascularization in *in vitro* and *in vivo*, is crucial for the advancement of tissue engineering applications.

The most significant challenges that prohibit the expansion of electrospinning applications are increasing scaffold thickness and pore size. If these two problems were to be solved, electrospun nanofibers would prove to be even more useful than they are currently. Electrospun scaffolds can also be used as a component of a composite scaffold, similar to those developed by Sahoo et al. (Sahoo et al., 2006; Sahoo et al., 2007; Sahoo et al., 2010a; Sahoo et al., 2010b) or incorporated with hydrogel scaffolds to create layered structures. More detailed understanding of how the process parameters interact with one another, especially the electric field, may provide ways to better control fiber deposition. Thus, basic research into the physics of electrospinning will be essential in overcoming the challenges outlined above. For example, in a recent study by Ziabari et al. the simultaneous effects of solution concentration, spinning distance, applied voltage, and flow rate on mean fiber diameter and standard deviation of fiber diameter were investigated (Ziabari et al., 2010). Perhaps with more basic research, we will be able to apply electrospun nanofibers as a controlled fiber jet such that the fibers can be precisely deposited with micron-scale or sub-micron accuracy. Such a capability would be immensely useful for tissue engineering scaffold design.

6. Acknowledgments

This work was supported by the Department of Energy (DE-FG02-09ER64711), Telemedicine and Advanced Technology Research Center (TATRC) at the U.S. Army Medical Research and Material Command (USAMRMC) through award W81XWH-07-1-0718, and Orthopaedic Trauma Research Program (USAMRAA OTRP07-07128091) of Department of Defense.

7. References

Badami, A.S. et al., 2006. Effect of fiber diameter on spreading, proliferation, and differentiation of osteoblastic cells on electrospun poly(lactic acid) substrates. *Biomaterials*, 27(4), pp.596-606.

Baker, B.M. et al., 2008. The potential to improve cell infiltration in composite fiber-aligned electrospun scaffolds by the selective removal of sacrificial fibers. *Biomaterials*, 29(15), pp.2348-2358.

Behonick, D.J. & Werb, Z., 2003. A bit of give and take: the relationship between the extracellular matrix and the developing chondrocyte. *Mechanisms of Development*, 120(11), pp.1327-1336.

Chan-Park, M.B. et al., 2009. Biomimetic control of vascular smooth muscle cell morphology and phenotype for functional tissue-engineered small-diameter blood vessels. *Journal of Biomedical Materials Research. Part A*, 88(4), pp.1104-1121.

Choi, J.S. et al., 2008. The influence of electrospun aligned poly(ε-caprolactone)/collagen nanofiber meshes on the formation of self-aligned skeletal muscle myotubes. *Biomaterials*, 29(19), pp.2899-2906.

Deitzel, J.M. et al., 2001. The effect of processing variables on the morphology of electrospun nanofibers and textiles. *Polymer*, 42, pp.261-272.

Doshi, J. & Reneker, D.H., 1995. Journal of Electrostatics. *Journal of Electrostatics*, 35(2-3), pp.151-160.

Ghasemi-Mobarakeh, L. et al., 2008. Electrospun poly(ε-caprolactone)/gelatin nanofibrous scaffolds for nerve tissue engineering. *Biomaterials*, 29(34), pp.4532-4539.

Greiner, A. & Wendorff, J.H., 2007. Electrospinning: a fascinating method for the preparation of ultrathin fibers. *Angewandte Chemie*, 46(30), pp.5670-5703.

Gupta, P. et al., 2005. Electrospinning of linear homopolymers of poly(methyl methacrylate): exploring relationships between fiber formation, viscosity, molecular weight and concentration in a good solvent. *Polymer*, 46, pp.4799-4810.

Gupta, D. et al., 2009. Nanostructured biocomposite substrates by electrospinning and electrospraying for the mineralization of osteoblasts. *Biomaterials*, 30(11), pp.2085-2094.

Hunziker, E.B., 2002. Articular cartilage repair: basic science and clinical progress. A review of the current status and prospects. *Osteoarthritis and Cartilage*, 10(6), pp.432-463.

Inoguchi, H. et al., 2006. Mechanical responses of a compliant electrospun poly(L-lactide-*co*-ε-caprolactone) small-diameter vascular graft. Biomaterials, 27(8), 1470-8.

Jang, J.-H., Castano, O. & Kim, H.-W., 2009. Electrospun materials as potential platforms for bone tissue engineering. *Advanced Drug Delivery Reviews*, 61(12), pp.1065-1083.

Ju, Y.M. et al., 2010. Bilayered scaffold for engineering cellularized blood vessels. *Biomaterials*, 31(15), pp.4313-4321.

Kameoka, J. et al., 2003. A scanning tip electrospinning source for deposition of oriented nanofibres. *Nanotechnology*, 14(10), pp.1124-1129.

Kumbar, S.G. et al., 2008. Electrospun poly(lactic acid-*co*-glycolic acid) scaffolds for skin tissue engineering. *Biomaterials*, 29(30), pp.4100-4107.

Ladd, M.R. et al., 2010. Co-electrospun dual scaffolding system with potential for muscle-tendon junction tissue engineering. *Biomaterials*, 32(6), pp.1549-1559.

Lee, C.H. et al., 2005. Nanofiber alignment and direction of mechanical strain affect the ECM production of human ACL fibroblast. *Biomaterials*, 26(11), pp.1261-1270.

Lee, S.J. et al., 2007. In vitro evaluation of electrospun nanofiber scaffolds for vascular graft application. *Journal of Biomedical Materials Research, Part A*, 83(4), p.999–1008.

Lee, S.J. et al., 2008. Development of a composite vascular scaffolding system that withstands physiological vascular conditions. *Biomaterials*, 29(19), pp.2891-2898.

Li, C. et al., 2006. Electrospun silk-BMP-2 scaffolds for bone tissue engineering. *Biomaterials*, 27(16), pp.3115-3124.

Li, D., Wang, Y. & Xia, Y., 2003. Electrospinning of Polymeric and Ceramic Nanofibers as Uniaxially Aligned Arrays. *Nano Letters*, 3(8), pp.1167-1171.

Li, W.-J. et al., 2005. A three-dimensional nanofibrous scaffold for cartilage tissue engineering using human mesenchymal stem cells. *Biomaterials*, 26(6), pp.599-609.

Li, X. et al., 2009. Nanofiber scaffolds with gradations in mineral content for mimicking the tendon-to-bone insertion site. *Nano Letters*, 9(7), pp.2763-2768.

Liao, S. et al., 2008. Processing nanoengineered scaffolds through electrospinning and mineralization suitable for biomimetic bone tissue engineering. *Journal of the Mechanical Behavior of Biomedical Materials*, 1(3), pp.252-260.

Lodish, H. et al., 2008. *Molecular Cell Biology* 6th ed., New York, New York: W. H. Freeman and Company. pp.820-833.

McCullen, S.D. et al., 2010. In situ collagen polymerization of layered cell-seeded electrospun scaffolds for bone tissue engineering applications. Tissue Engineering. Part C, 16(5), pp.1095-1105.

Megelski, S. et al., 2002. Micro- and nanostructured surface morphology on electrospun polymer fibers. *Macromolecules*, 35(22), pp.8456-8466.

Moffat, K.L. et al., 2009. Novel nanofiber-based scaffold for rotator cuff repair and augmentation. *Tissue Engineering. Part A*, 15(1), pp.115-126.

Moon, D.G. et al., 2008. Cyclic mechanical preconditioning improves engineered muscle contraction. *Tissue Engineering. Part A*, 14(4), pp.473-482.

Moore, K.L. & Dalley, A.F., 1999. *Clinically Oriented Anatomy* 4th ed., Baltimore, MD: Lippincott Williams & Wilkins. pp.12-14.

Nam, J. et al., 2007. Improved cellular infiltration in electrospun fiber via engineered porosity. *Tissue Engineering*, 13(9), pp.2249-2257.

Nie, H. & Wang, C.-H., 2007. Fabrication and characterization of PLGA/HAp composite scaffolds for delivery of BMP-2 plasmid DNA. *Journal of Controlled Release* 120(1-2), pp.111-121.

Pham, Q.P., Sharma, U. & Mikos, A.G., 2006. Electrospinning of polymeric nanofibers for tissue engineering applications: a review. *Tissue Engineering*, 12(5), pp.1197-1211.

Plotkin, L.I. et al., 1999. Prevention of osteocyte and osteoblast apoptosis by bisphosphonates and calcitonin. *Journal of Clinical Investigation*, 104(10), pp.1363-1374.

Powell, H.M. & Boyce, S.T., 2009. Engineered human skin fabricated using electrospun collagen-PCL blends: morphogenesis and mechanical properties. *Tissue Engineering. Part A*, 15(8), pp.2177-2187.

Prabhakaran, M.P., Venugopal, J.R. & Ramakrishna, Seeram, 2009. Mesenchymal stem cell differentiation to neuronal cells on electrospun nanofibrous substrates for nerve tissue engineering. *Biomaterials*, 30(28), pp.4996-5003.

Priya, S.G., Jungvid, H. & Kumar, A., 2008. Skin tissue engineering for tissue repair and regeneration. *Tissue Engineering. Part B, Reviews*, 14(1), pp.105-118.

Reneker, D. H., & Chun, I., 1996. Nanometre diameter fibres of polymer, produced by electrospinning. *Nanotechnology*, 7, pp.216-223.

Ross, M.H., Kaye, G.I. & Pawlina, W., 2003. Integumentary System. In *Histology A Text and Atlas*. Baltimore, MD: Lippincott Williams & Wilkins, pp. 400-433.

Sahoo, S. et al., 2006. Characterization of a novel polymeric scaffold for potential application in tendon/ligament tissue engineering. *Tissue Engineering*, 12(1), pp.91-99.

Sahoo, S., Cho-Hong, J.G. & Siew-Lok, T., 2007. Development of hybrid polymer scaffolds for potential applications in ligament and tendon tissue engineering. *Biomedical Materials (Bristol, England)*, 2(3), pp.169-173.

Sahoo, S. et al., 2010a. Bioactive nanofibers for fibroblastic differentiation of mesenchymal precursor cells for ligament/tendon tissue engineering applications. *Differentiation; Research in Biological Diversity*, 79(2), pp.102-110.

Sahoo, S., Toh, Siew Lok & Goh, J.C.H., 2010b. A bFGF-releasing silk/PLGA-based biohybrid scaffold for ligament/tendon tissue engineering using mesenchymal progenitor cells. *Biomaterials*, 31(11), pp.2990-2998.

Shin, H.J. et al., 2006. Electrospun PLGA nanofiber scaffolds for articular cartilage reconstruction: mechanical stability, degradation and cellular responses under mechanical stimulation in vitro. *Journal of Biomaterials Science. Polymer Edition*, 17(1-2), pp.103-119.

Sill, T.J. & von Recum, H. a, 2008. Electrospinning: applications in drug delivery and tissue engineering. *Biomaterials*, 29(13), pp.1989-2006.

Spalazzi, J.P. et al., 2008. Mechanoactive scaffold induces tendon remodeling and expression of fibrocartilage markers. *Clinical Orthopaedics and Related Research*, 466(8), pp.1938-1948.

Stankus, J.J. et al., 2006. Microintegrating smooth muscle cells into a biodegradable, elastomeric fiber matrix. *Biomaterials*, 27(5), pp.735-744.

Stoop, R., 2008. Smart biomaterials for tissue engineering of cartilage. *Injury*, 39 Suppl 1, pp.S77-87.

Sun, T. et al., 2005. Self-organization of skin cells in three-dimensional electrospun polystyrene scaffolds. *Tissue Engineering*, 11(7-8), pp.1023-1033.

Teo, W.E. & Ramakrishna, S, 2005. Electrospun fibre bundle made of aligned nanofibres over two fixed points. *Nanotechnology*, 16(9), pp.1878-1884.

Theron, A., Zussman, E. & Yarin, A., 2001. Electrostatic field-assisted alignment of electrospun nanofibres. *Nanotechnology*, 12, pp.384-390.

Thorvaldsson, A. et al., 2008. Electrospinning of highly porous scaffolds for cartilage regeneration. *Biomacromolecules*, 9(3), pp.1044-1049.

Tillman, B.W. et al., 2009. The in vivo stability of electrospun polycaprolactone-collagen scaffolds in vascular reconstruction. *Biomaterials*, 30(4), pp.583-588.

Vaquette, C. et al., 2010. Aligned poly(L-lactic-*co*-ε-caprolactone) electrospun microfibers and knitted structure: a novel composite scaffold for ligament tissue engineering. *Journal of Biomedical Materials Research. Part A*, 94(4), pp.1270-1282.

Wong, D.J., Chang, H.Y. & Biology, E., 2009. Skin tissue engineering. *StemBook*, pp.1-9.

Xu, C., 2004. Aligned biodegradable nanofibrous structure: a potential scaffold for blood vessel engineering. *Biomaterials*, 25(5), pp.877-886.

Yang, S. et al., 2001. The design of scaffolds for use in tissue engineering. Part I. Traditional factors. *Tissue Engineering*, 7(6), pp.679-689.

Yang, Y. et al., 2009. Electrospun Composite Mats of Poly[(D,L-lactide)-*co*-glycolide] and Collagen with High Porosity as Potential Scaffolds for Skin Tissue Engineering. *Macromolecular Materials and Engineering*, 294(9), pp.611-619.

Zhou, Y. et al., 2008. Electrospun water-soluble carboxyethyl chitosan/poly(vinyl alcohol) nanofibrous membrane as potential wound dressing for skin regeneration. *Biomacromolecules*, 9(1), pp.349-354.

Ziabari, M., Mottaghitalab, V. & Haghi, A.K., 2010. A new approach for optimization of electrospun nanofiber formation process. *Methodology*, 27(1), pp.340-354.

17

Three-Dimensional Nanofiber Scaffolds for Regenerative Medicine

Bit Na Lee, Jae Ho Kim, Heung Jae Chun[1] and Moon Suk Kim
Department of Molecular Science and Technology, Ajou University, Suwon,
[1]Institute of Cell & Tissue Engineering, Catholic University,
Korea

1. Introduction

Tissue engineering is an interdisciplinary technology of the basic concept of bioscience and biotechnology. It is aiming to make tissues that can replace or regenerate diseased tissues and organs after understanding of correlation between the structure and function of normal biological tissues. The regenerated tissues made in such way are improving, reviving, restoring, or substituting the functions of human body as well as maintaining the functions after transplanted into human body.[1]

The three elements of tissue engineering for regenerating the biological tissues are cell, growth factor, and scaffold. When the number of cells has been reduced due to some troubles of cell proliferation in the damaged tissue, cell can regenerate the damaged area by insertion of external cells, and the growth factor controls the growth and differentiation of cells. Scaffold is used to assist the growth and proliferation of cells. Scaffold helps cells growing and functioning in normal condition. If scaffold does not play its role properly, the replace or regenerate diseased tissues and organs must not success. Recent researches, therefore, have been focused on the development of scaffold that is influencing the proliferation and differentiation of cells.

Scaffold must have, particularly, the similar form of extracellular matrix that supports cells, which should have the following characteristics.[1,2] (1) Scaffold must connect tissue and blood vessel with each other by the appropriate size pores. (2) Scaffold must be able to adjust the biodegradability and bioabsorbability. (3) Scaffold must have the chemical surface where cells can achieve adherence, differentiation, and proliferation. (4) Scaffold must not induce other reverse functions or side effects. (5) Scaffold must be formed in various shapes and sizes, and it must be easy for the penetration or manipulation of various materials inside the scaffold.

Nanofiber is getting noticed the most intensively among the scaffolds with above characteristics (**Fig.1**).[3] Nanofiber is expected to overcome the limitation of conventional materials. So, it will be adopted in the new field with lots of advantages of high surface area per unit volume, high porosity, numerous fibers in the unit area, micro space created between fibers, and its flexibility. Nanofiber can be produced through phase separation, self-assembly method, electrospinning method, etc. Nanofiber can be used in various biomedical application such as high-functional filter and wound healing material, reinforced fiber of composite biomaterials, and scaffold for tissue engineering. Among them, the nanofiber as a scaffold for tissue engineering can provide similar environment as collagen of the extracellular matrix (ECM). It is advantageous for the cell adhesion when cultivating the cells.

This chapter focuses on three-dimensional nanofiber scaffolds, and highlights their potential applications for regenerative medicine. The first section reviews the fabrication of nanofiber scaffolds that provide an optimal microenvironment for cell proliferation, migration, differentiation, and guidance for the reconstruction or replacement of damaged or diseased tissues and organs. The following section focuses on natural and synthetic biodegradable biomaterials that have been applied as nanofiber scaffolds. The last section focuses on the preclinical applications of nanofiber scaffolds in regenerative medicine.

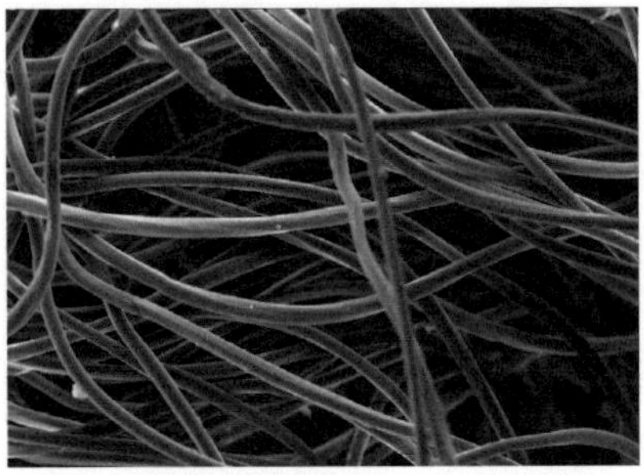

Fig. 1. Scanning electron micrograph (SEM) image of nanofibers.

2. Nanofiber manufacturing technology

Nanofiber manufacturing technology can be largely divided into 3 technologies of phase separation, self-assembly method, and electrospinning method. The nanofiber manufactured by phase separation and self-assembly method shows the limitation as a scaffold for the applications for tissue engineering. On the other hand, the nanofiber manufactured by electrospinning method shows various characteristics, which are suitable for the tissue engineering. This section introduces the nanofiber productions by phase separation, self-assembly method, and electrospinning method.

2.1 Phase separation

Phase separation is the porous polymer membrane forming technique used for years. Phase separation can control the pore structure of nanofiber by using two or more materials of different physical characteristics, and porous fiber is obtained when using polymer and highly volatile solvent. So, pore size can be changed by control of the volatility of solvent. Also it is possible to manufacture the nanofiber of which hydrophilic property has been adjusted, as pore structure can be changed by the interaction of solvent and water molecules in the air. However, there happens rapid phase separation between solvent and solute, when using the volatile solvent, due to the radical solidification of polymer with the volatilization of solvent. So it is not easy to control the concentration of polymer solution. It also has a problem that mass production of nanofiber is difficult as it can be applied only for limited numbers of polymer.

2.2 Self-assembly method

Self-assembly means that each component forms orderly structure voluntarily by the noncovalent bond. The universal method to make nanofiber is synthesizing the Peptide

Amphiphile (PA). When attaching PAs consisted with dialkyl chain (tail part of hydrophobicity) to the N-α amino group in the end of peptide chain, the peptides become similar to the base sequence of collagen amino acid of human ECM. However, self-assembly method is limited only for several polymer arrays (two block copolymers, three block copolymers, peptides-amphiphilic three block copolymers, and dendrimers). Another problem is that its mass production is not easy because of complicated manufacturing process and low productivity.

2.3 Electrospinning method

Electrospinning had been proposed in 1930s together with electrospraying method. But it has not been commercialized since its development as its application was limited (**Fig.2**). Electrospinning method gets spotlighted again later in mid of 1990s when Reneker succeeded in the manufacturing of nanofiber with various polymers after simplifying the electrospinning device.[4] The porosity, thickness, and components of nanofiber can be adjusted with simple experimental equipment in the electrospinning method. So it has been able to produce continuous nanofiber not only with polymer but also with ceramics at low process cost. It can produce nonwoven fabric type nanofiber at the same time of spinning, so its spinning time is short and its construction is simpler than general spinning facilities. Also, nanofiber can be produced with various polymers, and the spinning is available with just a little amount of polymer. The biggest feature of electrospinning method is that its nanofiber has the similar structure of ECM in terms of morphology. ECM forming collagen is consisted with micro fibril of 50~100 nm, and the ECM-similar nanofiber can be produced by using the electrospinning method.

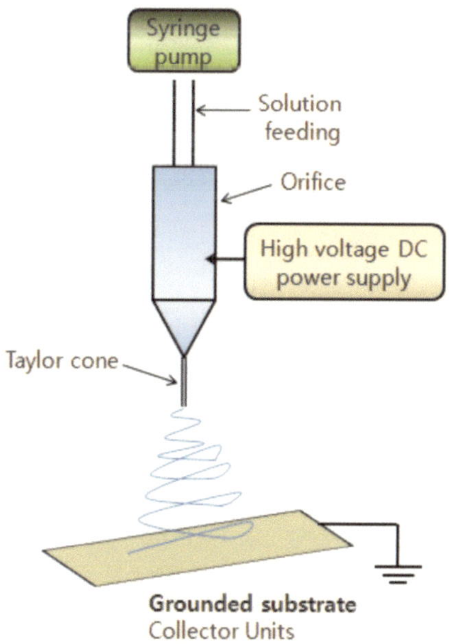

Fig. 2. Schematic image of the electrospinning method.

Reviewing the electrospinning principle, polymer fine bubbles are forming the taylor cone by the mutual repulsion of induced electric charges when applying high electric field to the polymer solution hanging on the end of a charged jet. Then, polymer solution is emitted when electric repulsion is higher than surface tension. Solvent is volatilized while emitted solution is in the air, then fibers of 50~100 nm diameter are laminated in 3-dimensional network structure to form the mesh structure. The nanofiber made in this way has high ratio of surface area compared to its volume, and it has high porosity to constitute favorable environment for cells to live in. As complex polymers can be used to produce nanofiber, it has been possible to manufacture the nanofiber that has overcome the limit of properties of existing polymers.

3. Nanofiber manufacturing materials

The scaffold is transplanted into the living body for treatment, so it is necessary to pay a lot of attention in selecting the materials. Both natural polymer and synthetic polymer have been used for treatment as the highly absorbing materials for over 30 years. It is important in recent tissue engineering to make scaffold that decomposes easily in the body without intermediate products or other side effects. Natural polymer is frail in its property but it is superior in the cell affinity and cell compatibility. Synthetic polymer can adjust the property and degradation period easily, but it is less biocompatible as there are no molecules to which cells can adhere. This section will describe the characteristics of each material applied for the production of nanofiber.

3.1 Natural biomaterials

As natural polymer has similar structure to the macromolecules in the body, it is used a lot as the biomaterial. For the production of nanofiber, chitosan, alginate, and elastin as well as collagen are used representatively (**Fig.3**).

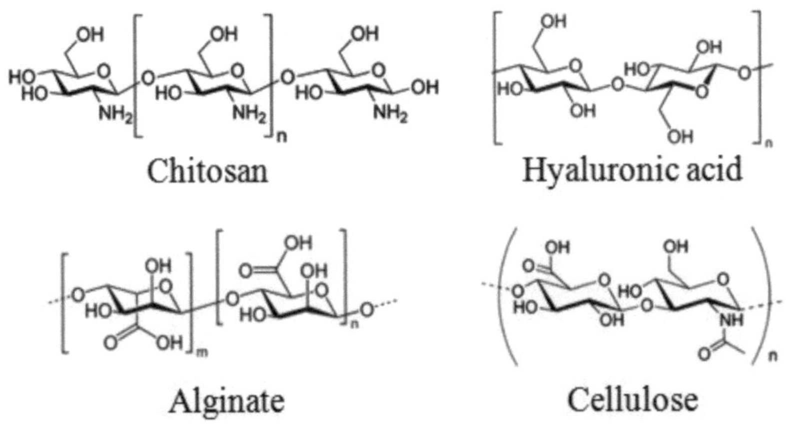

Fig. 3. Structures of natural biomaterials.

Collagen

Collagen occupies 20~30% of total protein in vertebrate animals, which is a main component of ECM. At least no less than 22 kinds of collagens are existing in the human body to keep up the tissues or organs and to maintain the figure of human body. Collagen has higher cell affinity and less immune reaction, moreover, as it contains lots of chemical inducers which are related to the adhesion, proliferation, and differentiation of cells. Collagen is also reported to have effectiveness in improving the physiology such as cell proliferation and organ formation, promotion of wound healing, and so on. It is disclosed that collagen nanofiber is inducing the biological tissue switching effect of nanocomposites in the connective tissues, accordingly, its applications are getting wider.

Chitosan

Chitosan, a naturally abundant antibacterial polymer, is found in the microbe cell walls and in the exoskeleton of crustaceans. Chitosan is manufactured by removing acetyl out of chitin, a kind of polysaccharide. The chitosan based nanofiber production is under research as it has less stimulation, excellent biocompatibility and biodegradability, and blood coagulating function. Also the chitosan nanofiber is being researched as antibacterial wound dressings in the tissue engineering.

Alginate

Alginate is one of natural polymers, which is safe without toxicity. Alginate is produced in fiber form and processed in the forms of woven fabric, nonwoven fabric, and composite material for the treatment of wound area. Calcium ion is exchanged with sodium ion in the body fluids, when alginate nanofiber contacts with wound exudation, so it is used widely for the absorbent wound dressing material. As alginate is an electrolyte of very high conductivity and high viscosity, however, it is not easy for electrospinning. So, it must be electrospinning processed after mixing with water-soluble substances.

Elastin

Elastin is a major component of elastic tissues such as blood vessel, lungs, ligament, and skin, which plays an important role to maintain the elasticity of tissues. Elastin is being proposed now as elastin nanofiber in 3-dimensional structure to produce excellent bioabsorbability, although it had not been easy to form and process elastin due to its stiff cross-linked structure.

3.2 Synthetic biomaterials

Synthetic polymer is being developed for over 30 years in various forms of surgical suture or screw form, mesh structure, etc. Synthetic polymer is manufactured according to the unique characteristics of each material, and it is being produced to minimize the immune reactions in the tissue engineering (**Fig.4**). As the biodegradation period of nanofiber made of synthetic polymer can be adjusted, its commercialization is being progressed partially as a tissue engineering scaffold (**Table 1**).

Materials	Product names	Degradation period
Polycaprolactone (PCL)	MONOCRYL	> 20 months
Poly L-lactic acid (PLLA)	BioScrew, PL-FIX	20~60 months
Polydioxanone (PDO)	PDS BIOSYN	6 months
Polyglycolic acid (PGA)	DEXON	1~4 months
Poly (lactic-co-glycolic acid) (PLGA)	VICRYL (90% Glycolide 10% Lactide)	2 months

Table 1. Degradable periods of synthetic polymer.[5]

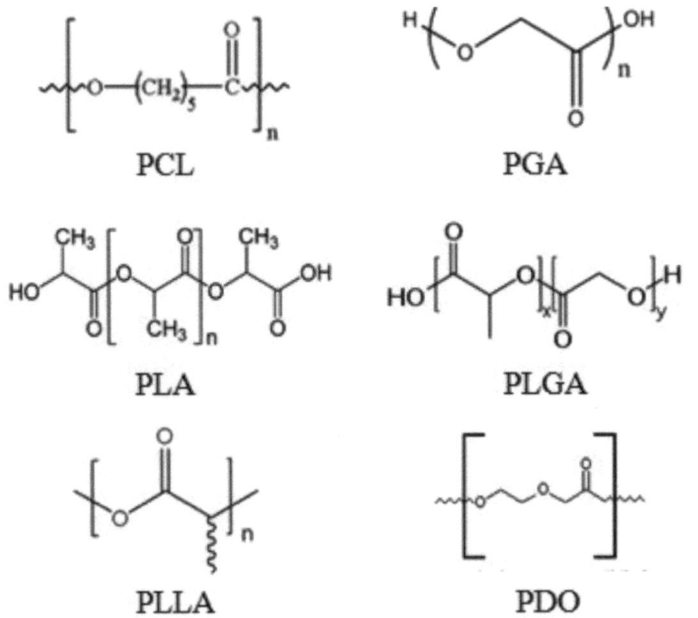

Fig. 4. Structures of synthetic biomaterials.

Polycaprolactone (PCL)

PCL is a polyester polymer of high elasticity without toxicity, which is superior in biocompatibility but its degradation period is rather slow for 1~2 years. It is aliphatic polyester with repeating 5 nonpolar methylene groups and 1 polar ester group, which is hydrophobic as of numerous carbons in its structure. It has disadvantages that initial protein absorption capacity is low, as the surface of PCL nanofiber is hydrophobic, the cell adhesion is slow, and the cell differentiation and tissue regeneration is slow. But, it is used as a good absorbent material for either soft or hard tissues as its properties can be adjusted.

Polyglycolic acid (PGA)

PGA is biodegradable aliphatic polyester, which had been developed for surgical suture in 1970s. Its biological absorptiveness can be predicted, and it takes 2~4 weeks for degradation as it is hydrophilic in the body. Its strength is reduced by 60%, for the first 2 weeks, also the pH and crystallinity are lowered down due to the generation of hydrolysis. The PGA nanofiber has high strength and elasticity initially, which is biodegraded rapidly by the diffusion of water and hydrolysis at the body temperature. However, it may cause undesirable reactions in the tissue as the pH of topical area is increasing rapidly while degradation speed is getting faster. To solve such a problem, researches are progressing for better biocompatibility by improving cell's adhesion capability. The PGA nanofiber is pretreated with acid in order to hydrolyze the ester bond and to expose the carboxylic acid and alcohol base.

Polylactic acid (PLA)

PLA is a synthetic polymer with steric hindrance, more hydrophobic than PGA, as it contains methyl base. It is highly soluble in the organic solvents and slow in hydrolysis, and its degradation period is as long as 30~50 weeks. There are L-type and D-type stereoisomers in PLA. PLLA (Poly L-lactic acid), which is an L-type stereoisomer, has a merit of excellent mechanical property in the polymerized form of lactic acid that is synthesized in the living body. PLLA nanofiber has problems as it is hydrophobic, fragile, and slow in degradation speed. But it has been proved, as a scaffold imitating collagen, to be highly capable in the cell adhesion and differentiation.

Poly (lactic-co-glycolic acid) (PLGA)

PLGA, which is a copolymer of polyglycolide and polylactide, is a synthetic polymer approved by FDA. It is widely used as the biomaterial for porous scaffold, drug delivery system, etc. in tissue engineering. PLGA is harmless to human body as it is changed to lactic acid and glycolic acid by hydrolysis in the living body and changed to carbon dioxide and water when discharging out of the body. As it is highly biocompatible, biodegradable, and processable, it is used widely in tissue engineering and drug delivery system. PLGA is a synthetic polymer with various degradation periods when the amount of monomers is adjusted. The PLGA nanofiber is being applied the most widely as a scaffold imitating biological tissues or drug delivery system, and biosensors. Especially, PLGA is spreading its applications after controlling its properties by adjusting the amount of PLA and PGA.

Polydioxanone (PDO)

PDO is biodegradable polyester of 55% crystallinity, which has been developed originally as a biodegradable surgical suture. Its degradation speed in the body is 6 months, intermediate period between PGA and PLA. PDO is superior in flexibility as it contains ester oxygen in the chains of monomer. PDO nanofiber is suitable for the biomaterial as it carries the property between minimum elasticity coefficient of collagen and maximum elasticity coefficient of elastin.

4. Nanofiber applications for tissue engineering

For the applications of nanofiber scaffold to the human body, it must provide processability and appropriate conditions for adhesion, proliferation, and differentiation of cells in the tissue as well as the properties of conventional scaffolds. As the nanofiber satisfying such requisites is widely applied to the artificial tissues such as skin, blood vessel, bone, etc., wound dressings, and drug delivery system, this section is going to describe the applications of nanofiber.

4.1 Artificial skin

Skin is the biggest tissue that is covering the surface of human body. It prevents loss of moisture, adjusts the body temperature, blocks bacterial invasion, and protects human body from radiant rays and ultraviolet rays. Skin grafting technology has been developed to regenerate the skin tissue when it had been damaged by a burn, external wound, carcinoma resection, skin disease, etc. It is proved that nanofiber is greatly effective as a tegaderm in the adhesion and proliferation of human skin cells, and it has been reported that nanofiber can be used as an artificial skin.

4.2 Artificial blood vessel

Scaffold of excellent biocompatibility must be used in the vascular grafting, and it must satisfy several specific requirements such as mechanical elasticity and durability that can put up with repeated inflations and compressions. The nanofiber produced by electrospinning can imitate similarly the component, structure, and mechanical characteristics of blood vessel. It is reported that the adhesion, proliferation, and differentiation of cells have been improved when cultivating the nonstriated muscle cells with the scaffold made of copolymer P(LLA-CL) of PLLA and PCL. Its result has shown that P(LLA-CL) nanofiber is potential to be an ideal artificial blood vessel.

4.3 Artificial bone

In order to regenerate the bone in tissue engineering, the human regeneration mechanism promoting the bone formation and restoration should be increased or the tissue similar to the bone in living body must be developed. For the past several decades, autograft or allograft has been used for the damaged bones by disease or traumatic injury. However, there was limitation in such bone transplantations. So bioactive materials such as hydroxyapatite or tricalcium phosphate have been used for its substitution, or glass and ceramic have been used in dentistry or in orthopedics as alternative bones. Bioactivity, tissue integration, and mechanical strength are requested for such materials. Various nanofiber scaffolds have been developed recently as bone substitutions to fulfill such requirements. Ramakrishna et al. has made the scaffold by electrospinning of collagen, which has similar structure of extracellular matrix, and hydroxyapatite at the mixing ratio of 1:1. It is proved, when osteoblast was cultured on the nanofiber that had been electrospinning processed with hydroxyapatite, that it has been mineralized as of high differentiation of cells and high concentration of calcium and phosphorus.

4.4 Artificial cartilage

Peculiarly, there is no blood vessel, nerves and lymphoid tissue in the articular cartilage. There is no inflammation reaction when damaged, therefore, and it is difficult to supplement the cells to recover the damage. So, it is highly limited to recover or regenerate the articular cartilage when it has been damaged. Nanofiber is applied to make hyaline articular cartilage tissue, therefore, by cultivating articular cartilage cells or adipose-derived stem cells in the appropriate 3-dimensional scaffold. Nanofiber scaffold promotes the generation of ECM by the transplanted cells. The nanofiber is also able to adjust the chondrogenesis of human cartilage cells and that nanofiber can be applied widely in the chondral resurfacing by adjusting the shape and size of nanofiber.

4.5 Wound dressing

Dermis is exposed as epithelial tissue is peeled off in the affected part of skin tissue by wound such as external injury, burns, diabetes, and clogging of venous blood flow. Wound dressing is the first-aid kit to stop bleeding, which promotes the injury protection as blood or serum out of the wound is penetrating the wound dressings. If the wound dressing is constituted with nanofiber of micro diameter, the body fluids such as blood cannot penetrate the nanofiber, and so blood flow stays in the wound and coagulated. As there is no blood coagulation in the nanofiber, therefore, it is easy to remove the nanofiber. Also it is easier to exchange the oxygen and evaporate the water-vapor between the wound surface and air when porosity is higher. As nanofiber is a mat of very tiny diameter and pores with high specific surface area, its moisturization and breathability is good enough, it protects wound from germs and prevents body fluids from penetrating, and it is easy to remove. So its application for various wound dressings is also being reviewed actively.

4.6 Drug delivery system

Drug delivery system is being developed to reduce the number of medications, to enhance the drug effectiveness and stability by adjusting the initial emissions through continuous emission of constant amount of drug to the medicating area. Numbers of researches are reported recently for using nanofiber produced with biodegradable polymers in the drug delivery systems. The drug delivery system made of nanofiber carries out the drug delivery function with fewer side effects by installing physical barriers using its wide surface area. Also the fibrosis of biocompatible polymers is easy and it does not need to raise the temperature when electrospinning. It is advantageous as drug is not decomposed in the nanofiber.

5. Conclusion and prospect

Nanofiber made of biomaterials in various methods is getting noticed for its very wide application potential as a scaffold imitating the ECM. Nanofiber shows significant effect in the adhesion, proliferation, and differentiation of cells as of its wide surface area and high porosity. Moreover, nanofiber is researched for the scaffold for the musculoskeletal tissues such as bone, spine, ligament, and skeletal muscle and for the tissues of skin, nerves, and blood vessel. It is also expanding its applications up to adjusting the delivery of drug, protein, and DNA. It is expected obviously that nanofiber will be used as an important scaffold in tissue engineering based on these researches. However, researches must be accompanied simultaneously for productivity and safety of nanofiber and securing application safety. It is also for easy biological application of nanofiber and for assessment of long-term regeneration capability in long-term view. Nevertheless, interdisciplinary researches of the experts in the fields of nanotechnology, biomaterials, medicines, and clinics are currently progressing so as to apply nanofiber clinically in the near future.

6. References

[1] A Atala, R Lanza, JA Thomson, R Nerem, Principles of Regenerative Medicine, 2nd Ed., Academic Press, San Diego, 2010.

[2] WM Saltzman, Tissue Engineering: Engineering Principles for the Design of Replacement Organs and Tissues, Oxford University Press, Oxford, 2004.

[3] K Gonsalves, C Halberstadt, CT Laurencin, L Nair, Biomedical Nanostructures, Wiley-Interscience, New Jersey, 2008.

[4] DH Reneker, I Chun, Nanometre diameter fibres of polymer produced by electrospinning, Nanotechnology, 7, 216, 1996.

[5] PA Walker, KR Aroom, F Jimenez, SK Shah, MT Harting, BS Gill, CS Jr Cox. Advances in Progenitor Cell Therapy Using Scaffolding Constructs for Central Nervous System Injury, Stem Cell Rev., 5, 283-300, 2009.

18

Incorporation of DNA into Electrospun Nanofibrous Scaffolds: Fundamental Characterization Studies and Gene Delivery

Michael Hadjiargyrou
Stony Brook University
USA

1. Introduction

The ability to successfully deliver exogenous DNA into cells has provided the scientific community with a plethora of opportunities into the realm of basic, clinical, and translational research. It has afforded us the means to study the structure and function of genes and their products (i.e. proteins), correct genetic or acquired disorders, and generate tremendous amounts of data which researchers around the world continuously work to build upon in order to generate the next "magic bullet". Despite such progress, gene delivery continues to suffer from various limitations, predominantly, low transfection efficiencies when using non-viral approaches (i.e. naked DNA, cationic lipids/liposomes, synthetic and natural polymers, etc.) (Al-Dosari & Gao, 2009) and unacceptable levels of toxicity and immunogenicity when using the more traditional viral vectors. To address these limitations, researchers have focused on a number of innovative alternatives, one of which involves the incorporation of DNA into electrospun nanofibrous, non-woven and biodegradable scaffolds.

Electrospinning is a variation of the electrospray process and enables the experimenter to formulate scaffolds with specified mechanical, biological, chemical and kinetic properties, all readily controlled by alterations of polymer solution composition (i.e. polymer molecular weight and concentration, viscosity, salt concentration, conductivity, surface tension) and processing parameters (i.e. temperature, humidity, electric field strength, distance between spinneret and collector, feed rate and flow rate) (Chiu et al., 2005; Bhardwaj & Kundu, 2010). The resulting nanofibrous scaffolds possess desirable attributes such as high surface area to volume ratio and numerous interconnected pores that mimic the topology of the extracellular matrix (ECM). Obviously these features are a requirement for the transport of oxygen, nutrients, and wastes through the scaffold, as well as supporting robust cell adhesion and viability. Further, the flexible nanofibers and their pliability is beneficial for cell migration throughout the 3D scaffold. As such, electrospinning represents a truly rational and versatile approach for the construction of custom-tailored tissue engineering scaffolds, especially those capable of supporting cell/tissue growth, as well as deliver a range of bioactive molecules, including, DNA, proteins, drugs, etc. These "functionalized" or "biomimetic" scaffolds represent a new paradigm in the design and implementation of tissue engineering strategies.

Since the initial study of Fang & Renneker (1997), demonstrating the possibility of electrospinning pure DNA nanofibers, the use of this technique for a variety of applications has truly exploded. Research has been reported in areas as diverse as scaffolds for tissue regeneration/regenerative medicine (Martins et al., 2007), drug delivery systems (Hadjiargyrou & Chiu, 2008), wound dressing materials (Khil et al., 2003; Kang et al., 2010; Chen & Chian, 2010; Cai et al., 2010), biocatalysis (Nair et al., 2007), cardiovascular grafts (Sell et al., 2009), prevention of post-surgical adhesions (Zong et al., 2004; Bölgen et al., 2007), filtration (Kattamuri et al., 2005), energy and environmental applications (Thavasi et al., 2008), biosensors (Manesh et al., 2007; Kowalczyk et al., 2008; Ding et al., 2010), and protective clothing (Gibson et al., 1998 ; Seungsin & Obendorf, 2006). In fact, there are over 1,650 reports indexed in PubMed for a keyword search of "electrospun or electrospinning". As many of these electrospinning applications are covered in other parts of this book, this chapter will strictly focus on the current literature and latest research on the use of electrospun polymeric scaffolds for fundamental characterization studies on the incorporation and behavior of DNA as well as their application as gene delivery systems.

2. Fundamental studies: Incorporation and characterization of DNA into polymeric nanofibers

The initial and simple study of electrospinning DNA into nanofibers was reported in 1997 by Fang and Reneker, who used commercially available calf thymus genomic DNA that was diluted in a range of 0.3 - 1.5% in a mixture of 70% water and 30% ethanol and electrospun at 12-14 kV onto porous copper grids. The resulting pure DNA fibers had diameters as small as 30 nm that spanned the 50 μm pores of the grids. In addition, the DNA fibers contained beads with diameters from 80 - 300 nm that were a few microns apart on each DNA nanofiber, resulting from DNA retraction into droplets. Lastly, the authors presented a high magnification image of a single DNA fiber that was 62 nm in diameter and speculated that ~600 DNA molecules could pass through each cross section to form the fiber (Fang and Reneker, 1997).

A similar study was conducted using salmon testis DNA with a range of 200-90 kb fragments that was diluted in a range of 0.5 - 2.5 wt% solution in a mixture of 70% water and 30% ethanol and electrospun using a range of voltages (15 - 30 kV) onto mica substrates (Takahashi et al., 2005). The resulting DNA fibers were examined with atomic force microscopy (AFM) and the authors reported the presence of four different types of fibers; twisted shaped, twisted and stretched, curved and straight. Further, the height and length of the fibers ranged from 1.8 nm and 1 μm, respectively, but these slightly varied with the different % DNA solutions and voltages used. The authors concluded that the optimum parameter values for generating ultrathin nanofibers were 1.5 wt% for DNA concentration and 20 kV for the applied voltage.

Aside from electrospinning pure DNA in solution, other investigations were carried out using DNA mixed with various synthetic polymers. For example, one study mixed fluorescently labelled bacteriophage lambda DNA molecules with polyethylene oxide (PEO) and produced polymer nanofibers that contained isolated and stretched DNA molecules (Bellan et al., 2005). These DNA nanofibers had diameters that ranged between 100-350 nm and lengths between 1 and 19 μm, as measured by AFM (Fig. 1). Making use of the fluorescently labelled DNA, this study also examined the relaxation time of the DNA in a viscous solution (bulk PEO) in order to understand the fluid dynamic nature of the DNA.

The authors found that the DNA length vs. time yielded a time constant ranging from 2.1 - 19 seconds, thus adding new knowledge on the dynamic behaviour of the DNA in the PEO solution while in the electrospinning jet.

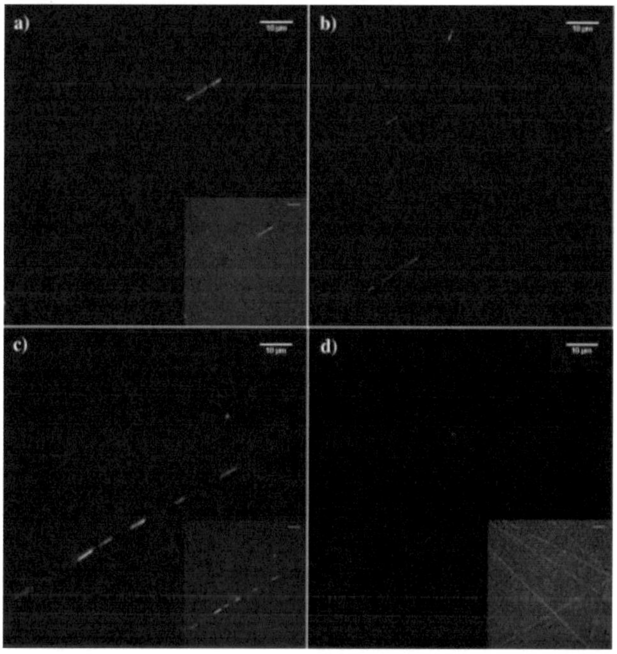

Fig. 1. Fluorescently labelled DNA/PEO nanofibers. Images a-c show the presence of the fluorescently labelled DNA within the PEO nanofibers. Image d shows control nanofibers without any DNA. Reprinted with permission from Bellan et al., 2005, American Chemical Society.

The same group also electrospun fluorescently labelled DNA and deposited it over photolithographically patterned trenches and gold electrodes and found that the nanofibers displayed diameters of ~27 nm (Bellan et al., 2007). Further, they imaged the nanofibers via a number of imaging techniques, including fluorescence microscopy, scanning electron microscopy (SEM), and AFM. Using AFM, the Young's modulus of a single suspended electrospun DNA fiber was determined by a three point bending test to be 15 ± 2 GPa. In their conclusion, the authors suggest that this approach could conceivably be used to manufacture nanosensors based on such DNA nanofibers, as well as aid in the development of self-assembled DNA devices.

An additional study was reported in the literature that utilized a mixture of DNA/PEO to electrospin nanofibers (Liu et al, 2007). The authors extensively evaluated the basic electrospinning solution properties such as conductivity, surface tension and viscosity (Table 1) and determined that the ionic conductivity of the solution increased significantly with the addition of DNA. In contrast, there was only a slight increase in the ionic conductivity with increasing amounts of PEO, which also decreased the surface tension. On the other hand, the viscosity increased with the addition of either DNA or PEO and as such the authors concluded that solutions containing both DNA and PEO had ideal properties for electrospinning. The electrospun DNA/PEO nanofibers displayed diameters of 50–250 nm that became thinner with decreased feed rate (tested a range between 50-200 μL/min), increased tip-to-collector distance (10-25 cm range) and thicker with low voltage (5-20 kv range) applied during electrospinning (Liu et al., 2007). As a result of these analyses, it was concluded that these DNA/PEO ultrathin fibers create a 3D porous structure with future potential applications in biosensing as well as cell culturing.

Solution no.	DNA (g)	PEO (g)	Amount of water (g)	Conductivity (S/cm)[a]	Surface tension (mN/m)[b]	Viscosity (mPa s)[c]
1	2	0	98	32.2	64.21	28.2
2	4	0	96	66.7	70.15	49.4
3	6	0	94	74.1	74.51	369.6
4	8	0	92	79.1	77.79	1068.8
5	10	0	90	104.0	74.26	63766.5
6	0	2	98	1.67	66.68	91.82
7	0	4	96	1.75	65.39	233.07
8	0	6	94	1.79	62.12	369.62
9	3	3	94	38.8	64.84	233.07
10	4	4	92	41.4	63.74	1104.14
11	5	5	90	53.6	61.58	2589.68
12	3	7	90	28.8	62.62	2337.77
13	4	6	90	41.9	64.11	2537.88
14	6	4	90	59.2	63.38	1998.76
15	7	3	90	80.4	61.14	8899.07

[a]Determined by a model 20 pH/conductivity meter (Denver Instrument).
[b]Determined by a KSV contact angle analyzer (goniometer, KSV instruments Ltd).
[c]Determined by a DV-II viscometer, shear rate using 20 s^{-1} for viscosity testing.
Reprinted with permission from Liu et al., 2007, Elsevier.

Table 1. Properties of aqueous solutions (100 mL) containing DNA/PEO.

Taking further advantage of the unique optical and electronic properties of DNA and thinking along the lines of using DNA as a candidate for device applications (electronics, filtration, sensors), Ner and colleagues (2008), complexed salmon sperm DNA (molecular weight [MW] 500 kDa) with a cationic surfactant (cetyltrimethylammonium chloride, CTMA), as well as a hemicyanine chromophore and electrospun nanofibers. These nanofibers had an average diameter between 250-350 nm and demonstrated amplified emission of the chromophore when compared to thin films of identical composition. When the material properties of these nonwoven DNA-CTMA fiber meshes were tested, it was found that the average initial modulus was 1.44 MPa with a breaking strength of 37.4 kPa, orders of magnitude lower than those of electrospun fluorescently labelled DNA (15 ± 2 GPa), as previously reported (Bellan et al., 2007). The reason for this difference maybe the presence of the CTMA in the DNA fibers, which were capable of only 11% elongation prior to failure, indicating stress-stiffening that may be due to the particular alignment of the DNA strands along the stress axis or the lateral inter-laminar sliding within the fibers. Although the incorporation of dye did not affect the fibers, it did enhance fluorescence that resulted from both the fiber morphology (5–6-fold amplification) and specific interactions (groove-binding) between the chromophore and DNA (18–21-fold amplification). Based on their data the authors suggest that electrospun DNA nanofibers can serve as advanced functional biomaterial for molecular electronics (Ner et al., 2008).

3. Gene delivery: Transfection efficiency

The aforementioned fundamental studies all dealt mostly with processing parameters and characterization of the DNA nanofibers, either electrospun on with pure DNA or in a mixture with various polymers. Moving past the stage of initial characterization and translating into biological feasibility, the issue of transfection efficiency is key for non-viral gene delivery. This was addressed in a number of studies reported in the literature. Thus, I have divided these manuscripts that focus strictly in generating electrospun scaffolds as gene delivery systems into two categories; those that focus on *in vitro* cell based studies and those that report on implanted scaffolds *in vivo* using various animal models.

3.1 *In vitro* studies

Beyond the critical fundamental studies on DNA and polymer interactions, the first demonstration of the ability of electrospun scaffolds to serve as non-viral gene delivery systems using plasmid DNA (pDNA) was conducted in my laboratory in 2003. We prepared polymer solutions of poly-lactide-co-glycolide (PLGA) mixed with either, 10, 12 or 15% of a poly(D,L-lactide)-poly(ethylene glycol) (PLA-PEG-PLA) triblock copolymer and with pDNA, electrospun multiple scaffolds and evaluated the morphology, material properties, pDNA incorporation and release, and transfection efficiency (Luu et al., 2003). These studies showed that increasing the concentration of tri-block copolymer resulted in the formation of larger diameter (~2.5 - 5.0 μm with the 15% triblock) electrospun fibers in comparison to those observed with lower concentrations of block copolymer (250 875 nm and 375-1.1 μm for the 10 and 12% triblock, respectively) (Fig. 2).

Despite the variation in fiber diameters, no significant differences in material properties were detected. On average these electrospun scaffolds displayed a tensile modulus of about 35 MPa, 45% tensile strength and ultimate stress of 3.3 MPa, matching those of skin and cartilage. We also showed that the addition of pDNA to the fibers did not change the ultimate strain but did significantly increase both the tensile modulus (by ~10 MPa) and ultimate stress (by ~1 MPa).

Characterizing the incorporation and release of pDNA showed that ~68-80% of the pDNA that was incorporated, was released in two phases over the 20 day study-period. The first phase constituted of an initial burst, where ~18-36% of the pDNA was released within 15 min of immersion of the scaffold into a saline buffer solution and the second phase was a slow and sustained release over the next 20 days. The initial release burst of pDNA indicated that a substantial amount of the electrospun DNA remained on the surface of the nanofibers with the rest embedded within the nanofibers. The integrity of the electrospun pDNA was demonstrated via agarose gel electrophoresis which revealed that the pDNA was indeed structurally intact and more importantly, its bioactivity was demonstrated by directed transfection in cell culture. Sections of the pDNA containing scaffolds placed in culture medium above a monolayer of pre-osteoblastic (MC3T3) cells were able to release active pDNA and transfect cells with the encoded reporter gene (β-galactosidase). The transgene was found to be expressed, albeit, at a very low efficiency in comparison to lipid mediated transfection (Fig. 3). Interestingly, when the released pDNA was complexed with a lipid mediator (Fugene6), the transfection efficiency only reached 20% that of the positive control (Fig. 3), but even 20% is considered a low level of transfection.

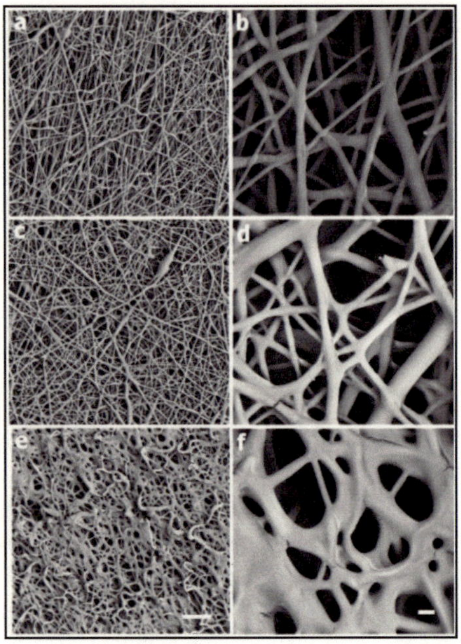

Fig. 2. Morphology of electrospun scaffolds. Highly porous scaffolds shown under low magnification on left, and high magnification on right. 10% block copolymer formulations (a, b) exhibited thin, fairly uniform fibers (250 to 875 nm). 12% block copolymer formulations (c, d) showed larger fibers and a larger distribution (375 nm to 1.1 mm). 15% block copolymer solutions (e, f) displayed thick, plate-like fibers of roughly 2.5 to 5 mm diameter at the junctions. Scale bar for (a), (c), (e) (left) 10 mm and for (b), (d), (f) (right) 1 μm. Reprinted with permission from Luu et al., 2003, Elsevier.

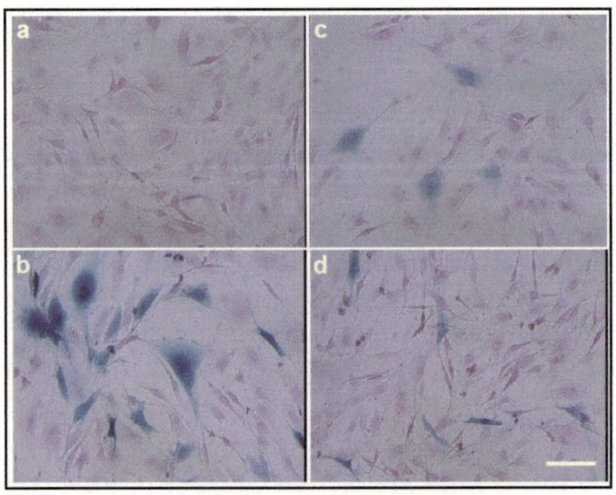

Fig. 3. Bioactivity of released DNA. (a) Naked DNA (2 μg) added directly to cell medium. (b) Cells transfected with complexed control DNA (2 μg) with a lipid-based transfection reagent (Fugene6). (c) Two 1.5 x 1cm sections of scaffold were directly incubated with cells for 4 h, then removed. (d) Released DNA from scaffold (2 μg) complexed with Fugene6. Following transfection (48 hr), cells were fixed and stained for β-galactosidase. Scale bar 100 μm. Reprinted with permission from Luu et al., 2003, Elsevier.

In subsequent experiments we were able to increase the transfection efficiency substantially by plating the cells directly on the electrospun pDNA/PLGA/PLA-PEG-PLA scaffolds (Fig. 4, Liang et al., 2005). This increase in transfection efficiency probably occurred because the cells were physically connected to the surface of the nanofibers where a substantial amount of pDNA was present. As a result, the degradation of the pDNA that probably occurred in our previous study (Luu et al., 2003) where the scaffolds were placed in the culture medium floating above the cells and the pDNA was released into the medium, was eliminated. Unfortunately, we did not conduct a confocal analysis of the scaffolds and thus we were not able to determine the exact transfection efficiency (transfected cells/total cells). Regardless, from the cells on surface shown in these images (Fig. 4), the transfection efficiency appears to be greater than 65%.

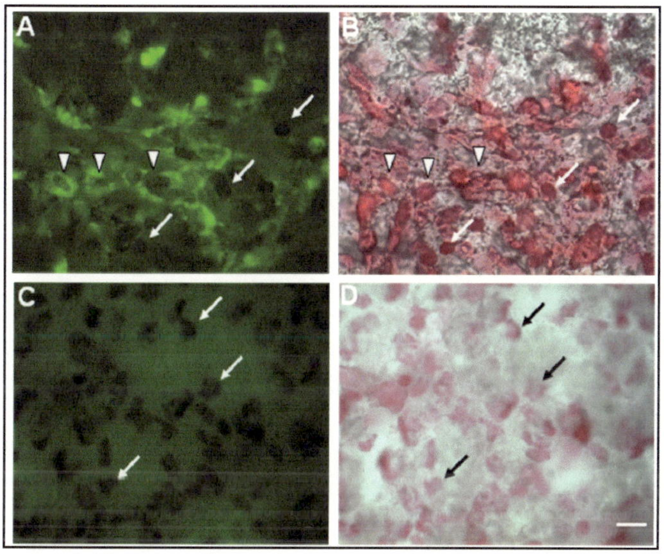

Fig. 4. Transfection of MC3T3 cells by nanofibrous electrospun scaffolds. Panels A-D show the increased transfection by plating MC3T3 cells directly on GFP pDNA/PLGA/10% block copolymer scaffolds. (A) Fluorescent image of cells 24 hr post-plating. Arrowheads indicate representative transfected GFP-expressing cells and arrows indicate non-transfected cells. (B) Light micrograph of cells (shown in A) stained with nuclear fast red. Arrowheads and arrows indicate identical cells in A. (C) Fluorescent image of cells on a control scaffold containing no DNA and indicating no fluorescent (green) cells (arrow). (D) Light micrograph of cells (shown in C) stained with nuclear fast red. Arrow indicates identical cells in C. This transfection efficiency reflected that of cells present only on the exterior surface of the scaffold and not for all cells within the scaffold. Scale bar: 10 μm. Reprinted with permission from Liang et al., 2005, Oxford University Press.

We also investigated the interactions between the polymers and pDNA using laser light scattering (LLS). Specifically, we determined that in a solution of N,N-dimethylformamide (DMF), the organic solvent used to dissolve the PLGA polymer, pDNA condenses into sphere-like structures and in the presence of the triblock copolymer, the formation of a micellar structure occurs with DNA being encapsulated by a shell made from the PLA-PEG-PLA triblock copolymer. When electrospun, these micelles are deposited randomly on and into the PLGA fibers, with a significant amount located on the surface (Fig. 5), which accounts for the observed rapid burst release of pDNA (Luu et al., 2003; Liang et al., 2005). Finally, it was also established that the presence of the polylactide shell encapsulating the pDNA, protects it during electrospinning, as in its absence, the pDNA was completely degraded.

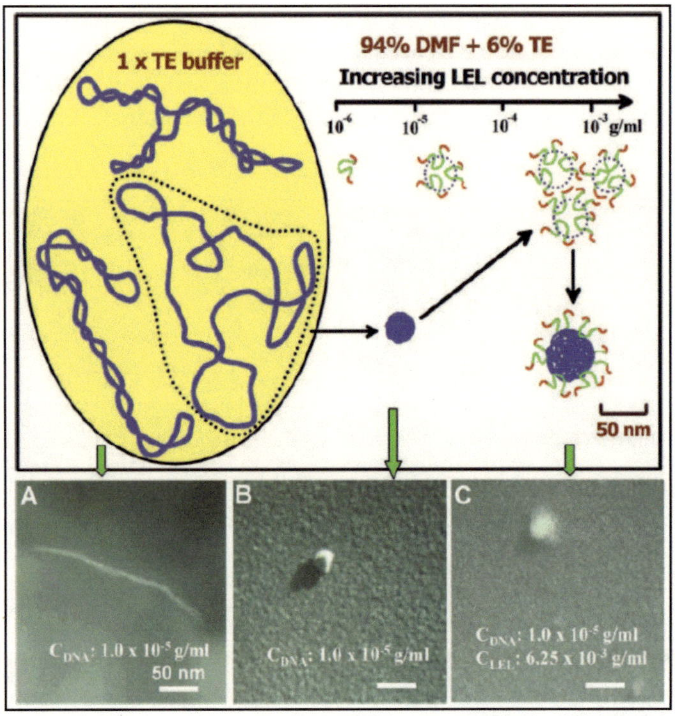

Fig. 5. Schematic presenting the condensed pDNA, the aggregation of LEL and the encapsulation of DNA by LEL. (A–C) Shows transmission electron microscopy (TEM) images of pDNA in 1x Tris-EDTA (TE) buffer, DNA in 94% DMF + 6% TE and encapsulated DNA in 94% DMF + 6% TE, respectively. Reprinted with permission from Liang et al., 2005, Oxford University Press.

Following our initial reports on the fabrication and use of electrospun scaffolds for gene delivery, a number of other studies were published that further expanded on the field, some with impressive increases in pDNA transfection efficiency. For example, Nie and Wang (2007) fabricated PLGA/hydroxyapatite (HAp) composite electrospun scaffolds using different HAp contents (0, 5, and 10%) in three compositions: 1) using only naked pDNA; 2) encapsulation of pDNA/chitosan nanoparticles; and 3) encapsulation of pDNA/chitosan nanoparticles mixed with a PLGA/HAp solution. The scaffolds generated were non-woven, with nano to micron fiber structures composed predominantly of PLGA with the HAp dispersed within the fibers (Fig. 6, white arrows). Scaffolds with the encapsulated pDNA/chitosan nanoparticles mixed with a PLGA/10%HAp showed the largest fiber diameters (100 nm). In addition, these scaffolds exhibited a much higher tensile strength (>4 times) than those with either naked pDNA or encapsulated with pDNA/chitosan nanoparticles (Nie and Wang, 2007).

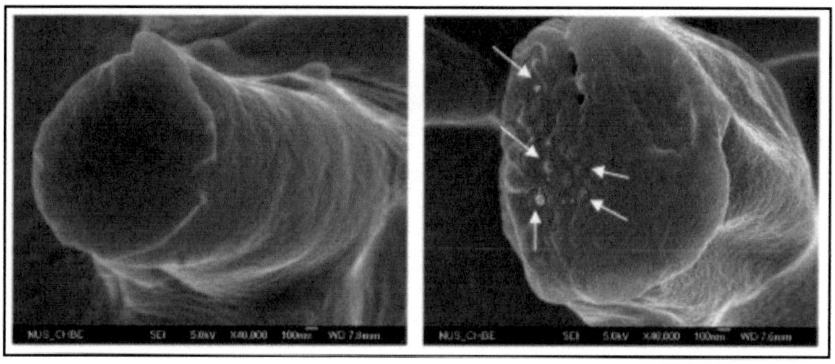

Fig. 6. The morphology observed in fiber cross sections containing pDNA/chitosan nanoparticles (left panel) and pDNA/chitosan nanoparticles mixed with a PLGA/HAp solution (right panel). This image illustrates clearly the encapsulated pDNA/chitosan nanoparticles (arrows) at the cut section of the fiber. Reprinted with permission from Nie and Wang, 2007, Elsevier.

These structural differences also translated into differences in pDNA release kinetics and transfection efficiencies. For the composite scaffold with just naked pDNA, an initial burst release of ~80-85% pDNA was observed for the first 7-9 days with the remaining pDNA released in the following 5 days. The scaffold with the encapsulated pDNA/chitosan nanoparticles did not show a burst release, but rather a slight delayed release with an overall cumulative release of 100% after ~24-30 days. In contrast to both of these scaffold types (just naked pDNA; pDNA/chitosan nanoparticles), those with encapsulated pDNA/chitosan nanoparticles mixed with a PLGA/HAp solution displayed a more linear and longer release curve with a 95% cumulative DNA release in 45-55 days. Further, for all three different scaffold types, the released DNA appeared to be structurally intact, as monitored by gel electrophoresis. When the authors measured the transfection efficiency of their scaffolds using human mesenchymal stem cells (plated directly on the scaffolds), they show that the second type of scaffold (encapsulated with DNA/chitosan nanoparticles) increased the *in vitro* expression of the delivered transgene (BMP-2), but unfortunately, the exact % transfection efficiency was not quantified. In addition, it was reported that these scaffolds resulted in reduced cell viability over time, which was attributed to the invasive transfection of the pDNA/chitosan nanoparticles as a result of the initial burst release (Nie and Wang, 2007). These results highlight the problem of balancing the need for high transfection efficiency with maintenance of cell viability. Thus, a scaffold that can do both, namely increase transfection efficiency while maintaining robust cell viability, would serve as an ideal gene delivery system.

A different method was utilized to generate electrospun pDNA containing scaffolds; surface immobilization of PEI/pDNA on electrospun PLA scaffolds using a layer by layer approach as a means of effectively controlling the functionality of the scaffold (Sakai et al., 2008). In this particular study, pDNA (pGL3 encoding luciferase) was incorporated on individual nanofibers via successive immersion into a solution of PEI and increasing concentration of pDNA. Results showed that the concentration of pDNA deposited on the electrospun nanofibers can be increased with increasing layers. Surprisingly, when tested with COS cells plated directly onto scaffolds, the scaffold with the highest plasmid concentration displayed decreased transfection efficiency as compared to one with a lower concentration (Fig. 7). Unfortunately, this study did not report transfection efficiency as a total percentage of cells.

Rather, they reported data normalized to the quantity of cellular protein which only approximated the number of transfected cells. Lastly, the authors also show that with longer incubations, the overall transfection efficiency did increase. Taken together, this study clearly showed that transgene expression can be controlled by changing both the concentration of DNA (via the number of immersions), as well as the incubation time with the cells (Sakai et al., 2008).

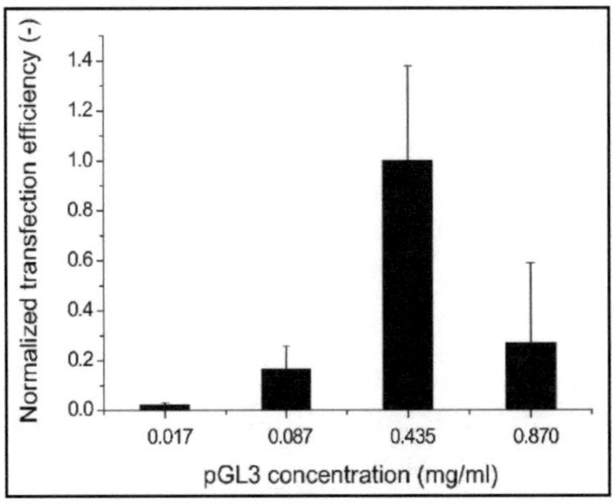

Fig. 7. Effect of pGL3 concentration in the immersion solution on transfection efficiency of PLA-(PEI/pGL3)$_2$PEI fibrous mat after 30 h of cultivation. Bars represent mean +/- SD (n= 4 – 10). Reprinted with permission from Sakai et al., 2008, John Wiley and Sons.

Gene delivery was also investigated using nanofibrous scaffolds that were generated using coaxial electrospinning. These scaffolds have a sheath/core fiber morphology whereby two immiscible polymer solutions are used to fabricate the individual fibers. Coupled to coaxial electrospinning, this study investigated the effect of four parameters: a) Poly(εcaprolactone) (PCL) concentration; b) PEG MW; c) PEG concentration; and d) DNA concentration (Saraf et al., 2010). By varying the amounts of these four components, the authors fabricated eight groups of scaffolds and then directly compared them by examining scaffold morphology, DNA incorporation and release and transfection efficiency. The various scaffolds exhibited fiber diameter distribution of ~200 nm to 4 μm, with higher fiber diameters obtained with increasing PCL concentration, followed by increased PEG MW, PEG concentration and higher pDNA concentration. More importantly, the investigators evaluated pDNA release over 60 days and observed a pattern that could be divided into four groups: a) burst release (0 - 24 hrs); b) Phase 1 (2 - 10 days); c) Phase 2 (11 - 28 days); and d) Phase 3 (35 - 60 days). Even though there were large differences between the groups for the burst release of pDNA, none of the fabricated scaffolds achieved release profiles with Phases 1-3 (Fig. 8).

The average transfection efficiency (in rat fibroblasts) over 60 days for the eight groups of coaxial electrospun scaffolds, was ~15%. Despite the low level of transfection, in comparison to scaffolds carrying pDNA alone without rhodamine-tagged PEI-HA, all experimental groups showed significance increases in transfection efficiency (~15% vs. ~5%, respectively). It is important to note that despite the rapid release kinetics, with some coaxial electrospun scaffolds, transfection efficiencies approached ~40%, a substantial level for a non-viral approach. Lastly, the cells remained viable for the duration of the 60 day transfection experiment (Saraf et al., 2010) which represents a significant improvement given the lower viability results reported in the Nie and Wang study (Nie and Wang, 2007).

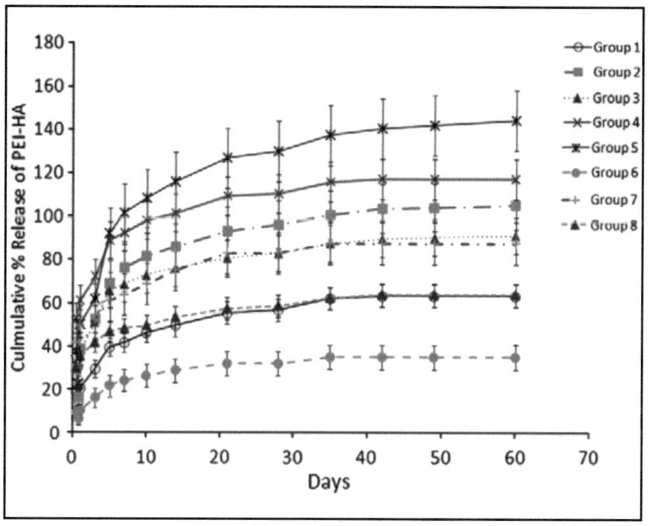

Fig. 8. Cumulative release of rhodamine-tagged PEI-HA (r-PEI-HA) from electrospun coaxial fiber meshes at 37 °C in PBS with agitation at 115 rpm. Error bars represent standard deviation for n=4. Reprinted with permission from Saraf et al., 2010, Elsevier.

A very recent study further investigated the relationship between DNA transfection via electrospun scaffolds and cell viability. Yang et al. (2011) utilized emulsion electrospinning to fabricate core-sheath structured fibers with a core loading of pDNA or pDNA polyplexes inside the fibers with either PEI or PEG. Four specific formulations of electrospun scaffolds were prepared with fibers consisting of: 1) pDNA/poly(DL-lactide)-PEG (PELA); 2) pDNA/PELA-PEI; 3) pDNA-PEI/PELA; and 4) pDNA-PEI/PELA-PEG. Morphological studies revealed that in comparison to the electrospun pDNA-loaded fibers (~500 - 600 nm diameter), the emulsion electrospun PELA fibers displayed a much wider size distribution, ranging between 700 and 1000 nm. In addition, the release rate also varied, with the pDNA/PELA fibers showing a two-phase profile, with an initial burst (~22% of pDNA) followed by a sustained one (~67%) over 3 weeks. On the other hand, adding PEI led to a higher burst release (~35%) of pDNA but lower cumulative release over the 3 weeks. Even lower burst and cumulative pDNA release resulted from the pDNA-PEI/PELA scaffold. Interestingly, pDNA-PEI/PELA scaffolds containing different MW PEG led to faster and more sustained release as a result of the molecular chains of high MW PEG forming larger channels thereby accelerating dissolution (Yang et al., 2011).

Similar to our studies (Luu et al., 2003; Liang et al., 2005), Yang et al (2011) also examined the integrity and bioactivity of the released pDNA and found it to be conformationally intact (attributed to the protective effect of core-sheath fibers on the encapsulated pDNA), as well as bioactive since it retained its activity for transfection and expression of the transgene (GFP) in NIH3T3 fibroblasts. But again the exact transfection efficiency was not reported. Instead, data were presented as relative light units, normalized to total cellular protein. Lastly, despite good cell attachment on some of their scaffolds, the authors did report cytotoxic effects on cells and poor adhesion on scaffolds that contained higher MW PEG, as well as with variable concentration of the pDNA polyplexes. As such, the study concludes that pDNA-polyplex loaded scaffolds containing 10% PEG show the best performance in balancing transfection efficiency and cell viability (Yang et al., 2011).

3.2 *In vivo* studies

Despite the presence of numerous manuscripts reporting on the transfection efficiency of DNA-loaded electrospun scaffolds using *in vitro* studies, only two very recent studies have published results dealing with *in vivo* applications of the electrospun scaffolds to deliver DNA. In addition, my laboratory has also conducted a study (Zhao et al., 2008) on bone regeneration (using a critical size drill defect) with electrospun scaffolds (PLGA mixed with a PLA-PEG diblock copolymer and GFP pDNA) and some of our results are also presented here.

Prior to the two recent studies, my laboratory performed experiments to test the efficacy of electrospun scaffolds to serve as gene delivery biomaterials (Zhao et al., 2008). Composite scaffolds of PLGA mixed with a 10% PLA-PEG diblock copolymer and GFP pDNA were prepared, cut in 5 x 1 cm strips and sterilized by brief ethanol immersion. Once in ethanol, the scaffolds shrank uniformly and considerably (~40 - 50%) and were then rolled into a 4 mm plugs (Fig. 9A). Once dried, we observed that these scaffold plugs were hard, similar in consistency to plastic. Using the unilateral 4 mm tibial plateau critical size defect rat model, the scaffold plugs were embedded in the defect and held in place by a pair of sutures. After 1, 2 and 3 weeks, the tibias were harvested and we analyzed the integration of the scaffold into the tissue as well as the GFP expression by the surrounding cells. Gross examination of a 3 week implanted electrospun scaffold indicates that it did not integrate well within the tissue (Fig. 9B). This was verified histologically, with a clear demarcation between the scaffold and surrounding tissue (Fig. 9C). In addition, we did not detect any GFP expression in the surrounding cells (data not shown). Based on these results we conclude that the scaffold sterilization employed in this study (100% ethanol) causes the release of a substantial amount of DNA that is present on the surface of the nanofibers, as observed in our previous studies (Luu et al., 2003; Liang et al., 2005) and the scaffolds to substantially shrink and thus alter their structural morphology, with porosity decreasing substantially and reducing the scaffolds advantageous high surface to volume area. The ethanol also likely increased the hydrophobicity of the scaffolds, which may have prevented the cells from attaching and thus decreased the overall scaffold biocompatibility. Thus, these results, are inconclusive in answering the question of whether these electrospun scaffolds can successfully be used as tissue engineering scaffolds (as prepared in this study) to deliver DNA and the experiments need to be repeated following development of a more suitable means of scaffold sterilization.

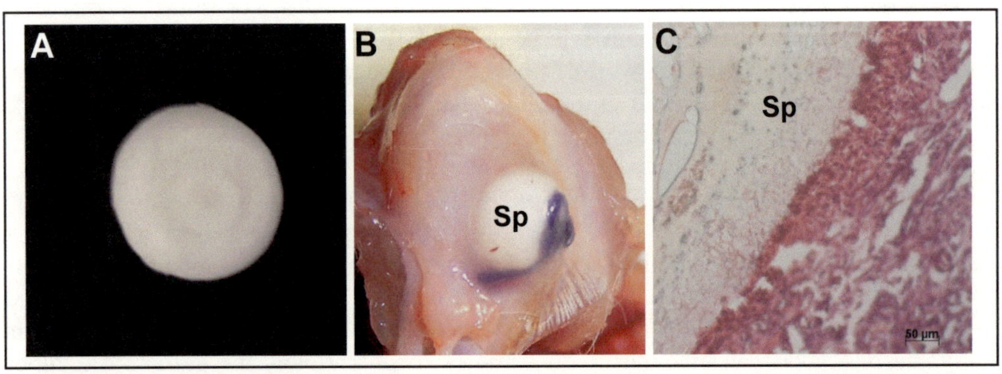

Fig. 9. (A) 4 x 4 mm electrospun Scaffold "Plug"; (B) 3 week post-implantation scaffold plug within rat tibial critical size defect (purple material is the suture used to hold scaffold within defect); (C) Tissue section showing the clear boundary between the implanted scaffold and surrounding tissue. Sp, scaffold plug.

The study by Zhang et al. (2010) generated three different kinds of electrospun scaffolds; one containing pure PCL, another with PCL/PEI and the third with PCL/PEI-PEG. The pDNA was added following fabrication of the electrospun scaffolds by immobilization (absorption) onto the nanofibers. Prior to the *in vivo* experiments, the authors tested the incorporation and release of DNA, as well as cell viability and transfection efficiency. Results indicated that approximately 15% and 50% of DNA was absorbed by the pure PCL and PCL/PEI scaffolds, respectively (the PCL/PEI-PEG also showed 50% maximum absorption). Despite the differences in DNA absorption both types of scaffold showed ~60% DNA release over 72 hrs. The transfection efficiency of these electropsun scaffolds was determined using two different cell types, a human embryonic kidney cell line (HEK293) and rat primary mesenchymal stem cells (MSCs). In HEK293 cells, the transfection efficiencies using PCL/PEI-PEG and PCL/PEI were ~70% and 38%, for each scaffold, respectively. Not surprisingly, with MSCs, the transfection efficiencies were lower at ~44.5% and 19%, for the PCL/PEI-PEG and PCL/PEI electrospun scaffolds, respectively. Further, the viability of these cells were also determined on both scaffolds and it was found that the PEG modification of PEI improved the biocompatibility, but still cell viability only reached ~50-80% of cells plated on PCL scaffolds. Following the *in vitro* characterization, the authors implanted the three different types of pDNA loaded scaffolds subcutaneously in mice and results showed that both pDNA containing PCL/PEI and PCL/PEI-PEG scaffolds successfully transfected surrounding tissue cells, with the highest measured fluorescent intensity obtained with PCL/PEI-PEG (Fig. 10, Zhang et al., 2010). As such, the absolute number of transfected cells and thus the exact effectiveness of these scaffolds to transfect with high efficiency *in vivo* remains to be determined.

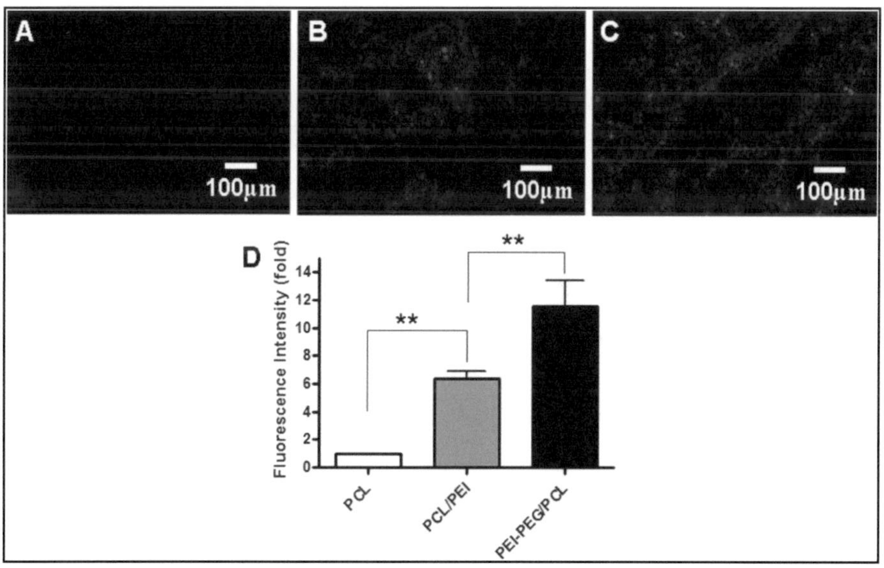

**Statistical significance at $p < 0.01$. Reprinted with permission from Zhang et al., 2010, John Wiley and Sons.

Fig. 10. Representative fluorescence images of *in vivo* transfection 14 days after subcutaneous implantation. (A) PCL/pDNA. (B) PCL/PEI/pDNA. (C) PCL/PEI-PEG/pDNA. (D) Averaged fluorescence intensity from three independent experiments.

In the second study, Kim and Yoo (2010) devised a clever approach to deliver DNA for the treatment of diabetic ulcers. Specifically, they reasoned that because diabetic ulcers contain high concentrations of matrix metalloproteinases (MMPs), they functionalized their PEI based electrospun scaffolds to contain an MMP-cleavable linkage. The idea was very simple; in the presence of MMPs, the linkerPEI (LPEI)/pDNA complex would be cleaved from the nanofibrous scaffolds, thereby transfecting the surrounding cells. This synthesis scheme is outlined in Fig. 11. In addition, the authors tested the incorporation of various amounts of DNA, its release and transfection efficiency *in vitro* prior to the *in vivo* experiments.

Various amounts of pDNA were loaded in the LPEI nanofibers and demonstrated that the DNA incorporation efficiency significantly increased with additional amounts of pDNA (from 17.4% to 79.5%) and this was attributed to the high cationic charge densities present on the nanofiber surface with respect to the anionic charge densities of the DNA. When the authors investigated pDNA release in the presence or absence of MMP-2, they found that depending on the amount of pDNA, MMP-2 treatment resulted in ~40 - 80% being released from the scaffolds (in the absence of MMP-2, the amount of released pDNA was below 30%). Further, the transfection efficiency of the released pDNA was found to be dependent on the charge ratio of the incorporated pDNA into the scaffolds rather than the actual amount of pDNA, but the data do not reveal what percent of the cells were transfected. Instead, it represents a comparison between the released pDNA with that of LPEI/DNA scaffolds.

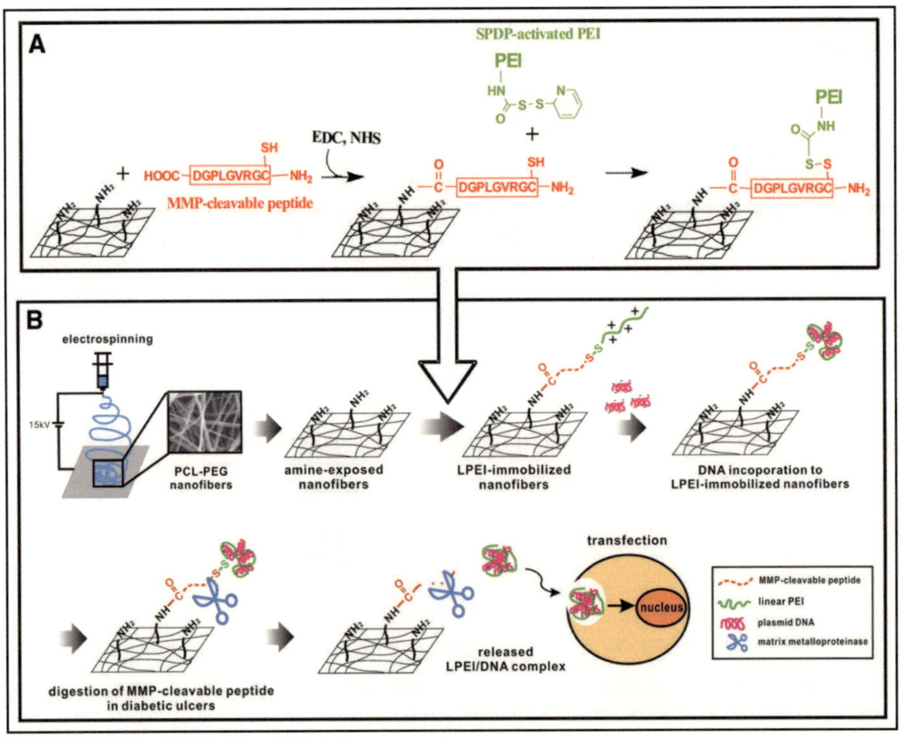

Fig. 11. (A) Synthetic scheme of LPEI immobilization on the surface of nanofibrous matrix via an MMP-cleavable linkage. (B) Schematic diagram of MMP-responsive electrospun nanofibrous matrix for gene delivery. Reprinted with permission from Kim and Yoo, 2010, Elsevier.

To investigate the performance of these MMP-responsive scaffolds (PCL-PEG and LPEI) in their ability to deliver GFP pDNA *in vivo*, a diabetic dorsal dermal ulcer mouse model was utilized. Results from these experiments showed that the highest level of GFP expression was detected in the dermal wounds treated with LPEI 3 days post-treatment. In contrast, very little GFP expression was detected with either naked DNA or with the PCL-PEG (no LPEI) scaffolds (Fig. 12). Similar results were also obtained with application of the scaffolds to normal, non-diabetic dermal wounds, although GFP expression with the LPEI containing scaffolds was increased in the diabetic wounds, as compared to the normal wounds (Fig. 12). Based on these results, the authors conclude that these MMP responsive DNA based electrospun scaffolds can be potentially applied for local gene delivery for the treatment of diabetic ulcers.

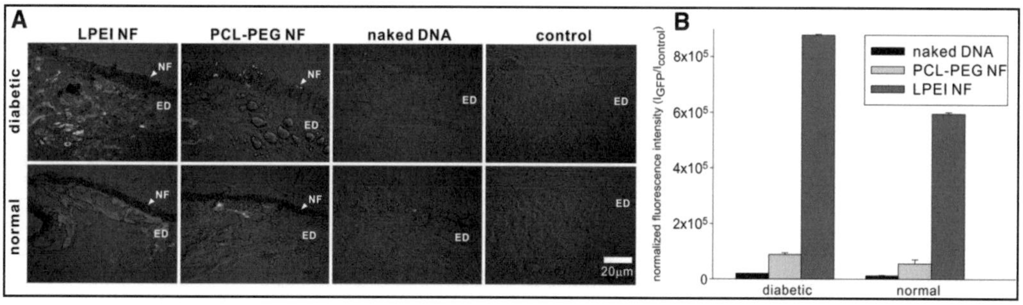

Fig. 12. Cross-sectional images of diabetic and normal wound tissue administered with DNA-incorporated nanofibrous matrix (NFs) at day 3. DNA/LPEI-immobilized nanofibrous matrix [LPEI NF] (NF 16) via a MMP-cleavable linker; DNA/PCL-PEG nanofibrous matrix without LPEI [PCL-PEG NF]. NF and ED indicate nanofibrous matrix and epidermis, respectively. GFP was visualized by confocal microscopy. (B) Image-analysis results of quantifying fluorescence intensity. Image-analysis methods based on an intensity based accumulation of the computer software (Image-Pro 6.0) was employed to quantify each image and average values and standard deviation were obtained (n=3). Reprinted with permission from Kim and Yoo, 2010, Elsevier.

4. Conclusion

The application of electrospinning in the fundamental studies of DNA and polymer behavior and gene delivery remains an active area of research. Although successful studies have been reported on the interactions between DNA and various polymers, the incorporation of DNA within nanofibers and its release, the structural characterization of the nanofibrous scaffolds and more importantly, demonstrated high transfection efficiency in both *in vivo* and *in vitro* settings, to date no definitive data supports the notion that electrospun scaffolds can serve as effective non-viral gene delivery systems that rival the efficiency of viral vectors. As such, there is a continual need to explore the nature of these electrospun scaffolds in order to be able to fine tune: a) the type(s) and combination(s) of polymers; b) the amounts of incorporated pDNA; c) the electrospinning parameters; d) the release rate of pDNA; e) the structural integrity of the released pDNA; and most of all, f) quantifiable and high transfection efficiency both *in vitro* and *in vivo*. Greater insight on the interactions between DNA and various polymers, understanding of the formation of complex structures and their properties, the process of electrospinning, as well as DNA transfer into cells, can all play a significant role in enabling us to generate effective gene delivery systems with higher transfection efficiencies than currently achievable. If successful, the field of gene therapy will be enhanced and more widely embraced than it has been so far as a result of the shortcomings of viral gene delivery systems that are widely used to deliver the therapeutic genes. This is even more important now since regulatory molecules such as microRNAs and RNA interference (RNAi) are beginning to find their way to the clinic. Thus, electrospun scaffolds with all of the functional versatility could prove to be the next "magic bullet" in the realm of molecular medicine.

5. Acknowledgment

I would like to thank Drs. David Komatsu and Yen Kim Luu for critically reading the document and for their insightful comments.

6. References

Al-Dosari, M.S. & Gao, X. Nonviral gene delivery: principle, limitations, and recent progress. *American Association of Pharmaceutical Scientists Journal*, Vol.11, No.4, (December 2009), pp. 671-81.

Bellan, L.M., Cross, J.D, Strychalski, E.A., Moran-Mirabal, J. & Craighead, H.G. Individually resolved DNA molecules stretched and embedded in electrospun polymer nanofibers. *Nano Letters*, Vol.6, No.11, (November 2006), pp. 2526-2530.

Bellan, L.M., Strychalski, E.A. & Craighead, H.G. Electrospun DNA nanofibers. *Journal of Vacuum Science & Technolology B* Vol.25, No.6 (September 2008), pp. 2255-2257.

Bhardwaj, N. & Kundu SC. Electrospinning: a fascinating fiber fabrication technique. *Biotechnology Advances* Vol.28, No.3, (January 2010), pp. 325-347.

Bölgen, N., Vargel, I., Korkusuz, P., Menceloğlu, Y.Z. & Pişkin, E. In vivo performance of antibiotic embedded electrospun PCL membranes for prevention of abdominal adhesions. *Journal of Biomedical Materials Research B Applied Biomaterials*, Vol.81, No.2 (May 2007), pp. 530-543.

Cai, Z.X., Mo, X.M., Zhang, K.H., Fan, L.P., Yin, A.L., He, C.L. & Wang, H.S. Fabrication of Chitosan/Silk Fibroin Composite Nanofibers for Wound-dressing Applications. *International Journal of Molecular Sciences*. Vol.11, No.9, (September 2010), pp. 3529-3539.

Chen, J.P. & Chiang, Y. Bioactive electrospun silver nanoparticles-containing polyurethane nanofibers as wound dressings. *Journal of Nanoscience and Nanotechnology*, Vol.10, No.11, (November 2010) , pp. 7560-7564.

Chiu, J., Luu, Y.K., Fang, D., Hsiao, B.S., Chu, B. & Hadjiargyrou, M. Electrospun nanofibrous scaffolds for biomedical applications. *Journal of Biomedical Nanotechnology*, Vol.1, No.2, (June 2005) pp. 115 -132.

Ding, Y., Wang, Y., Su, L., Bellagamba, M., Zhang, H. & Lei, Y. Electrospun Co_3O_4 nanofibers for sensitive and selective glucose detection. *Biosensors and Bioelectronics*, Vol.26, No.2, (October 2010), pp. 542-548.

Fang, X. & Reneker DH. (1997) DNA Fibers by Electrospinning. *Journal of Macromolecular Sciences – Physics B*, Vol.36, No.2, (1997) pp. 169-173.

Gibson, P., Schreuder-Gibson, H. & Pentheny, C. Electrospinning technology: direct application of tailorable ultrathin membranes. *Journal of Coated Fabrics*, Vol.28, (July 1998), pp. 63-72.

Hadjiargyrou. M. and Chiu, J.B. Enhanced composite electrospun nanofiber scaffolds for use in drug delivery. *Expert Opinion on Drug Delivery*, Vol.5, No.10, (October 2008), pp. 1093-1106.

Kang, Y.O., Yoon, I.S., Lee, S.Y., Kim, D.D., Lee, S.J., Park, W.H. & Hudson, S.M. Chitosan-coated poly(vinyl alcohol) nanofibers for wound dressings. *Journal of Biomedical Materials Research B Applied Biomaterials*, Vol.92, No.2, (February 2010), pp. 568-576.

Kattamuri, N., Shin, J.H., Kang, B., Lee, C.G., Lee, J.K. & Sung, C. Development and surface characterization of positively charged filters. *Journal of Materials Science*, Vol.40, No.17, (September 2005), pp. 4531-4539.

Khil, M.S., Cha, D.I., Kim, H.Y., Kim, I.S. & Bhattarai, N. Electrospun nanofibrous polyurethane membrane as wound dressing. *Journal of Biomedical Materials Research B Applied Biomaterials*, Vol.67, No.2, (November 2003), pp. 675-679.

Kim, H.S. & Yoo, H.S. MMPs-responsive release of DNA from electrospun nanofibrous matrix for local gene therapy: in vitro and in vivo evaluation. *Journal of Controlled Release*, Vol.145, No.3, (August 2010), pp. 264-271.

Kowalczyk, T., Nowicka, A., Elbaum, D. & Kowalewski, T.A. Electrospinning of bovine serum albumin. Optimization and the use for production of biosensors. *Biomacromolecules*, Vol. 9, No.7, (July 2008), pp. 2087-2090.

Liang, D., Luu, Y.K., Kim, K., Hsiao, B.S., Hadjiargyrou, M. & Chu, B. In Vitro Non-viral Gene Delivery with Nano-fibrous Scaffolds. *Nucleic Acids Research*, Vol.33, No.19 (November 2005), pp. e170.

Liu, Y., Chen, J., Misoska, V., & Wallace, G.G. Preparation of novel ultrafine fibers based on DNA and poly(ethylene oxide) by electrospinning from aqueous solutions. *Reactive and Functional Polymers*, Vol.67, No.5, (May 2007), pp. 461-467.

Luu, Y.K., Kim, K., Hsiao, B.S., Chu, B. & Hadjiargyrou, M. Development of a nanostructured DNA delivery scaffold via electrospinning of PLGA and PLA-PEG block copolymers. *Journal of Controlled Release*, Vol.89, No.2, (April 2003), pp. 341-353.

Manesh, K.M., Santhosh, P., Gopalan, A. & Lee, K.P. Electrospun poly(vinylidene fluoride)/poly(aminophenylboronic acid) composite nanofibrous membrane as a novel glucose sensor. *Analytical Biochemistry*, Vol.360, No.2, (January 2007), pp. 189-195.

Martins, A., Araújo, JV., Reis, RL. & Neves, NM. Electrospun nanostructured scaffolds for tissue engineering applications. *Nanomedicine*, Vol.2, No.6, (December 2007) pp. 929-942.

Nair, S., Kim, J., Crawford, B. & Kim, S.H. Improving biocatalytic activity of enzyme-loaded nanofibers by dispersing entangled nanofiber structure. *Biomacromolecules*, Vol.8, No.4, (April 2007), pp. 1266-1270.

Ner, Y., Grote, J.G., Stuart, J.A. & Sotzing, G.A. Enhanced fluorescence in electrospun dye-doped DNA nanofibers. *Soft Matter*, Vol.4, No.7, (May 2008), pp. 1448-1453.

Nie, H. & Wang, C.H. Fabrication and characterization of PLGA/HAp composite scaffolds for delivery of BMP-2 plasmid DNA. *Journal of Controlled Release*, Vol.120, No.1-2, (July 2007), pp. 111-121.

Sakai, S., Yamada, Y., Yamaguchi, T., Ciach, T. & Kawakami, K. Surface immobilization of poly(ethyleneimine) and plasmid DNA on electrospun poly(L-lactic acid) fibrous mats using a layer-by-layer approach for gene delivery. *Journal of Biomedical Matererials Research A*, Vo.88, No.2, (February 2009), pp. 281-287.

Saraf, A., Baggett, L.S., Raphael, R.M., Kasper, F.K. & Mikos, A.G. Regulated non-viral gene delivery from coaxial electrospun fiber mesh scaffolds. *Journal of Controlled Release*, Vol. 143, No.1, (April 2010), pp. 95-103.

Sell, S.A., McClure, M.J., Garg, K., Wolfe, P.S. & Bowlin, G.L. Electrospinning of collagen/biopolymers for regenerative medicine and cardiovascular tissue engineering. *Advanced Drug Delivery Reviews*, Vol.61, No.12, (October 2009), pp. 1007-1019.

Lee, S.S. & Obendorf, K. Developing protective textile materials as barriers to liquid penetration using melt-electrospinning. *Journal of Applied Polymer Science*, Vol.102, No.4, (November 2006), pp. 3430–3437.

Takahashi, T., Taniguchi, M. & Kawai T. Fabrication of DNA Nanofibers on a Planar Surface by Electrospinning. *Japanese Journal of Applied Physics*, Vol.44, No.27, (June 2005) pp. L6860-L862.

Thavasi, V., Singh G., & Ramakrishna S. Electrospun nanofibers in energy and environmental applications. *Energy & Environmental Science*, Vol.1, No.2, (July 2008), pp. 205-221.

Yang, Y., Li, X., Cheng, L., He, S., Zou, J., Chen, F. & Zhang, Z. Core-sheath structured fibers with pDNA polyplex loadings for the optimal release profile and transfection efficiency as potential tissue engineering scaffolds. *Acta Biomaterialia.* (February 2011) [Epub ahead of print]

Zhang, J., Duan ,Y., Wei, D., Wang, L., Wang, H., Gu, Z. & Kong, D. Co-electrospun fibrous scaffold-adsorbed DNA for substrate-mediated gene delivery. *Journal of Biomedical Matererials Research A*, Vol.96, No.1, (January 2011), pp. 212-220.

Zhao, X., Hsiao, B.S., Chu, B. & Hadjiargyrou, M. Functionalization of a PLGA/PEG-PLA Composite Electrospun Scaffold With rhBMP-2 Plasmid DNA for Bone Regeneration. *Journal of Bone and Mineral Research*, Vol 23, Supplement (September 2008), pp. S158.

Zong, X., Li, S., Chen, E., Garlick, B., Kim, K.S., Fang, D., Chiu, J., Zimmerman, T., Brathwaite, C., Hsiao, B.S. & Chu, B. Prevention of postsurgery-induced abdominal adhesions by electrospun bioabsorbable nanofibrous poly(lactide-co-glycolide)-based membranes. *Annals of Surgery*, Vol.240, No.5, (November 2004), pp. 910-915.

19

Nanocomposites for Vehicle Structural Applications

James Njuguna, Francesco Silva and Sophia Sachse
School of Applied Sciences, Cranfield University,
Bedfordshire,
UK

1. Introduction

Fiber-reinforced composites are a type of engineering material that has exhibited high strength–weight and modulus–weight ratios, even compared with some metallic materials. In the last two decades, some studies have shown the potential improvement in properties and performances of fiber reinforced polymer matrix materials in which nano and micro-scale particles were incorporated. This technology of nano and micro-scale particle reinforcement can be categorized into inorganic layered clay technology, single walled and multi-walled carbon nanotube, carbon nanofiber technology, and metal particle technology. To date, nanoparticle reinforcement of fiber reinforced composites has been shown to be a possibility, but much work remains to be performed in order to understand how nanoreinforcement results in major changes in material properties. The understanding of these phenomena will facilitate their extension to the reinforcement of more complicated anisotropic structures and advanced polymeric composite systems.

Large quantity of nanomaterials (such as carbon nanotubes, nanofibers, SiO_2 and montmorillonite) is presently available due to the establishment of well-developed manufacturing technologies such as chemical vapor deposition method, ball milling and electrospinning. With ease to manufacture in bulk, fiber-reinforced polymer nanocomposites are finding increasingly more practical applications (for example, in the manufacture of composite structures microelectronics). The improvements that are being identified for the high performance structures and payloads are related with the type of property that is expected to be modified primarily mechanical, thermal and electrical properties. In reality, the high performance structures and payloads are multi-functional elements with various design drivers that cannot be separated, due to the vehicular limited size and weight. High performance structure design imposes a number of restrictions on the properties of materials to be used. It also follows that lighter, thinner, stronger and cheaper structures are very important goals of materials science and engineering nowadays. The restricted parameters could be seen as a subset of the design parameters. Formulation and establishment of the design parameters build up a process of certification.

Fiber-reinforced composites are very sensitive to intrinsic damage, such as delamination (in particular), matrix cracking and fatigue damage. Several approaches are adopted to tackle these, which include: (i) improving the fracture toughness of the ply interfaces of the fiber-matrix including resin blends and (ii) reducing the mismatch of elastic properties (and stress concentrations) at the interfaces between the plies. However, these materials lack other required functional properties like high electrical/thermal conductivities to account for electrostatic dissipation and lightening strike protection. Currently, it is believed that the best route to achieve multi-functional properties in a polymer is to blend it with nanoscale fillers. The fundamental basis behind this idea is the three main characteristics that define the performance of polymer nanocomposites:

i. nanoscopic confinement of matrix polymer chains;
ii. nanoscale inorganic constituents and variation in properties; that is, many studies have reported that mechanical, conductive, optical, magnetic, biological and electronic properties of several inorganic nanoparticles significantly change as their size is reduced from macro- to micro- and nano-scale levels and
iii. nanoparticle arrangement and creation of large polymer/particle interfacial area.

One of the technology drawbacks is that the mechanical reinforcement by the application of nanoparticles as a structural element in polymers is quite difficult to realize and still remains a challenging task. An efficient exploitation of the nanoparticles properties in order to improve the materials performance is generally related to the degree of dispersion, impregnation with matrix and to the interfacial adhesion. The advantage of nano-scaled compared to micro-scaled fillers is their enormous surface-to-volume ratio, which can act as interface for stress-transfer. The methods published so far on the improvement of mechanical properties of polymer composites have mainly focused on the optimization of the manufacturing process of the composites, i.e. with the use of prepared nanofillers. The property and performance enhancements made possible by nanoparticle reinforcement may be of great utility for carbon or glass fiber-epoxy composites that are used for the high performance and high temperatures applications such as aerospace engine components and nacelle, storage of cryogenic liquids, and motorsports. Then again, precautionary measures should be observed in high temperatures since the structure and properties of these materials can change radically when they are exposed to extreme temperatures, especially in a cyclical fashion.

Better dispersion of nanosized, silicate-based filler in matrice systems is expected to yield improved materials properties in several areas. Various mechanical properties, specifically improved fracture toughness, as well as improved flame-retardant effects are of interest. A key objective of the ongoing worldwide research is investigating whether a nano-modified matrix yields improved delamination resistance in a fiber-reinforced laminate compared to a laminate with neat polymer as matrix material. It is necessary, however, to evaluate the damage resistance characteristics of three-phase nanocomposite laminates for the enhancement of their applicability to the structural elements. Further, as lamented earlier on, another complication is that the microstructure of semicrystalline polymer matrices is influenced not only by the processing history but also by the presence of nanofillers. The addition of various types of carbon nanotubes, nanofibers or nanoclays to polymers has already been observed to influence the crystallization kinetics and resulting morphology. Such changes in matrix morphology need to be considered when evaluating the nanocomposite performance with regard to the intrinsic filler properties. The effects of nanoparticles on such oriented polymer systems, although significant, have not yet been fully established. In addition, as mentioned earlier on, it should be noted that the presence of additives such as coloring pigments has been shown to influence matrix morphology during fiber spinning, whilst there is the whole technology of nucleating agents which are deliberately added to influence crystalline microstructure.

2. Strength and stiffness properties

Material used for crashworthy structural application, traditionally have been metals, due to their plastic deformation characteristics that enable them to absorb impact energy in a controlled manner. Polymer composite materials, contrasting metals, do not typically exhibit plastic deformation. However, their stress-strain relationships may show signs of other types of nonlinearities, but they are superior to metals for specific energy absorption. Nanoreinforced polymers have focused the attention, because of their potential to exhibit impressive enhancements of material properties compared to the pure polymers (Sun et al., 2009). For example, enhancement of materials strength and stiffness properties compared with the unreinforced material. The most important parameter which affects these properties is the load transfer between the composite phases, realized via shear stress induced in the interface region. If high-modulus filler is added into low-modulus matrix, the stress in the composite is transferred into the stiffer phase, causing higher loads to be carried by the filler. That is why efficiency of the stress transfer across composite phases depends on quality of the interfacial region. This includes interfacial strength and stiffness (Zhang et al., 1993). If weak coupling exists between the phases, then the load transfer is poor and the properties of the composite are dominated by the matrix. Enhancement of interfacial properties causes that the load is transferred more efficiently, leading to an increase in composite modulus and strength.

An important influence on the properties of the interfacial region is the aspect ratio of the reinforcement. The higher the aspect ratio of the reinforcement is, the bigger the load carrying area, and the higher the composite modulus and strength. This phenomenon explains the unique properties of the materials reinforced with nanoparticles. Large number of small particles in nanocomposite increases the available interfacial area leading to significant enhancement of the mechanical properties. The effectiveness of nanofillers in reinforcing polymeric materials depends on several factors, which are defined and summarized here after.

2.1 Filler size

It is well known that incorporation of rigid micro-fillers usually increases composite stiffness, but at the cost of reduced ductility. This is caused by the stress concentration regions that exist in close proximity of the reinforcement. In case of nanofillers the stress concentrations are significantly reduced. Therefore, composite ductility can be maintained at the constant level or even improved, in relation to the net polymer.

The size effect of TiO_2 particles on mechanical properties of the epoxy resin was investigated by Ng *et al.* (Ng et al., 1999). The authors studied tensile properties of the epoxy material reinforced with nano (32nm) and micro (0.24μm) particles, at constant 10% weight fraction. The obtained results show, that incorporation of micro-particles increased the modulus of the epoxy resin, but decreased its strain to failure. On the other hand, incorporation of nanoparticles increased both, the modulus and the strain to failure. The effects of nanoparticles on mechanical properties of polyurethane/clay (PU/MMT) nanocomposite foams have also been investigated in the literature. It is well known that H-bond formation among urethane groups, greatly contributes to the strength and modulus of PUs. The overall performance of PU nanocomposite foams depends on the competition between the positive effects of clay on polymer reinforcement and foam morphology, and the negative effects on H-bond formation and network structure.

Effect of particles size on the mechanical properties of polyurethane foams was studied by Javni *et al.* (Javni et al., 2002). Incorporation of nanosized filler was found to increase the compression strength of the foam, and to decrease its rebound resilience. On the other hand, the addition of micro-sized fillers was found to lower the hardness and compression strength, at the same time leading to an increase in rebound resilience. In another development, (Njuguna *et al. 2008)* fabricated and characterised a series of nanophased hybrid sandwich composites, based on polyurethane/montmorillonite (PU/MMT). Polyaddition reaction of the polyol premix with 4,4′-diphenylmethane diisocyanate was applied to obtain nanophased polyurethane foams, which were then used for fabrication of sandwich panels. It has been found that the incorporation of MMT resulted in higher number of PU cells with smaller dimensions and higher anisotropy index. The obtained materials exhibited improved parameters in terms of thermal insulation properties. Importantly, these foams can also be selectively stiffened to meet specific requirements.

2.2 Volume fraction of fillers

Another important factor which can have significant impact on the stress-strain behavior of nanocomposites is the filler content. In contrast to microparticles, which reduce composite strength at high filler content, the nanoparticles offer an increase in both modulus and strength, with rising fraction of the filler. However, bad dispersion of particles or existence of agglomeration regions may cause stress concentrations, which initiate local cracks, resulting in reduced ductility of the composite. This phenomenon is especially augmented at high filler loadings where agglomerations are more likely to occur (Javni et al., 2002).

The enhancement of mechanical properties due to the high filler content is associated with inter-particle distance. If the distance between particles is below the critical value, then a three dimensional network of interphase region is created across the particles and the matrix. Within this region the properties of the composite can be assumed as a gradual transition from that of the particle to that of the matrix. Value of the inter-particle distance τ is closely related to the concentration φ and average size of the particles d, what can be described using equation (1), assuming ideal dispersion of the particles (Zhang et al., 2006).

$$\tau = d \cdot \left[\left(\frac{\pi}{6\varphi} \right)^{1/3} - 1 \right] \tag{1}$$

Zhang *et al.* (Zhang et al., 2006) found, that if inter-particle distance is smaller than the average particle size d, then composite performances increase significantly, as it is presented in Figure 1. Similar conclusions were drawn by Wetzel *et al.* (Wetzel et al., 2006) who compared experimental fracture toughness with theoretical data presented in (Evans, 1972).

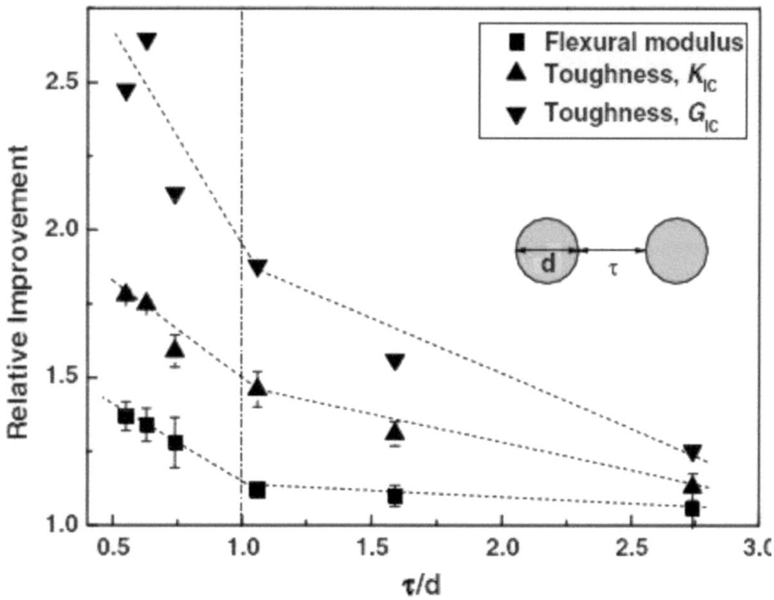

Fig. 1. Correlation between inter-particle distance and mechanical properties of SiO_2 /epoxy nanocomposites (Zhang et al., 2006).

2.3 Filler shape

The surface-area-to-volume ratio of the filler has the biggest influence on the mechanical properties of a nanocomposite, which is closely related with its shape. Figure 2 shows three typical geometries, which can be used as reinforcement in nanocomposites. Each of them possesses different aspect ratio, which can be expressed using equations (2-4) (Crosby & Lee, 2007).

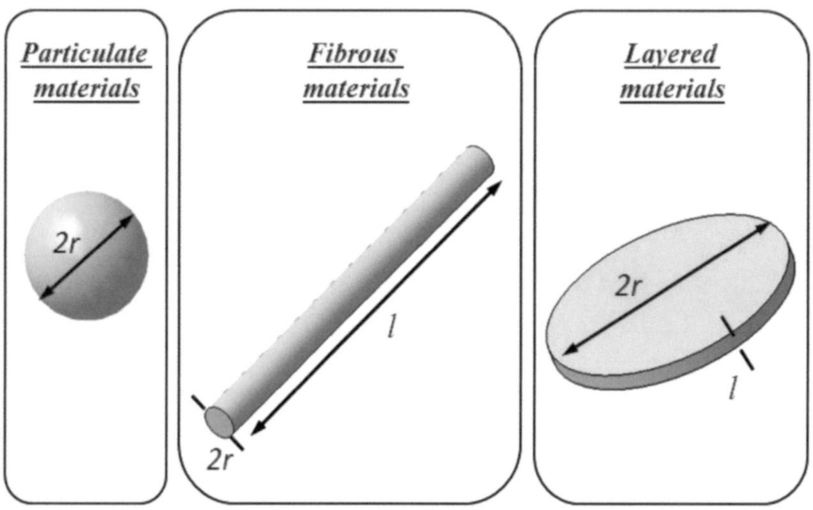

Fig. 2. Various reinforcement geometries.

- Particulate materials

$$\frac{A_s}{V_s} = \frac{4\pi r^2}{4/3\pi r^3} = \frac{3}{r} \tag{2}$$

- Fibrous materials

$$\frac{A_f}{V_f} = \frac{2\pi r^2 + 2\pi rl}{\pi r^2 l} = \frac{2}{r} + \frac{2}{l} \qquad l > r \tag{3}$$

- Layered materials

$$\frac{A_l}{V_l} = \frac{2\pi r^2 + 2\pi rl}{\pi r^2 l} = \frac{2}{r} + \frac{2}{l} \qquad r > l \tag{4}$$

If we compare surface-area-to-volume ratios of the particulate and cylindrical materials we obtain the following relation:

$$\frac{SV_p}{SV_c} = \frac{3}{2(1 + r/l)} \tag{5}$$

If the filler is in the form of plates (*r*>*l*) or short rods (*l*<2*r*), then its surface-area-to-volume ratio is bigger than that of a spherical particle. On the other hand, spherical particles have bigger surface-area-to-volume ratio in relation to long fibers (*l*>2*r*). Influence of r/l ratio on aspect ratio of the filler is shown in Figure 3.

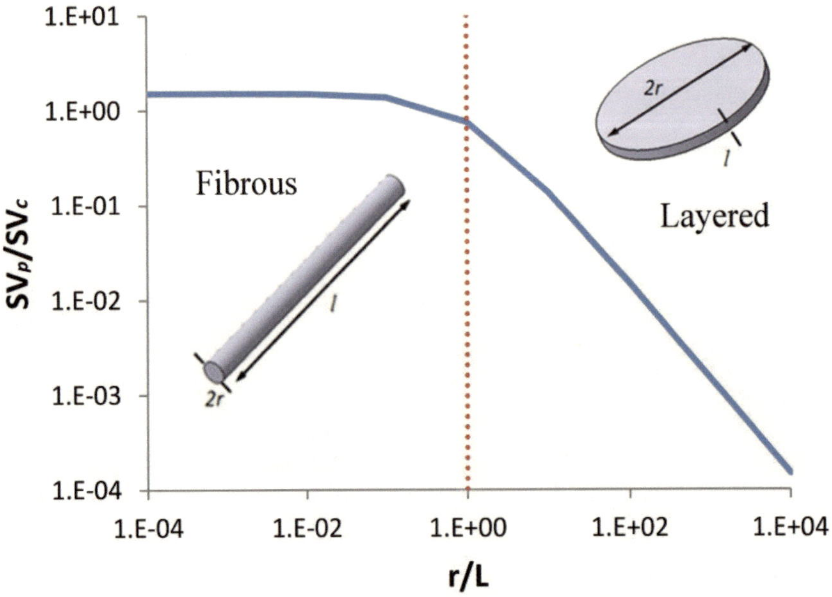

Fig. 3. (SV_p) to (SV_c) ratio as a function of particle radius (r) to length (l) ratio.

Summarizing, we can note that the best surface properties can be achieved using plate geometries. However, the mechanical properties of nanocomposites depend on various factors, i.e. size, dispersion and volume fraction of the filler. That is why it is difficult to directly compare different shapes of fillers in experiment. For example, cylindrical fillers are difficult to disperse in a homogeneous manner, at high volume fraction. Therefore, mechanical properties of the non-homogeneously dispersed filler of the one shape are not directly comparable with properly dispersed filler of the other shape, even at a constant volume fraction.

3. Impact resistance and energy absorption performance

Advanced vehicle structures must be able to withstand severe impact loads, at the same time providing safety of the occupants. That is why structural materials, used for crashworthy applications for instance, must be characterized by the energy absorption capability. In order to ensure survivability of the accident, structure has to dissipate energy in a controlled manner. This is limited by the two factors: induced decelerations (controlled by the strength of structure) and maintaining of a survival space for occupants during a crash. Traditionally metallic materials have been applied for the crashworthy structures, due to their ability to sustain plastic deformations. In contrast, composite materials do not exhibit plastic deformations as they are usually brittle. However, if they are properly designed they can absorb high amounts of impact energy by the progressive crushing and delamination.

For light weight constructions, among various nanocomposites, much attention has been paid to polymer/silica nanocomposites, because of the enhanced mechanical properties, high thermal stability and high flame-retardants. According to (Lux Research, 2004; Business Communications Company [BCC], 2006) nanoclays are accounting approximately 70% of the total volume of nanomaterials commercially used. In the automotive and packaging sector nanoclays are widely used, mainly due to their low cost and availability. PMMA-epoxy-nanoclay composites (Hyun Park & Jana, 2003), polypropylene-nanoclay composites (Galgali et al., 2004), polyvinylidene fluoride-nanoclay nanocomposites (Dillon et al. 2006) and nanoclay-modified rigid polyurethane foam (Widy & Macosko, 2005) have shown exhibit improved properties when compared to their bulk polymer constituents and conventional macro-composite counterparts. Numerous studies have reported the improvement of energy absorption of nanoclay/polymer nanocomposites (Chen & Evans, 2008; Sterky et al., 2010; Iqbal et al., 2009; Njuguna et al., 2008). For example, John *et al.* (John et al., 2010) have shown that the incorporation of 2 and 4 vol. % of nanoclay respectively improves the tensile modulus of cyanate ester syntactic foams by 6 and 80%.

Damages results from low-velocity impact events weaken the structure of composite materials, due to a continuous service load (Figure 4).

Furthermore, the impact may generate different types of flaws before full perforation, i.e. sub-surface delamination, matrix cracks, fiber debonding or fracture, indentation and barely visible impact damage (BVID). Over time, these effects can induce variations in the mechanical properties of such composites structures (the primary effect of a delamination is to change the local value of the bending stiffness and of the transverse-shear stiffness), leading to possible catastrophic failure conditions (Capezzuto et al., 2010).

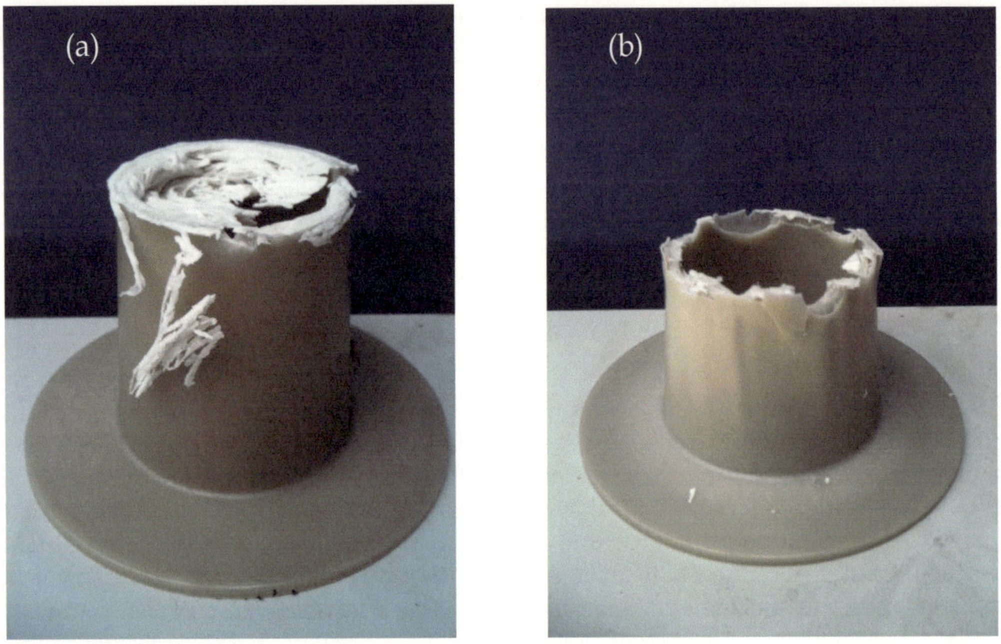

Fig. 4. 1kJ impact on (a) Glass-fiber/nanoclay reinforced Polypropylene Glass-fiber/nanosilica reinforced Polyamide 6.

Numerous studies have investigated the impact behaviour of polymer nanocomposite and are listed in Table 1.

It has been reported that the energy absorption capability and related properties of polymer matrices can be engineered by adding nano-scale fillers. For example, rigid nanosized particles such as SiO_2, TiO_2, $CaSiO_3$, $Al_2 O_3$ powder, carbon nanotubes (CNTs) and clay nanoplatelets have been used, and some important findings are summarised in this section.

3.1 Effects of particle stiffness

Viana (Viana, 2006) has shown that the particle stiffness influenced the properties of the polymer matrix. While soft/elastic fillers improved the impact toughness, at the same time those fillers reduced the modulus of elasticity of the polymer blend. However, increasing the amount of hard/rigid fillers improved both the impact toughness and the modulus of elasticity. For instance, the addition of nanoclay (less than 5 wt. %) into a polyurethane foam matrix, significantly improved the failure strength and energy absorption of the foam, with over a 50% increase in the impact load carrying capacity when compared to the unenforced matrix (Mohamme, 2006). Another example is the addition of rigid $CaCO_3$ (Ø 600 nm, 0.2 vol. %) in high density polyethylene (HDPE), which improved the impact strength by more than 200 % (Liu, 2002).

Such exorbitant improvements are not typically observed for composites reinforced with conventional micro-particles. Subramaniyan *et al.* (Subramaniyan & Sun, 2007) reported that core shell rubber (CSR) nanoparticles having a soft rubber core and a glassy shell improved the fracture toughness of an epoxy vinyl ester resin significantly more than nano-clay particles, having the same weight fraction. However, hybrid blends of CSR and nanoclay

Matrix material	Nanofiller	Ref.
Carbon fiber reinforced plastics	Multi-wall carbon nanotubes	(Kostopoulos et al., 2010)
Polystyrene	Cu_2S nanoparticles	(Capezzuto et al., 2010)
Polystyrene	Ammonium-treated montmorillonite clay	(Chen & Evans, 2008)
Nylon 6	Organoclay	(Lim et al., 2010)
Polyvinyl chloride	Organically modified montmorillonite	(Sterky et al., 2010)
Polyoxymethylene	Zinc oxide	(Wacharawichanant et al., 2008)
Epoxy matrix	Poly(acrylonitrile-co-butadiene-co-styrene) (ABS), clay (layered nanofiller) and nano-TiO_2	(Mirmohseni & Zavareh, 2010)
Carbon fiber–epoxy	Octadeclyamine modified montmorillonite s	(Iqbal et al., 2009)
Polyacrylate (PA), polyimide(PI) and polypropylene (PP)	Silica	(Li, 2008)
DGEBA (diglycidyl ether of bisphenol A) epoxy resin	TiO_2	(Bittmann et al., 2010)

Table 1. Studies investigating impact behaviour of polymer nanocomposite.

were found to yield the best balance of toughness, modulus and strength. The same investigators highlighted that when the nanoclay particles were used to enhance the polymer matrix in a conventional glass fiber-reinforced composite, the interlaminar fracture toughness of the composite was less than that of the unreinforced composite. As a possible reason for this result the arrangement of the nanoclay particles along the fiber axis was suggested.

Two factors dominate the capacity of rigid particles for energy dissipation at high loading rates (Dubnikova et al., 2004):

- the ability of the dispersed particles to detach from the matrix and to initiate the matrix local shear yielding in the vicinity of the voids and
- the size of the voids.

Therefore, the optimal minimal rigid particle size for polymer toughening should assure two main requirements:

- be smaller than the critical size for polymer fracture and
- have a debonding stress which is small compared to the polymer matrix yield stress.

3.2 Effects of particle geometry

Typical fillers for the reinforcement of polymer matrices are particles e.g. silica or aluminum oxide particles, tubes e.g. nanofibers or nanotubes and plates e.g. nanoclay platelets.

Significant enhancement of impact strength of polymeric nanocomposites was achieved by adding amino-functionalized multi-wall carbon nanotubes (MWCNTs) or small amounts of single-wall carbon nanotubes (SWCNTs) (Donclero & Gorga, 2006). Furthermore, the impact toughness of PMMA (polymethyl-methacrylate) has been improved considerably by the addition of CNTs and the toughness and modulus of MWCNTs reinforced PP exhibited a maximum at 1 wt.% CNTs (Donclero & Gorga, 2006). Additionally, the impact toughness and stiffness of CNT-reinforced polymer matrix have been found to be functions of the Young's modulus of the nanotubes (Kireitseu et al., 2008). The impact toughness of polymers containing inorganic nanofillers such as MMT (montmorillonite) nanoclay, based polymer composites was found to decrease (Viana, 2006), whereas adding Al_2O_3 nano-whiskers, glass fibers and wollastonite in a polymer resin improved the fracture toughness, KI_C, compared with that of the unreinforced resin (Yilmaz & Korkmaz, 2007)

3.3 Effects of inter-particle distance and volume fraction of fillers

The toughening of particle modified semi-crystalline polymers is related to the inter-particle distance *s*, independently of the type of added particles (Figure 5). The distance *s* relates both to the concentration *u*, and the average size d of the particles. Zhang *et al.* (Zhang et al., 2006) reported that the toughness improves significantly when the inter-particle distance *s* becomes smaller than the particle diameter, *d*. Qi *et al.* (Qi et al., 2006) showed that the fracture toughness and modulus of DGEBA epoxy resin increased with increasing volume fraction of nano-clay particles.

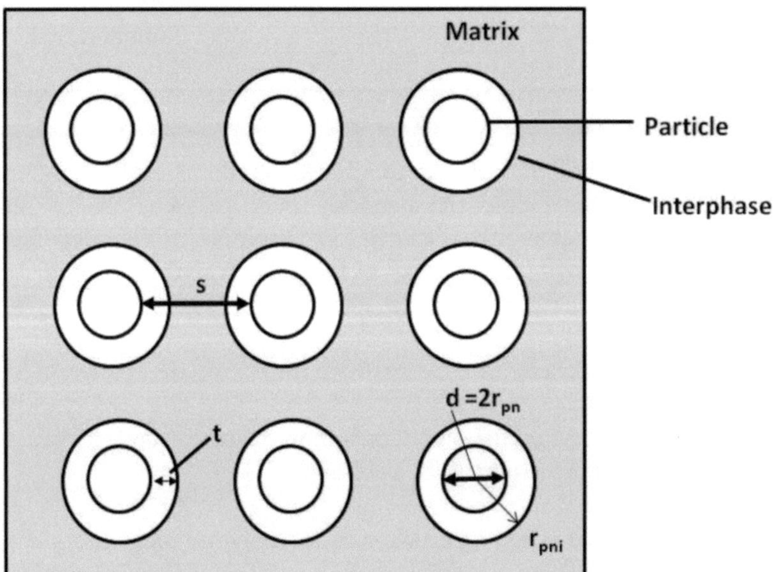

Fig. 5. Geometric parameters in nanoparticle-reinforced composites.

For nanophased hybrid sandwich composites based on polyurethane/montmorillonite (PU/MMT) shows that nanophased sandwich structures are capable of withstanding higher peak loads than those made of neat polyurethane foam cores when subject to low-velocity impact despite their lower density (Fig. 6).

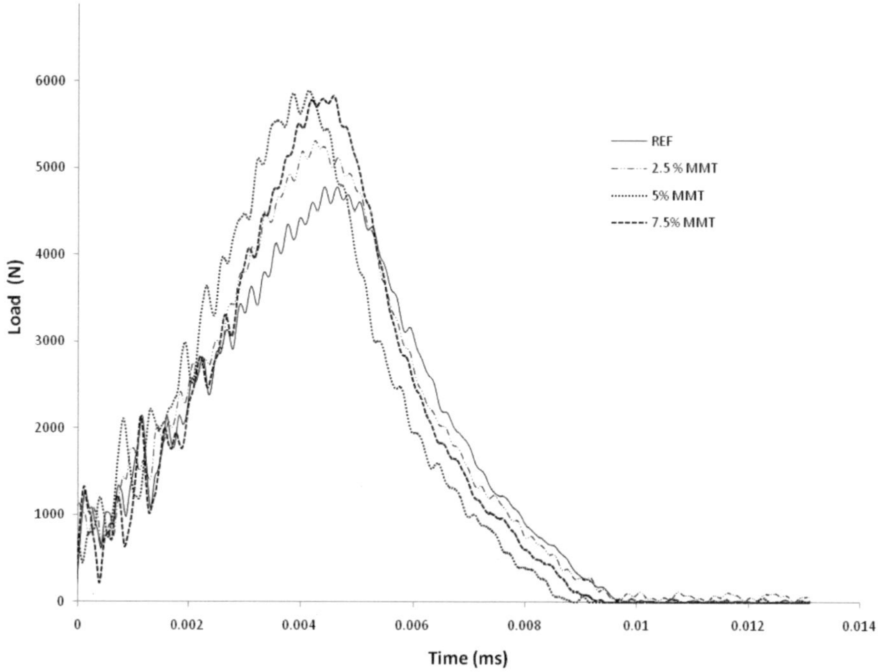

Fig. 6. Load vs time graph obtained from impact test.

This is especially significant for multi-impact recurrences within the threshold loads and energies studied by Njuguna *et al.* (Njuguna et al. 2010). Feasible applications for these lightweight structures are in energy absorbing structures or as inserts in hollow structures.

3.4 Effects of particle size

To investigate different toughening or energy absorption effectivenesses for fillers with micro- and nano-scale size, *d*, Wacharawichanant *et al.* (Wacharawichanant et al., 2008) measured tensile and impact properties of POM/ZnO (71 nm) and compared them with those of composites having micron- size ZnO (0.25 μm). It was shown that the nanocomposite has a higher mechanical properties e.g. tensile strength, Young's modulus and impact strength than the microcomposite. Fu *et al.* (Fu et al., 2008) investigated the mechanical behavior of polypropylene (PP)/$CaCO_3$ nanocomposites; the particle size varied from 10 nm, 80 nm, 1.3 to 58 μm. The study has shown that the composite strength increases with decreasing particle size due to higher total surface area for a given particle loading. This indicated that the strength increased with increasing surface area of the filled particles through a more efficient stress transfer mechanism. Furthermore, the Izod toughness was also found to increase with decreasing particle size.

3.5 Effects of stochastic variation of particle size and strength

Due to individual stochastic variation in the strength, different carbon nanotubes may have direct effect on the fracture mechanisms as in micro-based fiber composites. Xiao *et al.* (Xiao et al., 2008) measured the relation between individual CNT and CNT bundle strengths and proposed a method for determining the tensile strength distribution of individual CNTs or

CNT sub-bundles from experimental measurements on CNT bundles. The Weibull shape and scale parameters of the tensile strength distribution of CNTs were found to have wide variability of strength.

Tomar *et al.* (Tomar & Zhou, 2005) highlighted, that a microstructure less prone to fracture showed higher variations in fracture response when compared to the one which offers least resistance to crack propagation. Additionally, for a particular micro-structural morphology, the levels of variations in the crack surface area generated and the variations in the energy release rate were of the same order as the levels of variations in constituent properties.

3.6 Effects of polymer matrix and fillers types

The energy absorbent of nanocomposites depends on the polymer-clay affinity. For example, leads the reinforcement of polyacrylate (PA), polyimide (PI) and polypropylene (PP), with silica to striking variation in impact toughness behavior under identical processing conditions. The maximum stress intensity factor of PA-based enhanced significantly as compared with PI and PP based composites. Additionally the steady state fracture toughness of PA composite is approximately 45% and 25% higher than pure polyacrylate and both PP-based and PI-based hybrid specimens (Lin, 2008).

3.7 Effects of interfacial adhesion

Hybrid organic–inorganic composites usually exhibit enhanced mechanical properties compared to those of their separate components. The dispersion of the inorganic components into the inorganic matrices affects the mechanical performance of the materials. Hence, homogeneous despaired filler components are desirable. The employment of inorganic bulk fillers that can be exfoliated into well organized nano size fillers in the polymer matrices is one method to get well dispersed inorganic reinforced organic matrix nanocomponent. For example, Park *et al.* (Park, 2009) confirmed that silane-treated MMT could be successfully intercalated and dispersed in the epoxy matrix. The experimental results show that silane coupling agent (SCA) treatments cause an increase in both the specific component (due to the epoxide ring and amine group) and the dispersive component (due to the increase of the specific molecular volume). It was considered that the presence of SCAs leads to an increase of the interfacial adhesion between the MMTs and epoxy resin, which results from the improvement of the polar functional groups or of the specific component of the surface free energy of the MMTs. This enhances the mechanical interfacial strength between the MMTs and the epoxy matrix.

Hamming *et al.* (Hamming et al., 2009) studied the quality of dispersion and the quality of the interfacial interaction between TiO_2 nanoparticles and host polymer on properties such as glass transition temperature (T_g), elastic modulus and loss modulus. T_g is an attractive target property because of its high sensitivity to chain mobility and its use as a benchmark in other studies. The results showed that as the degree of dispersion improved (finer scale agglomerations) the T_g was depressed for samples of unmodified TiO_2 nanoparticles in PMMA and increased for samples of modified TiO_2 nanoparticles in PMMA. These results indicate that the quality of dispersion shifted T_g. However, the direction of shifts depended on the interfacial interaction of the nanoparticles with the matrix. This study indicated that the bulk properties of nanocomposites were highly sensitive to the quality of the interfacial interaction and quality of dispersion of the nanoparticles.

3.8 Distribution status of fillers

A number of experiments have shown that fracture toughness improved with addition of clay nano-platelets to epoxy when the clay nano-platelets were not fully exfoliated, and intercalated clay nano-platelets were present (Xidas & Triantafyllidis, 2010). Xidas *et al.* (Xidas & Triantafyllidis, 2010) showed that quaternary alkylammonium ion-modified organoclays (C18 alkylammonium ion and hydrogenated tallow ammonium ions) provided highly intercalated structures, which led to significant increase of tensile strength and strain at break.

4. Thermal properties

The deterioration of materials depends on the duration and the extent of interaction with the environment. Degradation of polymers includes all changes in chemical structure and physical properties of polymers due to external chemical or physical stresses caused by chemical reactions, involving bond scissions in the backbone of the macromolecules that lead to materials with characteristics different (usually worsened) from those of the starting material (Fig. 7). As a consequence of degradation, the resulting smaller fragments do not contribute effectively to the mechanical properties, the article becomes brittle and the life of the material becomes limited. Thus, any polymer or its (nano) composite, which is to be used in outdoor applications, must be highly resistant to all environmental conditions.

Research indicates that the modified epoxy nanocomposites possess better flame retardance than conventional composites. By the Kissinger method, the activation energies of thermo-oxidative degradation for the epoxy nanocomposites are less than those of the pure epoxy in the first stage of thermo-oxidative degradation. However, the activation energies of thermo-oxidative degradation for epoxy nanocomposites are generally higher than those of the pure epoxy in the second stage of thermo-oxidative degradation. For example, the main mechanism of layered silicate is a barrier formation influencing the flame spread in developing fires. Several minor mechanisms are significant, but important fire properties such as flammability or fire load are hardly influenced. Hence combinations with aluminum hydroxide and organo-phosphorus flame-retardants need to be evaluated. It has been shown that carbon nanotubes could surpass nanoclays as effective flame-retardant additives if the carbon-based nanoparticles (single- and multi-walled nanotubes, as well as carbon nanofibers) form a jammed network structure in the polymer matrix, such that the material as a whole behaves rheological like a gel (Kashiwagi et al., 2005).

The thermal degradation of nanoclay nanocomposites depends on the clay loading, and structure and the nature of the ambient gas. Leszczyńska *et al.* (Leszczyńska et al., 2007) have reviewed thermal stability of various polymer matrices improved by montmorillonite clay, and their influencing factors in detail. For the majority of polymers, owing to their hydrophobic character, the clay must be modified with a surfactant in order to make the gallery space sufficiently organophilic to permit it to interact with the polymer. In fact, several factors were found to govern the thermal stability of nanocomposite materials, such as intrinsic thermal resistance of polymer matrix, nanofiller content, chemical constitution of organic modifier and chemical character of polar compatibilizers as well as an access of oxygen to composite material during heating. For surface modification of clay, the surfactant is usually described as an 'onium' salt, but in fact ammonium salts are most commonly used. The quaternary ammonium ion is nominally chosen to compatibilize the layered silicate with a given polymer resin. However, the molecular structure (length and number of alkyl chains and unsaturation) is also the determining factor of the thermal stability of the polymer/MMT nanocomposites.

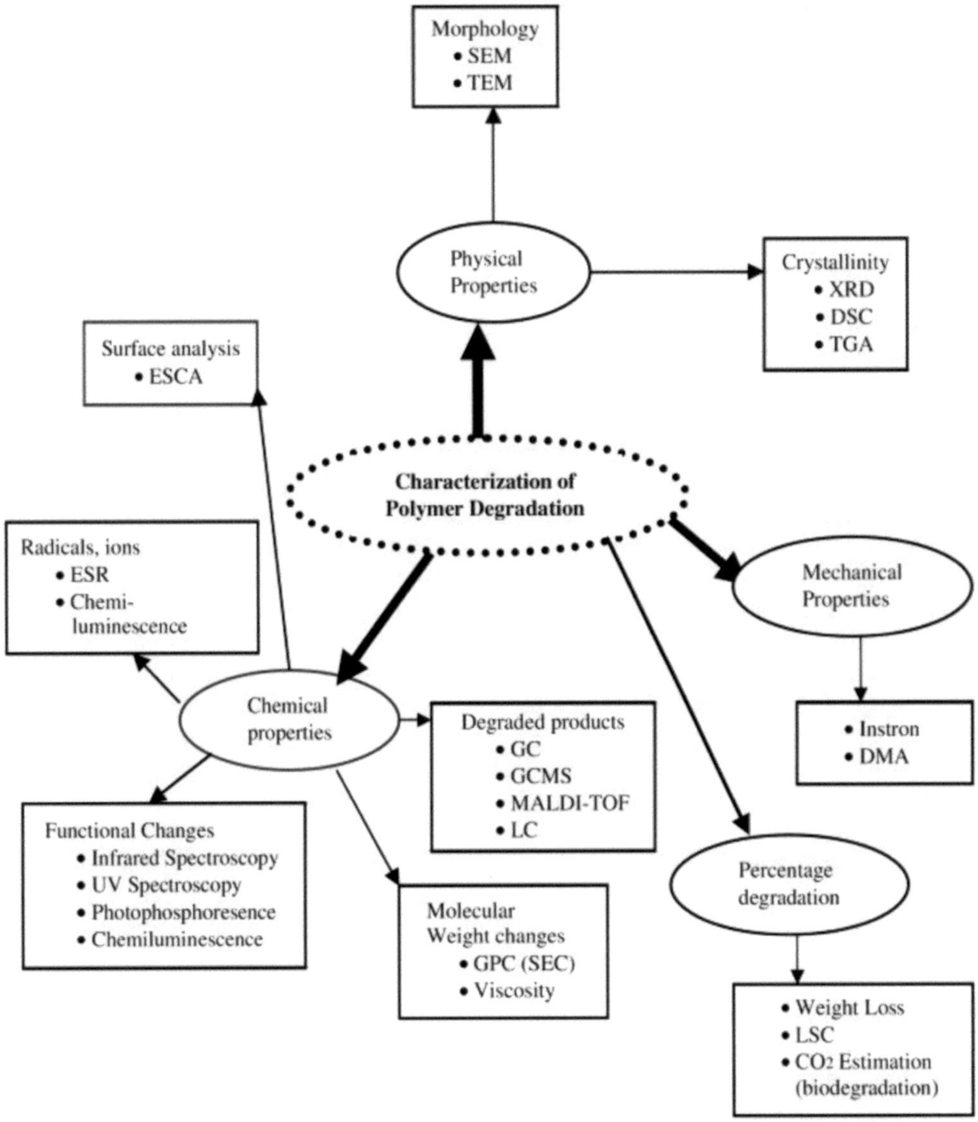

Fig. 7. Schematic representation of different degradation and stabilization processes in polymers (Pandey et al., 2005).

The possible mechanism for degradation of modifiers in silicates by among others Hwu *et al.* (Hwu et al., 2002), Leszczyńska et al. (Leszczyńska et al., 2007) and research has shown that surfactants degrade between 200 and 500 °C. The amount of surfactant lost during thermogravimetric analysis of various organoclays indicates that surfactants with multiple alkyl tails have greater thermal stability than those with a single alkyl tail. It has been proposed that the organic modifiers start decomposing at temperature around 200 °C, and the small molecular weight organics are released first while the high-molecular weight organic species are still trapped by organic layered silicate matrix. With the increase of temperature, the high-molecular organic polymer chains may still exist between the

interlayers until the temperature is high enough to lead to their further decomposition. The incorporation of silicate layers with high-aspect ratio decomposed/charred material on the clay surface act as carbonaceous insulators. The silicate has an excellent barrier property that prevents against permeation of various degraded gaseous products.
The addition of clay enhances the performance by acting as a superior insulator and mass transport barrier to the volatile products generated during decomposition. The clay acts as a heat barrier, which could enhance the overall thermal stability of the system, as well as assisting in the formation of char during thermal decomposition. In the case of the nanocomposite, the temperature at which volatilization occurs, increases as compared to the case of the micro-composite. Moreover, the thermal oxidation process of the polymer is strongly slowed down in the nanocomposite with high-char yield both by a physical barrier effect, enhanced by ablative reassembling of the silicate, and by a chemical catalytic action due to the silicate and to the strongly acid sites created by thermal decomposition of the protonated amine silicate modifier.
The polymers that show good fire retardancy upon nanocomposite formation exhibit significant intermolecular reactions, such as inter-chain aminolysis/acidolysis, radical recombination and hydrogen abstraction. In the case of the polymers that degrade through a radical pathway, the relative stability of the radical is the most important factor for the prediction of the effect that nanocomposite formation has on the reduction in the peak heat release rate. The more stable the radical produced by the polymer is; the better the fire retardancy is, as measured by the reduction in the peak heat release rate, of the polymer/clay nanocomposites.
The presence of MWNTs improves the thermal stability of PA6 under air obviously, but has little effect on the thermal degradation behavior of PA6 under nitrogen atmosphere. The thermal degradation mechanism of PA-6 has been proposed by Levchik *et al.* (Levchik et al., 1999). Vander Hart *et al.* (VanderHart et al., 2001.) observed that in the presence of clay the α-phase of PA 6 transforms into the γ-phase. The effect of modifier on the degradation of nanocomposite was studied by ^{13}C NMR. In the presence of modifier (dihydrogenated-tallow ammonium ion) the nylon nanocomposite begins to degrade at 240 °C, whereas the virgin polymer does not. They concluded that the organic modifier is less stable. The combination of shear stress and temperature may lead to extensive degradation of the modifier and the extent of clay dispersion may not depend on the modifier.
Davis *et al.* (Davis et al., 2003) have studied the thermal stability of injection moulded PA-6 nanocomposites by ^{13}C NMR. The virgin PA 6 and its nanocomposites were injection moulded at 300 °C. PA 6 did not degrade at processing temperature, whereas there was significant decrease in molecular weight in nanocomposites in the same conditions. It was observed that the degradation might depend upon the percentage of water in the nanocomposites, which might cause hydrolytic cleavage. Fornes *et al.* (Fornes et al., 2003) found color formation with polymer matrix degradation after twin-screw extrusion of polymer nanocomposites. The researchers reported that degradation of the nanocomposites depends on the type of nylon-6 materials as well as the chemical structure of the surfactant in the organically modified MMT (OMMT). Hydroxy-ethyl groups in the surfactant, as opposed to methyl groups, and tallow substituents, as opposed to hydrogenated tallow substituents, produced more colors in nanocomposite, which was related to unsaturation in the alkyl ammonium surfactant causing considerable polymer degradation. The kinetic parameters of PA-6 clay nanocomposite decomposition have also been studied.

5. Future trends

5.1 Numerical modeling for material properties

It is well known that interactions between nanofiller and matrix have a significant impact on the mechanical behavior of nanocomposites (Ajayan et al., 2003). That is why modeling of these materials becomes a challenging task, due to the wide range of length scales that need to be considered in the analysis. The most common used numerical methods for modeling nanocomposites include: Molecular Dynamics (MD); micro-mechanical, macro-mechanical and multiscale modeling. MD methods consider interaction at atomistic scale, and apply principles of quantum mechanics and molecular dynamics. These methods allow the determination of mechanical properties at nanoscale by solving equations of motion of interacting atoms within inter-atomistic potentials (Rapaport, 2004). MD simulations have been successfully applied in investigation of crack propagation and fracture of individual single and multi-walled carbon nanotubes (CNTs) (Belytschko et al., 2002; Namilae, 2005) and predicting the elastic properties of CNTs (Cornwell & Wille, 1997). However, these techniques are computationally expensive, and therefore, they are limited to studies of individual particles and polymer chains (Fisher, 2006). Link to the modeling of larger structures is currently unavailable. For a larger scale structures, continuum mechanics and micromechanics based finite element (FE) methods can be applied. General description and main challenges associated with these techniques are presented hereafter.

5.1.1 Macroscopic modeling

Macroscopic modeling is a simplified approach where micro or nano-scopic interactions between matrix and reinforcement are neglected. Properties of the composite are homogenized before the analysis, producing anisotropic and homogenous continuum (Aboudi, 1990). Effective properties like stress-strain curve and yield point are then obtained by using mechanical testing.

A big challenge when composite materials are considered is modeling of yield and failure under multiaxial loading. Most of the existing models, implemented into commercial FE packages, are developed for metals, and therefore, they are based on von-Misses criterion. In this approach yielding occurs at the same point in tension and compression, and it is specified by the value of effective stress. These assumptions are not necessarily correct for polymer composites. Contrary to metals, polymers yielding and failure are dependent on hydrostatic pressure, which is omitted in von Misses theory. Several experiments confirmed that hydrostatic part of stress tensor increases polymers yielding (Rabinowitz et al., 1970; Pae & Mears, 1968; Pugh et al., 1971) and failure (Ol'khovik, 1983) stress. That is why several modifications of Tresca and von Misses criteria were proposed to properly predict polymers behavior in the literature (Bowden & Jukes, 1972; Quinson et al., 1997; Sternstein & Ongchin 1969). These formulations take into account both hydrostatic and deviatoric components of the stress tensor. The yield stress is then calculated as a linear or nonlinear function of the hydrostatic pressure. In case of polymers under low pressure the linear representation is in good agreement with the experimental results, while the nonlinear characteristic is valid for the polymers under high pressure (Altenbach & Tushtev, 2001).

While selecting the material model for polymer composites it is also important to consider existence of nonlinear elastic region and strain rate effect. Some polymers exhibit significant rate dependency of modulus while others do not. That brings additional complications to modeling process, as not all material models are capable to represent these effects.

Application of the rate independent model to rate dependent material leads to an important error in stiffness prediction (Lobo, 2007).

5.1.2 Micro and multiscale modeling

In micro-scale modeling, each phase of a composite is considered separately. Loading and boundary conditions are applied both to the matrix and the reinforcement. The overall macroscopic properties of the material are then averaged and calculated as a function of the constituents properties, using a homogenization method or direct FE representation of Representative Volume Element (RVE). In contrast to the macroscopic approach multiscale modeling is able to provide information about stresses in phases, and it allows the definition of per-phase failure. Moreover, with this formulation, influence of the filler content on the composite properties can be predicted and investigated without performing an expensive material testing, for each configuration. An important drawback of this method is the larger computation time, in relation to the macroscopic approach, where material properties are calculated 'off line' (Aboudi, 1988).

However, the main principle of micro-mechanical modeling is the concept of RVE. The RVE is defined as a volume V of heterogeneous material, which must be large enough to correctly reflect statistical fluctuations of the composite properties. It must ensure an accurate prediction of the effective properties like Young's modulus or Poison's ratio (Kanit et al., 2003). Furthermore, it must contain all the data which can influence the composite behavior e.g. volume fraction, size and distribution of the filler.

In order to predict mechanical properties of the composite materials, various homogenization models can be applied. The most widely used models are Mori-Tanaka (Mori & Tanaka, 1973) and Halpin-Tsai (Ashton et al., 1969; Halpin, 1969) which are based on Eshelby's equation (Eshelby 1957; Eshelby 1961). Good description of these models can be found in (Tucker & Liang, 1999).

In order to predict the response of composite structures, taking into account its microstructure, a multiscale analysis can be applied. In this approach calculations are conducted at both macro and micro levels. At micro level, a finite number of constituents are defined, and each is described by its own constitutive model and material properties. The constitutive relation and the effective properties of composite at macro level are then calculated using a homogenization method. Once the effective properties are known a conventional FE analysis can be performed, giving macroscopic strain tensor ε at each point x, which at the micro level is viewed as a centre of the RVE. Subsequently, knowing the macro values of stress and strain tensor it is possible to compute approximated values at the micro level, in each phase.

Micromechanical models have been applied in several studies to predict elastic properties of nanocomposites (Luo & Daniel, 2003; Fornes et al., 2003; Wu et al. 2004; Wilkinson et al., 2007; Chavarria & Paul, 2004) as well as damage and failure (Fornes et al., 2003; Chen et al., 2003; Chen et al., 2007; Boutaleb et al., 2009). Luo *et al.* (Luo & Daniel, 2003) studied properties enhancement of epoxy matrix reinforced with silicate clay particles. A three-phase model, including the epoxy matrix, the exfoliated clay nano-layers and the nano-layer clusters was developed to account for partial exfoliation of nano-layers. The Mori-Tanaka method was applied to predict elastic properties of the material. Various parameters including the exfoliation ratio, clay layer and cluster aspect ratios, intra-gallery modulus, matrix modulus and matrix Poisson's ratio were taken into account. The exfoliation ratio

was investigated using TEM micrographs and the estimated value (10%) was included in the model. Predicted values of elastic properties were found to be in close agreement with the experimental results. That is why influence of various parameters on mechanical properties of the composite was further studied using the same method.
Boutaleb *et al.* (Boutaleb et al., 2009) extended micromechanical model based on Eshelby's formulation to predict stiffness and yield stress of silica reinforced polymer nanocomposites. The presented model took into account an interphase region around the nanoparticles and calculated its elastic modulus as gradual transition from that of the naoparticle to that of the matrix. The third phase centred around the particle is assumed to be inhomogeneous. Each phase was described by the own elastic stiffness tensor C. The dispersion of the filler and the interphase region in the matrix was assumed to be random. The yield function of the composite was calculated using the averaging procedure derived by Ju *et al.* (Ju & Sun, 2001). The numerical results obtained by the authors are in good agreement with the experiment. A dominating role of the interphase region on the composite yield stress is observed. For the small particle size the stiffness of the interphase is relatively strong. Increase in size of the particle leads to a softening of the interphase region. This indicates that process of debonding is dominating in this case.

5.2 Natural fiber-reinforced nanocomposites

At present the awareness of environmental pollution has increased. Carbon footprint, global warming and greenhouse gases are the main keywords related to this topic and the industrial and social attitude towards a "cleaner" future is omnipresent. Therefore, one of the most promising structures as component in nanocomposites are cellulosic nanofibrillars, due to their renewable and environmentally being nature and their outstanding mechanical properties. Cellulose nanofibers are obtained from various sources such as flax bast fibers, hemp fibers, kraft pulp, and rutabaga, by chemical treatments followed by innovative mechanical techniques. The nanofibers thus obtained have diameters between 5 and 60 nm. However, the key issues of concern for applications in which cellulosic nanocomposites are either intended or destined for long term use remains the durability. Although cellulosic nanocomposites proven to have superior properties, it remains a challenge to guarantee to the end user that such properties are reliably achieve during industrial scale processing. Furthermore, the degradation of properties over time cannot be assessed and secured at present (Hubbe et al., 2008).

5.3 Age and durability performance

Durability of the nanofiber layers can be affected by many different characteristics of the ultimate material composite. Therefore, each step towards the fabrication of the final composite material must be considered. Furthermore, durability depends to a great extent on structural elements and phenomena which are effective at micro- but most important at nanoscale.
Pramoda *et al.* (Pramoda et al., 2003) observed that the temperature of onset of degradation for PA-6 and 2.5% clay filled nanocomposites was higher than other compositions (neat polymer, 5% and 7.5%), respectively. Gilman *et al.* (Gilman et al., 1997) proposed that with the higher loading of clay, temperatures of onset of degradation remain unchanged, which was attributed to agglomeration in nanocomposites. The presence of organoclay (for PA6 – 2.5 wt. % clay nanocomposite) increased the activation energy for degradation, E_a,

compared to the neat PA-6 under N_2. The major evolved gas products were cyclic monomers, hydrocarbons, CO_2, CO, NH_3 and H_2O for PA-6 and PA-6/clay nanocomposites. During flammability measurements on the calorimeter in conjunction with an FTIR spectrometer in real time, changes in the condensed phase of PA-6 and a PA-6/clay nanocomposite have revealed that the spectra obtained during the burning are of sufficiently high quality to show the progression of the material in contact with the probe from molten polymer to thermal degradation products. The spectral features are consistent with the evolution of formation of caprolactam as a result of depolymerization.

Compared with pure PA6, the PA6/clay nanocomposite has higher activation energy, lower thermal decomposition rate constants and better thermal decomposition stability (VanderHart et al., 2001). The activation energy (E_a) of PA6/clay nanocomposites containing 1.2% and 3.5% clay are 174.47 and 309.175 kJ/mol, respectively, and that of pure PA6 is only 143 kJ/mol. Using ^{13}C NMR, they observed that in the presence of modifier (dihydrogenated-tallow ammonium ion) the nylon nanocomposite began to degrade at 240 °C, whereas the virgin polymer did not. They concluded that the organic modifier was less stable. The combination of shear stress and temperature may lead to extensive degradation of the modifier and the extent of clay dispersion may not depend on the modifier.

In an intumescent EVA-based formulation, using PA-6 clay nanocomposite instead of pure PA6 (carbonization agent) has been shown to improve the fire properties of the intumescent blend. Using clay as "classical" filler enabled the same level of FR performance to be obtained in the first step of the combustion as when directly using exfoliated clay in PA-6. But in the second half of the combustion, the clay destabilized the system and increased the flammability. Moreover, a kinetic modeling of the degradation of the EVA-based formulations showed that adding clay to the blend enables same mode of degradation and the same invariant parameters as for the polyamide-6 clay nanocomposite containing intumescent blend. The increase in the flammability by the K-10 in the second half of the combustion showed the advantages of using nanoclay rather than micronclay in an intumescent system (Levchik et al., 1999).

Organically modified clay-reinforced polyamide 6 was subjected to accelerated heat aging to estimate its long-term thermo-oxidative stability and useful lifetime compared to the virgin material (Kiliaris et al., 2009). Changes in molecular weight and thermal and mechanical properties were monitored and connected to the polymer modification encountered during aging. Generally, the strong interaction between the matrix and the clay filler renders the polymer chains, mainly that adjacent to silicates, highly restrained mechanically, enabling a significant portion of an applied force to be transferred to the higher modulus silicates. This mechanism explains the enhancement of tensile modulus that the non-aged clay-reinforced PA6 exhibited (1320 MPa) with regard to the neat polymer (1190 MPa), as shown in Figure 8.

The effects of hydrothermal ageing on the thermo-mechanical properties of high-performance epoxy and its nanocomposites are also reported in the literature (Njuguna et al., 2010). It was found that the storage modulus and relaxation behavior were strongly affected by water uptake, while the fracture toughness and Young's modulus were less influenced. Dependence of tensile strength and strain at break on water uptake was found to be different in neat epoxy and epoxy–clay systems. Further improvement of the flame retardancy using combinations of the nanofiller and traditional FR-additives (e.g. aluminum trihydrate) was observed. The nanocomposites based on nanofillers and aluminum trihydrate passed the UL 1666 riser test for fire-resistant electrical cables (Beyer, 2005).

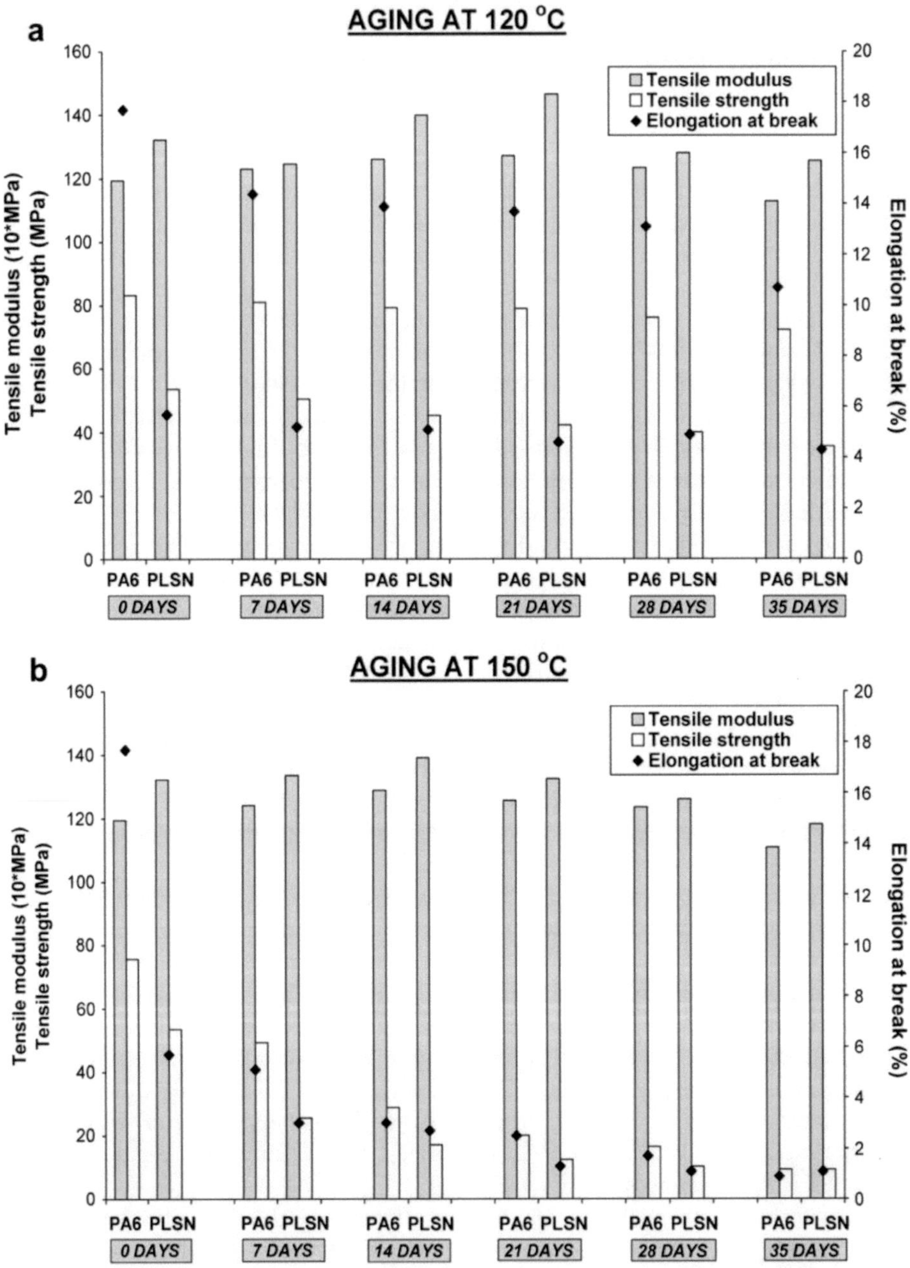

Fig. 8. Tensile properties of PA6 and PA6 nanocomposite, oven-aged at (a) 120 °C and (b) 150 °C (Kiliaris et al., 2009).

There are two factors which have opposite influences on the thermal stability of epoxy–clay nanocomposites. First factor is that the addition of clay to epoxy decreases the curing reactivity of epoxy resin. Lower reactivity of the resin generally results in lower cross-linking density of the cured resin and the longer polymer chains among the cross-linking points. It is known that a longer polymer chain is less stable thermally than a shorter chain, so both the nanocomposites are easier to degrade than the pristine epoxy resin. Secondly, silicate layers have good barrier to gases such as oxygen and nitrogen, they can insulate the underlying materials and slow the mass loss rate of decomposition products. Moreover, exfoliated nanocomposites have better barrier properties and thermal stability than intercalated ones. In the case of intercalated nanocomposites (10 wt. % clay), the first factor is dominant, whereas for an exfoliated nanocomposites (2 wt.% clay), the second factor is dominant. Becker *et al.* (Becker et al., 2004) have found that the water uptake (in aquatic environment) was considerably reduced in epoxy nanocomposites with a particular clay loading percentage.

5.4 Electrical properties

In the conventional fiber-reinforced polymers (FRPs), fiber-orientation is usually in-plane (x- and y-direction) resulting in fiber-dominated material properties in these directions, whereas matrix dominates in the z-direction. Therefore, FRPs are very sensitive to intrinsic damage, such as delamination (in particular), matrix cracking and fatigue damage. Moreover, their usage as structural materials in aerospace applications has evolved from less than 5% by structural weight to 50% (mainly due to the necessity to reduce weight and gain fuel savings), which leads to many issues related to their functionalities like vibration/acoustic damping, tribological behavior, and hygroscopic properties. Therefore, for enhanced structural stability, durability and performance, it is necessary to address and accommodate these undesirable issues.

Several approaches have been adopted to tackle these problems independently, which include (a) using epoxy/elastomer blends at ply interfaces, (b) incorporating interleave viscoelastic layers, (c) reducing the mismatch of elastic properties (and stress concentrations) at the interfaces between the laminated plies, and (d) using carbon filaments in place of interleaves. These approaches have their own positives and negatives as these materials are 'functional' rather than 'structural' and their high cost, low stiffness/stiffness, compatibility/adhesion with the FRP laminates, or poor processability always result in a compromised situation.

A specific area of interest is tuning the interleave material to achieve the desired multi-functionalities or to have an enhanced effect on the desired properties compared to FRPs without interleaf layers.

For the concept of non-destructive evaluation, damage sensitivity and reinforcing effect of carbon nanocomposites could be obtained from the electrical volume and contact resistivity measurement with acoustic emissions techniques. Adding conductive particles to an isolating polymer can result in an electrically conductive composite, if the particle concentration exceeds the percolation threshold, which is the particle volume fraction required to form a conductive network through the bulk polymer. Because of their aspect ratio and strong tendency to reagglomerate, CNTs are very effective fillers concerning the implication of electrical conductivity into polymeric matrices. Percolation thresholds as low as 0.0025 wt. % have been reported using MWNTs in an epoxy matrix (Sandler et al., 2003). The percolation threshold for the materials and the dispersion process used in this work was found to be below 0.1 wt. % (Gojny et al., 2005). Conductivities of up to 2×10^{-2} S/m could be achieved with unfunctionalised CNTs. The functionalisation of CNTs with amino-groups usually increases the percolation threshold and decreases the maximum achievable

conductivity, because the graphitic structure of the nanotubes becomes more defected during the functionalisation process.
Furthermore, the average nanotube length is reduced, which explains the increasing percolation threshold (Gojny et al., 2005). Park *et al.* (Park, 2003) explored this concept by applying electro-micromechanical technique to obtain the fiber damage and the reinforcing effect of carbon nanocomposites with their content. The sensitivity for fiber damage such as fiber fracture, matrix deformation and fiber tension was highest for 2.0 vol. % CNT composites. They suggested that damage sensitivity by electrical resistance measurement might be related to closely three-dimensional network structure. For CNT composites, mechanical properties and apparent modulus indicating the reinforcing effect increased with CNT content. The researchers confirmed that, apparent modulus measurement by electro-micromechanical test could be applied to evaluate mechanical properties of fiber reinforced composites. Reinforcing effect of 2.0 vol. % CNT obtained from mechanical properties and apparent modulus measurements was the highest. In 2.0 vol. % CNT case, percolation structure was observed when compared to 0.1 and 0.5 vol. % CNT cases. Morphological trends were found consistent with the result of damage sensitivity based on electrical properties.
The optimal use of sensor technology is in itself crucial in the design, manufacturing, maintenance, and proper functioning of a number of aerospace and defense related adaptive structures and other strategic equipment. Electrode materials with carbon nanotubes resulted in better behavior than traditional carbon electrodes including good conducting ability and high chemical stability. Another possible application entails the inherent multifunctionality of CNT-based materials that lead to designs which are self-sensing - sensor skins capable of probing the environment around the vehicle could be designed so that they are part of the vehicle itself. Their multifunctionality arises from the ability of the nanotubes to be either metallic or semi-conducting based on their chirality. Due to this property, SWNT have been used to fabricate several nanoscale devices, such as field-effect transistors and molecular logic devices. Furthermore, simply changing the environment around a nanotube can change its conducting behavior. This phenomenon is being exploited in order to create sensors capable of measuring several parameters related to vehicle structural health (i.e. strain, pressure, temperature, etc.). In the context of aerospace systems, innovative lighter weight and smaller volume sensors will allow carrying out enhanced real-time prognostic health monitoring and diagnosis of aerospace and military structures, strategic tactical transportation and weapon systems. However, as lamented earlier on, there are concerns that presently used maintenance techniques cannot meet the growing demand for high reliability and readiness. New advanced built-in diagnostic techniques have to be developed which can perform damage diagnosis automatically, avoiding human error; provide advanced warning of structural failure to pilots or operators; minimize unnecessary downtime for scheduled maintenance; maintain reliability and improve safety of aging structures; reduce maintenance costs; and enhance combat readiness.
Structural Health Monitoring (SHM) is the process of implementing a damage detection and characterization strategy for engineering structures. Here damage is defined as changes to the material and/or geometric properties of a structural system, including changes to the boundary conditions and system connectivity, which adversely affect the system's performance. The SHM process involves the observation of a system over time using periodically sampled dynamic response measurements (Hiem et al., 2004).

5.5 Health and safety aspects of nanoparticles

Numerous studies have focused on studying the potential health risks of the ultrafine- and nanoparticles, in the past few years (Maynard & Aitken R, 2007; European Agency for Safety and Health at Work [EU-OSHA], 2009). The European Union funded several projects dealing with the toxicological effect of nanoparticles. The NANOTOX (Impart- Nanotox, 2010) and NEPHH (Nanomaterial Related Enviormental Pollution and Health Hazard Throughout their Life cycle [NEPHH], 2010) projects both deal with the toxicological impact of nanoparticles on human health and the environment. To do so, the properties of different nanoparticles must be characterized and a sufficient amount must be sampled for toxicological investigations. A major risk of those projects is that the nanoparticles cannot be detected, and hence cannot be examined on their toxicological impact. Particles in those size ranges are suspected to enter the human cells more easily than micro sized particles, via respiratory, dermal or oral absorption. The high biological activity can be explained by the large surface-to-volume ratio; hence many types of nanoparticles have shown toxic impact (European Agency for Safety and Health at Work [EU-OSHA], 2009; Dreher, 2004). Despite limited knowledge about the risks associated with the exposure to and the release of nanoparticles in the environment, the development of mass production of products and materials has begun (Business Communications Company [BCC], 2006).

Crucial for the assessment of the potential environmental and health risks of ultrafine- and nanoparticles are the understanding of the exposure mechanism. In relation to toxicological studies, investigation of all physical and chemical parameters of engineered nanoparticles would be ideal, but represents a major workload. A significant number of parameters can therefore be retained as a minimum for successfully conducting meaningful toxicological studies (Murdock et al., 2008). These would include:

- Particle size and distribution,
- Specific surface area,
- Crystalline structure,
- Surface reactivity and composition,
- Purity.

Therefore, attention will have to be paid to the characterization of the different properties and hence the need for feasible and reliable measurement and characterization equipment for nanosized particles is fundamental.

At present, a standard procedure to measure airborne particles in workplaces and environment is to measure the aerosol mass concentration. The simplest approach is to use a filter-based sampler comprising some form of inertial particle pre-selector. This conventional pump based filter sampling of aerosols is not the best solution for exposure assessment for airborne nanostructures. There are two reasons for that: Firstly, mass concentration, is not necessarily well-suited to the toxicity assessment of inhaled nanoparticles. Toxicity studies show that particle toxicity increases as they become smaller (Oberdörster et al., 2005); Secondly none of the existing instruments used for monitoring give specific information about particle concentration below 1 µm aerodynamic diameter (European Agency for Safety and Health at Work [EU-OSHA], 2009).

More relevant parameters for describing airborne nanostructures are criteria to relating to their size and shape. Some of the equipments developed around these indicators allow a continuous measurement. The sections 6.2 and 6.3 of ISO Standard TR 27628 include a classification of each instrument based on continuous measurement of size, number and surface area parameters for the collected aerosol. These parameters X, can be described as a function of the time *t*, as shown in equation (6):

$$X = f(t) \quad (6)$$

X involves the whole aerosol or one of its particle size distribution fractions (European Agency for Safety and Health at Work [EU-OSHA], 2009). Table 2 summaries the currently available instruments on the market, to measure airborne nanoparticles.

Instrument Type	Measured Parameter (X)	Size Range (μm)	Concentration Range (particles/cm³)	Response Time (s)	Sample Flow Rate (l/min)
CPC	Number Concentration	0.003-.025	10^4-10^3	4	0.3-3
SMPS	Size Distribution, Number Concentration	0.0025-1	1-10^8	30 to 600	0.2-4
FMPS	Size Distribution, Number Concentration	0.0056-.56	10^6	1	10
ELPI	Size Distribution, Number Concentration	0.03 -10	n/a	< 5	10 or 30
APS	Aerodynamic size distribution	0.5 to 20	10^3-10^4	n/a	5
TEOM	Mass Concentration	2.5-10	1.5 μg/m³	0.5	0.5–5

APS, Aerodynamic Particle Sizer; CPC, Condensation Particle Counter; FMPS, Fast Mobility Particle Sizer; SMPS, Scanning Mobility Particle Sizer; ELPI, Electrical Low Pressure Impactor; TEOM, Tapered Element Oscillating Microbalance

Table 2. Instrument for measuring of Airborne Particles.

The instruments available at present on the market are capable to characterize airborne nanoparticles. The instruments are adequate to characterize nanoparticle properties such as particle size and concentration, as well as mass concentration. However, those instruments have several limitations and drawbacks, as low volume flows, background noises and particle losses. Furthermore, the lack of standardization makes the comparison of results obtained in different studies, which use different instruments, rather challenging. The high financial investment and the required specialist skills additional limit the different instruments. These different considerations require foreseeing a universal instrument for aerosol measuring, which would have the ability to record various parameters permitting optimum, simultaneous evaluation of (Maynard & Aitken, 2007):

- aerosol numerical, surface and mass concentration;

- an ultrafine fraction (which indeed requires definition as well as conventional respirable, inhalable and thoracic fractions who are defined in standard ISO 7708);
- Rapidly collected data and their storage for later use.

A universal instrument should be portable. At present all the instruments monitoring continuously a single parameter can be used static, because of their weight. Furthermore, concentration enriched systems, for direct exposure of cells with airborne particles needs further development in order to avoid sampling of particles with ESPs or filter, which leads to contamination and might change the original physical properties of the airborne particles.

For instance, low velocity impact of PU/MMT nanocomposites generated nanoparticles, which can be seen in the increasing of total number of particles for the period after the impact (Fig. 9). The particle increased for the first 15min after impact, after this period particle concentration decreases until it reached a normal ambient level again. Decrease of particle concentration is caused by three factors: Firstly, larger particles deposit due to gravitation; Secondly, airborne particles agglomerate to larger particles which is demonstrated by the increase of particles size over the time period; And thirdly, decrease of particle concentration is caused by the volume flow of the SMPS+C (0.3 l/min).

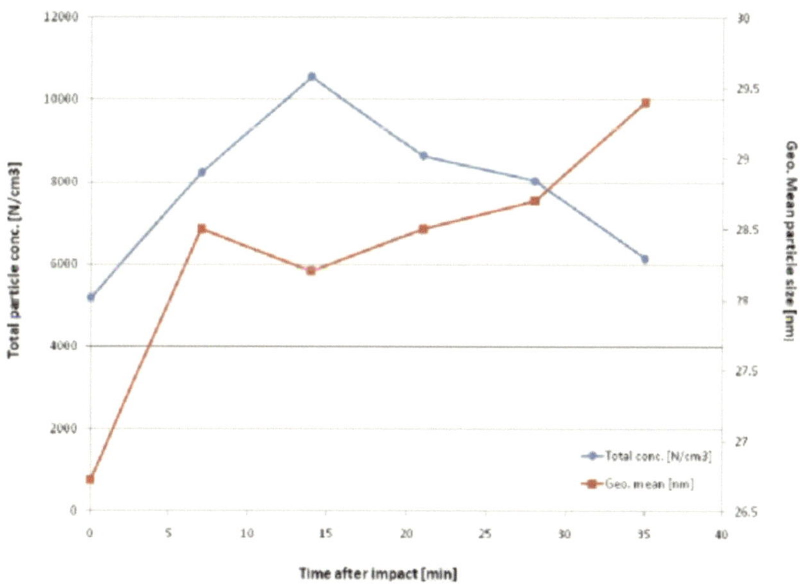

Fig. 9. Geometric mean particle size and total dependency of time after impact.

6. Applications in vehicle structures

Toyota researchers reported in the early 1990s, work based on true nylon-6-clay thermoplastic nanocomposites technology. Their results showed significant improvement of mechanical properties, with small amount (4.2wt%) filler loading. The strength increased more the 50%, the modulus doubled and additional improvement of thermal properties were obsereved (Oriakhi, C.O. 1998; Kozima, 1993). Since then, polymer nanocomposites are manufactured commercially for diverse structural vehicles applications. The consumption of clay-based polymer nanocomposites will increase to 181,094 metric tons and 692.3 million USD by 2014. (Business Communications Company [BCC], 2006). For example, one of the leading automotive manufacturers is using 300000 kg of nanoclay composites annually for various automotive exterior part/panel applications at present (Keles et al., 2009). However, other nanofillers have found applications for exterior vehicle panels. Carbon nanofibres have been used in automotive exterior panels for weight reduction, in bumpers and fenders. Figure 10 shows the potential application of for nanoclay and carbon nanofibers in automotive applications. As shown in figure 10 the opportunities of clay Nanocomposite lay in the increased stiffness and reduced density, which will lead to thinner wall stock and lower part weight. Additionally improved scratch resistance and reduced thermal sensitivity can be achieved. The incorporation of carbon nanofibers in sheet moulded compounded body panels, improves surface quality and can be used for improved toughness and thinner panels.

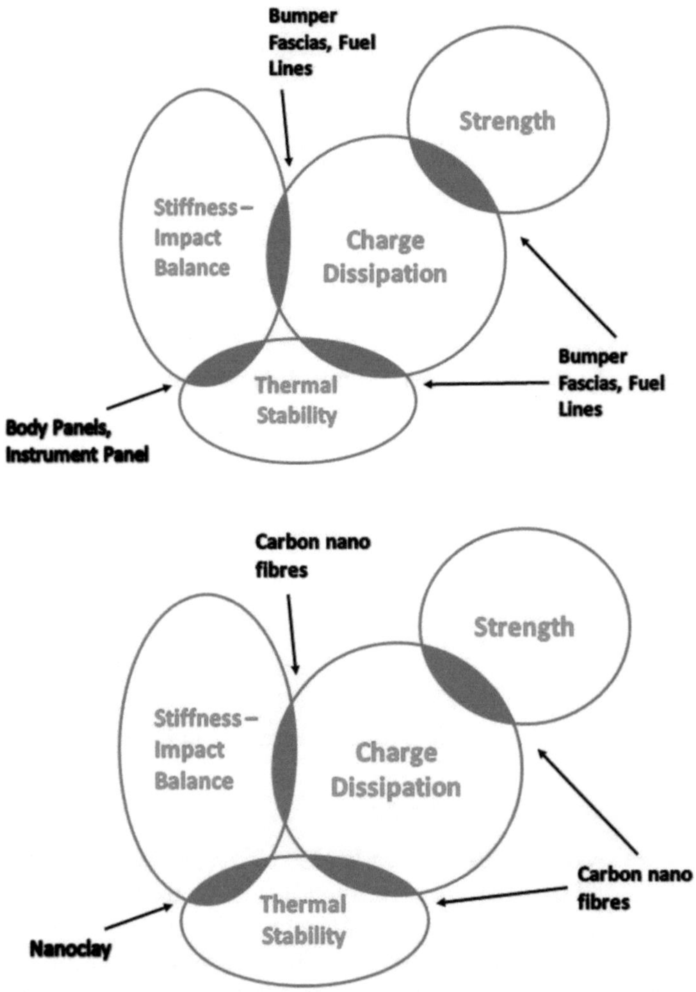

Fig. 10. Potential Automotive Applications of Nanocomposites.

Nevertheless, the nature of the constrained region as the mechanism of reinforcement has yet to be satisfactorily explained and property improvements have yet to be confirmed on commercially available nanocomposites using standard processing techniques. In automotive applications such as side mouldings, trim and panels in General Motor's vehicles, nanoclay compounds have advantages including reduced weight, a wider processing window, improved colourability and improved scratch and mar resistance. A consequential research focal point that has emerged relates to the known deleterious effects of the automobile exhaust air pollutant NO_x on nylon-6, in efforts to protect the nanocomposite from degradation (Njuguna et al., 2009).

7. Conclusions

Numerous studies have demonstrated that nanofiller have potential in improving both stiffness and energy absorption of polymer and/or conventional fiber-reinforced polymer composites. Tube crushing experiments have been widely used for measuring energy absorption in conventional structural composites over large displacement ranges. Energy absorption of nanocomposites as characterized by quasi-static fracture or low-speed impact energy absorption is caused by different energy absorption mechanisms in nanocomposites. The influence of some key control parameters such as shape and size of nanofillers, mechanical properties of nanofillers and matrix materials, interfacial adhesion, interphase characteristics, as well as the volume fraction and dispersion of fillers in the matrix. The complex mechanisms and numbers of control parameters make it difficult to obtain an ideal nanocomposite.

Due to numerous potential applications such as crashworthy transportation vehicle structures, protective armor, noise and vibration control, fracture-resistant structures, and multifunctional materials and structures, further research on energy absorption capacity of nanocomposites is crucial. Overall there is essential requirement to investigate the durability of these nanocomposites in different environmental conditions to extend the applicability of these hybrid materials. Further, the current limited knowledge of the degradation mechanism of polymer nanocomposites has led to development of more efficient stabilizers for improving the product performance on one hand and on the other, development of sensitizers to produce degradable polymers and to preserve the environment.

Further, through nanoparticles reinforcement, an electrically conductive matrix could provide enhanced feasibilities including stress–strain monitoring or damage detection. The application of an electrical field is known to orient the nanoparticles in the in-field-direction, which may result in an increased efficiency of the z-reinforcement of the laminates. As a further benefit, the electrical conductivity in z-direction should be increased with this approach. Nanoparticles and especially CNTs provide a high potential for the modification of polymers. They are very effective fillers regarding mechanical properties, especially toughness. Besides, they allow the implication of functional properties, which are connected to their electrical conductivity, into polymeric matrices. The electro-micromechanical technique had been studied as an economical non-destructive evaluation method for damage sensing, characterization of interfacial properties, and non-destructive behavior because conductive fiber can act as a sensor in itself as well as a reinforcing fiber.

The degradation of polymers has harmful aspects as well as beneficial aspects. If unchecked it can play havoc with a polymer nanocomposites performance, if uncontrolled it can lead to safety hazards of fire and toxicity, but if properly harnessed it can be used for producing new and better materials. It will be worthwhile to focus further studies with nanoparticulates such as clay, carbon nanotubes, metal oxides, metal salts in the aspect of degradation in an environment where they are to be utilized in specific aerospace and military applications. Fundamental to the success of nanotechnology is its perceived safety by the public. Many concerns have been aired concerning the use of manufactured nanoparticles.

8. Acknowledgment

The authors would like to acknowledge the European Commission financial support through the FP7 Project- CP-FP; Project Reference No.: 228536– 2.

9. References

Aboudi, J. (1988), "Micromechanical analysis of the strength of unidirectional fiber composites", Composites Science and Technology, vol. 33, no. 2, pp. 79-96.

Aboudi, J. (1990), "The nonlinear behavior of unidirectional and laminated composites-A micromechanical approach", Journal of Reinforced Plastics and Composites, vol. 9, no. 1, pp. 13.

Ajayan, P. M., Schadler, L. S., Braun, P. V., Picu, C. and Keblinski, P. (2003), Nanocomposite science and technology, Wiley Online Library.

Altenbach, H. and Tushtev, K. (2001), "A new static failure criterion for isotropic polymers", Mechanics of Composite Materials, vol. 37, no. 5, pp. 475-482.

Ashton, J., Halpin, J. C. and Petit, P. H. (1969), Primer on composite materials: analysis, Technomic Pub. Co. Stamford, Conn.

Business Communications Company [BCC] Research Report (2006) Nanocomposites, Nanoparticles, Nanoclay, and Nanotubes. Report Code: NAN021C.

Becker, O., Varley, R. and Simon, G. (2004), "Thermal stability and water uptake of high performance epoxy layered silicate nanocomposites", European Polymer Journal, vol. 40, no. 1, pp. 187-195.

Belytschko, T., Xiao, S., Schatz, G. and Ruoff, R. (2002), "Atomistic simulations of nanotube fracture", Physical Review B, vol. 65, no. 23, pp. 235430.

Beyer, G. (2005), "Flame retardancy of polymers by nanocomposites: a new concept", in High performance fillers 2005, Rapra Technology Ltd., Shrewsbury, UK, , pp. P5-P7.

Bittmann B., Haupert F., Schlarb A.K. (2010), "Preparation of TiO_2/epoxy nanocomposites by ultrasonic dispersion and their structure property relationship", Ultrason. Sonochem., In Press

Boutaleb, S., Zaïri, F., Mesbah, A., Naït-Abdelaziz, M., Gloaguen, J., Boukharouba, T. and Lefebvre, J. (2009), "Micromechanics-based modelling of stiffness and yield stress for silica/polymer nanocomposites", International Journal of Solids and Structure., vol. 46, no 7-8, pp1716-1726

Bowden, P. and Jukes, J. (1972), "The plastic flow of isotropic polymers", Journal of Materials Science, vol. 7, no. 1, pp. 52-63.

Capezzuto F., Ciampa F., Carotenuto G., Meo M., Milella E., Nicolais F. (2010) "A smart multifunctional polymer nanocomposites layer for the estimation of low-velocity impact damage in composite structure", Composite Structures, vol. 92,pp. 1913-1919.

Chavarria, F. and Paul, D. (2004), "Comparison of nanocomposites based on nylon 6 and nylon 66", Polymer, vol. 45, no. 25, pp. 8501-8515.

Chen B. and Evans J.R.G. (2008) "Impact and tensile energies of fracture in polymer-clay nanocomposites", Polymer, vol. 49, pp. 5113-5118.

Chen, J., Huang, Z. and Mai, Y. W. (2003), "Constitutive relation of particulate-reinforced viscoelastic composite materials with debonded microvoids", Acta Materialia, vol. 51, no. 12, pp. 3375-3384.

Chen, J., Huang, Z. and Zhu, J. (2007), "Size effect of particles on the damage dissipation in nanocomposites", Composites Science and Technology, vol. 67, no. 14, pp. 2990-2996.

Cornwell, C. and Wille, L. (1997), "Elastic properties of single-walled carbon nanotubes in compression", Solid State Communications, vol. 101, no. 8, pp. 555-558.

Crosby, A. J. and Lee, J. Y. (2007), "Polymer nanocomposites: the "nano" effect on mechanical properties", Polymer reviews, vol. 47, no. 2, pp. 217-229.

Davis, R. D., Gilman, J. W. and VanderHart, D. L. (2003), "Processing degradation of polyamide 6/montmorillonite clay nanocomposites and clay organic modifier", Polymer Degradation and Stability, vol. 79, no. 1, pp. 111-121.

Deher K. L. (2004) "Health and environmental impact of nanotechnology: toxicological assessment of manufactured nanoparticles", Toxicological Science, vol.77, pp 3-5.

Dillon D. R., Tenneti K. K, Li C. Y., Ko F. K., Sics I., Hsiao B. S., (2006) "On the structure and morphology of polyvinylidene fluoride-nanoclay nanocomposites", Polymer, vol. 47, pp. 1678-1688.

Donclero W.E., Gorga R.E. (2006) "Morphological and mechanical properties of carbonnanotube/polymer composites via melt compounding", Journal of Polymer Science, Part B: Polymer Physics vol. 44, pp. 864–78.

Dubnikova I.L., Berezina S.M., Antonov A.V. (2004) "Effect of rigid particle size on the toughness of filled polypropylene", Journal of Applied Polymer Science, vol. 94, pp. 1917–26.

Eshelby, J. (1957), "The determination of the elastic field of an ellipsoidal inclusion, and related problems", Proceedings of the Royal Society of London.Series A.Mathematical and Physical Sciences, vol. 241, no. 1226, pp. 376.

Eshelby, J. (1961), "Elastic inclusions and inhomogeneities", Progress in solid mechanics, vol. 2, no. 1, pp. 89-140.

European Agency for Safety and Health at Work [EU-OSHA] (2009) European Risk Observatory-Workplace exposure to nanoparticles.

Evans, A. (1972), "The strength of brittle materials containing second phase dispersions", Philosophical Magazine, vol. 26, no. 6, pp. 1327-1344.

Fisher FT, (2006), "Nanomechanics of nanoreinforced polymers.", Handbook of theoretical and computational nanoscience. American Scientific Publishers, pp. 253-360.

Fornes, T. and Paul, D. (2003), "Modeling properties of nylon 6/clay nanocomposites using composite theories", Polymer, vol. 44, no. 17, pp. 4993-5013.

Fornes, T. D., Yoon, P. J. and Paul, D. R. (2003), "Polymer matrix degradation and color formation in melt processed nylon 6/clay nanocomposites", Polymer, vol. 44, no. 24, pp. 7545-7556.

Fu S.-Y., Feng X.-Q, Lauke B, Mai Y.-W. (2008) "Effects of particle size, particle/matrix interface adhesion and particle loading on mechanical properties of particulate-polymer composites", Composites Part B, vol. 39, pp. 933-961.

Galgali G., Agarwal S., Ashish L. (2004) "Effect of clay orientation on the tensile modulus of polypropylene-nanoclay composites", Polymer, vol. 45, pp. 6059-6069.

Gilman, J. W., Kashiwagi, T. and Lichtenhan, J. D. (1997), "Nanocomposites: A revolutionary new flame retardant approach", SAMPE Journal, vol. 33, no. 4, pp. 40-46.

Halpin, J. (1969), "Stiffness and expansion estimates for oriented short fiber composites", Journal of Composite Materials, vol. 3, no. 4, pp. 732.

Hamming L. M., Qiao R., Messersmith P.B., Catherine Brinson L. (2009) "Effects of dispersion and interfacial modification on the macroscale properties of TiO_2 polymer-matrix nanocomposites" Composite Science Technology, vol. 69, pp. 1880-188.

Hiem, C., Drager, S., Flynn, C., Renz, T. and Burns, D. (2004), "Advanced computer technology for novel information processing paradigms", Journal of Aerospace Computing, Information, and Communication, vol. 1, no. 7, pp. 308-317.

Hwu, J. M., Jiang, G. J., Gao, Z. M., Xie, W. and Pan, W. P. (2002), "The characterization of organic modified clay and clay-filled PMMA nanocomposite", Journal of Applied Polymer Science, vol. 83, no. 8, pp. 1702-1710.

Hyun Park J., Jana S. C., (2003) "The relationship between nano- and micro-structures and mechanical properties in PMMA–epoxy–nanoclay composites", Polymer, vol. 44, pp. 2091-2100.

Impart- Nanotox: Investigative Support for the Elucidation of the Toxicological Impact of Nanoparticles on Human Health and the Environment (2010) http://www.impart-nanotox.org. Accessed 22 July 2010.

Iqbal K., Khan S.-U., Munir A., Kim J.-K., (2009) "Impact damage resistance of CFRP with nanoclayfilled epoxy matrix", Composite Science Technology, vol. 69,pp. 1949-1957.

Javni, I., Zhang, W., Karajkov, V., Petrovic, Z. and Divjakovic, V. (2002), "Effect of nano-and micro-silica fillers on polyurethane foam properties", Journal of Cellular Plastics, vol. 38, no. 3, pp. 229.

John B., Reghunadhan Nair C.P., Ninan K.N., (2010) "Effect of nanoclay on the mechanical, dynamic mechanical and thermal properties of cyanate ester syntactic foams", Material Science Engineering A, vol. 527, pp. 5435–5443

Ju, J. and Sun, L. (2001), "Effective elastoplastic behavior of metal matrix composites containing randomly located aligned spheroidal inhomogeneities. Part I: Micromechanics-based formulation", International Journal of Solids and Structures, vol. 38, pp. 183-201.

Kanit, T., Forest, S., Galliet, I., Mounoury, V. and Jeulin, D. (2003), "Determination of the size of the representative volume element for random composites: statistical and numerical approach", International Journal of Solids and Structures, vol. 40,.pp.3647–3679.

Kashiwagi, T., Du, F., Douglas, J. F., Winey, K. I., Harris Jr., R. H. and Shields, J. R. (2005), "Nanoparticle networks reduce the flammability of polymer nanocomposites.", Nature materials., vol. 4, no. 12, pp. 928-933.

Kiliaris, P., Papaspyrides, C. D. and Pfaendner, R. (2009), "Influence of accelerated aging on clay-reinforced polyamide 6", Polymer Degradation and Stability, vol. 94, no. 3, pp. 389-396.

Kireitseu M., Hui D., Tomlinson G. (2008) "Advanced shock-resistant and vibration damping of nanoparticle-reinforced composite material", Composite Part B: Engineering. vol, 39, pp. 128–38.

Kostopoulos V., Baltopoulos A.,Karapappas P., Vavouliotis A., Paipetis A. (2010) "Impact and after-impact properties of carbon fibre reinforced composites enhanced with multi-wall carbon nanotubes", Composite. Science Technology, vol. 70, pp. 553-563.

Leszczyńska, A., Njuguna, J., Pielichowski, K. and Banerjee, J. R. (2007), "Polymer/montmorillonite nanocomposites with improved thermal properties: Part I. Factors influencing thermal stability and mechanisms of thermal stability improvement", Thermochimica Acta, vol. 454, pp. 1-22.

Levchik, S. V., Weil, E. D. and Lewin, M. (1999), "Thermal decomposition of aliphatic nylons", Polymer International, vol. 48, no. 6-7, pp. 532-557.

Lim S.-H., Dasari A., Wang G.-T., Yu Z.Z., Mai Y.-W, Yuan Q., Liu S.,Yong M. S., (2010) "Impact fracture behaviour of nylon 6-based ternary nanocomposites", Composites Part B, vol. 41, pp. 67-65.

Lin J. C. (2008) "Investigation of impact behaviour of various silica reinforced polymeric matrix nanocmposites", Composite Structure, vol. 84, pp. 125-131.

Liu Z.H., Kwok K.W., Li R.K.Y., Choy C.L. (2002) "Effects of coupling agent and morphology on the impact strength of high density polyethylene/$CaCO_3$ composites", Polymer, vol. 43, pp. 2501-2506.

Lobo, H. (2007), "Methodology for selection of material models for plastics impact simulation", 6th European LS-DYNA Users' Conference.

Luo, J. J. and Daniel, I. M. (2003), "Characterization and modeling of mechanical behavior of polymer/clay nanocomposites", Composites Science and Technology, vol. 63, no. 11, pp. 1607-1616.

Lux Research, (2004) "The Nanotech Report"

Martin A. Hubbe, Orlando J. Rojas, Lucian A. Lucia, Mohini Sain (2008). "Cellulosic nanocomposites: A review", BioResources, vol. 3, no. 3, pp. 929–980.

Maynard A. D., Aitken R. J. (2007) Assessing exposure to airborne nanomaterials: Current abilities and future requirements. Nanotoxicology, vol. 1, pp. 26-41.

Mirmohseni A., Zavareh S. (2010) "Preparation and characterization of an epoxy nanocomposite toughened by a combination of thermoplastic, layered and particulate nano-fillers", Materials and Design, vol. 31, pp. 2699-2706.

Mohammed A. A., Hosur M. V., Jeelani S. (2006) "Processing and characterization of nanophased polyurethane foams", Cellular Polymers, vol. 25, pp. 293-306.

Monteiro-Riviere, N. A., Nemanich, R. J., Inman, A. O., Wang, Y. Y. and Riviere, J. E. (2005), "Multi-walled carbon nanotube interactions with human epidermal keratinocytes", Toxicology Letters, vol. 155, no. 3, pp. 377-384.

Mori, T. and Tanaka, K. (1973), "Average stress in matrix and average elastic energy of materials with misfitting inclusions", Acta Metallurgica, vol. 21, no. 5, pp. 571-574.

Murdock R.C., Braydich-Stolle L., Schrand A.M., Schlager J.J., Hussain S. M. (2008) "Characterization of nanomaterial dispersion in solution prior to in vitro exposure using dynamic light scattering technique", Toxicological Sciences, vol. 101, pp. 239-253.

Namilae, S. (2005), "Multiscale model to study the effect of interfaces in carbon nanotube-based composites", Journal of Engineering materials and technology, vol. 127, pp. 222.

Nanomaterial Related Enviormental Pollution and Health Hazard Throughout their Life cycle [NEPHH] (2010) www.nephh-fp7.eu. Accessed 22 July 2010.

Ng, C., Schadler, L. and Siegel, R. (1999), "Synthesis and mechanical properties of TiO_2-epoxy nanocomposites", Nanostructured Materials, vol. 12, no. 1-4, pp. 507-510.

Njuguna J., Pielichowski K., Desai S., (2008) "Nanofiller-reinforced polymer nanocomposites", Polymers for Advanced Technologies, vol. 19, pp. 947-959.

Njuguna, J., Peña I., Zhu H., Rocks S. A., Blázquez M. and Desai S. A., (2009) "Opportunities and environmental health challenges facing integration of polymer nanocomposites technologies for automotive applications" International Journal of Polymers and Technologies, vol. 2, pp. 117-126

Njuguna, J., Michalowski,S., Pielichowski, K., Kayvantash, K., and Walton, A. C. (2010), Fabrication, characterisation and low-velocity impact on hybrid sandwich composites with polyurethane/layered silicate foam cores, Polymer Composites, vol. 32,pp. 6–13.

Njuguna, J. P., K. Pielichowski (2010), "Ageing and performance predictions of polymer nanocomposites for exterior defence and aerospace applications", 11-12 March 2010, Hamburg, Germany, Smithers Rapra Technology.

Oberdörster G., Oberdörster E., Oberdörster J. (2005) "Nanotoxicology: an ermerging disciplineevolving from studies of ultrafine particles", Environment Health Perspectives, vol. 113, pp. 173-179.

Ol'khovik, O. (1983), "Apparatus for testing the strength of polymers in a three-dimensional stressed state", Mechanics of Composite Materials, vol. 19, no. 2, pp. 270-275.

Pae, K. and Mears, D. (1968), "The effects of high pressure on mechanical behavior and properties of polytetrafluoroethylene and polyethylene", Journal of Polymer Science Part B: Polymer Letters, vol. 6, no. 4, pp. 269-273.

Pandey, J. K., Raghunatha Reddy, K., Pratheep Kumar, A. and Singh, R. P. (2005), "An overview on the degradability of polymer nanocomposites", Polymer Degradation and Stability, vol. 88, no. 2, pp. 234-250.

Park S.- J., Kim B.-J., Seo D.-I., Rhee K.-Y., Lyu Y.-Y.(2009) "Effects of a silane treatment on the mechanical interfacial properties of montmorillonite/epoxy nanocomposites" Material Science Engineering A, vol. 523,pp. 74-78.

Pramoda, K. P., Liu, T., Liu, Z., He, C. and Sue, H. J. (2003), "Thermal degradation behavior of polyamide 6/clay nanocomposites", Polymer Degradation and Stability, vol. 81, no. 1, pp. 47-56.

Pugh, H. L. D., Chandler, E., Holliday, L. and Mann, J. (1971), "The effect of hydrostatic pressure on the tensile properties of plastics", Polymer Engineering & Science, vol. 11, no. 6, pp. 463-473.

Qi B., Zhang Q.X., Bannister M., Mai Y.W. (2006) "Investigation of the mechanical properties of DGEBA-based epoxy resin with nanoclay additives", Composite Structurs, vol. 75, pp. 514–9.

Quinson, R., Perez, J., Rink, M. and Pavan, A. (1997), "Yield criteria for amorphous glassy polymers", Journal of Materials Science, vol. 32, no. 5, pp. 1371-1379.
Rabinowitz, S., Ward, I. and Parry, J. S. C. (1970), "The effect of hydrostatic pressure on the shear yield behaviour of polymers", Journal of Materials Science, vol. 5, no. 1, pp. 29-39.
Rapaport, D. C. (2004), The art of molecular dynamics simulation, Cambridge Univ Pr.
Sandler, J. K. W., Kirk, J. E., Kinloch, I. A., Shaffer, M. S. P. and Windle, A. H. (2003), "Ultra-low electrical percolation threshold in carbon-nanotube-epoxy composites", Polymer, vol. 44, no. 19, pp. 5893-5899.
Sterky K., Jacobsen H., Jakubowicz I., Yarahmadi N., Hjertberg T., (2010) "Influence of processing technique on morphology and mechanical properties of PVC nanocomposites", European Polymer Journal, vol. 46, pp. 1203-1209.
Sternstein, S. and Ongchin, L. (1969), "Yield criteria for plastic deformation of glassy high polymers in general stress fields", Polymer preprints, vol. 10, no. 2, pp. 1117–1124.
Subramaniyan A.K., Sun C.T. (2007) "Toughening polymeric composites using nanoclay:crack tip scale effects on fracture toughness", Composites Part A, vol. 38, pp. 34–43.
Sun L., Gibson R. F., Gordaninejad F., Suhr J., (2009) "Energy absorption capability of nanocomposites:A review", Composites Science and Technology, vol. 69, pp. 2392-2409.
Tomar V., Zhou M. (2005)"Deterministic and stochastic analyses of fracture processes in a brittle microstructure system", Engineering Fracture Mechanics, vol. 72, pp. 1920–41.
Tucker, C. and Liang, E. (1999), "Stiffness predictions for unidirectional short-fiber composites: review and evaluation", Composites Science and Technology(UK), vol. 59, no. 5, pp. 655-671.
VanderHart, D. L., Asano, A. and Gilman, J. W. (2001), "Solid-state NMR investigation of paramagnetic nylon-6 clay nanocomposites. 2. Measurement of clay dispersion, crystal stratification, and stability of organic modifiers", Chemistry of Materials, vol. 13 , pp 3781–3795.
Viana J.C. (2006) "Polymeric materials for impact and energy dissipation", Plastics Rubber and Composites Processing and Applications, vol. 35, pp. 260-267.
Wacharawichanant S., Thongyai S., Phutthaphan A.,Eiamsam-ang C., (2008) "Effect of particle sizes of zinc oxide on mechanical, thermal and morphological properties of polyoxymethylene/zinc oxide nanocomposites", Polymer Test, vol. 27, pp. 971-976.
Wetzel, B., Rosso, P., Haupert, F. and Friedrich, K. (2006), "Epoxy nanocomposites-fracture and toughening mechanisms", Engineering Fracture Mechanics, vol. 73, no. 16, pp. 2375-2398.
Widy T., Macosko C. W., (2005) "Nanoclay-modified rigid polyurethane foam", Journal of Macromolecular Science, Part B Physics., vol. 44, pp. 897-908.
Wilkinson, A., Man, Z., Stanford, J., Matikainen, P., Clemens, M., Lees, G. and Liauw, C. (2007), "Tensile properties of melt intercalated polyamide 6-Montmorillonite nanocomposites", Composites Science and Technology, vol. 67, no. 15-16, pp. 3360-3368.

Wu, Y. P., Jia, Q. X., Yu, D. S. and Zhang, L. Q. (2004), "Modeling Young's modulus of rubber-clay nanocomposites using composite theories", Polymer Testing, vol. 23, no. 8, pp. 903-909.

Xiao T., Ren Y., Liao K., Wu P., Li F., Cheng H.M. (2008) "Determination of tensile strength distribution of nanotubes from testing of nanotube bundles", Composites Science and Technology, vol. 68, pp. 2937-2942.

Xidas P. I., Triantafyllidis K. S. (2010) "Effect of the type of alkylammonium ion clay modifier on the structure and thermal/mechanical properties of glassy and rubbery epoxy-clay nanocomposites", European Polymer Journal, vol. 46, pp. 404-417.

Yilmaz C., Korkmaz T. (2007) "The reinforcement effect of nano and microfillers on fracture toughness of two provisional resin materials", Material Design, vol. 28, pp. 2063–70.

Zhang H., Zhang Z., Friedrich K., Eger C. (2006) "Property improvements of in situ epoxy nanocomposites with reduced interparticle distance at high nanosilica content", Acta. Materialia, vol. 54, pp. 1833–42.

Zhang, M., Zeng, H., Zhang, L. and Lin, G. (1993), "Fracture characteristics of discontinuous carbon fibre-reinforced PPS and PES-C composites", Polymers & Polymer Composites, vol. 1, no. 5, pp. 357-365.

20

Filtration and Catalytic Behaviors of Titanium Silicate-1 Supported on Carbon Nanofibers for Cyclohexanone Ammoximation

Qian Zhao[1,2], Shiyuan Zhang[1], Ping Li[1], Weikang Yuan[1], Alex Chikin Yip[3] and Xijun Hu[3]
[1]*State Key Laboratory of Chemical Engineering, East China University of Science and Technology, Shanghai*
[2]*Department of Chemical Engineering and Technology, Hebei University of Technology, Tianjin*
[3]*Department of Chemical Engineering, Hong Kong University of Science and Technology, Hong Kong*
P. R. China

1. Introduction

Titanium silicalite-1 (TS-1), a MFI structure zeolite, is exclusively used as the catalyst in cyclohexanone ammoximation for its extremely high catalytic activity and selectivity.[1,2] It has been proved that the active sites inside the microporous channels of TS-1 are easily accessible to reactants when the particle size is controlled in a range of 200-300 nm. However, the separation of the nanosized TS-1 from liquid reaction systems by filtration is very difficult in industrial applications.[3] To deal with this problem, efforts have been made to immobilize TS-1 crystallites on supports with relatively large dimensions so as to facilitate the filtration.[4-7] Nevertheless, the dissolution of oxide supports such as $A1_2O_3$ and SiO_2 during TS-1 synthesis can result in a structural change of TS-1 crystallites and consequently a low utilization efficiency of H_2O_2 in reaction.[4-6] Carbon materials such as active carbon and graphite can be excellent candidates for support because of their chemical inertness to the reaction. But the separation performance of the TS-1/carbon composites has rarely been reported yet.[7]
Carbon nanofibers (CNFs)[8-11] have been attracting extensive interests owing to their large external surface and few micropores, chemical inertness and hybrid electronic structure. In particular, their high mechanical strength and long filamentary modality are beneficial for them to tolerating vigorous agitation in slurry systems. Furthermore, they can be facilely filtrated from the slurry.[8] Therefore, CNFs was deliberately selected as the support for TS-1 immobilizing.[12]
In this paper, TS-1/CNFs composite catalysts were synthesized by in situ crystallization and simple mechanical blending methods, respectively. The catalytic behaviors and filtration performance of TS-1/CNFs were examined in cyclohexanone ammoximation. The structure of catalyst is characterized using techniques including SEM, XRD, FT-IR, UV-Vis, Raman and XRF spectroscopy. The aim of the study is to develop a composite TS-1/CNFs catalyst

which can be easily separated from the slurry system without compromising the catalytic activity and the selectivity.

2. Experimental

2.1 Synthesis

2.1.1 Synthesis of CNFs

CNFs were grown on a Fe catalyst at 600 °C by a chemical vapor deposition (CVD) method using CO as carbon source.[12] The CNFs as grown were then immersed in a 4 M HCl solution at 60 °C under stirring for 2 h so as to dissolve Fe particles. Subsequently, the CNFs suspending was filtered and the CNFs on the filter medium were washed with copious deionized water. After triply filtering and washing, the purified CNFs were finally dried at 120 °C overnight. The final CNFs contained less than 0.1 wt % Fe as detected by means of inductively coupled plasma spectroscopy (ICP, Optima 2100 DV, PerkinElmer Instruments).

2.1.2 Synthesis of TS-1

Firstly, tetraethyl orthosilicate (TEOS) was added to tetrapropyl ammonium hydroxide (TPAOH) with vigorously stirring. Then, a tetrabutyl orthotitanate (TBOT) solution (in isopropanol) was brought dropwise into the above mixture. The resulting solution was aged at 80 °C for 5 h. A typical molar composition of the final gel was set to SiO_2: TiO_2: TPAOH: H_2O=1:0.025:0.3:20, corresponding to a Si/Ti molar ratio of 40. Next, the gel was transferred into an autoclave for crystallization at 175 °C for 72 h. After that, the cooled product was filtered and the solid product on the filter medium was washed with deionized water. Finally, the solid product was dried at 120 °C for 12 h, followed by calcining in air or Ar atmosphere at 550 °C for 6 h to eliminate TPAOH.

2.1.3 Preparation of blended TS-1/CNFs composite

TS-1 crystallites and the CNFs powder were mechanically mixed with deionized water at room temperature. After 24 h agitation, the mixture was then filtered and the solid TS-1/CNFs was dried at 120 °C for 12 h. The weight ratio of TS-1 to CNFs in the TS-1/CNFs composite was kept at 2.

2.1.4 Preparation of in situ crystallized TS-1/CNFs composite

The procedure for the preparation of in situ crystallized TS-1/CNFs composite is almost the same as that for the synthesis of TS-1, with the exception of the addition of CNFs into the colloid before hydrothermal treatment at 175 °C. TS-1/CNFs were calcined in Ar at 550 °C for 6 h. The weight ratio of TS-1 to CNFs was also regulated to 2. For the synthesis of composites with different Si/Ti ratios, the amount of TiO_2 in the gel was adjusted on the basis of SiO_2.

2.2 Characterization

Material morphology observations were executed on a JEOL JSM-6360LV scanning electron microscope (SEM). X-ray diffraction (XRD) patterns of material solid phase composition were recorded in a 2θ range 5° ~ 80° using a Rigaku D/Max 2550VB/PC diffractometer equipped with a Cu/K_α radiation source (λ=0.154056 nm) and operated at a voltage of 40 kV and a current of 100 mA. Infrared (IR) spectra of material were acquired on a Bruker Equinox-55 FT-IR spectrometer (KBr plate). Diffuse reflectance ultraviolet-visible spectrometry (UV-Vis, Varian Cary-500 spectrometer) was performed in a range 200~800 nm using Si-1 zeolite as the reference. Laser Raman spectra (Renishaw inVia-Reflex spectrometer) were taken with an Ar^+ laser beam tuned at 514 nm radiation wavelength. Material composition was detected on an X-ray fluorescence spectrometer (XRF-1800, Shimadzu Corporation).

2.3 Evaluation of catalytic performance of catalyst

Batch-wise cyclohexanone ammoximation was performed to evaluate the catalytic activity and selectivity. The reaction was carried out in a three-neck flask fitted with a stirrer, a condenser and a feed pump. The catalyst of 1.125 g TS-1/CNFs composite (TS-1: CNFs = 2 in weight) or 0.750 g TS-1 or 0.375 g CNFs was dispersed into a mixture of 5 g cyclohexanone, 12.5 ml t-butanol and 12.5 ml H_2O; and the flask was warmed in a water bath at 80 °C. With turbulently agitating the mixture, NH_3 (28 wt %) was injected into the flask with a dose every 30 min and H_2O_2 (30 wt %) was injected dropwise through the feed pump within 1.5 h. The total molar ratio of cyclohexanone: H_2O_2: NH_3 was 1: 1.4: 2. After 2 h reaction, the cooled reaction slurry was then filtered and the filtrate was extracted with toluene. Both cyclohexanone oxime and the unreacted cyclohexanone in the extractive were analyzed by a gas chromatography (GC, HP-6890) equipped with a HP-5 capillary column. 1,3,5-trimethylbenzene was used as an internal standard substance.

The catalytic activity is defined as the conversion of cyclohexanone after reaction based on the weight of cyclohexanone in the feed, and the selectivity is the weight of cyclohexanone oxime produced on the basis of cyclohexanone converted.

2.4 Evaluation of filtration performance of catalyst

The filtration of the reaction slurry was managed with a constant velocity using an injecting pump while the particles in the slurry were passing through a composite fiber membrane with uniform pores of 1.2 μm. The filtration resistance was measured with a pressure gauge. The turbidity of filtrate, an indication of the amount of filterable particles, was tested by a nephelometer (turbidity unit: NTU, number of transfer unit).

3. Results and discussion

3.1 Catalytic behaviors and filtration performance of CNFs, TS-1 and TS-1/CNFs

As listed in Table 1, 47.3% of cyclohexanone was converted but no cyclohexanone oxime was detected while using pure CNFs as the catalyst after 2 h reaction. In contrast, over the TS-1 crystallites, cyclohexanone was almost completely converted and the selectivity to oxime reached to ca. 95%. A slight decline in oxime selectivity was observed on the TS-1 calcined in Ar than in air, probably due to the template residues emerging in inactive atmosphere which partly block the channels of TS-1.[13,14] By comparing separately the blended TS-1/CNFs composite and the in situ crystallized one with pure TS-1 in the catalytic performance, we can see that the former exhibited a higher activity and selectivity while the latter partly deactivated. The reasons will be anatomized hereinafter.

The filtration result of the slurry containing CNFs after reaction (Table 1) indicates almost no CNFs appearing in the filtrate. This is probably attributed to the unique morphology of

CNFs (Fig. 1a). CNFs have a long filamentous appearance with ten μm in length and 250 nm in diameter. The curly and entangled CNF filaments agglomerated into granules with macroporous pores (Fig. 1a), and the granule size was distributed up to 100 μm (Fig. 1b). The morphology of CNF granules remained in the reaction slurry having undergone 2 h vigorous stirring (Fig. 1c). The particle size distributions of the CNFs as grown and after ultrasonic treating for 2.5 h were measured by means of laser granularity analysis (Fig. 2). The result shows that the dominant particle size around 10 μm was retained after treating. It is evident that the compact agglomeration of CNFs to large dimensions renders CNFs to be facilely intercepted by filter membrane and thus to be separated from the slurry after reaction.

Catalytic material		Catalytic activity and selectivity		Filtration result after reaction	
Description	Si/Ti in gel	Cyclohexanone conversion/%	Cyclohexanone oxime selectivity/%	Turbidity of filtrate /NTU	State of filtrate
CNFs	-	47.3	0	1	Clear
TS-1 calcined in air	40	99.8	94.7	107	Plenty of white particles suspending
TS-1 calcined in Ar	40	99.8	89.5	115	Plenty of white particles suspending
Blended TS-1/CNFs	40	99.7	92.4	48	Few white particles suspending
Blended TS-1/CNFs after CNFs removal	40	98.9	94.9	112	Plenty of white particles suspending
In-situ crystallized TS-1/CNFs	40	49.7	7.7	15	Clear
In-situ crystallized TS-1/CNFs after CNFs removal	40	49.0	8.0	124	Plenty of white particles suspending
In-situ crystallized TS-1/CNFs	10	96.6	83.2	11	Clear
In-situ crystallized TS-1/CNFs	5	92.1	78.4	4	Clear

Table 1. Catalytic and filtration performance of CNFs, TS-1 and different TS-1/CNFs composites.

The filtration results of the TS-1 samples calcined in air and Ar were also listed in Table 1. The turbidity of the filtrate was relatively high for both samples because of many white particles suspending in the filtrate. The TS-1 crystallites with tiny dimension and isolated sphere morphology (Fig. 1d) undoubtedly cause the penetration of white TS-1 particles into the filtrate.

Both TS-1/CNFs composites showed better filtration performance than pure TS-1 (Table 1). The low filtrate turbidity of the composites especially the in situ crystallized one indicates the rarity of TS-1 or CNFs particles in the filtrate. Clearly, the TS-1 crystallites were scarcely detached from the composite during reaction, implying strong interactions between TS-1 and CNFs.

SEM images of both composites (Fig. 1e-f) illustrate large granules in congeries modality. Nevertheless, isolated TS-1 crystallites are difficult to be found, consistent with the observation of filtration performance. In the high magnification SEM images (Fig. 1g-h), the filamentous CNFs can be easily distinguished from the small TS-1 particles by length. Spherical TS-1 crystallites are uniformly distributed among the network of intertwining CNFs in both composites. The TS-1 particles formed through in situ crystallized manner are of 200~300 nm in diameter, which is comparable to that of pure TS-1 assembled in the blended composite. As pointed out in our previous study,[15] a surface electrostatic effect may exist between TS-1 and CNFs, playing an important role in cohesion of two materials. Furthermore, intensified molecular forces resulting from nano-size effect of two materials may also contribute to the enhanced adhesion of individual TS-1 particles on the surface of CNF filaments. For the in situ crystallized TS-1/CNFs composite, the surface oxygenous functional groups on CNFs[16] may favor the chemical interaction between CNFs and in situ generated TS-1 crystallites via oxygen bridges. Thus, the intra-action among the in situ crystallized composite may be stronger than that of the blended one.

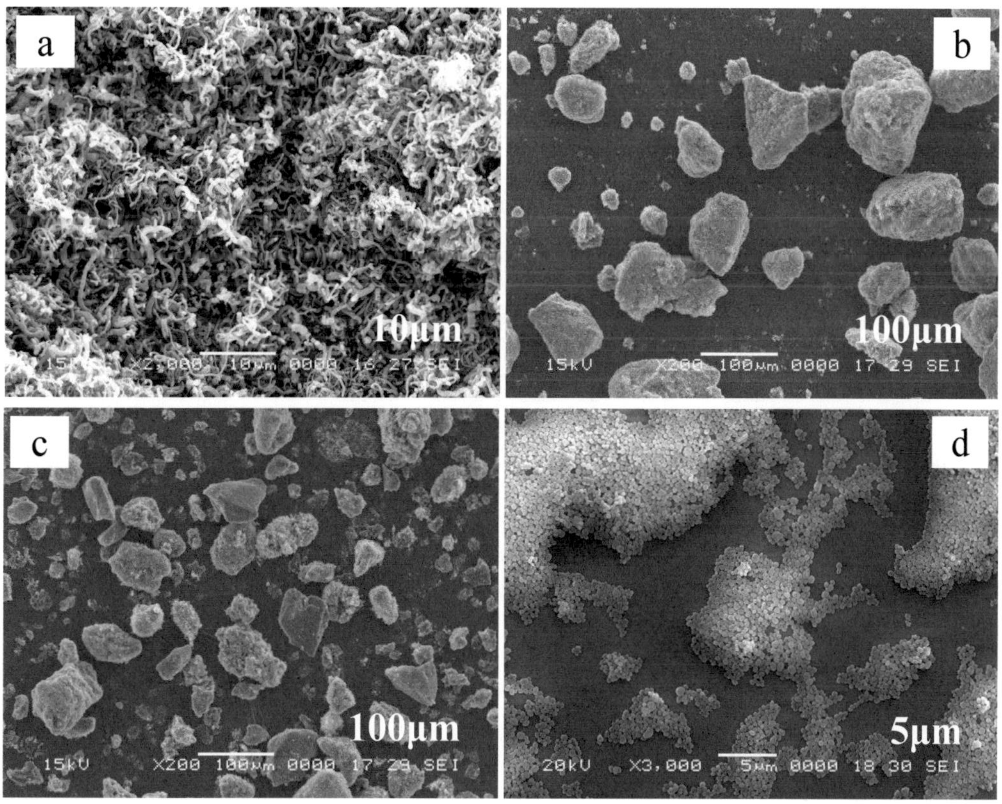

Fig. 1. Continued

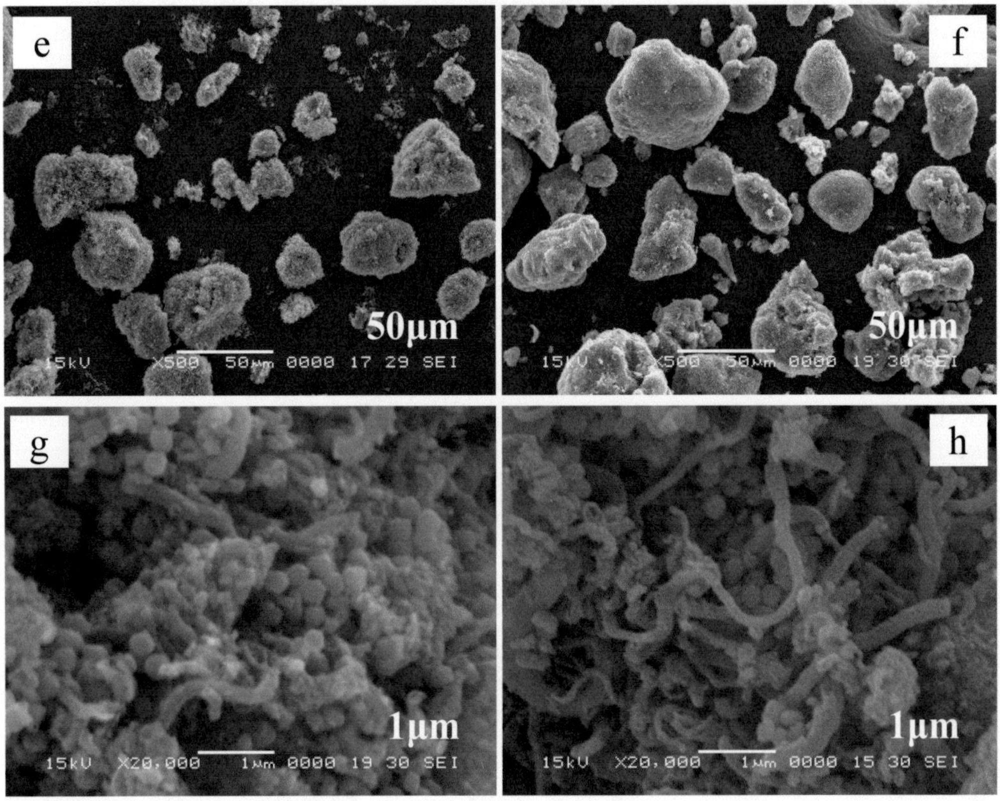

Fig. 1. SEM images of CNFs, TS-1 and TS-1/CNFs composites.
(a) CNFs; (b) CNF agglomerates; (c) CNF agglomerates after reaction; (d) TS-1; (e) Blended TS-1/CNFs composite agglomerates; (f) In situ crystallized TS-1/CNFs composite agglomerates; (g) Blended TS-1/CNFs composite (high magnification); (h) In situ crystallized TS-1/CNFs composite (high magnification).

The change in the resistance with time during slurry filtration under a constant velocity is demonstrated in Fig. 3. Extremely low resistance was undergone while filtrating the slurry containing CNFs. This manifests that the CNFs forming agglomerates with large size and macroporous pores can prevent them from blocking the filter membrane and the cake formed afterwards.[17,18] From the curve of TS-1, we can see that the filtration resistance increases rapidly right from the beginning owing to the blockage of the filter membrane by the tiny TS-1 crystallites. In comparison with the case involving pure TS-1, a significant decrease in the resistance was found during the filtration of slurry containing TS-1/CNFs composite, especially the in situ crystallized one. It is obvious that the filtration behavior of the composites is more similar to that of CNFs rather than TS-1. As having been known, a filter aid can render the cake configured highly porous, thereby facilitating the filtration process.[17] In view of this, the role of CNFs playing in the composites is a kind of filter aid in addition to the support of TS-1. Taking into consideration the network structure of the composites and the substantial affinity between two components, the excellent filtration performance of both TS-1/CNFs composites can be understandable.

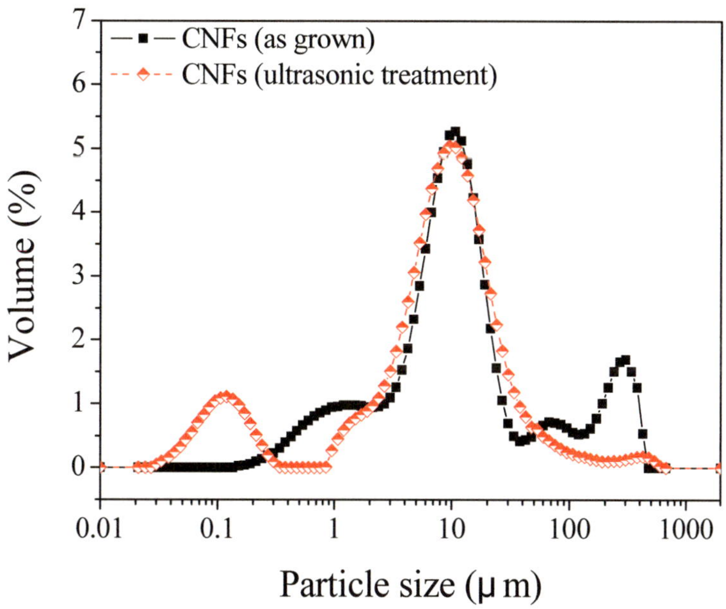

Fig. 2. Particle size distribution of CNFs as grown and after ultrasonic treatment.

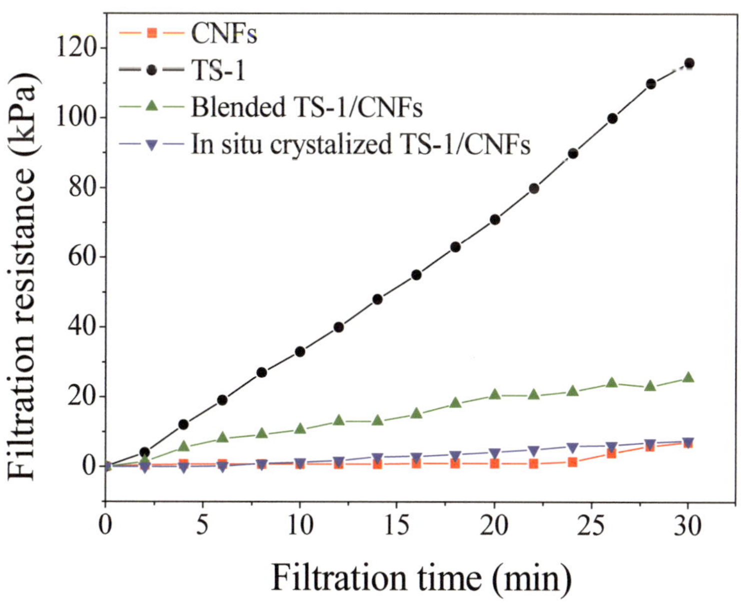

Fig. 3. Change in filtration resistance of different materials with filtration time.

3.2 TS-1 structures of the blended and the in situ crystallized TS-1/CNFs composites

As described above, the catalytic behaviors of pure TS-1 are basically preserved in the TS-1/CNFs composite prepared by simple mechanical blending. Only a slight decrease in selectivity was detected, being ascribed to the unselective decomposition of H_2O_2 induced by CNFs. Hence, we deduce that the interference of CNFs in the reaction is very weak. On the other hand, it is known that the catalytic behaviors of TS-1 depend on the content of lattice Ti in tetrahedral coordination, and the incorporation of Ti into the framework of zeolite generally occurs in the crystallization process.[19,20] However, the non-framework Ti species, such as octahedral Ti species and Ti oxides, are inactive for this type of selective oxidation reaction.[20-24] Therefore, with respect to the in situ crystallized TS-1/CNFs composite, its inferior catalytic performance is perhaps due to the change in Ti coordination states.

The structure of TS-1 for both composite samples was studied using XRD, FT-IR, UV-Vis, and Raman spectroscopy. However, owing to the strong optical adsorption and disturbance of the coexisting CNFs, the characteristics of TS-1 in the composite were difficult to identify.[12] In our measurements, in order to improve the quality of spectroscopy, the CNFs in the composite were removed by burning off carefully in air with a ramp rate of 2 °C/min to 550 °C for 6 h. In fact, oxidative calcination at high temperatures was generally adopted in TS-1 synthesis for template elimination,[19,25,26] in spite of the fact that the decrease in defect sites of titanium silicate material and the ejection of Ti framework species occur with the rise of temperature.[13,27]

XRD patterns show (Fig. 4) the diffraction peaks appearing at 2θ = 7.8°, 8.8°, 23.1°, 23.9°, 24.4° and 29.2°. These peaks are the characteristics of the typical MFI structure of zeolite, and two single reflection peaks at 24.4° and 29.2°, respectively, are an indicative of the presence of Ti in the framework.[28] TiO_2 (anatase) may not exist because the corresponding peaks (2θ = 25.4°, 37.8°, and 48.0°) weren't detected. Moreover, almost no difference between two XRD patterns of both samples implies a considerable homology of the TS-1 crystalline structure in two composites.

The FT-IR spectra (Fig. 5) show a strong band at 1100 cm^{-1} assigned to asymmetric Si-O-Si stretching vibration, a symmetric vibration at 800 cm^{-1} and a bending vibration at 450 cm^{-1}. The second strong band at 550 cm^{-1} is related to the lattice vibration of zeolite MFI structure.[12] The band at 960 cm^{-1} is a diagnostic character of Ti incorporation to zeolite framework and quantitatively associated with Ti content in tetrahedral coordination.[19,29] The infrared spectra of two TS-1 samples derived from different composites are nearly identical with the exception of the intensity of 960 cm^{-1} band. The in situ crystallized TS-1 sample has lower intensity of 960 cm^{-1} band than the blended one.

In UV-Vis spectrum, a band around 210 nm is generally assigned to the isolated tetrahedral Ti species in the framework.[30,31] Anatase TiO_2 can be identified by a band around 330 nm.[21,32] In addition, the band at 280 nm is assigned to the partially condensed hexa-coordinated Ti species (extra-framework Ti).[24,33] After removing CNFs, the UV-Vis spectra of TS-1/CNFs were recorded with referring Si-1 (Fig. 6). The TS-1 of the blended composite exhibits a main band at 220 nm with a weak band at 280 nm. No specific band around 330 nm can be perceived. In contrast, the bands at 220, 280 and 330 nm in the spectrum recorded for the in situ crystallized TS-1 indicates the formation of extra-framework Ti and anatase-like species.

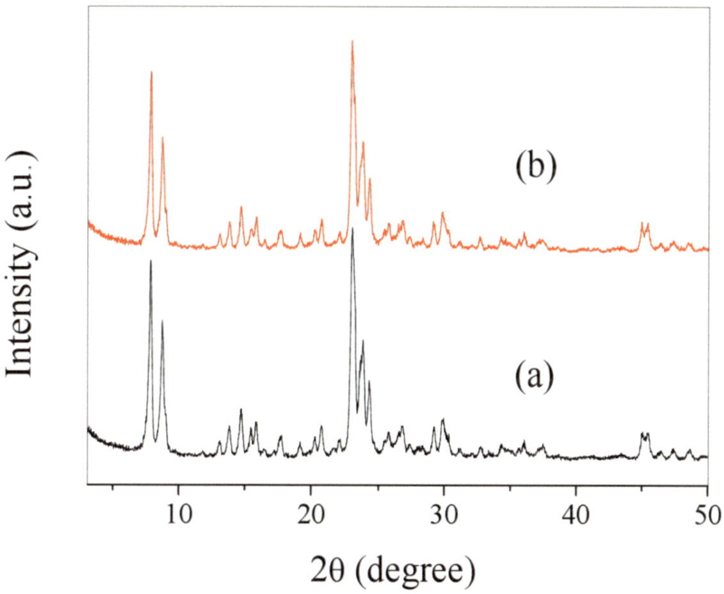

Fig. 4. XRD patterns of TS-1/CNFs composites after removing CNFs.
(a) Blended TS-1/CNFs (Si/Ti=40); (b) In-situ crystallized TS-1/CNFs (Si/Ti=40).

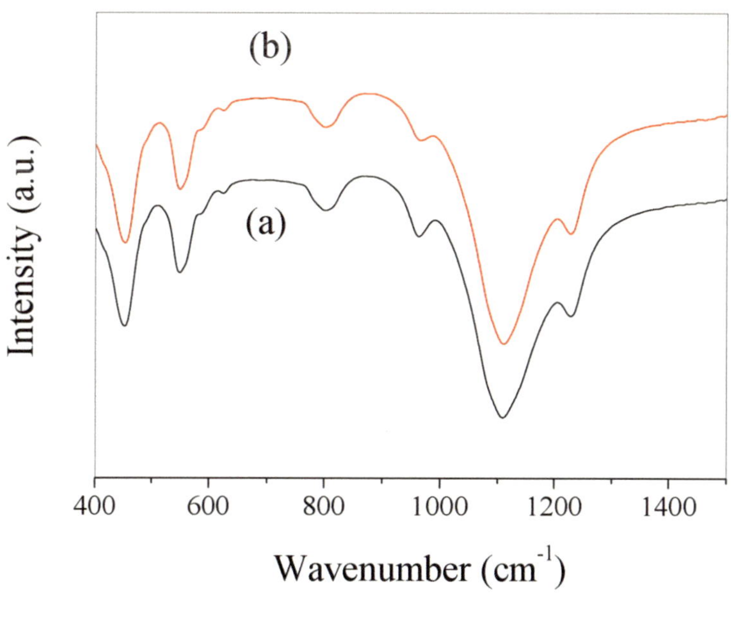

Fig. 5. FT-IR spectra of TS-1/CNFs composites after removing CNFs.
(a) Blended TS-1/CNFs (Si/Ti=40); (b) In-situ crystallized TS-1/CNFs (Si/Ti=40).

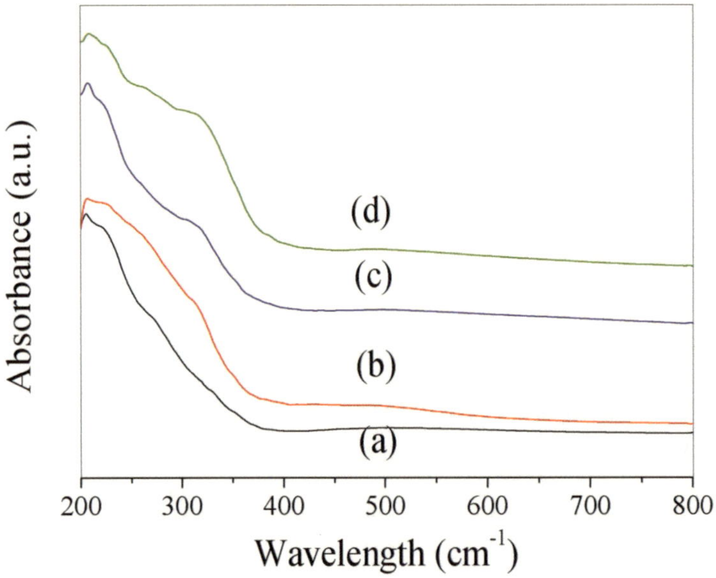

Fig. 6. UV-Vis spectra of TS-1/CNF composites after removing CNFs.
(a) Blended TS-1/CNFs (Si/Ti=40); (b) In-situ crystallized TS-1/CNFs (Si/Ti=40);
(c) In-situ crystallized TS-1/CNFs (Si/Ti=10); (d) In-situ crystallized TS-1/CNFs (Si/Ti=5).

The Raman spectrum (Fig. 7) of TS-1 in the blended TS-1/CNFs shows bands at 290, 375, 467, 800-830, 960, 1090 and 1128 cm^{-1}, respectively, corresponding to the typical structure of TS-1.[34] The bands at 960 and 1128 cm^{-1}, assigned to the asymmetric and symmetric stretching vibrations of TiO_4 tetrahedron, may be resulted from Ti incorporation to zeolite framework.[29,34,35] In the Raman spectrum of the in situ crystallized TS-1/CNFs, besides the typical bands of TS-1 described above, distinct bands are present at 397, 515 and 638 cm^{-1}, which are characteristics of anatase particles.[35]

It has been reported that there are numerous simultaneous and independent equilibria of nucleation and crystal growth during TS-1 crystallization. Meanwhile, there is a competition of Ti ions bonding to the framework with those aggregating to metal oxides. The presence of structure-breaking agents in the synthesis gel can impair the sufficiency of Ti incorporation.[21,24] In the present work, the CNFs with positive-charged surface might induce the sequential adsorption and condensation of negative Ti-OH groups onto them,[17] hence disturbing the oriental isomorphous substitution of Ti for Si. As a result, the non-framework Ti species formed and the lattice Ti content decreased, leading to inferior catalytic performance of the in situ crystallized TS-1/CNFs composite.

3.3 Catalytic and filtration performance of in situ crystallized TS-1/CNFs composites with different Si/Ti ratios

As listed in Table 1, both cyclohexanone conversion and oxime selectivity were increased with a decrease in the molar ratio of Si/Ti in gel from 40 to 10. However, the catalytic activity and selectivity were decreased with a continuous decrease in Si/Ti. Excellent filtration performance of the in situ crystallized TS-1/CNFs composite was still maintained with varying the Si/Ti ratio.

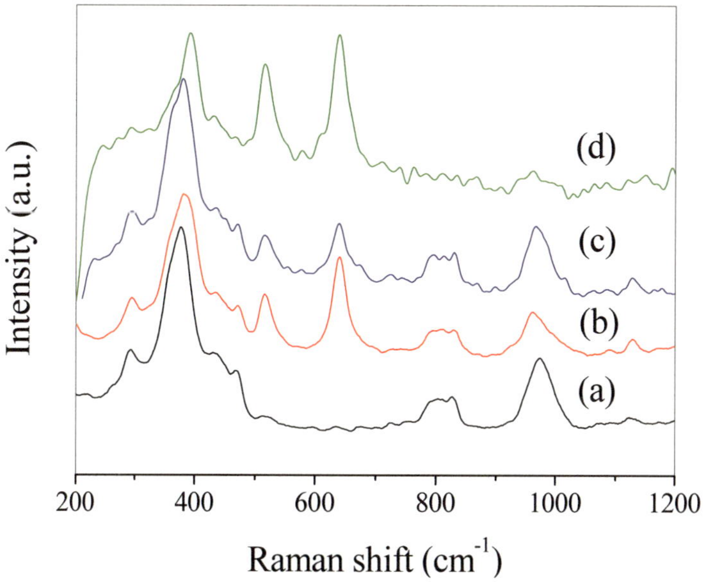

Fig. 7. Raman spectra of TS-1/CNF composites after removing CNFs.
(a) Blended TS-1/CNFs (Si/Ti=40); (b) In-situ crystallized TS-1/CNFs (Si/Ti=40);
(c) In-situ crystallized TS-1/CNFs (Si/Ti=10); (d) In-situ crystallized TS-1/CNFs (Si/Ti=5).

From the results of UV-Vis (Fig. 6) and Raman (Fig. 7) spectroscopy, it can be convinced that the MFI structure of TS-1 and the inclusion of Ti species in lattice are existent indeed in the TS-1/CNFs composites crystallized under different Si/Ti ratios. Meanwhile, the bands at 280 nm and 330 nm on the UV-Vis spectra in Fig. 6c-d are evident especially for the sample of Si/Ti=5, suggesting the generation of non-framework Ti species. The Raman bands of anatase particles are also observable at 397, 515 and 638 cm^{-1} in Fig. 7c-d.

In order to ascertain the lattice Ti content in each TS-1 samples with different Si/Ti ratios, the intensity of 960 cm^{-1} band on FT-IR spectra was normalized on the basis of the 455 cm^{-1} band, which is assigned to structure-insensitive Si-O bending mode of tetrahedral SiO_4 units.[19] The results of I_{960}/I_{455} were given in Table 2, where the value of Si/Ti ratio in solid TS-1 particles has been derived from XRF results.

Sample	Si/Ti in gel	Si/Ti in solid	I_{960}/I_{455}
Blended TS-1/CNFs after CNFs removal	40	44.9	0.38
In-situ crystallized TS-1/CNFs after CNFs removal	40	61.7	0.30
In-situ crystallized TS-1/CNFs after CNFs removal	10	42.8	0.49
In-situ crystallized TS-1/CNFs after CNFs removal	5	21.3	0.36

Table 2. Intensity ratio of I_{960}/I_{455} in TS-1/CNFs composites with different Si/Ti ratios.

As listed in Table 2, the Si/Ti ratios in solid are higher than those in gel, in accordance with the general rule followed in TS-1 synthesis.[36] The value of Si/Ti in solid for the in situ crystallized sample is larger than that for the blended one with the same Si/Ti ratio in gel at 40. Clearly, the addition of CNFs into the gel has hindered the inclusion of Ti ions into the solid TS-1 particles. With respect to the ratio of I_{960}/I_{455}, lower value for the in situ crystallized TS-1 was found than that of the blended one, indicating less content of Ti in the lattice of the in situ crystallized TS-1.[19] With increasing the Si/Ti ratio in gel to 10 for the synthesis of in situ crystallized TS-1, a higher ratio of I_{960}/I_{455} was attained, even exceeding the corresponding value of the blended TS-1 whose Si/Ti in gel is equal to 40. The I_{960}/I_{455} decreased at the Si/Ti ratio of 5.

Obviously, the incorporation of Ti into the lattice of TS-1 is affected by CNFs that is the primary reason for the decrease of the catalytic activity and selectivity. Furthermore, the formation of non-framework Ti species can also render the catalytic property worse. However, it should be pointed out that although the presence of CNFs in the synthesis gel can alter to some extent the proper circumstance for gestating perfect TS-1 crystallites, there still would be measures capable of modifying the synthesis system. For example, it has been reported that a kind of Ti-rich TS-1 catalysts free of extra-framework Ti species was developed by using different ammonium salts as crystallization-mediating agent and a great increase in the catalytic activities in some cases was achieved as expected.[19] In view of the underlying principle proposed in the example, an attempt at counteracting the undesirable effect of CNFs while synthesizing TS-1 is undertaking in our laboratory.

4. Conclusions

TS-1/CNFs composites synthesized by in situ crystallization and mechanical blending methods show excellent filtrating performance. The large dimensions and macroporous structure of the composites constructed by firmly integrating TS-1 crystallites with CNFs can significantly improve the filtration efficiency. However, the introduction of CNFs into the synthesis gel before the crystallization of TS-1 can inhibit the incorporation of Ti ions into the lattice of TS-1, resulting in low catalytic activity and selectivity of the TS-1/CNFs composite catalysts in cyclohexanone ammoximation. Furthermore, the presence of CNFs during TS-1 crystallization especially in the case of high Si/Ti ratio in the gel can give rise to the generation of TiO_2 (anatase) particles and hexa-coordinated titanium species, which can deteriorate the catalytic performance of TS-1. In contrast, the catalytic behavior of pure TS-1 is basically preserved in the blended TS-1/CNFs composite catalyst. Therefore, the blended composite catalyst seems to be more promising than the in situ crystallized one towards industrial applications.

5. Acknowledgements

The authors are grateful to the financial supports from the project of NSFC-20518001/RGC-N_HKUST620/05, international cooperation project of Chinese Ministry of Science and Technology (2007DFC61690), 111 project of Chinese Education Ministry (B08021), and opening project of State Key Laboratory of Chemical Engineering (SKL-ChE-08C05).

6. References

Perego, C.; Carati, A.; Ingallina, P.; Mantegazza, M. A.; Bellussi, G. Production of titanium containing molecular sieves and their application in catalysis. *Appl. Catal., A* 2001, *221*, 63-72.

Saxena, S.; Basak, J.; Hardia, N.; Dixit, R.; Bhadauria, S.; Dwivedi, R.; Prasad, R.; Soni, A.; Okram, G. S.; Gupta, A. Ammoximation of cyclohexanone over nanoporous TS-1 using UHP as an oxidant. *Chem. Eng. J.* 2007, *132*, 61-66.

Zhong, Z. X.; Xing, W. H.; Liu, X.; Jin, W. Q.; Xu, N. P. Fouling and regeneration of ceramic membranes used in recovering titanium silicalite-1 catalysts. *J. Membr. Sci.* 2007, *301*, 67-75.

Li, G.; Jin, C. Z.; Wang, X. S. The synthesis and application of a titanium silicalite composite catalyst. CN Patent 1 554 483, 2004.

Schodel, R.; Birke, P.; Geyer, R.; Kraak, P.; Muller, W.; Neubauer, H. D.; Pester, R.; Vogt, F.; Wendlandt, K. P. Oxidation catalysts, US Patent 5 736 479, 1998.

Yip, A. C. K.; Lam, F. L. Y.; Hu, X.; Li, P.; Yuan, W. K. Study on the synthesis of clay-based titanium silicalite-1 catalytic composite. *Ind. Eng. Chem. Res.* 2009, *48*, 5266-5275.

Birke, P.; Kraak, P.; Pester, R.; Schodel, R.; Vogt, F. Carbon supported TS-1 catalyst. *Stud. Surf. Sci. Catal.* 1994, *83*, 425-432.

De Jong, K. P.; Geus, J. W. Carbon nanofibers: catalytic synthesis and applications. *Catal. Rev. Sci. Eng.* 2000, *42*, 481-510.

Chinthaginjala, J. K.; Seshan, K.; Lefferts, L. Preparation and application of carbon-nanofiber based microstructured materials as catalyst supports. *Ind. Eng. Chem. Res.* 2007, *46*, 3968-3978.

Tessonnier, J. P.; Rosenthal, D.; Hansen, T. W.; Hess, C.; Schuster, M. E.; Blume, R.; Girgsdies, F.; Pfander, N.; Timpe, O.; Su, D. S.; Schlogl, R. Analysis of the structure and chemical properties of some commercial carbon nanostructures. *Carbon* 2009, *47*, 1779-1798.

Pinilla, J. L.; Lázaro, M. J.; Suelves, I.; Moliner, R.; Palacios, J. M. Characterization of nanofibrous carbon produced at pilot-scale in a fluidized bed reactor by methane decomposition. *Chem. Eng. J.* 2010, *156*, 170-176.

Zhao, Q.; Li, P.; Li, D. Q.; Zhou, X. G.; Yuan, W. K.; Hu, X. J. Synthesis and characterization of titanium silicate-1 supported on carbon nanofiber. *Microporous and Mesoporous Mater.* 2008, *108*, 311-317.

Fejes, P.; Nagy, J. B.; Halasz, J.; Oszko, A. Heat-treatment of isomorphously substituted ZSM-5 zeolites and its structural consequences: An X-ray diffraction, ^{29}Si MAS-NMR, XPS and FT-IR spectroscopy study. *Appl. Catal., A* 1998, *175*, 89-104.

Shibata, M.; Gabelica, Z. Synthesis of MFI titanosilicates from methylamine-TPABr media. *Zeolites* 1997, *19*, 246-252.

Zhao, Q.; Li, P.; Li, D. Q.; Zhang, L. M.; Zhou, X. G.; Yuan, W. K.; Hu, X. J. Filtration and catalytic performance of blended TS-1/CNF composite catalyst. *J. Chem. Ind. Eng. (China)* 2008, *59*, 2000-2006.

Li, P.; Zhao, T. J.; Zhou, J. H.; Sui, Z. J.; Dai, Y. C.; Yuan, W. K. Characterization of carbon nanofiber composites synthesized by shaping process. *Carbon* 2005, *43*, 2701-2710.

Heertjes, P. M.; Zuideveld, P. L. Clarification of liquids using filter aids. *Powder Tech.* 1978, *19*, 17-64.

Wenmakers, P. W. A. M.; van der Schaaf, J.; Kuster, B. F. M.; Schouten, J. C. Enhanced liquid-solid mass transfer by carbon nanofibers on solid foam as catalyst support. *Chem. Eng. Sci.* 2010, *65*, 247-254.

Fan, W.; Duan, R. G.; Yokoi, T.; Wu, P.; Kubota, Y.; Tatsumi, T. Synthesis, crystallization mechanism, and catalytic properties of titanium-rich TS-1 free of extraframework titanium species. *J. Am. Chem. Soc.* 2008, *130*, 10150-10164.

Serrano, D. P.; Uguina, M. A.; Ovejero, G.; van Grieken, R.; Camacho, M. Evidence of solid-solid transfomations during the TS-1 crystallization from amorphous wetness impregnated SiO-TiO_2 xerogels. *Microporous Mater.* 1996, *7*, 309-321.

Fan, W.; Fan, B.; Shen, X.; Li, J.; Wu, P.; Kubota, Y.; Tatsumi, T. Effect of ammonium salts on the synthesis and catalytic properties of TS-1. *Microporous Mesoporous Mater.* 2009, *122*, 301-308.

Li, G.; Wang, X. S.; Guo, X. W.; Liu, S.; Zhao, Q.; Bao, X. H.; Lin, L.W. Titanium species in titanium silicalite TS-1 prepared by hydrothermal method. *Mater. Chem. Phys.* 2001, *71*, 195–201.

Zecchina, A.; Bordiga, S.; Lamberti, C.; Ricchiardi, G.; Scarano, D.; Petrini, G.; Leofanti, G.; Mantegazza, M. Structural characterization of Ti centres in Ti-silicalite and reaction mechanisms in cyclohexanone ammoximation. *Catal. Today* 1996, *32*, 97-106.

Khouw, C. B.; Davis, M. E. Catalytic activity of titanium silicalites synthesized in the presence of alkali metal and alkaline earth ions. *J. Catal.* 1995, *151*, 77-86.

Van der Pol, A. J. H. P.; Van Hooff, J. H. C. Parameters affecting the synthesis of titanium silicalite-1. *Appl. Catal., A* 1992, *92*, 93-111.

Van der Pol, A. J. H. P.; Verduyn, A. J.; Van Hooff, J. H. C. Why are some titanium silicalite-1active and others not? *Appl. Catal., A* 1992, *92*, 113-130.

Bhaumik, A.; Samanta, S.; Mal, N. K. Highly active disordered extra large pore titanium silicate. *Microporous Mesoporous Mater.* 2004, *68*, 29-35.

Taramasso, M.; Perego, G.; Notari, B. Preparation of porous crystalline synthetic material comprised of silicon and titanium oxides. US Patent 4 410 501, 1983.

Ricchiardi, G.; Damin, A.; Bordiga, S.; Lamberti, C.; Spano, G.; Rivetti, F.; Zecchina, A. Vibrational structure of titanium silicate catalysts. A spectroscopic and theoretical study. *J. Am. Chem. Soc.* 2001, *123*, 11409-11419.

Reddy, J. S.; Kumar, R. Synthesis, characterization, and catalytic properties of a titanium silicate, TS-2, with MEL structure. *J. Catal.* 1991, *130*, 440-446.

Wang, X. S.; Guo, X. W. Synthesis, characterization and catalytic properties of low cost titanium silicalite. *Catal. Today* 1999, *51*, 177-186.

Huybrechts, D. R. C.; Buskens, P. L.; Jacobs, P. A. Physicochemical and catalytic properties of titanium silicalites. *J. Mol. Catal.* 1992, *71*, 129-147.

Blasco, T.; Camblor, M. A.; Corma, A.; Perez-Pariente, J. The state of Ti in titanoaluminosilicates isomorphous with zeolite β. *J. Am. Chem. Soc.* 1993, *115*, 11806-11813.

Deo, G.; Turek, A. M.; Wachs, I. E.; Huybrechts, D. R. C.; Jacobs, P. A. Characterization of titania silicates. *Zeolites* 1993, *13*, 365-373.

Gao, H. X.; Lu, W. K.; Chen, Q. L. Characterization of titanium silicalite-1 prepared from aqueous $TiCl_3$. *Microporous Mesoporous Mater.* 2000, *34*, 307-315.

Gontier, S.; Tuel, A. Synthesis of titanium silicalite-1 using amorphous SiO_2 as silicon source. *Zeolites* 1996, *16*, 184-195.

21

Au/TiO₂ Hierarchical Nanofibers Heterostructure: Controllable Synthesis and Enhanced Photocatalytic Performances

Chao Pan[1] and Li Dong[2]
[1]*College of Science, Dalian Ocean University, Dalian, Liaoning*
[2]*College of Science, Dalian Nationalities University, Dalian, Liaoning*
[1,2]*China*

1. Introduction

Titanium dioxide, because of its unique optical, electrical and chemical attributes, is a widely studied material. It is well-known for its use in photocatalysis, gas sensors and dye sensitized solar cells and batteries. (Fujishima & Honda, 1972,Garzella et al., 2000,Meng et al., 2003,Ohko et al., 1998) Titanium dioxide (TiO_2) was extensively studied for its applications in high performance photovoltaics. Now, the greatest challenge in TiO_2-based photocatalysis is obtaining a highly efficient catalyst. Many efforts have been made to enhance the activity and it is found that the morphology is one of the important factors influencing the activity, which may lead to interesting shape-dependent properties. (Li et al., 2007)

In various morphologies, the hierarchical 3D architectures have been regarded as the more interesting structures due to the greater number of active sites than other 2D or 1D architectures.(Wu et al., 2006,Hu et al., 2007) Therefore, Preparation of high-quality hierarchical TiO_2 nanostructure of desired morphology is of great fundamental and technological interest. Up to now, various methods, including hydrothermal,(Sinha et al., 2009) solvothermal,(Zhu et al., 2009) vapor deposition methods, (Wu & Yu, 2004)as well as electrochemical approaches,(Zhang et al., 2010) have been used to synthesize the TiO_2 materials with different morphologies and architectures, such as nanotubes,(Su et al., 2008) hollow nanospheres,(Li et al., 2006) nanofibers,(Pan et al., 2007) nanorods,(Xu et al., 2008) nanocolumn,(Li et al., 2008) and dandelion-structured spheres.(Bai et al., 2008) Though there are many effective ways to enhance the photocatalytic activity, this approach also has some disadvantage. In particular, direct fabrication of complex hierarchical nanostructures with controlled morphologies, crystalline orientations, grain size, and surface architectures is one of the most challenging topics because of their promising functions. To overcome the above inconvenience, many attempts have been made to effectively fabricate TiO_2 photocatalysts, for example, by electrospun Nanofiber-based hierarchically organized TiO_2 nanocomposite.

Electrospinning is a process by which high static voltages are used to produce nano and microscale fibers, with the fiber diameter in the range from less than 10 nm to over several micrometers. The variety of materials and fibrous structures that can be electrospun allow for the incorporation and optimization of various functions to the nanofiber, either during spinning or through post-spinning modifications.(Teo & Ramakrishna, 2009,Pan et al., 2010) Recent efforts have made this technique a new platform for fabricating complex nanostructures having controllable hierarchical features. For example, a number of groups have demonstrated that electrospinning allows one to maneuver the secondary structures of individual fibers as well as to increase their structural complexity.(Ostermann et al., 2006,T.S.He et al., 2010,Cao et al., 2010,Wang et al., 2008)

Here we demonstrate that electrospinning and sputter coating technique can be combined to provide a simple route to formed the 3D hierarchical structural Au/TiO_2 nanofibers. The structure of samples prepared at different Au depositing time were investigated. The relationship between photocatalytic activity and the structure is discussed in detailed.

2. Experimental

2.1 Materials

Poly (styrene-co-methacrylie aicd) (SMAI) was synthesized by our group. Titanium tetraisopropoxide ($Ti(OC_4H_9)_4$), N,N-dimethyl formamide (DMF), tetrahydrofuran(THF) and acetic acid (HAc) were purchased from Shanghai Guanghua reagent Plant. These chemicals were used without further purification. Distilled water was used as hydrolysis agent. DW-P403-1AC High Voltage supply was obtained from Tianjin Dongwen High Voltage Power Supply Go. A TS2-60 multi-syringe pump was obtained from Baoding Longer Precision Pump Co., ltd.

2.2 Preparation of Au/TiO_2 nanofibers

2.2.1 Synthesis of copolymer SMAI

A brief of synthesis process is as follows: The boiling medium polymerization techniquewas used for the synthesis of a copolymer SMAI. 6 mL of MAA monomer, 600 mL water and 0.6 g KPS was added to the three-necked flask, and the stirring speed was set at 300 rpm. The mixture was raised to reflux, and the medium had boiled. The reaction was stopped after 2 h. A particular describe view the references.(Gu et al., 2007)

2.2.2 SMAI/ $Ti(OC_4H_9)_4$ nanofibres

In a typical procedure, 2.48 g of titanium tetraisopropoxide ($Ti(OC_4H_9)_4$) was mixed with 3 mL of acetic acid. After 30 min, this solution was removed from the glove box and added to 9.87 mL of THF/DMF=6:4 (wt) mixture containing 1 g of SMAI, followed by magnetic stirring for 1 h (with the solution held in a capped bottle). The mixture was immediately loaded into a plastic syringe (5ml) equipped with a 6 gauge stainless needle. The needle was connected to a high-voltage supply that is capable of generating DC voltages up to 15 kV. The feeding rate for the precursor solution was controlled using a syringe pump. A copper grid was placed 15 cm below the tip of the needle to collect the nanofibers. The as-spun nanofibers were left in air for 5 h to allow the hydrolysis of $Ti(OC_4H_9)_4$ to go to completion.

2.2.3 TiO_2 naofibers

The SMAI/ $Ti(OC_4H_9)_4$ fibers were heated at 130 °C for 2 h and calcined at 450 °C for 2 h in air at a heating rate of 0.5 °C •min^{-1} from 130 °C to 450 °C to completely eliminate the organics. After cooling to room temperature, TiO_2 fibers were obtained.

2.2.4 Au/TiO_2 nanofibers

A Cu plate with SMAI/ $Ti(OC_4H_9)_4$ composite nanofibers on it was clipped by a holder(Hitachi E-1010 Ion Sputter). The chamber was evacuated to 1.33 Pa (10^{-2}torr) by a mechanical pump, and the sputtering current was controlled with 15mA, for sputtering of Au started. The process time was adjustabled at 5 sec, 40 sec, 2min, and 3 min. Finally, the SMAI was selectively removed from these nanofibers by treating them in air at 450 °C for 1 h and then cooled down to room temperature. The morphology and diameter of Samples were measured with SEM (S-3000 of Hitachi).

2.3 Photocatalytic activity measurements

The photocatalytic activities of the TiO_2: Au hybrid nanofibers for the photodecomposition of acetaldehyde gas were investigated. This was carried out at room temperature. An O_2(20%)-N_2 gas mixture adjusted to a relative humidity of 50% was used to fill a 500 mL Pyrex glass photocatalytic reaction vessel. Then 8mL acetaldehyde (1 vol% in N_2, Takachiho, Japan) were injected into the reactor using a syringe. After equilibrium in the dark for 2 h, the sample was irradiated with UV light with an intensity of 1 mW cm^2, and the concentrations of CO_2 and acetaldehyde were followed by gas chromatography (GC-8A, Shimadzu). The sample size was 2.5×2.5 cm^2, and the weight of the Au-TiO_2 hybrid nanofibers was 5.3±0.1mg.

3. Results and discussion

3.1 Characterization of prepared TiO_2 nanofibers and Au/TiO_2 nanofibers

Figure 1a shows the SEM image of TiO_2/SMAI composite nanofibers electrospun from a DMF/THF solution. The average diameter of this sample was 470-500 nm. As the SMAI was selectively removed by burning the sample in air at 450 °C for 1 h, TiO_2 nanofibers with 200-300 nm and smooth surface were synthesized (Fig. 1b)The formation of the TiO_2 nanofibers is considered to involve the following steps during the fabrication. Firstly, when the $Ti(OC_4H_9)_4$ containing polymer solution was ejected from the orifice of the syringe, a composite fiber composed of SMAI and $Ti(OC_4H_9)_4$ forms. Then the $Ti(OC_4H_9)_4$ in the fiber was hydrolyzed to titania in the SMAI fibers by moisture in the air. The following calcination at 450 °C removes the polymer and crystallizes the titania. As a result, the TiO_2 fibers appear. Figure 1c shows the micrograph of TiO_2 nanofibers coated with Au and then annealed at 450 °C for 1hr. In the SEM micrographs, the sintering of Au nanoparticles takes place, resulting in organization. It is clear that the islands are formed by Au.

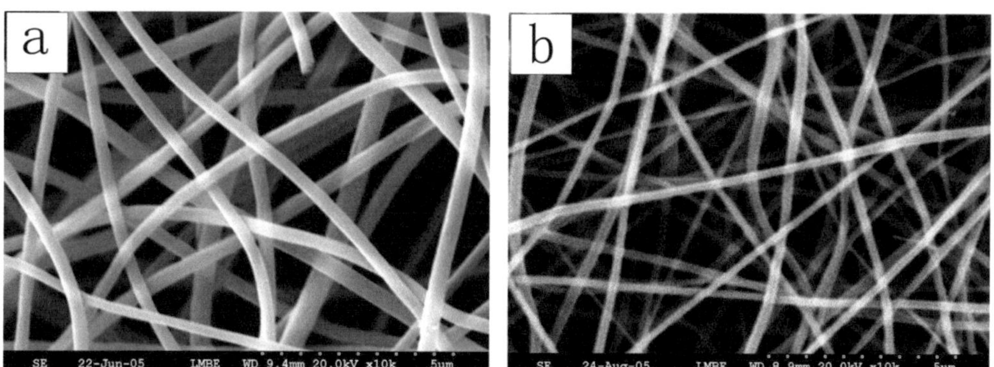

Fig. 1. (Continued)

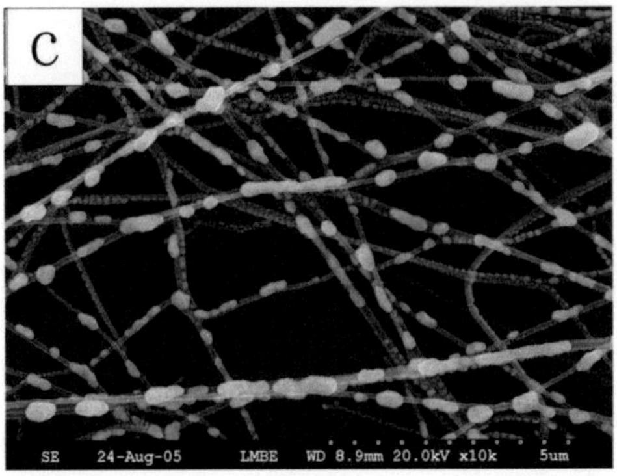

Fig. 1. (a) SEM image of TiO_2/SMAI nanofibers that were electrospun from an DMF/THF solution. (b) SEM image of the same sample after it had been calcined in air at 450 °C for 1 h. (c) SEM image of the same.

3.2 Characterization of the Au –TiO_2 nanofibersnanofibers surface

Figure 2 shows the XPS spectra for nanofibersnanofiberss of TiO_2 with Au. The peaks are marked with the figure to which they correspond. Figure 2a shows prominent peaks at

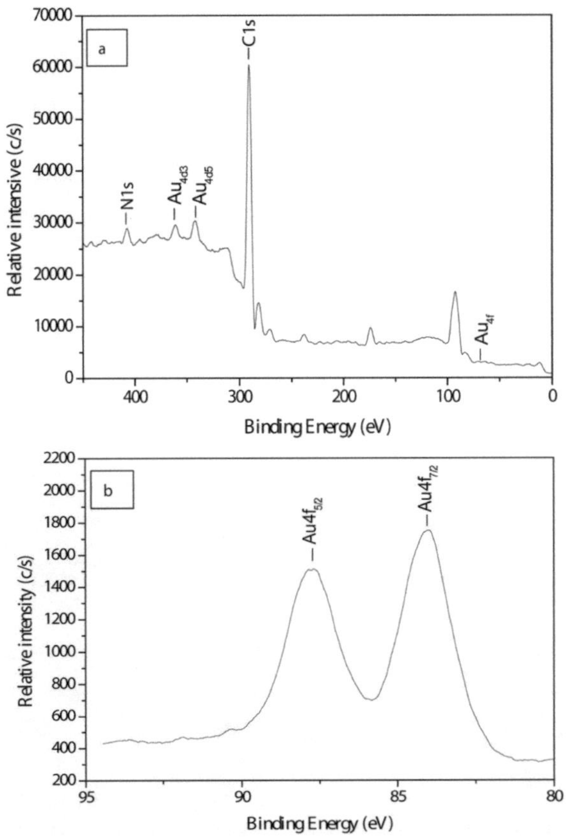

Fig. 2. a: Wide scanning XPS spectra; b: High energy resolution XPS of Au4f.

400.0 (N1s) and 284.8 eV (C1s). The peaks at 360 eV (Au4d3) and 345 eV (Au4d5), which are well visible, and small peaks at 84 eV (Au4f7/2) and 88 eV (Au4f5/2) indicate the presence of gold. (Database)The Au4f gold doublet size is on the order of the noise but is distinguishable at higher energy resolution (see 2b). Figure2b show the high energy resolution XPS spectra of the Au –TiO_2 nanofibers described above. The Au4f doublet at 84 (Au4f7/2) and 88 eV (Au4f5/2) is clearly resolved. The characteristic doublet at 84 (Au4f7/2) and 88 eV (Au4f5/2) together with the peaks at 360 (Au4d3) and 345 (Au4d5) eV, respectively, confirm the presence of metallic gold.(Ting et al., 1994) This indicates that that metallic gold was successfully converted into TiO_2 nanofibers surface.

3.3 The influence of the ion beam sputtering time on the morphology of fibers and the amount of Au loading

To investigate the influence of deposition time on the morphology of the nanofibers, 5s, 40s, 2min, and 3min of ion beam sputtering time were adjusted, respectively and the SEM images were shown in Figure 3.

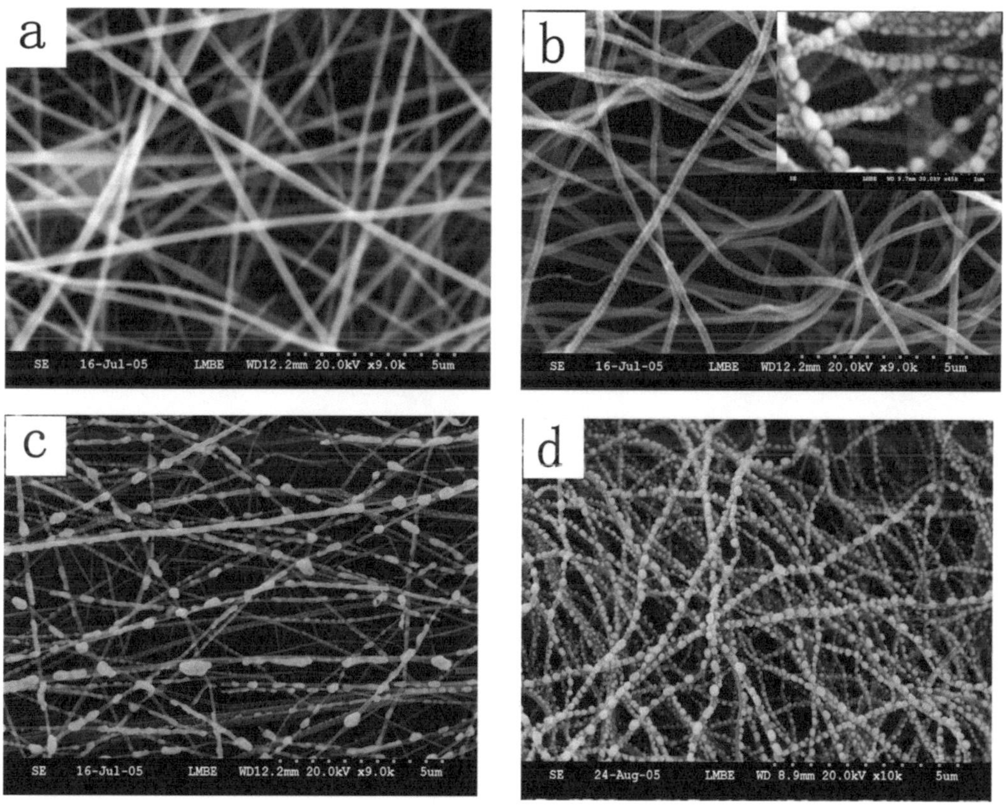

Fig. 3. SEM micrographs for Au sputtered for various lengths of time. (a) 5s, (b) 40s, (c) 2min and (d) 3min.

The electric conductivity of TiO_2 nanofibers was not good. As shown in Fig. 3a, there was a charging phenomenon because the content of Au loading after 5sec was not enough to conduct the electrons. The charging phenomenon was reduced with the increasing of Au deposition time, Figs. 3 (b) and (c). Meanwhile, some islands were formed on the surface. The Au islands became more with increasing of time. The Au islands were stacked with the clusters of the Au nanoparticles. The Au nanoparticles were implanted into TiO_2 nanofibers with a long sputtering time such as 3min (Fig. 3 d). The Au islands were in contact with each other, and heating process provided the energy for self-diffusion. Compared with Fig. 3b-c, there are fewer voids on the TiO_2 surface. Additionally, this clearly indicates that the morphology of the Au/TiO_2 nanofibers is dependent on the sputtering time. Simultaneity, we can see obviously that the content of Au loading increased with increasing of sputtering time.

The above results of Au deposition could be explained by a thin-film growth mode.(Kasuga et al., 1998,Lahav et al., 2003) As shown in Fig. 4, the period experienced two stages: (a) island, and (b) layer. When the smallest stable clusters are deposited on the TiO_2 surface and grow in three dimensions to form islands, it is called island growth mode, and the deposited atoms or molecules are more strongly bound to each other than to the TiO_2 surface. This usually happens in the system of Au on insulators. In our system, Au nanoparticles are stable to bind to each other, and TiO_2 nanofibers are insulating substrate. For the beginning there were many Au nanoparticles formed on the surface as show in Fig. 4(a), this is Au island form stage. Then the particles became larger and stacked with each other when the time of deposition was increased to 3 min as show in Fig. 4 (b). And the stacked layer became much thicker, this is Au layer form stage.

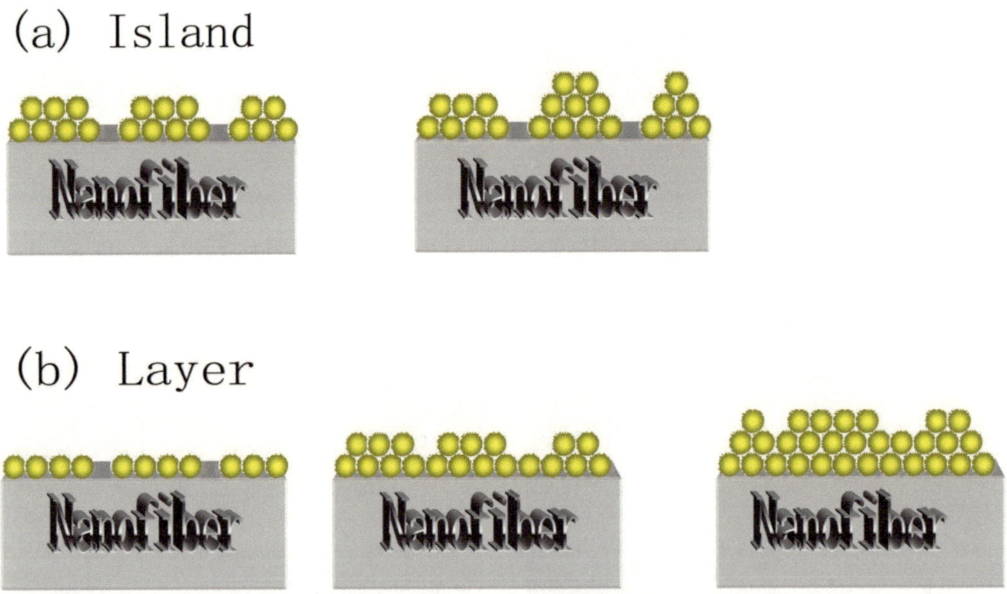

Fig. 4. Basic modes of Au thin-film growth: (a) island, and (b) layer.

3.4 Photocatalytic activity

We investigated the photocatalytic decomposition of acetaldehyde on the Au/TiO_2 (3min) complex nanofibers photocatalyst. A typical result for the photodecomposition of acetaldehyde is shown in Figure 5a.

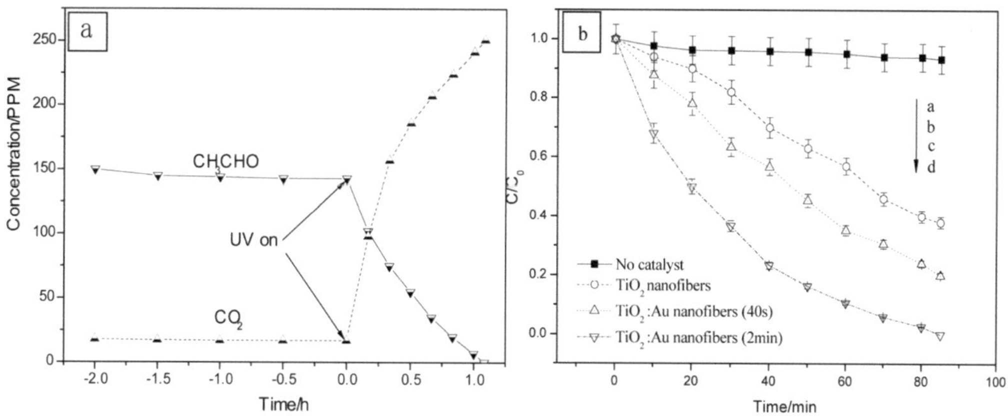

Fig. 5. a: Plot of concentration versus irradiation time for the degradation of acetaldehyde photocatalyzed by a TiO_2: Au nanofibers (3min). The sample area was 6.25 cm^2, and the light intensity was 1 mW cm^{-2}. b: Time courses for UV photodegradation of acetaldehyde. Curve a: Without photocatalyst; b: TiO_2 nanofibers; c: Au/TiO_2 nanofibers (40s); d: Au/TiO_2 nanofibers (2min).

The concentration of acetaldehyde dropped slightly in the dark but decreased quickly as a function of irradiation time. The decrease in acetaldehyde concentration could be fitted with a first-order kinetic model, which is a common phenomenon in diffusion-controlled photocatalytic processes.(Gibson et al., 2001) A parallel increase in the concentration of CO_2 was observed. There were no traces of acetaldehyde (in the photoreactor) after about 70 min reaction. Accordingly, the expected stoichiometric CO_2 concentration was generated. Indicating the complete mineralization of acetaldehyde on the TiO_2: Au fibrous photocatalyst under UV illumination. Figure 5b shows the change of concentration of acedaldehyde gas over time under UV irradiation in the presence of different TiO_2 materials. Control experiments indicated that the photocatalytic reaction hardly proceeded in the absence of TiO_2 (curve a). After 70 min, up to 54% (curve b) of the gaseous acetaldehyde was degraded in the presence of TiO_2 nanofibers. For the Au/TiO_2 nanofibers, we studied the influence of the Au depositing time on the photocatalytic decomposition of acetaldehyde. For 40s of the Au depositing time (curve c), after 70 min, up to 69.4% of the gaseous acetaldehyde was degraded. For 2 min of the Au depositing time, ca. 94% gaseous acetaldehyde was degraded to CO_2 and H_2O under the same conditions (curve d). This indicates that the Au depositing time influenced the photocatalytic effect, the speed of decomposition increased with the increasing of the Au depositing time. The photocatalytic activity of the Au/TiO_2 nanofibers was higher than TiO_2 nanofibers. The main reasons for that change are as following: Firstly, the open 3-D structure of the nanofibers photocatalyst has a beneficial effect on its activity.(Pan et al., 2009) Secondly, the Au cluster deposited on TiO_2 surface functions as the electron trap center, and the perimeters between Au and TiO_2 act as the active site for acetaldehyde decomposition. Thirdly, the effect of Au nanoparticles covered the surface of TiO_2 nanofibers enhanced as function of the sputtering time. This process inhibited ulteriorly the recombination rate of the photo-produced electrons and holes, as shown in Scheme 1. The three reasons complementary promoted the photocatalytic effect of the Au/TiO_2 nanofibers.

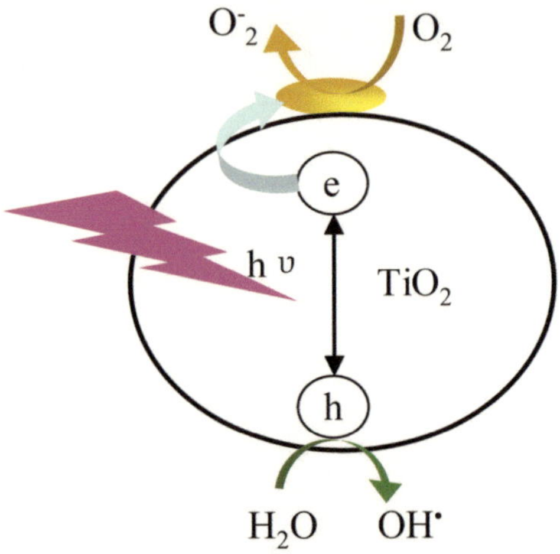

$$Au/TiO_2 \xrightarrow{h\upsilon} Au/TiO_2(e^- + h^+) \quad (1)$$

$$e^- + O_2 \rightarrow {}^{\bullet}O_2^- \quad (2)$$

$${}^{\bullet}O_2^- + H^+ \rightarrow HO_2^{\bullet}$$

(4)

$$h^+ + H_2O \rightarrow {}^{\bullet}OH$$

$$CH_3CHO + {}^{\bullet}OH \rightarrow CH_3C^{\bullet}O + H_2O \quad (5)$$

$$CH_3C^{\bullet}O + O_2 \rightarrow CH_3(CO)OO^{\bullet} \quad (6)$$

$$CH_3(CO)OO^{\bullet} + HO_2^{\bullet} \rightarrow CH_3(CO)OH \quad (7)$$

$$CH_3(CO)OH + {}^{\bullet}OH \rightarrow CO_2 + H_2O + CH_3^{\bullet} \quad (8)$$

Scheme 1. Interfacial charge transfer process of Au-doped TiO_2 nanocomposites under the irradiation of UV light.

4. Conclusions

In conclusion, Au/TiO_2 nanofibers were successfully produced by a combination of electrospinning and sputter coating technique. The Au morphology on the obtained TiO_2 nanofibers is dependent on the depositing time, the Au surface becomes more compact and the wall-structure seems more complete with increasing Au depositing time from 40s to 3min. The thin-film growth mechanism of Au on TiO_2 nanofibers can be explained in terms of islands mode which is usually applied to the systems of metal on insulator. Photocatalytic results showed that the photocatalytic activity of the Au/TiO_2 nanofibers was higher than TiO_2 fibers. In addition, the interconnected Au/TiO_2 nanofibers structure also opens up the opportunity of Dye-sensitized Solar Cells and photocatalytic functions in the same material.
Acknoledgements: This work was supported by National Nature Science Foundation of China (Grant No. 50773010, Grant No. 10705007).

5. References

A. Fujishima,K. Honda. Electronchemical photolysis of water at a semiconductor electrode. *Nature* 238, 37-38 (1972).

C. Garzella, E. Comini, E. Tempeti, C. Frigerib,G. Sberveglieri. TiO_2 thin films by a novel sol–gel processing for gas sensor applications. *Sensors & Actuators B-chem.* 68, 189-196 (2000).

Q. B. Meng, K. Takahashi, X. T. Zhang, I. Sutanto, T. N. Rao, O. Sato,A. Fujishima. Fabrication of an Efficient Solid-State Dye-Sensitized Solar Cell. *Langmuir* 19, 3572-3574 (2003).

Y. Ohko, D. A. Tryk, K. Hashimoto,A. Fujishima. Autoxidation of acetaldehyde initiated by TiO_2 photocatalysis under weak UV illumination. *J. Phys. Chem. B* 102, 2699-2704 (1998).

H. X. Li, Z. F. Bian, J. Zhu, D. Q. Zhang, G. S. Li, Y. N. Huo, H. Li,Y. F. Lu. Mesoporous Titania Spheres with Tunable Chamber Stucture and Enhanced Photocatalytic Activity. *J. Am. Chem. Soc.* 129, 8406-8407 (2007).

C. Z. Wu, Y. Xie, L. Y. Lei, S. Q. Hu,C. Z. OuYang. Synthesis of New-Phased VOOH Hollow "Dandelions" and Their Application in Lithium-Ion Batteries. *Adv. Mater* 18, 1727-1732 (2006).

X. L. Hu, J. C. Yu,J. M. Gong. Fast Production of Self-Assembled a-Fe_2O_3 Hierarchical Nanoarchitectures. *J. Phys. Chem. C.* 111, 11180-11185 (2007).

A. K. Sinha, S. Jana, S. Pande, S. Sarkar, M. Pradhan, M. Basu, S. Saha, A. Pal,T. Pal. New hydrothermal process for hierarchical TiO_2 nanostructures. *CrystEngComm* 11, 1210-1212 (2009).

J. Zhu, S. H. Wang, Z. F. Bian, C. L. Cai,H. X. Li. A facile synthesis of hierarchical flower-like TiO_2 with enhanced photocatalytic activity *Res Chem Intermed* 35, 769-777 (2009).

J. J. Wu,C. C. Yu. Aligned TiO_2 Nanorods and Nanowalls. *J. Phys. Chem. B* 108, 3377–3379 (2004).

J. Zhang, J. H. Bang, C. Tang,P. V. Kamat. Tailored TiO_2-$SrTiO_3$ Heterostructure Nanotube Arrays for Improved Photoelectrochemical Performance. *ACS Nano* 4, 387–395 (2010).

H. L. Su, Q. Dong, J. Han, D. Zhang,Q. X. Guo. Biogenic Synthesis and Photocatalysis of Pd-PdO Nanoclusters Reinforced Hierarchical TiO2 Films with Interwoven and Tubular Conformations. *Biomacromolecules* 9, 499-504 (2008).

X. X. Li, Y. J. Xiong, Z. Q. Li,Y. Xie. Large-Scale Fabrication of TiO_2 Hierarchical Hollow Spheres. *Inorg. Chem* 45, 3493-3495 (2006).

C. Pan, C. Sun, H. M. Wei, G. Z. Han, J. Z. Zhang, A. Fujishima,Z. Z. Gu. Bio-inspired titanium dioxide film with extremely stable super-amphilicity. *Materials Research Bulletin* 42, 1395-1401 (2007).

H. Xu, Z. Zheng, L. Z. Zhang, H. L. Zhang,F. Deng. Hierarchical chlorine-doped rutile TiO_2 spherical clusters of nanorods: Large-scale synthesis and high photocatalytic activity. *Journal of Solid State Chemistry* 181, 2516- 2522 (2008).

Y. Li, T. Sasaki, Y. Shimizu,N. Koshizaki. Hexagonal-Close-Packed, Hierarchical Amorphous TiO_2 Nanocolumn Arrays: Transferability, Enhanced Photocatalytic Activity, and Superamphiphilicity without UV Irradiation. *J. AM. CHEM. SOC.* 130, 14755-14762 (2008).

X. L. Bai, B. Xie, N. Pan, X. P. Wang,H. Q. Wang. Novel three-dimensional dandelion-like TiO_2 structure with high photocatalytic activity. *Journal of Solid State Chemistry* 181, 450-456 (2008).

W. E. Teo,S. Ramakrishna. Electrospun nanofibers as a platform for multifunctional, hierarchically organized nanocomposite. *Composites Science and Technology* 69, 1804-1817 (2009).

C. Pan, Y. G. Wang, J. S. Qiu, B. Qu,J. Wang. Fabrication of 3D Co_3O_4 Homoarchitectures via a Novel Template-assisted Coprecipitation Process. *Chemistry Letters* 39, 944-945 (2010).

R. Ostermann, D. Li, Y. D. Yin, J. T. McCann,Y. N. Xia. V_2O_5 Nanorods on TiO_2 Nanofibers: A New Class of Hierarchical Nanostructures Enabled by Electrospinning and Calcination. *Nano Lett.* 6, 1297-1302 (2006).

T.S.He, Z. F. Zhou, W. B. Xu, Y. Cao, Z. F. Shi,W. P. Pan. Visible-light photocatalytic activity of semiconductor composites supported by electrospun fiber. *Composites Science and Technology* 70, 1469-1475 (2010).

T. P. Cao, Y. J. Li, C. H. Wang, L. M. Wei, C. L. Shao,Y. C. Liu. Three-dimensional hierarchical CeO_2 nanowalls/TiO_2 nanofibers heterostructure and its high photocatalytic performance. *J. Sol-Gel. Sci. Technol.* 55, 105-110 (2010).

N. X. Wang, C. H. Sun, Y. Zhao, S. Y. Zhou, P. Chen,L. Jiang. Fabrication of three-dimensional ZnO/TiO_2 heteroarchitectures via a solution process. *J. Mater. Chem.* 18, 3909-3911 (2008).

Z. Z. Gu, H. H. Chen, S. Zhang, L. G. Sun, Z. Y. Xie,Y. Y. Ge. Rapid Synthesis of Monodisperse Polymer Spheres for Self-Assembled Photonic Crystals. *Colloids Surf. A.* 302, 312-319 (2007).

N. S. R. Database. 20 version 3.2 (Web Version) (http://srdata.nist.gov/xps).

Y. P. Ting, K. G. Neoh, E. T. Kang,K. L. Tan. Recovery of gold by electroless precipitation from acid solutions using polyaniline. *J. Chem. Technol. Biotechnol.* 59, 31-36 (1994).

T. Kasuga, M. Hiramatsu, A. Hoson, T. Sekino,K. Niihara. Formation of Titanium Oxide Nanotube. *Langmuir* 14, 3160-3163 (1998).

M. Lahav, T. Sehayek, A. Vaskevich,I.Rubinstein. Nanoparticle Nanotubes. *Angew. Chem. Int. Ed.* 42, 5576-5579 (2003).

P. Gibson, H. S. Gibson,D. Rivin. Transport properties of porous membranes based on electrospun nanofibers. *Colloids Surf. A* 187, 469-481 (2001).

C. Pan, L. Dong, L. Q. Ge, J. Wang,Z. Z. Gu. Highly Active TiO_2/Polyelectrolytes Hybrid Multilayered Hollow Nanofibrous Photocatalyst Prepared from Electrouspun Fibers Using Electrostatic Layer-by-Layer Technique. *J. Macromol. Sci., Part B: Phys.* 48, 92-105 (2009).